Wireless Communications Systems and Networks

Information Technology: Transmission, Processing, and Storage

Series Editor: Jack Keil Wolf
University of California at San Diego
La Jolla, California

Editorial Board: Robert J. McEliece
California Institute of Technology
Pasadena, California

John Proakis
Northeastern University
Boston, Massachusetts

William H. Tranter
Virginia Polytechnic Institute and State University
Blacksburg, Virginia

Coded Modulation Systems
John B. Anderson and Arne Svensson

Communication System Design Using DSP Algorithms: With Laboratory Experiments for the TMS320C6701 and TMS320C6711
Steven A. Tretter

A First Course in Information Theory
Raymond W. Yeung

Interference Avoidance Methods for Wireless Systems
Dimitrie C. Popescu and Christopher Rose

Nonuniform Sampling: Theory and Practice
Edited by Farokh Marvasti

Simulation of Communication Systems, Second Edition: Methodology, Modeling, and Techniques
Michel C. Jeruchim, Philip Balaban, and K. Sam Shanmugan

Stochastic Image Processing
Chee Sun Won and Robert M. Gray

Wireless Communications Systems and Networks
Mohsen Guizani

A Continuation Order Plan is available for this series. A continuation order will bring delivery of each new volume immediately upon publication. Volumes are billed only upon actual shipment. For further information please contact the publisher.

Wireless Communications Systems and Networks

Edited by

Mohsen Guizani

Western Michigan University
Kalamazoo, Michigan

Kluwer Academic / Plenum Publishers
New York, Boston, Dordrecht, London, Moscow

ISBN: 0-306-48190-1

©2004 Kluwer Academic/Plenum Publishers, New York
233 Spring Street, New York, New York 10013

http://www.wkap.nl/

10 9 8 7 6 5 4 3 2 1

A C.I.P. record for this book is available from the Library of Congress

Printed in the United States of America

This work is dedicated to

my parents, my wife Saida,

and my kids Nadra, Fatma, Maher, Zainab, Sara, and Safa

ACKNOWLEDGMENT

This work would have not been possible without the help and encouragement of so many people. These people inspired my knowledge and helped directly or indirectly with many aspects of the writing process, as well as ideas of topics of interest to include in this book. These include, Ibrahim Habib, City University of New York, Ted Rappaport, University of Texas-Austin, K. C. Chen, National Taiwan University, and H-H. Chen, National Sun-Yat Sen University in Taiwan, to name a few. I am also very thankful to my graduate students Ghassen Ben Brahim, Mohamed Bourouha, and Anupama Raju for their help as well as Kluwer editors Ana Bozicevic and Alex Greene who facilitated many processing issues to get this work finished at a fast pace.

I also would like to thank all the contributors to the chapters—their patience during this process was very much appreciated.

This work is dedicated to my parents, my wife Saida and my kids Nadra, Fatma, Maher, Zainab, Sara, and Safa for their patience and support during the time I spent working on this project.

PREFACE

Since the early 1990s, the wireless communications field has witnessed explosive growth. The wide range of applications and existing new technologies nowadays stimulated this enormous growth and encouraged wireless applications. The new wireless networks will support heterogeneous traffic, consisting of voice, video, and data (multimedia). This necessitated looking at new wireless generation technologies and enhance its capabilities. This includes new standards, new levels of Quality of Service (QoS), new sets of protocols and architectures, noise reduction, power control, performance enhancement, link and mobility management, nomadic and wireless networks security, and ad-hoc architectures. Many of these topics are covered in this textbook.

The aim of this book is research and development in the area of broadband wireless communications and sensor networks. It is intended for researchers that need to learn more and do research on these topics. But, it is assumed that the reader has some background about wireless communications and networking. In addition to background in each of the chapters, an in-depth analysis is presented to help our readers gain more R&D insights in any of these areas. The book is comprised of 22 chapters, written by a group of well-known experts in their respective fields. Many of them have great industrial experience mixed with proper academic background.

The first chapter discusses peer-to-peer networking in mobile communications based on SIP. J. Huber of Siemens, Germany, describes the evolution path of the Global System for Mobile Communications towards UMTS that integrates Internet technologies and provides a platform for the development of new mobile specific services. These services include new developments that will incorporate mobility and person-to-person communications with real-time video, sound, and voice, describing the evolutionary path in achieving this objective.

The second chapter, written by A. Boukerche from the University of Ottawa, Canada, discusses some protocols for data propagation in wireless sensor networks. These protocols cover large numbers of these sensors that can be deployed in areas of interest, such as inaccessible terrains or disaster places, and use self-organization and collaborative methods to form a network. Their wide range of applications is based on the possible use of various sensor types. Thus, sensor networks can be used for continuous sensing, event detection, location sending, as well as micro-sensing. This is a challenging technological and algorithmic task. Features, including the huge number of sensor devices involved, the severe power shortage, computational and memory limitations, their dense deployment and frequent failures, pose new design analysis and implementation challenges. This chapter aims to present certain important aspects of the design, deployment, and operation of sensor networks, as well as some evaluation of their performance analysis.

The third chapter, by A. Safwat of Queen's University, Canada, talks about the design of a novel architecture for 4G and 4G+ wireless networks called "A-CELL." The A-Cell relay is used to reduce power consumption and enhance coverage and throughput. A-Cell increases spatial reuse via directive antennas and GPS. The author derives an analytical model for A-Cell based on multi-dimensional Markov chains. The model is then used to formulate A-Cell call blocking. This is the first time that directive antennas are used as a means to reduce interference, conserve energy, and enhance spatial reuse in a multi-hop UMTS wireless network.

H. Holma and A. Toskala from Nokia, Finland discuss third generation WCDMA radio evolution. They discuss technologies including: 2 Mbps data rate with variable bit rate capability, support of multi-service, QoS differentiation, and efficient packet data. These technologies also contain downlink packet data operation enhancement, high speed downlink packet access (HSDPA). HSDPA utilizes Hybrid ARQ and higher order modulation for improving data spectral efficiency and for pushing bit rates beyond 10 Mbps. The further 3GPP releases will study the enhancements of packet data performance in uplink. Other important features in future 3GPP releases include advanced antenna technologies and WCDMA standard for new spectrum allocations. The chapter also describes the main solutions of 3GPP WCDMA standard in more detail.

Next, H-H. Chen from National Sun Yat-Sen University, Taiwan, discusses the evolution of CDMA from interference-limited to noise-limited. The chapter covers various new CDMA-based wireless standards—such as TD-SCDMA and LAS-CDMA, etc., as well as those on other commonly referred CDMA-based standards, such as IS-95A/B, cdma2000, UTMS-UTRA, W-CDMA, etc.— providing a rather comprehensive coverage of the evolution of CDMA technologies from interference-limited to noise-limited. The discussions on various features of state-of-the-art CDMA technologies lead to the introduction of a new concept on *isotropic CDMA air-link technology,* which consists of two related sectors: isotropic spreading codes and isotropic spreading modulation. This can offer a homogenous performance in both synchronous and asynchronous CDMA channels. Based on the discussion on conventional CDMA technologies and introduction of two novel CDMA architectures, the author concludes that the evolution of CDMA technologies will eventually make a CDMA system offer a noise-limited performance.

Then, B. Hashem from Nortel Networks of Ottawa, Canada, describes a power control implementation in third generation CDMA systems. The inner loop and the outer loop algorithms are explained whereby the power commands are sent at a high rate, and uncoded, resulting in high error rate in receiving them. The problem of errors in the power control commands becomes worse during soft handoff where the terminal communicates simultaneously with more than one base station. A few algorithms are introduced during soft handoff to increase the system capacity, like power control rate reduction and the adjustment loop. Adjusting the transmission rate is also discussed as another scheme to achieve the required quality.

F. Gunnarsson of Linköping University, Sweden, discusses the characteristics and fundamentals of power control in wireless networks. He describes power control fundamentals, including both theoretical and practical limitations. The relationship to session management, such as admission and congestion control, is also addressed. Concepts and algorithms are illustrated with simple examples and simulations.

L. Yang and M-S. Alouini of the University of Minnesota present their results on the average outage duration of wireless communication systems. More specifically they discuss generic results for the LCR and AOD: (1) of maximal ratio combining systems subject to CCI operation over independent identically distributed Rician and/or Nakagami fading environments when a minimum desired signal power requirement is specified for satisfactory reception; and (2) of various selection combining diversity scheme in presence of multiple CCI and with both minimum SIR and desired signal power constraints over independent, correlated, and/or unbalanced channels. Corresponding numerical examples and plots illustrating the mathematical formulation are also provided and discussed.

A transport level approach in enhancing TCP performance in wide area cellular wireless networks are discussed by E. Hossain and N. Parvez from the University of Manitoba, Canada. They present a comprehensive study on the performances of the basic TCP variants (e.g., TCP Tahoe, TCP Reno, TCP New-Reno, SACK TCP, FACK TCP) in wide-area cellular wireless networks. For the basic TCP variants, an in-depth analysis of the transport-level system dynamics is presented based on computer simulations using *ns-2*. Impacts of variations in wireless channel error characteristics, number of concurrent TCP flows, and wireless link bandwidth on the average TCP throughput and fairness performances are investigated. The maximum achievable throughput under window-based end-to-end transmission control is also evaluated and the throughput performances of TCP New-Reno, SACK TCP, and FACK TCP are compared against this ideal TCP throughput performance. To this end an overview of the major modifications to the basic TCP variants based on transport-level approaches to enhance TCP performance in wireless networks is presented.

The deployment of several real-time multimedia applications over the Internet has motivated a considerable research effort on the provision of multiple services on the Internet. In order to extend this work over wireless links however, we must also take into account the performance limitations of wireless media. G. Xylomenos and G. Polyzos from the Athens University of Economics and Business in Greece survey various related approaches and conclude that link layer schemes provide a universal and localized solution. Based on simulations of application performance, over many link layer schemes, they show that different approaches work best for different applications by concurrently supporting multiple link layer schemes. Simulations of multiple applications executing simultaneously show that this approach dramatically improves performance for all of them. They finally consider embedding this approach into a Quality of Service oriented Internet, discussing the traditional best-effort architecture, the Differentiated Services architecture, and an advanced dynamic service discovery architecture.

A value added service to broadband wireless network is the remote access virtual private network (VPN). The corporate legitimate portable users can connect to their users through a wireless network from different locations and get secure services as if they were connected to the corporate local area network. One of the main challenges is to block illegitimate wireless users' requests. Registration and authentication functions should be implemented with highly secured wireless connections. These functions are accomplished by tunnelling the user information, in a secured form, to the corporate authentication server through the Internet traffic. The Corporate Authentication Server then grants or denies the user access. M. Matalgah from the University of Mississippi, J. Qaddour, from Illinois Institute of Technology, and their colleagues O. Elkeelany and K. Sheikh from Sprint Inc. have designed a portability architecture for nomadic wireless Internet access users and security performance evaluation to solve these challenges. This chapter addresses various portability scenarios, architectures, implementation, and requirement issues for portable wireless Internet access systems. Moreover, performance evaluation and comparison are presented for the state-of-the-art security and authentication techniques.

V. W-S Feng and Y-B Lin, of National Chiao Tung University and Industrial Technology Research Institute (ITR) of Taiwan, describe the design and implementation of a softswitch for Third Generation UMTS Mobile All-IP Network developed by Computer and Communication Research Laboratory (CCL) of ITRI. This softswitch can be utilized as call agent (media gateway controller) for the third generation mobile all-IP network such as UMTS. The CCL Softswitch follows the reference architecture proposed by International Softswitch Consortium. In this approach, the Intelligent Network (IN) call model is implemented to interwork with existing IN devices. They design protocol adapter modules and service provider interface to ensure that multiple VoIP protocols can be supported without modifying the core of the softswitch.

Furthermore, the message flows of call setup, call transfer, and inter-softswitch call are described to show the feasibility of the softswitch.

Then, in Chapter 13, E. Hossain et al., from the University of Manitoba, Canada, discuss issues and approaches of clustering in mobile wireless ad hoc networks. In a multi-hop ad hoc wireless network, which changes its topology dynamically, efficient resource allocation, energy management, routing, and end-to-end throughput performance can be achieved through adaptive clustering of the mobile nodes. Impacts of clustering on radio source management and protocol performance in a multi-hop ad hoc network are described and a survey of the different clustering mechanisms is presented. A comparative performance analysis among the different clustering mechanisms based on the metrics such as cluster stability, load distribution, control signaling overhead, energy-awareness is performed.

The maximum capacity of a CDMA cellular system's radio interface depends on the time varying radio environment. This makes it hard to establish the amount of currently available capacity. The received interference power is the primarily resource in the uplink. Ability to predict how different resource management decisions affect this spatial quality is therefore of utmost importance. The uplink interference power is related to the uplink load through the pole equation. In this chapter, E. Lundin and F. Gunnarsson of Ericsson Research and Linköping University, Sweden, discuss both theoretical and practical aspects of uplink load estimation concentrating on concepts and algorithms.

In Chapter 15, end-to-end performance analysis of multi-hop transmissions over Rayleigh .fading channels is presented. Several types of relays for both regenerative and non-regenerative systems are considered. In addition, optimum power allocation over these hops is investigated as it is considered a scarce resource in the context of relayed transmission. Numerical results show that regenerative systems outperform non-regenerative systems especially at low average signal-to-noise ratios, or when the number of hops is large. The authors, M. Hasna and M-S. Alouini of Qatar University and the University of Minnesota, also show that power optimization enhances the system performance, especially if the links are highly unbalanced in terms of their average fading power or if the number of hops is large. Interestingly, they also show that non-regenerative systems with optimum power allocation can outperform regenerative systems with no power optimization.

Mobility management plays a significant role in current and future wireless mobile networks in effectively delivering services to the mobile users on the move. Many schemes have been proposed and investigated extensively in the past. However, most performance analyses were carried out either under simplistic assumptions on some time variables or via simulations. Recently, Y. Fang and W. Ma of the University of Florida have developed a new analytical approach to investigate the modeling and performance analysis for mobility management schemes under fairly general assumptions. In this chapter, they present the techniques they have developed for this approach and summarize major results obtained for a few mobility management schemes such as movement-based mobility management, pointer forwarding scheme (PFS), Two-level location management scheme, and two-location algorithm (TLA).

Pervasive computing environments present entirely new set of challenges because of the fact that data may be acquired and disseminated at various stages within the system. Therefore, novel caching mechanisms are needed that take into account demand-fetched and prefetched (or pulled), as well as broadcast (or pushed) data. In addition, cached maintenance algorithms should consider such features as heterogeneity, mobility, interoperability, proactivity, and transparency that are unique to pervasive environments. Pervasive computing applications require continual and autonomous availability of 'what I want' type of information acquisition and dissemination in a proactive yet unobtrusive way. Mobility and hererogeneity of pervasive environments make this problem even more challenging. Effective use of middleware techniques—such a caching—can overcome the dynamic nature of communication media and the limitations of resource-poor devices. These and other challenging issues in this area are discussed by M. Kumar and S. Das, of the University of Texas at Arlington, in Chapter 17. Das *et al.* also present issues of security in wireless mobile and sensor networks in Chapter 18. In particular, they discuss security issues and challenges in wireless mobile ad hoc and sensor networks as well as in pervasive computing infrastructures. They also describe security protocols for such environments.

H-H. Chen discusses various issues on pulse shaping waveform design for bandwidth efficient digital modulations in Chapter 19. He starts with an overview on the evolution of pulse shaping technologies from early digital modulation schemes to recently emerging new carrier modulations such as quadrature-overlapped modulations. The impact of the pulse shaping technologies on both bandwidth-efficiency and power-efficiency, which are two essential merit parameters of a digital modem, are also discussed with an emphasis on shaping pulse design methodology as well as performance analysis of a pulse-shaped digital modem. Several new pulse shaped QO modulation schemes, which adopt various novel pulse shaping waveforms generated by time-domain convolution method, are presented and their performance analysis will be carried out in comparison with other traditional digital modems, such as QPSK, OQPSK and MSK.

In chapter 20, E. Biglieri and G. Taricco of Politecnico di Torino, Italy, investigate transmission systems where more than one antenna is used at both ends of the radio link. The use of multiple transmit and receive antennas allows one to reach capacities that cannot be obtained with any other technique using present-day technology. After computing these capacities, the authors show how "space-time" codes can be designed and how sub-optimum architectures can be employed to simplify the receiver.

In Chapter 21, a group of researchers (R. Schober, L. Lampe, S. Pasupathy, and W. Gerstacker) from the University of British Columbia and the University of Toronto, Canada, as well as the University of Wrlangen-Nuernberg, Germany, designed and optimized matrix-symbol-based space-time block codes (STBC's) for transmission over fading intersymbol interference (ISI) channels. They show that STBC's employing diagonal code matrices exclusively facilitate the successful application of suboptimum equalization techniques for the practically important case when only a single receive antenna is available. Three different types of diagonal STBC's are optimized for fading ISI channels and their performances are compared for decision-feedback equalization (DFE) and decision-feedback sequence estimation (DFSE), respectively. The robustness of the designed codes against variations of the characteristics of the fading ISI channel and the dependence of the equalizer performance on the STBC structure is investigated.

Last, but not least, M. Bassiouni, W. Cui, and B. Zhou of the University of Central Florida, present a fast routing and recovery protocols in hybrid ad-hoc cellular networks. They present and evaluate efficient recovery and routing protocols for mobile routers in HACN (hybrid ad-hoc cellular network). In the dual backup recovery protocol, a standby router is dedicated to each primary router. A more flexible protocol, called the distributed recovery protocol, is obtained by relaxing this one to one mapping and scattering the backup routers geographically among the primary routers. The effectiveness of the distributed recovery protocol is demonstrated by simulation results. A simple analytical model is also presented for computing the blocking probability of new calls in the presence of router failures. Chapter 22 is concluded by presenting an efficient location–based routing protocol for HACN mobile routers.

Mohsen Guizani

CONTENTS

Chapter 1

PEER-TO-PEER NETWORKING IN MOBILE COMMUNICATIONS BASED ON SIP

Josef F. Huber
Siemens AG
ICM N PG SP NI
Hofmannstr. 51
D-81359 Munich
Germany
Tel.: +49 89 722 44564, Fax: +49 89 722 58929
josef-franz.huber@siemens.com

Abstract   The mobile Internet, ubiquitous computing and universal mobile communications are key words related to the integration of information technology and communications. Until today, the Internet has been dominated by non-realtime services based on client-server relations. As its continued developments encompass realtime voice and video, peer-to-peer communications will emerge. Consequently, the take-up of Internet technologies in mobile networks leads to new innovations there. This chapter describes the evolution path of mobile networks and services in the framework of the universal mobile telecommunication system UMTS.

Keywords: Mobile Internet, Next Generation Networks, IP Multimedia, SIP, 3G Services and Networks, UMTS, Ubiquitous Computing

1.   Why Mobile NGN's?

What are the targets of mobile next generation networks (NGN's)? This is the key question when we discuss the evolution of hardware and software technology in the field of mobile communications. One target could be computing anywhere, anytime, another one personalisation, or let us summarise and say ubiquitous computing.

The discussion about ubiquitous computing (UC, or ubicomp) has been going on for many years. UC describes the evolution of computing towards the so-called third era of computing. Mobile computing will be a main contributor to this development. The main aim of UC is to embed many small and highly specialised devices within the everyday environment in such a way that they operate seamlessly and become transparent to the person using them. They will operate either off-line or on-line. UC products aim to be everywhere (e. g., by being portable), to be small, and to be aware (of their environments, users, the contexts). Products and devices embodying these characteristics will provide a physical entity with complete freedom of movement as well as freedom of interaction [1].

The cornerstones of this vision are that computers as they are known today will be replaced by a multitude of networked computing devices embedded in our environments and that these devices will be invisible in the sense of not being perceived as computers. Wireless connectivity is a key contributor to this vision.

In order to facilitate UC, mobile NGN's have to fulfil such requirements like providing flexible bit rates and wide area coverage outdoor and indoor, as well as intelligent support encompassing location and situation of the computing devices.

2.   Enabling Technologies for Mobile NGN's

With the evolution from second generation (2G) to third generation (3G) mobile networks, wideband radio access and Internet-based protocols will prepare the way from a mobile handset today to a mobile multimedia device in the future. Providing wireless access to the wireline Internet brings, of course, more flexibility and facilitates penetration. New services come with new enabling functions – like mobility, personalisation, and localisation capabilities – which are the characteristics of 3G mobile systems. This is the motivation for the industry to evolve the wireline Internet to a mobile Internet with new capabilities and applications. Key enabling functions for the mobile NGN are as follows:

- *IP-transparency:* All elements involved in the end-to-end communications path have to support IP, both in the fixed and in the mobile parts of the network.
- *Mobility management:* It has to function in a globally networked environment for roaming.
- *Addressing:* It must allow every user a unique address capability, which is independent from the user's location.
- *Personalisation of information and positioning:* There must be means to provide such functionality.
- *Positioning:* The individual must be positioned to enable location-dependent services.

- *Security:* It has to be provided end-to-end for fixed and mobile devices.

Such functionality will make the mobile Internet different from a wireless access to the Internet, like wireless LANs (WLAN) offer today.

3. New Infrastructures

Many mobile operators are currently in the process of evolving their networks to 3G services, from current voice and data bearer services to high bit rate IP-based services. To do so, they must upgrade their networks from second generation 2G to evolved 2G (2.5G) and third generation systems (3G).

The major 2G mobile networks are based on four technologies: Global System for Mobile Communications (GSM), Universal Wireless Communications (UWC-136), Personal Digital Cellular (PDC) and Code Division Multiple Access (CDMA), also called cdma One. As illustrated, the generally accepted 3G migration path for the GSM technology is GPRS/EDGE/WCDMA, for cdmaOne it is CDMA2000 and for PDC it is WCDMA.

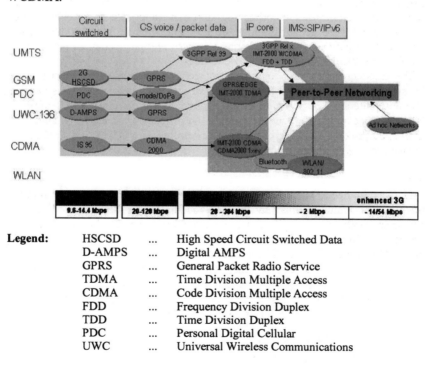

Legend:	HSCSD	...	High Speed Circuit Switched Data
	D-AMPS	...	Digital AMPS
	GPRS	...	General Packet Radio Service
	TDMA	...	Time Division Multiple Access
	CDMA	...	Code Division Multiple Access
	FDD	...	Frequency Division Duplex
	TDD	...	Time Division Duplex
	PDC	...	Personal Digital Cellular
	UWC	...	Universal Wireless Communications

Fig. 1   Migration of Mobile Networks

These same paths are also available for UWC-136. It is clear that the GSM operators and the majority of UWC-136 operators are preferring the GSM path to 3G. There are a number of reasons for example the present worldwide GSM footprint with a market share of beyond 75 % (1 bio. users in first half of 2004). Thus, it is obvious that the GSM technology path GPRS – EDGE – WCDMA will become the most widely accepted standard for 3G services.

The evolution begins with an upgrade of the GSM network to GPRS technology, which provides more effective mobile data capabilities.

GPRS

GPRS (General Packet Radio Service) is a 2.5G radio system, but a 3G system in terms of the core network. It enhances GSM data services significantly by providing genuine end-to-end packet-switched data connections, offers data transmission speeds up to 171.2 kbps (peak data rate) and supports the leading Internet protocols TCP/IP and X.25.

The integration of GPRS into GSM is a rather straightforward process. A subset of time slots on the air interface are defined for GPRS allowing scheduled packet data multiplexing of several mobile stations. The radio subsystem needs a minor modular upgrade associated with the packet control unit (PCU) to provide a routing path for packet data between the mobile terminal and gateway node. A minor software upgrade becomes necessary to employ the different channel coding schemes.

The GSM core network constructed for circuit-switched connections has to be extended with new packet data switching and gateway nodes, the so-called GGSN (Gateway GPRS Support Node) and SGSN (Serving GPRS Support Node). However this acquisition endures the migration towards 3G, since the high-speed packet switching core network provided by GPRS and EDGE can be used for UMTS almost completely.

EDGE

EDGE (Enhanced Data rates for Global Evolution) is an approved 3G radio transmission technique that can be deployed in existing spectrum of TDMA and GSM operators. EDGE reuses the GSM carrier bandwidth and time slot structure. Thus, EDGE-capable infrastructure and terminals are fully compatible with GSM and GPRS (EGPRS).

Due to adaptive modulation/coding schemes optimal bit rates are achieved for all channel qualities. The maximum user peak data rate that can be achieved in a 200 kHz carrier with the most sensitive modulation/coding scheme and combining of all 8 time slots is 473.6 kbps.

W-CDMA or UMTS/FDD

W-CDMA (Wideband Code Division Multiple Access) is the UMTS radio access (UTRA) technology for paired band operation. It operates in separated up- and downlink frequencies (FDD mode) and is based on the direct sequence (DS) spread spectrum (CDMA) technology using a chip rate of 3.84 Mcps within a 5 MHz frequency band. A higher bandwidth and higher spreading provide an increase of processing gain and a higher receiver multi-path resolution, which was a decisive point for the IMT-2000 bandwidth fixing.

W-CDMA fully supports both circuit and packet-switched high-bit-rate services and ensures the simultaneous operation of mixed services with an efficient packet mode. Moreover, W-CDMA supports highly variable user data rates based on the rate matching procedure, where data capacity among the users can change from frame to frame (frame length 10 ms).

The current W-CDMA standard is equipped with QPSK, a more robust modulation scheme than 8-PSK, which provides a peak data rate of 2 Mbps with good transmission quality in a large coverage area.

W-CDMA is the new radio transmission technology with a new radio access network (RAN) called UTRAN.

UMTS/TDD

With increasing shortage of spectrum it may be more likely that unpaired spectrum blocks can be cleared and, therefore, TDD systems may become more important in the near future. The IMT-2000 standard for TDD, called IMT-TC, defines two types of air interfaces; which are standardised as TDD hcr (high chip rate) and TDD lcr (low chip rate), respectively. Both technologies use a hybrid access technology combining TDMA and CDMA. The consequent harmonisation of the TDD transmission techniques with W-CDMA and the efficient usage of the existing unpaired bands make the TDD standard (IMT-TC) an interesting proposition for wireless operators.

HSDPA

All the advanced transmission technologies such as adaptive modulation and incremental redundancy are not exploited in W-CDMA yet. Therefore, the next logical evolutionary step of W-CDMA in terms of higher downlink data rates is being standardised in 3GPP with the working title HSDPA (High Speed Downlink Packet Access). HSDPA will improve the average throughput of the cell and, furthermore, the end user access speed of up to 14 Mbps on the downlink and similarly on the uplink side (EUDPA).

WLAN

The next step is the introduction of 3G services with WLAN integration for Hotspot coverage and indoor applications. All these technologies will provide the high capacity and bit rates, which are means to enable multimedia services up to 54 Mbps. From a switching point of view, however, they are not entirely sufficient for true multimedia operations, which are characterised by

- simultaneous handling of realtime video, voice and data in one user session

- combined management of real-time and non-real-time services with their respective QoS requirements
- multiple session handling and presence related services management

For these reasons standardisation work started to build a new network architecture based on IP for both – realtime voice and video and non-realtime data services. Such an approach takes us to the next generation mobile networks, which are described as part of the UMTS concept.

4.   Future Mobile Infrastructure Platforms

The start-up of UMTS networks follows the architectural approach from GSM: Realtime voice/video services will be handled via circuit switched channels, data services via packet switched channels. Both, circuit and packet switched services are separated from each other.

The first UMTS specification release was approved in 1999 (R99), then followed by Release 4 in 2000, Release 5 in 2002 and by Release 6 in 2003/2004. The work started from a network architecture comprising circuit switched and packet switched domains (see Fig. 2) in Release 99 followed by Release 4, which integrates TD-SCDMA radio access for the use of unpaired frequency bands.

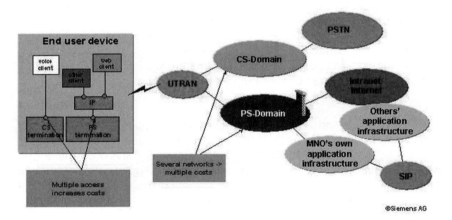

Fig. 2   UMTS – Hybrid Circuit/Packet Switched Network Architecture

Real-time or non-real-time services are related to these domains, thus they are kept separated. Video and voice services can therefore be combined with data applications only outside the network. This leads to difficulties in the Quality of Service (QoS) management, to complex billing and affects the user interface. IP-based communications are using IPv4, mobility management of a user is handled with the Mobile Application Part from GSM (GSM-MAP), thus based on E.164 addressing. E.164 addressing allows a unique address for the mobile user, however the user has to deal with two addressing schemes – E.164 and DNS/IPv4.

The first step into a common IP-based system for real-time and non-real-time services, for video, voice and data was done with Release 5. Key motivations for the new approach were the expectation of cost reduction and gains for multi-service transport and switching via integration of real-time and non-real-time applications. The new development is based completely on the IPv6 protocol rather than IPv4. The main components of Release 5 which are part of the IP Multimedia Subsystem (IMS), are shown in Fig. 3.

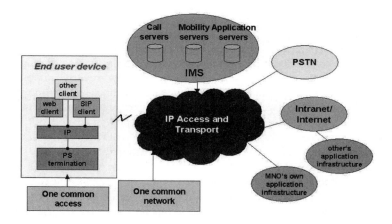

Fig. 3   UMTS Rel. 5: SIP is implemented inside the mobile network

Release 5, of course, comprises backward compatibility by allowing in parallel the circuit and packet domains of upgraded R99 and R4 networks. In the longer term IMS evolves to the extent, where it is the common means providing packet based transport for all services – voice and data – in the switching and the radio part as well as in the user terminal. The subsequent Release 6 finalizes the specifications for implementation of IMS.

4.1   Network Control Architecture based on SIP

The new vision in the mobile sector is the integration of realtime mobile voice/video communications with non-real-time data beyond the physical integration on the access and transport side. This means, as a consequence, that mobile radio and transport technology has to be merged with Internet technologies and further, that a new call control architecture for multimedia session management has to be introduced. The basis for both will be Internet protocols (IP) and the Session Initiation Protocol (SIP) [6]. Of course, a new architecture has to imply functions, which are beyond IP and SIP. For example, there are functions needed to manage

- Quality of Service: they are highly important for real-time voice and video
- Mobility Management: it has to guarantee roaming (national and international for the services the user has chosen)
- User Authentication and Terminal Identification: a unique address and equipment identification has to be provided
- Security: end-to-end security is an important factor
- Multiple Session Handling: the user should be able to activate/de-activate sub-sessions under one main session

Multimedia services require multiple sessions over one physical channel. In principle, such capability will be possible with packet-switched networks. They share available physical transmission capacity for several sessions, either for one user or among all users leading to more efficient capacity utilisation as compared to the traditional Circuit Switching. In order to allow combined real-time and non-real-time services in one physical channel, a common transport protocol is needed as well as the separation of user information and control. This leads to separate Service, Control and Transport in all network elements including end-user equipments. There are also other reasons for a new approach in the network, ranging from distributed functionality to more flexible and easier introduction of new features. The new approach is called Internet protocol Multimedia Subsystem (IMS) and enables Next Generation Mobile Networks to converge existing voice and data networks into one facilitating the voice-data communication business on a large scale. As shown in Fig. 2, the common protocol set is IP: in the horizontal direction (user to user/host) and in the vertical direction (session control, QoS and billing). It allows distributed functionality, network intelligence, packet-based radio access and backbone, open platforms and mediation technologies. The take up of the "Session Initiation Protocol" (SIP) into the UMTS IP Multimedia Subsystem (IMS) separates signalling and control tasks from end-to-end user information exchange enabling the user to manage several multimedia sessions simultaneously. In addition, the Internet Protocol IPv6 will handle mobility and allow unique addressing for every user [7].

The New Approach: Internet protocol Multimedia Subsystem (IMS)

Approved by the Third Generation Partnership Project (http://www.3gpp.org/About/about.htm) in March 2002 as part of UMTS Release 5, IMS provides packet-based transport for both data and voice services. The IP radio and core network represent the server and communicate with the user as the client for session control. It is a concept which encompasses end-user terminal, radio and core network and gateway to external networks. As Fig. 4 shows, the bearer plane links to an ISP and portal for content-based services from the Internet and switches users' voice and video connections. The control plane comprises all network control functions, including gateways to external packet- and circuit-switched networks. APIs in the service plane use the common object request broker architecture as the software standard to support network applications. The Internet protocol enables distributed functionality, network intelligence, packet-based radio access and backbone, open platforms, and mediation technologies. A similar solution using SIP is seen in Microsoft's Windows XP.NET framework functionality. However, it considers terminal (client) and server as entities outside the mobile network and does not provide network control functions. It also allows voice and video applications shared with messaging, sending files or to do applications sharing. Application sharing means that several users can work on the document, can have video conferencing together. With .NET, the network is assumed to be a transparent transport system. In contrast to IMS, QoS cannot be guaranteed, because the network has no knowledge about the detailed session requirements. IMS, however, brings the network into the position of an intelligent session management facility between users providing services support regarding QoS, location. Multiple sessions between several users with address translation from the telephone numbering scheme E.164 to DNS and vice versa will be handled via IMS. In IMS, the network represents the server for session set-up and clear and communicates with the user as client. Since SIP is used for session initiation and clear, the user application itself (e, g. between terminal and content server or user-to-user for voice, video) can be built on its own mechanism.

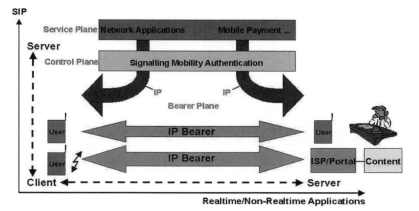

Fig. 4   IMS uses IP for User-to-Host and for Network Control Functions

4.2 The IMS Multimedia Call Model

It includes user service functions for multimedia session handling and management, for roaming, for end-to-end quality of service, for subsession control and provides support for inter-working with the Internet and ISDN/PSTN as well as with mobile networks (GSM, IS41, PDC). Key feature of IMS is the ability to set up **multiple sessions**, which include any mix of media (e. g. voice, video, text) or mix of real-time communications with non-real-time information, person-to-person, person-to-machine or machine-to-machine. Calls within calls (such as "private chat") are included in this. Fig. 5 shows examples of session arrangements of IMS.

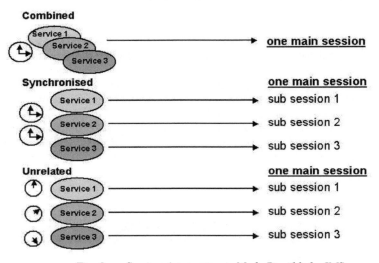

Fig. 5   *Session Arrangements Made Possible by IMS*

The integration of services is the ability to dynamically modify the media types active during a multimedia call session. The range of media types in use is dependent on the capabilities of the user terminal. In this way, IMS "integrates" into a single session what are today separate services. For the user, this single session capability means that it can multitask. For example it does not have to terminate a voice call (or place it on hold) in order to send a text message or video clip.

The integration of services also allows interaction between media types. Interaction is creating a new capability and user experience. As an example, a user could be browsing a Web site and then with a simple click, create a voice or video call. In this way, different services interact with each other to provide a seamless user experience.

Table 1   IMS Main Features

Feature	Protocols, SW Tools, APIs
Network control for bundled services	Session Initiation Protocol (SIP) stack, "always on"
One account for all services	Charging for one account
VoIP, Video over IP	SIP with extensions for mobility, security
Dynamic QoS control reducing delay, jitter	Resource Reservation Protocol RSVP Protocol Label Switching Protocol MPLS
Interworking E.164 – IPv6	ENUM generic address translation between URL and E.164 numbers
Inter-working Gateways PSTN, ISDN, GSM	Media Gateway to PSTN, ISDN, GSM, other PLMN SIP Proxy to Internet
Single sign-in access Personalised addressing	Java based IMS SIM Card (ISIM)
Presence Service	User attendance, user location recognition
Value-added services	API (Parlay, ...)

IMS services represent certainly a challenge to the mobile industry. Although part of the 3GPP Release 5 for UMTS, it will take some time to bring this functionality into the field. The first demos were already shown by Siemens and others at the GSM World Exhibition 2002 in Cannes/France, the introduction into the marketplace is foreseeable in the year 2005 following the 3G services deployments of Multimedia Messaging (MMS), rich voice, infotainment, Internet and intranet access and Location Based Services (LBS). Enhanced Messaging Services (EMS) will probably be substituted by MMS, as they are seen as an intermediate step from SMS to MMS.

4.3 IMS Addressing

Addressing is another reason, why IMS is an important issue in a converged mobile network-Internet environment. In the Internet, users who are hosted by computers, are addressed by their domain name (TLD). Instead – in a mobile network today - a user making a telephone call to another party is using an E.164 number. For multimedia connections between these two different addressing worlds, address translation has to take place according to the IETF extended numbering scheme called ENUM. It allows bi-directional address translation on a generic basis. The user will only have to click on the name and then the number or URL will automatically be chosen. In addition, the "Presence" service of IMS will further inform the caller, if his contact is available at that instant and, if not offering alternative means of contact. This

IMS service therefore enhances address translations for multimedia communications.

IPv6: this addressing scheme restores the paradigm of end-to-end address capability in the Internet. Therefore it is a must for mobile communication, where network address translation (NAT) does not work when it comes to unique addressing for every user. IPv6 supports mobility better than IPv4 by using two layer addressing. Also security is better supported.

4.4   Extending the Scope of Standardisation of the Mobile World and the Internet World

Of fundamental importance for mobile cellular systems and essential for world-wide roaming is global standardisation. Agreed on the ITU level (the IMT-2000 Framework Standards on the radio side), the UMTS specifications are developed in a joint standardisation work, called 3rd Generation Partnership Project (3GPP). The standards for the Internet have their roots in the development of the ARPANET and in the LAN area. They emerged to a global standard with the world-wide use of the Internet protocols in PCs and computers. IETF, W3C, ICANN and other IP-related standards organisations are working in this field. Their standards, which are coming from different roots, will be merged for UMTS. The Session Initiation Protocol Standard (SIP) is the main basis for a major work on the control level in 3GPP, called IP Multimedia Subsystem (IMS). It will provide IP call services, which are independent from applications [2, 3].

In mobile NGN's, when it comes to content related services, commercial software will play an increasing role. This leads to the integration of proprietary standards into middleware and end-user equipments and to the definition of open Application Program Interfaces (API). Examples are Java and the Java Virtual machine with ISO Java API, the MultOS API, Parlay, xhtml etc. The Open Mobile Alliance (OMA), which was founded in the year 2002 as a cross-industry standardisation group for mobile applications, is creating such standards (Fig. 6).

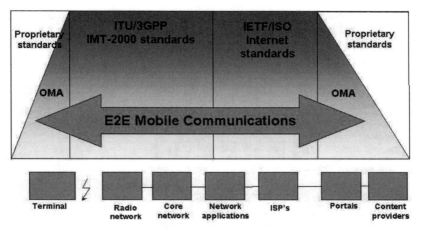

Fig. 6   Scope of UMTS Standardisation

5.   IMS Network Description

5.1   IMS Core Network

The development of the core network involves the provision of **new control platform** components (see Fig. 7). The **Call State Control Function (CSCF)** manages SIP session establishment and call control and forms the link to the **SIP Application Server (AS)**. The **Policy Control Function (PCF)** manages the Quality of Service policy. The **Multimedia Resource Function (MRF)** controls the multi-party conferencing features of SIP. A **Media Gateway Control Function (MGCF)** together with a **Transport Signalling Gateway (TSGW)** enable inter-working with ISDN based Circuit Switched networks including inter-working between IP addresses and E.164 numbering schemes. Also shown is the **Home Subscriber Server (HSS)** which is the packet (GPRS) equivalent of the HLR in GSM and which carries through from 2.5G/3G into IMS. In addition, an **MSC server** could be added, which is another form of soft switch enabling this same all IP packet switched Core Network to deliver GSM Circuit Switched services via a 2G or 3G RAN to conventional mobile terminals. The **Telecom Management** system (shown as OAM – Operations Administration & Maintenance on the diagram) including **Charging** applications and **Terminal Management** is also required.

The applications servers are placed in service plane. These are the **Service Switching Function (SSF)**, the **Service Content Provision (SCP)**, the **Presence Server (PC)** etc. API's care about flexibility using CORBA as the software standard to support service applications. The bearer plane is logically separated from the control platforms. The routers in packet technology perform their tasks in the transmission path, but are relatively unintelligent compared to a conventional ISDN based switch. They build the link to an ISP and Portal for content based services from the Internet or switch the voice, video connection to other users, directed from the control plane.

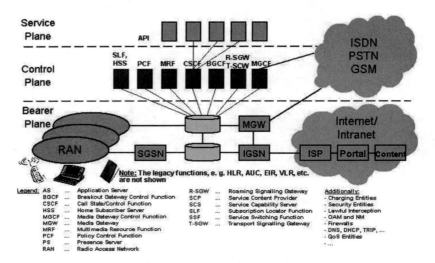

Fig. 7   Block Diagram for IMS in UMTS Release 5 and beyond

5.2 IMS Radio Access

The IMS radio access can be either GPRS based, or IP over ATM based or in future IP based. It goes into two radio access schemes: the GSM Enhanced Radio Access Network (GERAN) and the Universal Terrestrial Radio Access Network (UTRAN). The packet transport in the Radio Access Network (RAN) must meet the QoS requirements for non-real-time and real-time services.

Although the 3G radio access UTRAN has the high capacity to make multimedia IP over the air feasible, the overhead incurred when an IP bearer is used to transport basic speech over the air interface reduces spectrum efficiency. Enhancements to the standards include Robust Header Compression (ROHC), which will be developed for 3GPP by IETF. Capacity improvements which are under way are including High Speed Packet Downlink Access (HSPDA) with up to 14 Mbps bit rate.

6.   Service Evolution with IMS

Numerous articles, reports and documents are available that discuss mobile services and applications. Yet, in all this literature, there is no clear definition of the two terms. The labels "service" and "application" often seem to be interchangeable – even within the same document. A concept such as m-commerce will be classified as a service in one report and an application in the next. The terminology serves to confuse rather than clarify. The UMTS Forum, which is an international body of operators, manufacturers and regulators has done a number of studies in this context [3]. They resulted in the following definition:

Services are the Operator Portfolio choices to users:

Users select their preferred services based on the options in that product portfolio. Services are characterised by

- Clear Service Definition and Service Pricing model
- Roaming – Service Portability
- Interoperability between mobile devices, QoS

Applications are built on Service of choice, related to:

- Communications (voice, video)
- Content (access to information/3rd party content)
- Transactions (electronic banking)
- etc.

Defining the mobile services for creating a mobile data market starts with a top-down analysis of mobile applications, business models and user demand. Of course, the easiest way is to add mobility to existing fixed Internet applications e.g. mobile Internet access and mobile Intranet/Extranet access. The user base in the Internet is already quite large – about 500 mio. users – and wireless mobile access will certainly enrich the Internet based business. The limitations of mobile terminals, however, do not always allow to use the same services as they are on the fixed Internet. In addition, the mobile user – especially the mass market user has different needs in contrast to a stationary user at home or in the office.

This experience took the industry focus to the definition of new mobile oriented services, which will be the platforms for new mobile applications.

The UMTS Forum identified six mobile service categories in its market studies, which are either Internet access related or mobile specific services. They are shown in Fig. 8 indicating: Mobile Internet Access, Mobile Intranet/Extranet Access, Customised Infotainment (CI), Multimedia Messaging Service (MMS), Location Based Services (LBS) and Rich Voice. CI, MMS and LBS are shown in Fig. 5 as mobile specific services. Rich Voice is the only real-time service. It is, in effect, the enhanced voice service provided with the Video Telephony Service. Simple Voice is the term which describes the basic speech telephony services.

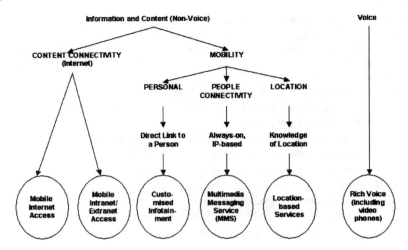

Fig. 8   *3G Services are Structured into Wireless Content*
            Connectivity and Mobility Services

IMS enables a qualitatively better end-user experience due to the increased service integration and interaction. Without IMS, similar services will still be possible, but will lack the ease-of-use that IMS can provide.

Using the service categories in Fig. 6, the following impacts can be seen:

Services that are made possible by IMS: Rich Voice services, which by definition are the integration of real-time voice with multimedia, will be impossible without IMS. Therefore IMS is wholly responsible for the revenues generated from the Rich Voice services category.

Services that are enhanced with IMS capability: IMS (largely through SIP and presence) provides increased service integration and interaction in all service categories. In addition, the ability to integrate non-real-time services (such as MMS) with real-time (Rich Voice) services in a single device improves ease-of-use and enhances the end-user experiences. Therefore, it is reasonable to expect that the deployment of IMS will have some positive impact on service adoption in all other service categories. For example:

Location-Based Services and Customised Infotainment: Services based on personalised access to content would see slight increase in service adoption due to the increased ability to engage interactively with other users while viewing content (such as gaming, navigation data, and advertising information) and conducting transactions.

"Mixed Media" Mobile Access (Business)

This service is also part of the Mobile Intranet/Extranet Access Service category and an enhancement to the Mobile VPN service. With Mixed Media capability, the Mobile VPN user will be able to respond to different types of media messages through any media type in the same session (e. g. respond to a voice mail message by sending an e-mail while listening – without disconnecting the voice call). The addition of "presence capability" and with the ability to access PIM/contact information kept in the corporate Intranet or on the user's fixed PC, the user is able to respond to messages through any media type and in a way that is acceptable to the receiving party.

"Multimedia Messaging" with Presence

This service is part of the Multimedia Messaging Service Category[1]. Short Messaging Services (SMS) are extremely popular with the youth in Western Europe, Japan, and other countries. The addition of 3G/UMTS capacity will allow these SMS users to also send and receive multimedia attachments such as downloaded video clips (e. g. movie trailers), audio clips, photos, user-created video clips, etc. The addition of IMS will allow an interactive element, much like in mobile gaming, where the user can collaborate and interact with other users while creating and/or viewing the multimedia message. For example,

- Teenage friends can take pictures of clothing while window-shopping at the mall, send it to their friends at another store, and then discuss the item on the phone while simultaneously viewing it.
- Similarly, several teenagers in different locations can simultaneously download a movie trailer video clip and discuss whether they want to see the movie while viewing the clip.
-

Dispatch Service (or Push-to-Talk Voice Conferencing)

A Dispatch Service can potentially be implemented very easily using SIP call processing. The term "Dispatch Service" comes from the type of two-way radio system used by dispatchers. Dispatchers are most immediately associated with the emergency services, Fire, Police and Ambulance as well as radio controlled taxi services and delivery services etc.

The main difference is that communication is one way at a time, it is initiated by pressing and holding a switch (usually on the microphone) rather than by dialling and the communication can be overheard by others in the community

---

[1] A consumer 3G service, that offers non-real-time, multimedia messaging with always-on capabilities allowing the provision of instant messaging. Targeted at closed user groups that can be services provider- or user-defined.

of interest. As well as "Dispatch Service", similar systems are known as "Push-to-Talk", "Two-Way Radio", and "Walkie Talkies".

Advanced Mobile Videophone Service

Advanced Mobile Videophone is a two-way, real-time, conversational video, with the same feature capabilities as voice service. As with the existing early stage mobile videophone service, IMS Advanced Mobile Videophone Service will be a person-to-person transmission, but will extend this to person-to-multiparty transmissions and enables additional functional capabilities such as the ability to independently initiate the video component and the ability to set up or receive other media calls.

Multimedia Group Broadcast

Multimedia Group Broadcast is a voice call, enhanced with multimedia elements, which can be either two-way or one-way, 1:n or n:n communications. With this service, multimedia content would be broadcast from a central point (server or live conference) to a work group. IMS Multimedia Group Broadcast adds the mobile environment to currently available conferencing options and provides a new mechanism for communication with employees that have no access to fixed Internet and/or fixed wireline telephony. When used as an extension of fixed group conferencing services, IMS Multimedia Group Broadcast would allow users with mobile devices to participate more fully in a Web conference or Web cast.

More information is available in [4, 5].

7.   Additional Requirements for IMS Deployments

IMS is positioned between two worlds: the traditional telecommunication infrastructure and the Internet. Each world has its own environment related to addressing, Quality of Service, network control, security and billing. Thus, a number of requirements are valid for IMS deployments:

- The Domain Name System DNS, based on IPv6
- ENUM address translation for the inter-operation between E.164 and Internet address schemes

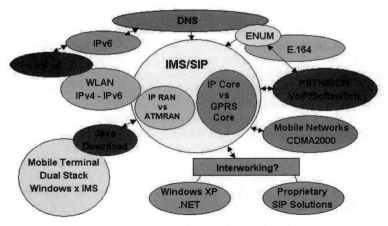

Fig. 9   Areas to be clarified for IMS

- Interworking with Voice over IP (VoIP), with the fixed telecoms networks and with the enterprise networks
- Interworking with SIP solutions in the Internet
- Java download to enable mobile terminals for IMS
- Interworking with WLAN based access
- Billing

8.   Conclusions

It has become obvious over the last few years, that low-cost, small, long-life, and low-powered microprocessor-driven devices will enforce ubiquitous computing. This development is additionally driven by advanced network architectures in the sectors of mobile network markets. The circuit-switched mobile networks evolved to a concentration of voice traffic towards large nodes, data traffic moved into packet-switched networks. In the Internet sector networking led to a decentralised and heterogeneous node-structure with a number of advantages for data services. The advantages of the circuit-switched networks are guaranteed bit rate and QoS which are important for voice; their disadvantages lie in the bit rate inflexibility and in the growing resource loss with increasing bit rates. The advantages of the Internet are applied now for mobile NGN's. They lie mainly in the flexibility regarding bit rate adaptions and conversions between servers and terminals; the disadvantages lie in the field of quality of service. With growing bitrates and capacity of packet-switched networks, it becomes feasible to take over transport and switching for realtime services like voice and video. This needs to be done with the SIP protocols. The Internet brought a lot of dynamics into more decentralised networking. In the future, there will be also networks that are organised on an ad hoc basis: these will be spontaneous dynamic networks complementing the further developed existing network infrastructures. The question is, will both work together and how? This question we cannot answer today.

Another question that affects ubiquitous computing in the future is the question of integration different wireless access technologies into a harmonised mobile environment. There are the Wireless Personal Area Networks (WPAN's), the Wireless Local Area Networks (WLAN's) and future active self-organised networks. The UMTS vision of the integration of radio zones and technologies remains a project for the future and will not really happen as long as there are no real requirements arising from the market. Initially this has already been a target for IMT-2000 – the result was different solutions from different standardisation organisations like IEEE, Bluetooth/SIG, and the regional standardisation bodies like ARIB, ETSI, ANSI, CWTS, TTA, etc. Meanwhile, it is obvious that the initial vision of a single common radio standard in all environments will not be fulfilled as long as no real demand drives the development in such directions. The upcoming convergence of digital broadcast radio technology with 3G mobile radio also does not look like it will bring a common solution. From a technical point of view, we will have to wait for software-defined radio.

Another development could be driven by mobile specific services: they may take away the mobile networks from fixed Internet. This possible development is indicated in Fig. 10. However, convergence remains an issue.

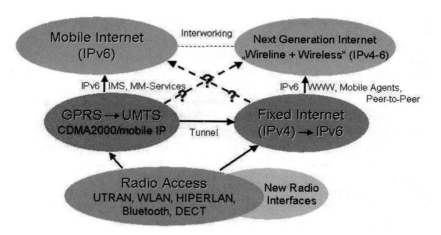

Fig. 10   NGN's: Wireless Internet or Mobile Internet?

The emerging IPv6 based session handling protocols are driven strongly by mobile operators and users which have mobile devices with Internet capability.

The mobile operators want a well defined session handling in order to do charging and billing. The user requires multimedia communication and expects assistance from the network for parallel sessions related to different services. For personalisation and situation-dependent services, user and position-information are needed and have to be combined with content.

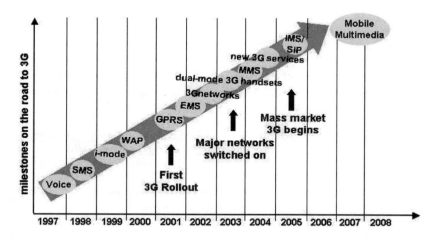

Fig. 11   3G Services Roadmap

Such tasks can only be performed by intelligence in the network. Fig. 11 shows the present roadmap for 3G mobile multimedia services. These services allow mobile access to fixed Internet services, which will continue to use comprehensive web content up to very high complexity and quality. In addition, they also support users with mobile specific applications with pre-selected information according to his personal profile and geographical situation. The IP Multimedia Subsystem will provide the possibility to combine a mix of services according to the users' need. It can be foreseen that IMS will be deployed in the timeframe between 2005 and 2006.

These mobile network developments will take us to the question in future: Will the mobile NGN representing a mobile Internet be a part of the fixed Internet or will it be a separate network environment with the same protocols, however with its own functionality, its own specific services, its own DNS and will it also play a role as an access means for fixed Internet/Intranet user access? The future will answer this question.

More information can be found in [1].

References

[1]   Book: UMTS and Mobile Computing
      Josef Franz Huber
      Alexander Joseph Huber, www.artechhouse.com

      Mail: Artech House Books, April 2002
      46 Gillingham Street
      London, SW1V 1AH, UK

[2]   UMTS Forum Report No. 10: Shaping the Mobile Multimedia Future –
      An Extended Vision from the UMTS Forum. UMTS Forum, 2000,
      www.umts-forum.org

[3]   UMTS Forum Report No. 13: The UMTS Third Generation Market –
      Phase II: Structuring the Service Revenue Opportunities. UMTS Forum,
      2000, www.umts-forum.org

[4]   UMTS Forum Report No. 20: IMS Service Vision for 3G Markets.
      UMTS Forum, May 2002, www.umts-forum.org

[5]   UMTS Forum Report No. 27: Strategic Considerations for IMS – the 3G
      Evolution. UMTS Forum, January 2003, www.umts-forum.org

[6]   Session Initiation Protocol (SIP). IETF June 2002
      http://ftp.rfc-editor.org/in-notes/rfc3261.txt

[7]   IEEE Computer Magazine: Toward the Mobile Internet. October 2002,
      pp100 - 102, http://computer.org

Chapter 2

PROTOCOLS FOR DATA PROPAGATION IN WIRELESS SENSOR NETWORKS

Azzedine Boukerche*

School of Information Technology and Engineering (SITE)
University of Ottawa
Ottawa, ONT, Canada
boukerch@site.uottawa.ca

Sotiris Nikoletseas[†]

Department of Computer Engineering and Informatics
University of Patras, and Computer Technology Institute (CTI)
Greece
nikole@cti.gr

Abstract

Recent dramatic developments in micro-electro-mechanical systems (MEMS), wireless communications and digital electronics have already led to the development of small in size, low-power, low-cost sensor devices. Such devices integrate sensing, data processing and communication capabilities.

Examining each such device individually might appear to have small utility, however the effective distributed co-ordination of large numbers of such devices may lead to the efficient accomplishment of large sensing tasks. Large numbers of sensors can be deployed in areas of interest (such as inaccessible terrains or disaster places) and use self-organization and collaborative methods to form a network. Their wide range of applications is based on the possible use of various sensor types. Thus,

*This work has been partially supported by Research Grants from the Canada Research Chair Program, Canada Foundation Innovation and Ontario Distinguished Researcher Award OIT/ODRA #201722.
†This work has been partially supported by the IST/FET Programme of the European Union under contract numbers IST-1999-14186 (**ALCOM-FT**) and IST-2001-33116 (**FLAGS**).

sensor networks can be used for continuous sensing, event detection, location sensing as well as micro-sensing.

We note however that the efficient and robust realization of such large, highly-dynamic, complex, non-conventional networking environments is a challenging technological and algorithmic task. Features including the huge number of sensor devices involved, the severe power, computational and memory limitations, their dense deployment and frequent failures, pose new design, analysis and implementation challenges.

This chapter aims at presenting certain important aspects of the design, deployment and operation of sensor networks. In particular, to provide a) a brief description of the technical specifications of state-of-the-art sensor devices b) a discussion of possible models used to abstract such networks c) a presentation of some characteristic protocols for data propagation in sensor networks, along with an evaluation of their performance analysis.

Keywords: Wireless sensors, smart dust, algorithms

1. Modeling Aspects of Wireless Sensor Networks

1.1 Technical Specifications of Current Micro-Sensor Devices

State-of-the-art sensor devices exhibit an enhanced integration of three basic components: the sensing part, the processing and communication unit and the battery. An important challenge is to incorporate the desired sensing, computing and communication capabilities within an extremely small volume, while maintaining very low power consumption.

Towards building a networking infrastructure for ultra-low energy, communication technologies are based on radio frequency (RF) or optical transmission. Each technique has relative advantages and disadvantages.

Several research groups and projects have recently begun to investigate distributed sensor networks. Without aiming at being exhaustive, we mention the following pioneering research and development activities:

- The "Smart Dust" Project at Berkeley, pursuing a volume goal at the scale of a few cubic millimeters, focusing on free-space optical transmission. They mention the following basic challenges for optical sensor networking: a) optical links require uninterrupted line-of-sight paths b) the directional transmission characteristics must be supported by system design and c) protocols should deal with the severe trade-offs between transmission range, energy per bit, directionalilty and transmission range.

- The "Wireless Integrated Network Sensors" (WINS) Project at UCLA, with similar goals, concentrating on RF communications over short distances. The project is developing low power MEMS-based devices that can sense, actuate and communicate.

- The "Ultralow Power Wireless Sensor" Project at MIT, with a primary focus on extremely low energy consumption, using RF communication. Their prototype will be able to support various transfer rates, from 1 bit/sec up to 1 megabit/sec, with power levels varying from 10 μW to 10 mW.

Towards presenting in some detail and discussing the capabilities and limitations of current sensor networks, we below provide the basic technical specifications of an existing system, developed in Berkeley (see e.g. [28]).

We note that other approaches and systems developed within the projects mentioned above might have served this goal equally well.

The basic hardware uses a fraction of a watt of power and consists of commercial components a square inch in size. The hardware design contains of a small, low-power radio and processor board (known as a mote processor/radio, or MPR, board) and one or more sensor boards (known as a mote sensor, or MTS, board). The combination of the two types of boards form a networkable wireless sensor.

The MPR board includes a processor, radio, A/D converter and battery. The processor runs at 4 MHz, has 128 KB of flash memory and 4 KB of SDRAM. In a given area, thousands of sensors may continuously report data, creating a heavy data load. Thus, the overall system is memory-constrained, but this feature is a rather common design challenge in any wireless sensor network.

The MPR modules also contain various sensors, which are available through a small connector that links the MPR and MTS modules. The sensor interface includes an A/D converter and two serial ports. This allows the MPR module to connect to a variety of MTS sensor modules, including MTS modules that use analog sensors as well as digital sensors.

The processor, radio, and a typical sensor consume about 100 mW in active mode. This figure should be compared with the 30 μA draw when all components are in "sleep" mode. Thus, well-designed schemes that allow particles to alternate between the two modes without delaying too much data delivery are particularly successful in saving energy.

The MTS sensor boards currently include light, temperature, two-axis acceleration, and magnetic sensors and $420mA$ transmitters. The wireless transmission is at $4Kbps$ rate and the transmission range may vary, depending also on the type of antenna used, including directional antennae. Researchers are also developing a GPS board and a multi-sensor board that incorporates a small speaker and light, temperature, magnetic, acceleration, and acoustic (microphone) sensing devices.

Table 1.1.   Sensor Device Specifications

CPU Speed	4 MHz
Memory	ROM: 128Kb FLASH
	SDRAM: 4Kb
	EEPROM: 4Kb
Power Supply	2X AA batteries
Power Consumption	0.75 mW
Processor Current Draw	5.5 mA (active current)
	< 20 μA (sleep mode)
Radio Current Draw	12 mA (transmit current)
	1.8 mA (receive current)
	< 1 μA (sleep mode)
Output Device	3 LEDs
I/O Port	Expansion Connected (51 pin)
	Serial port (Proprietary 16-pin)
Network	Wireless 4 Kbits/sec at 916 MHz (ISM band)
	Radio range depends on antennae configuration

An important challenge in the development of sensor devices and networks is in the software embedded in the sensors. The software runs the hardware and network, making sensor measurements, routing measurement data, and controlling power dissipation. In effect, it is the key ingredient that makes the wireless sensor network to properly operate.

To this end, a lot of research and development has gone into the design of a software environment that appropriately supports wireless sensors. The result is a very small operating system, which however allows the networking, power management, and sensor measurement details to be properly abstracted from the application development. The operating system also creates a standard method of developing applications and extending the hardware. Although tiny, this operating system is quite efficient, as shown by the small stack handling time.

For a summary of technical specifications, see Tables 1.1 and 1.2.

1.2   An Abstract Model
for Wireless Sensor Networks

Sensor networks are comprised of a vast number of ultra-small homogenous sensors, which we call *"grain" or "smart dust" particles*. Each grain particle is a fully-autonomous computing and communication de-

Table 1.2. Operating System Key Facts

Software Footprint	3.4 Kb
Transmission Cost	1 μJ/Bit
Inactive State	$< 25\ \mu A$
Peak Load	20 mA
Typical CPU Usage	$< 50\ \%$
Events Propagate Thru Stack	$< 40\ \mu S$

vice, characterized mainly by its available power supply (battery) and the energy cost of computation and transmission of data. Such particles (in our model) do not move.

Each particle is equipped with a set of monitors (sensors) for various conditions, such as light, pressure, humidity, temperature etc. Each particle has a *broadcast* (digital radio) *mode*, which can be also a directional transmission of varying angle α around a certain line (possibly using some special kind of antenna).

We adopt here (as a starting point) a two-dimensional (plane) framework: A *smart dust cloud* (a set of grain particles) is spread in an area (for a graphical presentation, see Fig. 1.1). Usually the deployment of particles is done in a rather random manner (such as when particles are dropped by an airplane over the area of interest. In variations of this basic model, we may include the possibility of a (more or less) structured deployment (possibly done by humans or robots).

Let n be the number of smart dust particles and let d (usually measured in numbers of *particles*/m^2) be the *density* of particles in the area. Let R be the maximum transmission range of each grain particle. The transmission range of each particle may vary during the network's lifetime.

There is a single point in the network area, which we call the sink S, that represents a control center where data should be propagated to. In variations of this basic model, there might be multiple sinks, which may be static or moving.

Furthermore, we assume that there may exist a set-up phase of the smart dust network, during which the smart cloud is dropped in the terrain of interest, when using special control messages (which are very short, cheap and transmitted only once) each smart dust particle is provided with the direction of S. In the case of a moving sink this set-up phase may run when significant changes in the sink's position occur. By assuming that each smart-dust particle has individually *a sense of direction*, (such as one provided by a common coordinates system) and using

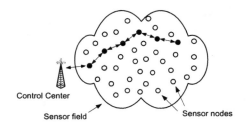

Figure 1.1. A Sensor Network

these control messages, each particle is aware of the "general" location of S.

This model, although quite simple to allow a rigorous analysis of protocol performance to develop, depicts accurately enough the technological specifications of real smart dust systems.

2.   Characteristic Applications

Very large numbers of sensor nodes can be deployed in areas of interest (such as inaccessible terrains or disaster places) and use *self-organization and collaborative methods* to form (in an ad-hoc way) a sensor network. The possible use of various sensor types (i.e. thermal, visual, seismic, acoustic, radar, magnetic, etc.) allows the network to monitor a wide variety of conditions (e.g. temperature, object presence and movement, humidity, pressure, noise levels etc.).

Thus, sensor networks can be used for continuous sensing, event detection, location sensing as well as micro-sensing. Hence, sensor networks have important applications, including:

(a) environmental applications (such as fire detection, flood detection, precision agriculture),

(b) health applications (like telemonitoring of human physiological data),

(c) home applications (e.g. smart environments and home automation) and

(d) military (like forces and equipment monitoring, battlefield surveillance, targeting, nuclear, biological and chemical attack detection).

For excellent surveys of wireless sensor networks see [1, 3] and also [10, 18].

3. New Challenges

The efficient and robust realization of such large, highly-dynamic, complex, non-conventional networking environments is *a challenging algorithmic and technological task*. Features including the huge number of sensor devices involved, the severe power, computational and memory limitations, their dense deployment and frequent failures, pose *new design, analysis and implementation aspects* which are essentially different not only with respect to distributed computing and systems approaches but also to ad-hoc networking techniques.

We emphasize the following characteristic differences between sensor networks and ad-hoc networks:

- The number of sensor particles in a sensor network is extremely large compared to that in a typical ad-hoc network.

- Sensor networks are typically prone to faults.

- Because of faults as well as energy limitations, sensor nodes may (permanently or temporarily) join or leave the network. This leads to highly dynamic network topology changes.

- The density of deployed devices in sensor networks is much higher than in ad-hoc networks.

- The limitations in energy, computational power and memory are much more severe in sensor networks.

Because of the above rather unique characteristics of sensor networks, efficient and robust distributed protocols and algorithms should exhibit the following critical properties:

Scalability
Distributed protocols for sensor networks should be highly scalable, in the sense that they should operate efficiently in extremely large networks composed of huge numbers of nodes. This feature calls for an urgent need to prove by analytical means and also validate (by large scale simulations) certain efficiency and robustness (and their trade-offs) guarantees for asymptotic network sizes.

Efficiency
Because of the severe energy limitations of sensor networks and also because of their time-critical application scenaria, protocols for sensor networks should be efficient, with respect to both energy and time.

Fault-tolerance
Sensor particles are prone to several types of faults and unavailabilities, and may become inoperative (permanently or temporarily). Various reasons for such faults include physical damage during either the deployment or the operation phase, permanent (or temporary) cease of operation in the case of power exhaustion (or energy saving schemes, respectively). The sensor network should be able to continue its proper operation for as long as possible despite the fact that certain nodes in it may fail.

4.   Data Propagation Protocols

Because of the complex nature of a sensor network (that integrates various aspects of communication and computing), protocols, algorithmic solutions and design schemes for all layers of the networking infrastructure are needed. Far from being exhaustive, we mention the need for frequency management solutions at the physical layer, for Medium Access Control (MAC) protocols to cope with multi-hop transmissions at the data link layer. The interested reader may use the excellent survey by Akyildiz et al [3] for a detailed discussion of design aspects of all layers of the networking infrastructure.

We focus in this chapter on algorithms and protocols for the network layer. We believe that a complementary use of rigorous analysis and large scale simulations is needed to fully investigate the performance of data propagation protocols in wireless sensor networks. In particular, asymptotic analysis may lead to provable efficiency and robustness guarantees towards the desired scalability of protocols for sensor networks that have extremely large size. On the other hand, simulation allows to investigate the effect of a great number of detailed technical specifications of real devices, a task that is difficult (if possible at all) for analytic techniques which, by their nature, use abstraction and model simplicity.

We below present some represenative state-of-the-art protocols and paradigms that try to avoid flooding the network, achieving good perforance (with respect to time and energy) and robustness.

The list of protocols discussed is not meant to be complete.

4.1   Directed Diffusion - a Data-centric Paradigm

4.1.1   The Model and the Protocol.   Directed Diffusion, a novel paradigm for wireless sensor networking introduced by Intanagonwiwat, Govindan and Estrin in [15], is data-centric in its nature, in the sense that all communication performed in the sensor network is for named data. Simply speaking, a network operated under this paradigm, obeys the following simple model: one or more sinks (possibly corresponding to human operators) broadcast questions of the following form: "How many objects of a certain kind move within a certain region X?". Sensors within area X should collect the information requested and subsequently collaborate with nearby sensors to forward the query result back to the sink(s).

Directed Diffusion manages the following four basic elements:

(a) Interest messages, i.e. queries specifying what the sinks are looking for. Such task descriptions are named by a list of attribute-value pairs. Thus, the task description specifies an interest for data matching the attributes.

(b) Interests are injected by the sinks and disseminated throughout the sensor network. During this process, gradients are set within the network in order to "collect" data matching the interest. More specifically, gradients are direction states created at each sensor that receives an interest pointing towards the sensor which the interest was received from.

(c) Clearly, this process may create multiple gradient paths towards the sinks. However, towards saving energy and avoid flooding the network, once a sink starts receiving multiple events, it reinforces one particular sensor to "draw down" real data, to achieve high quality tracking of targets and high data rates.

(d) Subsequently, data matching the interest is propagated towards the sinks using the established low-latency paths.

We should emphasize here that Directed Diffusion is a paradigm rather than a single protocol. It allows various design choices for its building components. To name a few, data propagation may occur along a single path, or be multi-path with selected quality or even randomized, and gradient reinforcement may vary with respect to when to reinforce.

4.1.2   Analytic Evaluation.   In [15, 16], the authors present an analysis of the cost of the protocol and also compare its performance to flooding. For simplicity, the analysis is performed on a idealized setting i.e. they assume that the network topology is a square grid, of N nodes. Transmission range is set in a way that each sensor can communicate directly (in one hop) with exactly eight neighbor sensors on the grid. A number of n sources of events are placed along the left edge of the grid, whereas all sinks are placed along the right edge.

Under this setting, the authors measure performance in terms of total cost of transmission and reception of one event from each source to all the sinks. They are able to prove an asymptotic $O(n\sqrt{N})$ cost for directed diffusion. Then they compare the performance of their technique to that of flooding. Recall that in flooding, sources of events broadcast data to all their neighbors in the network. It easy to see that the cost of the flooding scheme is $O(nN)$ i.e. much bigger compared to the cost of directed diffusion.

In [16], a packet-level simulation using ns-2 has been performed to comparatively study diffusion and flooding, and also investigate the impact of network dynamics (sensor failures) to performance. The experimental results obtained indicate that directed diffusion exhibits much better energy dissipation compared to flooding, while having good latency properties.

4.2   LEACH - A Cluster-based Approach

4.2.1   The LEACH Protocol.   The Leach (Low-Energy Adaptive Clustering Hierarchy) Protocol, introduced by Heinzelman, Chandrakasan and Balakrishnan in [13], is a routing protocol for sensor networks aiming at reducing energy dissipation and also increasing the network's lifetime, by evenly distribute energy consumption among the sensors in the network. Similar ideas were subsequently used by Chatzigiannakis and Nikoletseas in [8].

LEACH is a clustering, self-organizing scheme in the sense that the nodes in the network form (using randomization) local clusters, with one sensor per cluster being a cluster-head. As opposed to conventional clustering techniques, cluster-heads are not fixed throughout the network's lifetime. Instead, cluster-head positions rotate in a randomized way among all sensors of the network. This "load balancing" technique avoids exhausting the battery of a single sensor and thus increases the lifetime of the network.

Randomization, as a load balancing technique, is used in the following way: Each sensor may become a local cluster-head at any given time with a certain probability. After their election, cluster-heads inform all other sensors in the network about their status. Each sensor chooses which cluster-head it belongs to by trying to minimize the communication cost within a cluster. Typically, this is done by selecting the closest cluster-head as the one a sensor belongs to.

Once clusters are formed and cluster-heads become chosen, communication to the sink is done in a hierarchical way: The sensors in each cluster communicate with the cluster-head. The cluster-head collects data and transmits it to the sink in one hop. Thus, the largest part of communication is done locally and thus the cost it incurs is rather small. Only few nodes (the cluster-heads) transmit directly to the sink.

Such direct transmissions might be of long distance, thus leading to high energy costs. The fact, however, that cluster-heads are few in number and also periodically rotate, avoids exhausting the energy of any single node and leads to significant energy savings. In fact, one can estimate an optimal number of cluster-heads based on several parameters, including network topology and communication and computing restrictions.

In [13] the authors presented a set of simulation experiments and have demonstrated that a significant energy dissipation reduction (a factor of 7) compared to direct communication to the sink. Furthermore, LEACH exhibits significant improvements in system lifetime compared to direct communication and static clustering. This is due to the fact that (because of randomized cluster-head rotation) each sensor dies rather randomly and at almost the same rate. In particular, the authors in [13] simulate LEACH for a random network, certain radio and computation cost parameters, while varying the percentage of total nodes that are cluster heads. Their experiments show how the energy dissipation in the system varies as the percent of nodes that are cluster-heads is changed.

They also show that LEACH can achieve over a factor of 7 reduction in energy dissipation compared to direct communication with the base station, when using the optimal number of cluster-heads. The main energy saving of the LEACH protocol is due to combining lossy compression with the data routing. There is clearly a trade-off between the quality of the output and the amount of compression achieved. In this, case, some data from the individual signals is lost, but this results in a substantial reduction of the overall energy dissipation of the system.

4.3   STEM - A Topology Management Protocol

4.3.1   Protocol Description.
The STEM (Sparse Topology and Energy Management) Protocol, proposed by Shurgers, Tsiatsis, Ganeriwal and Srivastava in [27], tries to achieve energy saving by appropriately affecting the topology of the sensor network, with respect to two parameters: the path latency and the density of nodes. The topology management is achieved by aggressively putting nodes to "sleep" (i.e. switch their radio off) and wake-up only when they must propagate data. When a sensor enters a "sleep" state, it becomes disconnected from the network topology, which becomes sparser. Thus, STEM trade-offs latency for energy saving.

By further integrating with energy saving techniques that heavily rely on network density (taking advantage of the fact that nearby nodes are equivalent with respect to data routing, see GAF, SPAN protocols in [9], [32]) STEM achieves further energy savings.

More specifically, when a node "falls asleep", it ceases any communication with other nodes (since its radio is switched off), however its sensors and processor remain active (since they are much less energy consuming), thus it can still sense events in its monitoring area. An efficient topology management scheme should appropriately coordinate these sleeping periods of all the nodes, to save as much energy as possible while at the same time ensuring that data propagation to the sink can be done efficiently. Towards this goal, several topology management techniques (such as SPAN and GAF) try to exploit the density of the network to put nodes asleep while guaranteeing that nearby nodes will still be able to propagate data. It is clear that such techniques become more efficient in saving energy as the density of the network increases.

On the other hand, STEM exploits time rater than density. This approach is particularly successful in many practical scenaria when data propagation must in fact be done infrequently (such as in the case of a rare event). In such cases, the network is most of the time in a "monitoring state" (waiting for an event to occur). During this state nodes can be asleep. When the event happens, the network must enter a "transfer state", in the sense that data has to be propagated to the sink quickly. The node detecting the event must now forward information about the event to the sink. Thus it turns its radio on and tries to notify a next-hop node on a multi-hop path towards the sink. Note however that it may fail to find such a closer-to-the-sink node, since all such nodes may have turned off their radios.

To deal with this problem, STEM forces nodes to periodically turn their radios on for an appropriately short time, to be able to listen if some other node wants to communicate with them. The node that wants to communicate to a next-hop node is called "the initiator node" while the node that must wake up to participate in data propagation is called the "target node". When the target node is notified of the need to enter an information forwarding process towards the sink and wakes up, it becomes an initiator node for the next hop itself and so on up to the sink.

In order to avoid actual data transmissions to interfere with wake-up beackons, STEM uses two different frequencies for each of them (and a separate radio for each frequency). This is already supported by current technology. This dual radio frequency approach has many benefits. If we use only one radio and one frequency, jamming phenomena may arise and extra delays may occur. Furthermore, if we use one radio and two frequencies we have to interrupt data transmission periodically to listen for wake-up beackons. Even in the dual radio setup collisions may arise, so STEM takes extra care of this (such as that a node becomes awake in the case of collisions of wake-up beacons).

4.3.2   Theoretical Analysis of the STEM Protocol.   In [27], two important performance parameters are analyzed: a) the setup latency of a link b) the energy savings achieved by the protocol. The setup latency T_s of a link is defined as the time duration between the moment the initiator node sends out wake-up beacons and the moment that both nodes (initiator and target) turn on their data radio. In a typical setting, the two nodes are not synchronized, so we can assume that the initiator's first wake-up message is sent out at a time which is uniformly randomly distributed in the interval T (the time interval that the initiator node waits for a response from the target node before starting data transmission). Let B_1 be the transmit time of a beacon packet and let B_2 be the transmission time of a beacon acknowledgement. Let T_{Rx} be the time period a target node turns its radio on. (For the target node to receive at least one beacon, T_{Rx} should be appropriately long). Let T_B the inter-beacon interval. (Thus, in order for the target node to receive the second beacon as well, we must have $T_{Rx} \geq B_1 + T_B$).

In the case of no collisions in the wake up plane, and under reasonable assumptions that usually arise in practice, in [27] it is proved that the average set up latency of a link is:

$$\bar{T}_s = \frac{T + T_B}{2} + 2B_1 + B_2 - T_{Rx}$$

The above equation assumes that $T > T_{Rx}$. In the case of no sleeping periods at all (i.e. $T = T_{Rx}$) the above yields

$$\bar{T}_s = B_1 + B_2$$

In the case of collisions of wake-up beacons, it is clear that initiators will eventually time out after time T, so the setup latency becomes

$$\bar{T}_s = T$$

As far as energy saving is concerned, let E_0 be the total energy consumed during a time internal t, and E be the energy consumption during interval t when using STEM.

In [27], it is proved that

$$\frac{E}{E_0} = \frac{1}{\beta} + f_S \left(t_{burst} + \frac{T}{2} \right) + 2 \frac{P_{sleep}}{P}$$

where :

- T is the time interval an initiator waits for a target node's response before starting data transmission.

- $\beta = T/T_{Rx}$, where T_{Rx} is the time duration a target turns its radio on (so β is the inverse of the duty cycle in the wake up domain).

- f_S is the number of link setups per second (i.e. we call f_S the setup frequency).

- If t_{data} is the total time the radio is turned on in the data plane, then we split t_{data} in bursts of average duration t_{burst}, where a burst of data transfer requires one link setup. Consequently, the fraction of time the data-plane radio is on, is $\alpha = \frac{t_{data}}{t} = f_S\, t_{burst}$.

- P_{sleep} and P represent power consumption during sleeping periods and link setups, respectively.

The main conclusions reached in [27] using the above formula are the following:

- Clearly, energy saving increases with β i.e. by increasing period T. This however means larger link setup latencies.

- Energy savings are also large when the link set-up frequency f_S becomes smaller (since in this case fewer link setups are needed).

- Clearly, the energy savings become larger when sleeping time periods increase.

4.4   The Local Target Protocol (LTP)

4.4.1   The Model.   The LTP Protocol was introduced by Chatzigiannakis, Nikoletseas and Spirakis in [5]. The authors adopt a two-dimensional (plane) framework: A *smart dust cloud* (a set of particles) is spread in an area (for a graphical presentation, see Fig. 1.1).

Let d (usually measured in numbers of *particles*/m^2) be the *density* of particles in the area. Let $\mathcal{R}$ be the maximum (radio/laser) transmission range of each grain particle.

A *receiving wall* $\mathcal{W}$ is defined to be an infinite line in the smart-dust plane. Any particle transmission within range $\mathcal{R}$ from the wall $\mathcal{W}$ is received by $\mathcal{W}$. $\mathcal{W}$ is assumed to have very strong computing power, able to collect and analyze received data and has a constant power supply and so it has no energy constraints. The wall represents in fact the authorities (the fixed control center) who the realization of a crucial event should be reported to. The wall notion generalizes that of the sink and may correspond to multiple (and/or moving) sinks. Note that a wall of appropriately big (finite) length suffices.

The notion of multiple sinks which may be static or moving has also been studied in [29], where Triantafilloy, Ntarmos, Nikoletseas and Spirakis introduce "NanoPeer Words", merging notions from Peer-to-Peer Computing and Smart Dust.

Furthermore, there is a set-up phase of the smart dust network, during which the smart cloud is dropped in the terrain of interest, when using special control messages (which are very short, cheap and transmitted only once) each smart dust particle is provided with the direction of $\mathcal{W}$. By assuming that each smart-dust particle has individually *a sense of direction*, and using these control messages, each particle is aware of the general location of $\mathcal{W}$.

4.4.2 The Protocol. Let $d(p_i, p_j)$ the distance (along the corresponding vertical lines towards $\mathcal{W}$) of particles p_i, p_j and $d(p_i, \mathcal{W})$ the (vertical) distance of p_i from $\mathcal{W}$. Let *info($\mathcal{E}$)* the information about the realization of the crucial event $\mathcal{E}$ to be propagated. Let p the particle sensing the event and starting the execution of the protocol. In this protocol, each particle p' that has received *info($\mathcal{E}$)*, does the following:

- *Search Phase*: It uses a periodic low energy directional broadcast in order to discover a particle nearer to $\mathcal{W}$ than itself. (i.e. a particle p'' where $d(p'', \mathcal{W}) < d(p', \mathcal{W})$).

- *Direct Transmission Phase*: Then, p' sends *info($\mathcal{E}$)* to p''.

- *Backtrack Phase*: If consecutive repetitions of the *search phase* fail to discover a particle nearer to $\mathcal{W}$, then p' sends *info($\mathcal{E}$)* to the particle that it originally received the information from.

Note that one can estimate an a-priori upper bound on the number of repetitions of the search phase needed, by calculating the probability of success of each search phase, as a function of various parameters (such as density, search angle, transmission range). This bound can be used to decide when to backtrack.

For a graphical representation see figures 1.2, 1.3.

4.4.3 Theoretical Analysis. [5] first provides some basic definitions.

DEFINITION 1.1 Let $h_{opt}(p, \mathcal{W})$ be the (optimal) number of "hops" (direct, vertical to $\mathcal{W}$ transmissions) needed to reach the wall, in the *ideal* case in which particles always exist in pair-wise distances $\mathcal{R}$ on the vertical line from p to $\mathcal{W}$. Let Π be *a smart-dust propagation protocol*, using

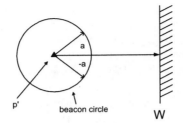

Figure 1.2.   Example of the Search Phase

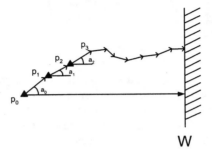

Figure 1.3.   Example of a Transmission

a transmission path of length $L(\Pi, p, \mathcal{W})$ to send info about event $\mathcal{E}$ to wall $\mathcal{W}$. Let $h(\Pi, p, \mathcal{W})$ be the actual number of hops (transmissions) taken to reach $\mathcal{W}$. The *"hops" efficiency* of protocol Π is the ratio

$$C_h = \frac{h(\Pi, p, \mathcal{W})}{h_{opt}(p, \mathcal{W})}$$

Clearly, the number of hops (transmissions) needed characterizes the energy consumption and the time needed to propagate the information $\mathcal{E}$ to the wall. Remark that $h_{opt} = \left\lceil \frac{d(p, \mathcal{W})}{\mathcal{R}} \right\rceil$, where $d(p, \mathcal{W})$ is the (vertical) distance of p from the wall $\mathcal{W}$.

In the case where the protocol Π is randomized, or in the case where the distribution of the particles in the cloud is a random distribution, the number of hops h and the efficiency ratio C_h are random variables and one wishes to study their expected values.

The reason behind these definitions is that when p (or any intermediate particle in the information propagation to $\mathcal{W}$) "looks around" for a particle as near to $\mathcal{W}$ as possible to pass its information about $\mathcal{E}$, it may not get any particle in the perfect direction of the line vertical to $\mathcal{W}$. This difficulty comes mainly from three causes: a) Due to the initial spreading of particles of the cloud in the area and because particles do not move, there might not be any particle in that direction. b) Particles of sufficient remaining battery power may not be currently available in the right direction. c) Particles may temporarily "sleep" (i.e. not listen to transmissions) in order to save battery power.

Note that any given distribution of particles in the smart dust cloud may not allow the ideal optimal number of hops to be achieved at all. In fact, the least possible number of hops depends on the input (the positions of the grain particles). [5] however, compares the efficiency of protocols to the ideal case. A comparison with the best achievable number of hops in each input case will of course give better efficiency ratios for protocols.

To enable a first step towards a rigorous analysis of smart dust protocols, [5] makes the following simplifying assumption: *The search phase always finds a p''* (of sufficiently high battery) in the semicircle of center the particle p' currently possessing the information about the event and radius R, in the direction towards $\mathcal{W}$. Note that this assumption on always finding a particle can be relaxed in the following ways: (a) by repetitions of the search phase until a particle is found. This makes sense if at least one particle exists but was sleeping during the failed searches, (b) by considering, instead of just the semicircle, a cyclic sector defined by circles of radiuses $\mathcal{R} - \Delta \mathcal{R}$, $\mathcal{R}$ and also take into account the density of the smart cloud, (c) if the protocol during a search phase ultimately fails to find a particle towards the wall, it may *backtrack*.

[5] also assumes that the position of p'' is uniform in the arc of angle $2a$ around the direct line from p' vertical to $\mathcal{W}$. Each data transmission (one hop) takes constant time t (so the "hops" and time efficiency of our protocols coincide in this case). It is also assumed that each target selection is stochastically *independent* of the others, in the sense that it is always drawn uniformly randomly in the arc $(-\alpha, \alpha)$.

The above assumptions may not be very realistic in practice, however, they can be relaxed and in any case allow to perform a first effort towards providing some concrete analytical results.

LEMMA 1.2 ([5]) *The expected "hops efficiency" of the local target protocol in the a-uniform case is*

$$E(C_h) \simeq \frac{\alpha}{\sin \alpha}$$

for large h_{opt}. Also

$$1 \leq E(C_h) \leq \frac{\pi}{2} \simeq 1.57$$

for $0 \le \alpha \le \frac{\pi}{2}$.

Proof: Due to the protocol, a sequence of points is generated, $p_0 = p, p_1, p_2, \ldots, p_{h-1}, p_h$ where p_{h-1} is a particle within $\mathcal{W}$'s range and p_h is part of the wall. Let α_i be the (positive or negative) angle of p_i with respect to p_{i-1}'s vertical line to $\mathcal{W}$. It is:

$$\sum_{i=1}^{h-1} d(p_{i-1}, p_i) \le d(p, \mathcal{W}) \le \sum_{i=1}^{h} d(p_{i-1}, p_i)$$

Since the (vertical) progress towards $\mathcal{W}$ is then $\Delta_i = d(p_{i-1}, p_i) = \mathcal{R} \cos \alpha_i$, we get:

$$\sum_{i=1}^{h-1} \cos \alpha_i \le h_{opt} \le \sum_{i=1}^{h} \cos \alpha_i$$

From Wald's equation for the expectation of a sum of a random number of independent random variables (see [26]), then

$$E(h-1) \cdot E(\cos \alpha_i) \le E(h_{opt}) = h_{opt} \le E(h) \cdot E(\cos \alpha_i)$$

Now, $\forall i$, $E(\cos \alpha_i) = \int_{-\alpha}^{\alpha} \cos x \frac{1}{2\alpha} dx = \frac{\sin \alpha}{\alpha}$. Thus

$$\frac{\alpha}{\sin \alpha} \le \frac{E(h)}{h_{opt}} = E(C_h) \le \frac{\alpha}{\sin \alpha} + \frac{1}{h_{opt}}$$

Assuming large values for h_{opt} (i.e. events happening far away from the wall, which is the most interesting case in practice since the detection and propagation difficulty increases with distance) we have (since for $0 \le \alpha \le \frac{\pi}{2}$ it is $1 \le \frac{\alpha}{\sin \alpha} \le \frac{\pi}{2}$) and the result follows.   ∎

4.4.4   Local Optimization: The Min-two Uniform Targets Protocol (M2TP).

[5] further assumes that the search phase always returns *two points* p'', p''' each uniform in $(-\alpha, \alpha)$ and that a modified protocol M2TP selects the best of the two points, with respect to the local (vertical) progress. This is in fact an optimized version of the Local Target Protocol.

In a similar way as in the proof of the previous lemma, the authors prove the following result:

LEMMA 1.3 ([5]) *The expected "hops efficiency" of the "min two uniform targets" protocol in the a-uniform case is*

$$E(C_h) \simeq \frac{\alpha^2}{2(1 - \cos \alpha)}$$

for $0 \le \alpha \le \frac{\pi}{2}$ and for large h.

Now remark that

$$\lim_{\alpha \to 0} E(C_h) = \lim_{\alpha \to 0} \frac{2\alpha}{2\sin a} = 1$$

and

$$\lim_{\alpha \to \frac{\pi}{2}} E(C_h) = \frac{(\pi/2)^2}{2(1 - 0)} = \frac{\pi^2}{8} \simeq 1.24$$

Thus, [5] proves the following:

LEMMA 1.4 ([5]) *The expected "hops" efficiency of the min-two uniform targets protocol is*

$$1 \le E(C_h) \le \frac{\pi^2}{8} \simeq 1.24$$

for large h and for $0 \le \alpha \le \frac{\pi}{2}$.

Remark that, with respect to the expected hops efficiency of the local target protocol, the min-two uniform targets protocol achieves, because of the one additional search, a relative gain which is $(\pi/2 - \pi^2/8)/(\pi/2) \simeq 21.5\%$. [5] also experimentally investigates the further gain of additional (i.e. $m > 2$) searches.

4.5 PFR - A Probabilistic Forwarding Protocol

4.5.1 The Model.

The PFR protocol was introduced by Chatzigiannakis, Dimitriou, Nikoletseas and Spirakis in [6]. They assume the case where particles are *randomly deployed* in a given area of interest. Such a placement may occur e.g. when throwing sensors from an airplane over an area.

Remark: *As a special case*, they consider the network being a lattice (or grid) deployment of sensors. This grid placement of grain particles is motivated by certain applications, where it is possible to have a pre-deployed sensor network, where sensors are put (possibly by a human or a robot) in a way that they form a *2-dimensional lattice*. Note indeed that such sensor networks, deployed in a structured way, might be useful in precise agriculture, where humans or robots may want to deploy the sensors in a lattice structure to monitor in a rather homogenous and uniform way certain conditions in the spatial area of interest. Certainly, exact terrain monitoring in military applications may also need some sort of a grid-like shaped sensor network. Note also that Akyildiz et al in a recent state of the art survey ([1]) do not exclude the pre-deployment possibility. Also, [15] explicitly refers to the lattice case. Moreover, as the authors of [15] state in an extended version of their work ([16]), they consider, for reasons of "analytic tractability", a square grid topology.

Let N be the number of deployed grain particles. There is a single point in the network area, which we call the sink S, and represents a control center where data should be propagated to.

We assume that each grain particle has the following abilities:

(i) It can estimate the direction of a received transmission (e.g. via the technology of direction-sensing antennae).

(ii) It can estimate the distance from a nearby particle that did the transmission (e.g. via estimation of the attenuation of the received signal).

(iii) It knows the direction towards the sink S. This can be implemented during a set-up phase, where the (very powerful in energy) sink broadcasts the information about itself to all particles.

(iv) All particles have a common co-ordinates system.

Notice that GPS information is not needed for our protocol. Also, there is no need to know the global structure of the network.

4.5.2   The Protocol.   The PFR protocol is inspired by the probabilistic multi-path design choice for the Directed Diffusion paradigm mentioned in [15]. Its basic idea of the protocol (introduced in [6]) lies in probabilistically favoring transmissions towards the sink within a *thin zone* of particles around the line connecting the particle sensing the event $\mathcal{E}$ and the sink (see Fig. 1.4). Note that transmission along this line is energy optimal. However it is not always possible to achieve this optimality, basically because certain sensors on this direct line might be inactive, either permanently (because their energy has been exhausted) or temporarily (because these sensors might enter a sleeping mode to save energy). Further reasons include (a) physical damage of sensors, (b) deliberate removal of some of them (possibly by an adversary in military applications), (c) changes in the position of the sensors due to a variety of reasons (weather conditions, human interaction etc). and (d) physical obstacles blocking communication.

The protocol evolves in two phases:

> **Phase 1: The "Front" Creation Phase.** Initially the protocol builds (by using a limited, in terms of rounds, flooding) a sufficiently large "front" of particles, in order to guarantee the survivability of the data propagation process. During this phase, each particle having received the data to be propagated, deterministically forwards them towards the sink. In particular, and for a sufficiently large number of steps $s = 180\sqrt{2}$, each particle broadcasts the information to all its neighbors, towards the sink. Remark that to implement this phase, and in

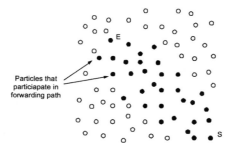

Figure 1.4.   Thin Zone of particles

particular to count the number of steps, we use a counter in each message. This counter needs at most $\lceil \log 180\sqrt{2} \rceil$ bits.

Phase 2: The Probabilistic Forwarding Phase. During this phase, each particle P possessing the information under propagation, calculates an angle ϕ by calling the subprotocol "ϕ-calculation" (see description below) and broadcasts $info(\mathcal{E})$ to all its neighbors with probability $\mathbf{P}_{fwd}$ (or it does not propagate any data with probability $1 - \mathbf{P}_{fwd}$) defined as follows:

$$\mathbf{P}_{fwd} = \begin{cases} 1 & \text{if } \phi \geq \phi_{threshold} \\ \frac{\phi}{\pi} & \text{otherwise} \end{cases}$$

where ϕ is the angle defined by the line EP and the line PS and $\phi_{threshold} = 134°$ (the selection reasons of this $\phi_{threshold}$ will become evident in Section 1.4.5.4).

In both phases, if a particle has already broadcast $info(\mathcal{E})$ and receives it again, it ignores it. Also the PFR protocol is presented for a single event tracing. Thus no multiple paths arise and packet sizes do not increase with time.

Remark that when $\phi = \pi$ then P lies on the line ES and vice-versa (and always transmits).

If the density of particles is appropriately large, then for a line ES there is (with high probability) a sequence of points "closely surrounding ES" whose angles ϕ are larger than $\phi_{threshold}$ and so that successive points are within transmission range. All such points broadcast and thus essentially they follow the line ES (see Fig. 1.4).

The ϕ-calculation subprotocol (see Fig. 1.5)
Let P_{prev} the particle that transmitted $info(E)$ to P.

(1) When P_{prev} broadcasts $info(E)$, it also attaches the info $|EP_{prev}|$ and the direction $\overrightarrow{P_{prev}E}$.

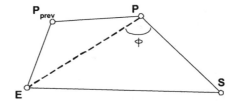

Figure 1.5. Angle ϕ calculation example

(2) P estimates the direction and length of line segment $P_{prev}P$, as described in the model.

(3) P now computes angle $(E\widehat{P_{prev}}P)$, and computes $|EP|$ and the direction of $\overrightarrow{PE}$ (this will be used in further transmission from P).

(4) P also computes angle $(\widehat{P_{prev}PE})$ and by subtracting it from $(\widehat{P_{prev}PS})$ it finds ϕ.

Notice the following:

(i) The direction and distance from activated sensors to E is inductively propagated (i.e. P becomes P_{prev} in the next phase).

(ii) The protocol needs only messages of length bounded by $\log A$, where A is some measure of the size of the network area, since (because of (i) above) there is no cumulative effect on message lengths.

Essentially, the protocol captures the intuitive, deterministic idea "if my distance from ES is small, then send, else do not send". [6] has chosen to enhance this idea by random decisions (above a threshold) to allow some local flooding to happen with small probability and thus to cope with local sensor failures.

4.5.3   Properties of PFR.   Any protocol Π solving the data propagation problem must satisfy the following three properties:

- **Correctness**. Π must guarantee that data arrives to the position S, given that the whole network exists and is operational.

- **Robustness**. Π must guarantee that data arrives at enough points in a small interval around S, in cases where part of the network has become inoperative.

- **Efficiency**. If Π activates k particles during its operation then Π should have a small ratio of the number of activated over the total number of particles $r = \frac{k}{N}$. Thus r is an energy efficiency measure of Π.

[6] shows that this is indeed the case for PFR.

Consider a partition of the network area into small squares of a fictitious grid G (see Fig. 1.6). Let the length of the side of each square be l. Let the number of squares be q. The area covered is bounded by ql^2. Assuming that we randomly throw in the area at least $\alpha q \log q = N$ particles (where $\alpha > 0$ a suitable constant), then the probability that a particular square is avoided tends to 0. So with very high probability (tending to 1) all squares get particles.

[6] conditions all the analysis on this event, call it F, of at least one particle in each square.

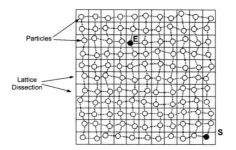

Figure 1.6. A Lattice Dissection G

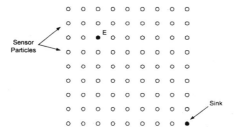

Figure 1.7. A Lattice Sensor Network

4.5.4 The Correctness of PFR.
Without loss of generality, we assume each square of the fictitious lattice G to have side length 1.

In [6] the authors prove the correctness of the PFR protocol, by using a geometric analysis. We below sketch their proof.

Consider any square Σ intersecting the ES line. By the occupancy argument above, there is with high probability a particle in this square. Clearly, the worst case is when the particle is located in one of the corners of Σ (since the two corners located most far away from the ES line have the smallest ϕ-angle among all positions in Σ).

By some geometric calculations, [6] finally proves that the angle ϕ of this particle is $\phi > 134^o$. But the initial square (i.e. that containing E)

always broadcasts and any intermediate intersecting square will be notified (by induction) and thus broadcast because of the argument above.
Thus the sink will be reached if the whole network is operational.

LEMMA 1.5 ([6]) PFR *succeeds with probability 1* in sending the information from E to S given the event F.

4.5.5   The Energy Efficiency PFR.   [6] considers the fictitious
lattice G of the network area and let the event F hold. There is (at
least) one particle inside each square. Now join all nearby particles of
each particle to it, thus by forming a new graph G' which is "lattice-
shaped" but its elementary "boxes" may not be orthogonal and may
have varied length. When G's squares become smaller and smaller, then
G' will look like G. Thus, for reasons of analytic tractability, in [6] the
authors assume that particles form a lattice (see Fig. 1.3). They also
assume length $l = 1$ in each square, for normalization purposes. Notice
however that when $l \to 0$ then "$G' \to G$" and thus all results in this
Section hold for any random deployment "in the limit".

The analysis of the energy efficiency considers particles that are active
but are as far as possible from ES. Thus the approximation is suitable
for remote particles.

[6] estimates an upper bound on the number of particles in an $n \times n$
(i.e. $N = n \times n$) lattice. If k is this number then $r = \frac{k}{n^2}$ ($0 < r \le 1$) is
the "energy efficiency ratio" of PFR.

More specifically, in [6] the authors prove the (very satisfactory) result
below. They consider the area around the ES line, whose particles
participate in the propagation process. The number of active particles
is thus, roughly speaking, captured by the size of this area, which in
turn is equal to $|ES|$ times the maximum distance from $|ES|$ (where
maximum is over all active particles).

This maximum distance is clearly a random variable. To calculate
the expectation and variance of this variable, the authors in [6] basically
"upper bound" the stochastic process of the distance from ES by a
random walk on the line, and subsequently "upper bound" this random
walk by a well-known stochastic process (i.e. the "discouraged arrivals"
birth and death Markovian process, see e.g. [20]). Thus they prove the
following:

THEOREM 1.6 ([6]) The energy efficiency of the PFR protocol is
$\Theta\left(\left(\frac{n_0}{n}\right)^2\right)$ where $n_0 = |ES|$ and $n = \sqrt{N}$, where N is the number of
particles in the network. For $n_0 = |ES| = o(n)$, this is $o(1)$.

4.5.6   The Robustness of PFR.   To prove the following robustness result, the authors in [6] consider particles "very near" to the *ES* line. Clearly, such particles have large ϕ-angles (i.e. $\phi > 134^o$). Thus, even in the case that some of these particles are not operating, the probability that none of those operating transmits (during the probabilistic phase 2) is very small. Thus, [6] proves the following.

LEMMA 1.7 ([6]) PFR manages to propagate the crucial data across lines parallel to *ES*, and of constant distance, with *fixed* nonzero probability (not depending on n, $|ES|$).

4.5.7   Comparative Study of LTP and PFR protocols.
In a recent paper ([7]), a large scale simulation is conducted to comparatively study the LTP and PFR protocols. The authors also propose three variations of the LTP protocol, with respect to different next particle selection criteria.

The comparison highlights the advantages and weaknesses of the two protocols. In the light of the above, they suggest a possible hybrid combination of the two protocols.

5.   Some Recent Work

In [11], Euthimiou, Nikoletseas and Rolim study the problem of *energy-balanced* data propagation in wireless sensor networks. The energy balance property guarantees that the average per sensor energy dissipation is the same for all sensors in the network, during the entire execution of the data propagation protocol. This property is important since it prolongs the network's lifetime by avoiding early energy depletion of sensors.

They propose a *new algorithm* that in each step decides whether to propagate data one-hop towards the final destination (the sink), or to send data directly to the sink. This randomized choice balances the (cheap) one-hop transimssions with the direct transimissions to the sink, which are more expensive but *"bypass" the sensors lying close to the sink*. Note that, in most protocols, these close to the sink sensors tend to be overused and die out early.

By a detailed analysis they *precisely estimate* the probabilities for each propagation choice in order to guarantee energy balance. The needed estimation can easily be performed by current sensors using simple to obtain information. Under some assumptions, they also derive a *closed form* for these probabilities.

The fact (shown by the analysis) that direct (expensive) transmissions to the sink are needed only rarely, shows that their protocol, besides energy-balanced, is *also energy efficient.*

In [2], the authors propose a new energy efficient and fault tolerant protocol for data propagation in smart dust networks, the Variable Transmission Range Protocol (VTRP). The basic idea of data propagation in VTRP is the varying range of data transmissions, ie. they allow the transmission range to increase in various ways. Thus data propagation in the protocol exhibits high fault-tolerance (by bypassing obstacles or faulty sensors) and increases network lifetime (since critical sensors, ie. close to the control center are not overused). As far as we know, it is the first time varying transmission range is used.

They *implement* the protocol and perform an *extensive experimental evaluation and comparison to a representative protocol* (LTP) of several important performance measures with a focus on energy consumption. The findings indeed demonstrate that the protocol achieves significant improvements in energy efficiency and network lifetime.

In [24], Nikoletseas et al a) propose *extended versions* of two data propagation protocols: the Sleep-Awake Probabilistic Forwarding Protocol (SW-PFR) and the Hierarchical Threshold sensitive Energy Efficient Network protocol (H-TEEN). These non-trivial extensions aim at improving the performance of the original protocols, by introducing *sleep-awake periods* in the PFR protocol to save energy, and introducing a *hierarchy of clustering* in the TEEN protocol to better cope with large networks areas. b) They have *implemented* the two protocols and performed an *extensive experimental comparison* (using simulation) of various important measures of their performance with a focus on energy consumption. c) They investigate in detail the *relative advantages and disadvantages* of each protocol and discuss and explain their behavior. d) In the light above they propose and discuss a possible *hybrid combination of the two protocols* towards optimizing certain goals.

Recently, Boukerche et. al [4] propose a novel and efficient energy-aware distributed heuristic, which they refer to as EAD, to build a special rooted broadcast tree with many leaves that is used to facilitate data-centric routing in wireless microsensor networks. EAD algorithm makes no assumption on local network topology, and is based on residual power. It makes use of a *neighboring broadcast scheduling* and *distributed competition among neighboring nodess.*

EAD basically computes a tree with many leaves. With the transceivers of all leaf nodes being turned off, the network lifetime can be greatly extended. In [4], Boukerche et. al. the implementation of EAD scheme, and present an extensive simulation experiments to study the its performance. The experimental results indicate clearly that EAD scheme outperforms previous schemes, such as LEACH among other protocols.

References

[1] I.F. Akyildiz, W. Su, Y. Sankarasubramaniam and E. Cayirci: Wireless sensor networks: a survey. In the Journal of Computer Networks, Volume 38, pp. 393-422, 2002.

[2] T. Antoniou, A. Boukerche, I. Chatzigiannakis, G. Mylonas and S. Nikoletseas: A New Energy Efficient and Fault-tolerant Protocol for Data Propagation in Smart Dust Networks using Varying Transmission Range. CTI Technical Report, 2003.

[3] I.F. Akyildiz, W. Su, Y. Sankarasubramaniam and E. Cayirci: A Survey on Sensor Networks. In the IEEE Communications Magazine, pp. 102-114, August 2002.

[4] A. Boukerche, X. Cheng, and J. Linus, Energy-Aware Data-Centric Routing in Microsensor Networks Proc. of ACM Modeling, Analysis and Simulation of Wireless and Mobile Systems, pp. 42-49, Sept 2003.

[5] I. Chatzigiannakis, S. Nikoletseas and P. Spirakis: Smart Dust Protocols for Local Detection and Propagation. Distinguished Paper. In *Proc. 2nd ACM Workshop on Principles of Mobile Computing – POMC'2002*. Also, accepted in the *ACM Mobile Networks (MONET) Journal, Special Issue on Algorithmic Solutions for Wireless, Mobile, Adhoc and Sensor Networks*, to appear in 2003.

[6] I. Chatzigiannakis, T. Dimitriou, S. Nikoletseas and P. Spirakis: A Probabilistic Algorithm for Efficient and Robust Data Propagation in Smart Dust Networks. CTI Technical Report, 2002.

[7] I. Chatzigiannakis, T. Dimitriou, M. Mavronicolas, S. Nikoletseas and P. Spirakis: A Comparative Study of Protocols for Efficient Data Propagation in Smart Dust Networks. In *Proc. International Conference on Parallel and Distributed Computing – EUPOPAR 2003*. Also accepted in the *Parallel Processing Letters (PPL) Journal*, to appear in 2003.

[8] I. Chatzigiannakis and S. Nikoletseas: A Sleep-Awake Protocol for Information Propagation in Smart Dust Networks. In *Proc.*

3rd Workshop on Mobile and Ad-Hoc Networks (WMAN)–IPDPS Workshops, IEEE Press, p. 225, 2003.

[9] B. Chen, K. Jamieson, H. Balakrishnan and R. Morris: SPAN: An energy efficient coordination algorithm for topology maintenance in ad-hoc wireless networks. In *Proc. ACM/IEEE International Conference on Mobile Computing* – MOBICOM'2001.

[10] D. Estrin, R. Govindan, J. Heidemann and S. Kumar: Next Century Challenges: Scalable Coordination in Sensor Networks. In *Proc. 5th ACM/IEEE International Conference on Mobile Computing* – MOBICOM'1999.

[11] H. Euthimiou, S. Nikoletseas and J. Rolim: Energy Balanced Data Propagation in Wireless Sensor Networks. CTI Technical Report, 2003.

[12] S.E.A. Hollar: COTS Dust. Msc. Thesis in Engineering-Mechanical Engineering, University of California, Berkeley, USA, 2000.

[13] W. R. Heinzelman, A. Chandrakasan and H. Balakrishnan: Energy-Efficient Communication Protocol for Wireless Microsensor Networks. In *Proc. 33rd Hawaii International Conference on System Sciences* – HICSS'2000.

[14] W. R. Heinzelman, J. Kulik and H. Balakrishnan: Adaptive Protocols for Information Dissemination in Wireless Sensor Networks. In *Proc. 5th ACM/IEEE International Conference on Mobile Computing* – MOBICOM'1999.

[15] C. Intanagonwiwat, R. Govindan and D. Estrin: Directed Diffusion: A Scalable and Robust Communication Paradigm for Sensor Networks. In *Proc. 6th ACM/IEEE International Conference on Mobile Computing* – MOBICOM'2000.

[16] C. Intanagonwiwat, R. Govindan, D. Estrin, J. Heidemann and F. Silva: Directed Diffusion for Wireless Sensor Networking. Extended version of [15].

[17] C. Intanagonwiwat, D. Estrin, R. Govindan and J. Heidemann: Impact of Network Density on Data Aggregation in Wireless Sensor Networks. Technical Report 01-750, University of Southern California Computer Science Department, November, 2001.

[18] J.M. Kahn, R.H. Katz and K.S.J. Pister: Next Century Challenges: Mobile Networking for Smart Dust. In *Proc. 5th ACM/IEEE International Conference on Mobile Computing*, pp. 271-278, September 1999.

[19] B. Karp: Geographic Routing for Wireless Networks. Ph.D. Dissertation, Harvard University, Cambridge, USA, 2000.

[20] L. Kleinrock: Queueing Systems, Theory, Vol. I, pp. 100. *John Wiley & Sons*, 1975.

[21] μ-Adaptive Multi-domain Power aware Sensors: http://www-mtl.mit.edu/research/icsystems/uamps, April, 2001.

[22] A. Manjeshwar and D.P. Agrawal: TEEN: A Routing Protocol for Enhanced Efficiency in Wireless Sensor Networks. In *Proc. 2nd International Workshop on Parallel and Distributed Computing Issues in Wireless Networks and Mobile Computing*, satellite workshop of *16th Annual International Parallel & Distributed Processing Symposium* – IPDPS'02.

[23] K. Mehlhorn and S. Näher: LEDA: A Platform for Combinatorial and Geometric Computing. *Cambridge University Press*, 1999.

[24] S. Nikoletseas, I. Chatzigiannakis, H. Euthimiou, A. Kinalis, A. Antoniou and G. Mylonas: Energy Efficient Protocols for Sensing Multiple Events in Smart Dust Networks. CTI Technical Report, 2003.

[25] C.E. Perkins: Ad Hoc Networking. *Addison-Wesley*, Boston, USA, January, 2001.

[26] S. M. Ross: Stochastic Processes, 2nd Edition. *John Wiley and Sons, Inc.*, 1995.

[27] C. Schurgers, V. Tsiatsis, S. Ganeriwal and M. Srivastava: Topology Management for Sensor Networks: Exploiting Latency and Density. In *Proc. MOBICOM 2002*.

[28] TinyOS: A Component-based OS for the Network Sensor Regime. http://webs.cs.berkeley.edu/tos/, October, 2002.

[29] P. Triantafilloy, N. Ntarmos, S. Nikoletseas and P. Spirakis: NanoPeer Networks and P2P Worlds. In *Proc. 3rd IEEE International Conference on Peer-to-Peer Computing* , 2003.

[30] W. Ye, J. Heidemann and D. Estrin: An Energy-Efficient MAC Protocol for Wireless Sensor Networks. In *Proc. 12th IEEE International Conference on Computer Networks* – INFOCOM'2002.

[31] Wireless Integrated Sensor Networks: http:/www.janet.ucla.edu/WINS/, April, 2001.

[32] Y. Xu, J. Heidemann, D. Estrin: Geography-informed energy conservation for ad-hoc routing. In *Proc. ACM/IEEE International Conference on Mobile Computing* – MOBICOM'2001.

Chapter 3

A-CELL: A NOVEL ARCHITECTURE FOR 4G AND 4G+ WIRELESS NETWORKS

Ahmed M. Safwat
Department of Electrical and Computer Engineering
Queen's University

Abstract: In this Chapter, we envision wireless multi-hopping as a complementary technology to conventional cellular networks.   Hybrid wireless networks consisting of mobile base stations are expected to play a vital role in enhancing future cellular communications.   However, numerous challenges pertaining to the wireless network and the user equipment are yet to be addressed.   We herein utilize multi-hop relaying as an overlay architecture for single-hop TDD W-CDMA cellular networks.   In our proposed architecture, namely Ad hoc-Cellular (A-Cell) relay, the inherently high node density in cellular networks is used to reduce power consumption, and enhance coverage and throughput.   A-Cell increases spatial reuse via directive antennas and GPS. We derive an analytical model for A-Cell based on multi-dimensional Markov chains. The model is then used to formulate A-Cell call blocking. To the author's best knowledge, this is the first time that directive antennas are used as a means to reduce interference, conserve energy and enhance spatial reuse in a multi-hop UMTS wireless network.

Key words: UMTS, TDD W-CDMA, wireless ad hoc networks, directive antennas, GPS

1.   INTRODUCTION

The complexity of the user equipment aside, the reduction in path loss as the distance between a pair of communicating nodes decreases results in considerable gains in performance, cost, and health (due to hazardous EM radiations).   This draws our attention to the significance of multi-hop relaying in rural and urban areas. Consequently, and against intuition, the mean delay decreases as the number of hops increases albeit the overheads associated with signaling (for route and channel acquisition). If so, multi-hop relaying renders the well-known coverage-capacity trade-off invalid. This was also among the key research areas for fourth generation (4G) wireless systems identified through a scenario-based approach in an effort led by a multidisciplinary group of the Swedish Personal Computing and Communications (PCC) program [6]. Various levels of wireless ad hoc networking models [22], [23] are expected to play an important role in future wireless communications. However, this is envisioned in the context of unlicensed bands (2.4, 5, 60 GHz) and autonomous operation. Thus, it is thought of as a competing rather than a complementary technology to conventional cellular communications. Besides, no solutions have been proposed for the erupting challenges to support hybrid ad hoc-cellular operation. User-deployed (fixed) base stations, similar to the WiFi approach requiring little to no site and placement planning, are also envisioned.

Multi-hop relaying may thus be considered as a free-of-charge overlay architecture for conventional cellular networks. Ideally, it utilizes the intrinsically high node density in cellular networks to reduce power consumption, and enhance coverage and throughput [1]. This becomes particularly useful in alleviating (and working around) shadowing in urban areas. However, power control is no longer fully centralized, and is thus more complex. Relaying can be done in one or more extra hops. As the number of hops increases, the network becomes more distributed, and the scarce wireless resources have to be controlled in a less centralized fashion by the base station(s) and the cellular nodes[1]. The benefits of two-hop relaying have been previously explored in [1] for noise-limited and interference-limited systems. In this Chapter, we address the problem of multi-hop relaying, among others, without constraining the number of hops and propose a novel architecture for multi-hop relaying using directive antennas and GPS. Multi-hop relaying has been extensively studied in the context of fully distributed wireless ad hoc networks [22], [23]. It was shown to be a non-trivial problem, especially in the presence of multiple objectives, such as maximizing the throughput while minimizing the control overhead. It is expected to be more tractable in multi-hop cellular networks due to the existence of intelligent central base stations.

Obviously, a mechanism must be developed to assign relaying channels whenever needed. While 2nd- and 3rd-generation (3G) wireless systems use Fixed Channel Assignment (FCA) strategies for Radio Resource Management (RRM), dynamic channel allocation policies may be employed in future multi-hop networks. We also need to investigate whether the relaying node and the relaying channel schemes are to be solved independently. For the sake of simplicity, both schemes have been

---

[1] The terms "node", "station" and "terminal" will be used interchangeably throughout this Chapter.

developed in isolation of one another. Intuitively, however, there can be mutual advantages for combining their respective decision-making processes. Excessive signaling is required for any optimal relaying channel assignment policy. Consequently, suboptimal schemes may be more appropriate in this context [21].

This chapter is organized as follows. Section 2 surveys the proposed architectures for future wireless networks and discusses their shortcomings in the context of enhancing capacity and power awareness. In Section 3, our novel A-Cell architecture is proposed. This is followed in Section 4 by a Markovian model for A-Cell. A-Cell's call blocking is derived in Section 5. Finally, Section 6 presents the Chapter summary and conclusions and a discussion on our future work.

2. RELATED WORK

Multi-hop relaying is different from the so-called self-organization for dynamically managing network resources in macro-, micro-, and pico-cellular environments [7]. Self-organization may be employed in single-hop and multi-hop networks alike, but with varying complexity. It can also be achieved through bunched networks by deploying fixed Remote Antenna Units (RAU) [7] (see Figure 1), which are synchronized on a timeslot basis. Nevertheless, no means of increasing spatial reuse through reassigning resources has been developed. In [1], [8], [3] algorithms based on path loss were utilized for routing purposes.

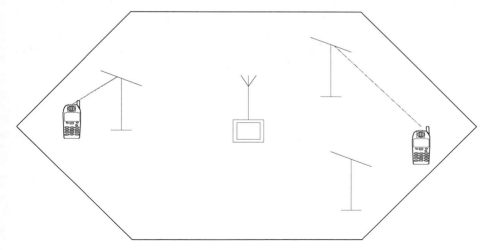

Figure 1. The bunch concept in a city center with a central unit and three remote antennas.

As shown in Figure 6, similar to RAU [7], in [5] a set of Ad hoc Relay Stations (ARSs) are used (at specific locations) besides the conventional infrastructure to relieve congestion in hot spots and achieve load balancing. The proposed scheme is called integrated Cellular and Ad hoc Relay (iCAR). The ISM band is used for inter-ARS, MT-ARS and ARS-MS operation. The authors obtain numerical results using steady-state analysis in a three-tier cellular network with the assumption that traffic intensity increases as we move outwards. It is shown that the call blocking probability increases as the area covered by the ARSs is enlarged. The analysis is rather oversimplified and neglects key PHY parameters and constraints (i.e., it is based on perfect channel conditions). In addition, there remain several open issues, one of which is the co-channel interference resulting from the possible simultaneous use of an ISM channel in the same cell or in adjacent cells. Thus, the achieved blocking probabilities will in effect be higher.

iCAR is very much similar to the Multi-hop Radio Access Cellular (MRAC) scheme [9]. A set of rooftop devices, similar to ARSs, are utilized for the sake of enhancing capacity and reducing power consumption. Only routes with no or one intermediate hop are permitted. The relaying node may either be a wireless station or a fixed rooftop device, and only Time Division Duplex (TDD) operation is considered to reduce the UE complexity. The performance of MRAC was evaluated with distinct isotropic antenna parameters (gain and height) and propagation models between the communicating nodes, such as free propagation, two-ray [10], and urban mobile propagation [11], [12]. This is then utilized in positioning the fixed router and cell layout. Although such architectures do not exploit multi-hop wireless networks up to their fullest potential, they represent an attractive and somewhat-conservative model for adopting new technologies to enhance coverage and throughput. This has been evidently demonstrated through Nokia's acquisition of rooftop (www.rooftop.com).

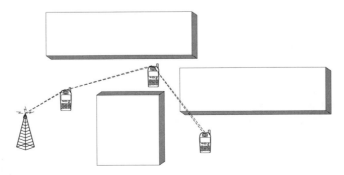

Figure 2. Intelligent relaying is used within a macro- or micro-cell in [7] for power conservation (and not coverage extension).

Opportunity-Driven Multiple Access (ODMA) has been proposed by the 3rd-Generation Partnership Project (3GPP) [13] to enhance coverage at the edges of a Universal Mobile Telecommunications System (UMTS) TDD network coverage area, while providing high access rates. ODMA was first introduced by the European Telecommunications Standards Institute (ETSI) in 1996. No further modifications to the TDD synchronization procedures and guard period are required due to the physical proximity of the communicating devices. Incorporating relaying into a Frequency Division Duplex system requires deploying last-hop gateway relay nodes (seeds). Even with this enhancement to cellular communications, ad hoc networks remain far more flexible since nodes may communicate with one another without being forced to go through the base station and no upper bound is imposed on the number of hops between the communicating parties. The former is due to billing complexities and the latter is to guarantee non-excessive signaling.

Three distinct routing strategies based on path loss and/or the location of the recipient are studied in [3]. Methods for single-hop, two-hop and interference-based ODMA are proposed. Interference-based ODMA allows the network to adapt to interference through re-routing. This is done, however, after an initial route has been established based on the total path loss to an immediate neighbor and from it to the target node. The latter scheme enhances capacity (defined as the number of supported calls/users). Nevertheless, as the user density increases, a star topology proves more useful from an energy perspective. This confirms the results reported in [4] in which the performance of a fully distributed ad hoc network is compared with that of a centralized single-cell network. The simulation study shows that wireless multi-hop ad hoc networks are superior in terms of capacity, end-to-end throughput, end-to-end delay and power consumption. Expectedly, it also shows that wireless multi-hop ad hoc networks suffer mainly from fairness and service interruption (caused by frequent network partitioning) problems. Thus, a simplistic model is proposed to switch between cellular and ad hoc mode in an infrastructure-based network. This method is a hybrid of both architectures with the instantaneous network throughput used as a toggle switch. Before adopting the conclusions drawn from the study, the performance of a full-scale network consisting of one or more cell clusters needs to be evaluated.

In [14], spatial diversity is used in a narrowband flat fading environment to overcome the effects of path loss and fading. This form of antenna and resource (e.g., battery) sharing is shown to outperform single-hop and multi-hop relaying in terms of power efficiency. The work in [14] can be extended to multiple intermediate nodes (relays). Nevertheless, Laneman et al do not address several keys problems. First, a criterion needs to be adopted through which the relay nodes are chosen. Secondly, there have been no means of provisioning extended coverage through one or multiple relays, which is a case that frequently arises at the edge of boundary cell sites. Thus, we infer that spatial diversity does not preclude the necessity of multi-hop relaying in urban and rural areas.

Balazs et al [15] modeled the capacity of a single-cell ODMA UMTS system based on the number of active wireless terminals, with which the solution grows exponentially. The model assumes an infinite number of W-CDMA channels and calling blocking occurs only when the sender cannot reach the destination in at most 10 hops (defined by the optional part of the standard). Moreover, load balancing has not been incorporated into ODMA. W-CDMA is used for the last link (between the final relay station and node B), whereas ODMA is used for all other links.

In [16], a routing protocol, namely Base-Centric Routing (BCR), is proposed for multi-hop cellular networks. The proposed protocol is a hybrid of table-driven and on-demand routing protocols commonly used in wireless ad hoc networks. Base stations use table-driven routing to track topological changes, while on-demand routing is used by the mobile stations to query base stations or the rest of the network, in case paths may not be obtained from the base. Along with introducing a new routing protocol, this work mainly investigates reducing the control overhead associated with routing via lowering the number of exchanged query changes. If considerable, such an overhead will degrade the throughput of a multi-hop wireless network and multi-hopping will be deemed inefficient. New messages for path and topology maintenance are used. HELLO messages broadcast periodically by the base enable the wireless stations to reach the base through the intermediate nodes that relayed the message. HELLO messages are uniquely identified in a manner similar to that used for distinguishing stale RREQs in DSR [2]. Link-state packets (LSP) provide the means for updating the base's topological database. Notably, LSPs are unicast packets sent to the base. Although it is proposed that an LSP may be broadcast at least once as triggered by the failure to receive the next hop's periodic beacon, several broadcasts may be required if such failure is triggered by more than one intermediate node. In essence, the difference between on-demand protocols in wireless ad hoc networks and BCR is basically due to the existence of the base station. Thus, unlike DSR and AODV, the mobile station is only required to query the base (using a PREQ packet) rather than the rest of the network. Obviously, this would not have been possible if the network consisted of all-mobile nodes. As mentioned earlier, the source node is permitted to broadcast the RREQ if no route is obtained. This phase then becomes similar to AODV. In addition, BROKEN_LINK and RERR messages are required for path maintenance and congestion mitigation.

Clearly, BCR is more appropriate for contention-based wireless multi-hop networks of small to moderate sizes. As such, it is applicable to multi-hop WiFi networks. The nodes may thus communicate with one another directly (i.e., without the intervention of the base station). This is not possible in a wide-area cellular network as a number of different issues ranging from medium access to billing become far more complex to address. Nevertheless, the capacity, defined as the number of simultaneous (successful) transmissions, is significantly enhanced. It is shown [16] that the capacity increases at a higher rate than the number of hops.

A multi-hop cellular architecture of relevance to static WiFi-based networks using virtual channel sensing, namely Multi-hop Cellular Network (MCN), is proposed in [17], [18]. The approach therein is to decrease the infrastructure-related cost through deploying fewer base stations than the single-hop case, or to reduce the transmission power. MCN-b and MCN-p correspond to the two variant architectures, respectively. As shown in Figures 3 and 4, in MCN-b fewer base stations are deployed and in MCN-p the transmission power is reduced for mobile nodes and base stations alike. The merits and limitations of MCN are examined, and the throughput is derived and is shown to be higher than in the single-hop case. Routing and handoff, however, remain particularly challenging.

A quantitative performance study is conducted in [4]. A hybrid wireless network (HWN) model is proposed in which the wireless stations, with the support of the base station, can toggle between the single-hop and the multi-hop modes of operation (by varying the transmission power) depending on the current network state and instantaneous throughput (see Figure 5). Thus, all the wireless stations must lie within the cellular boundaries to enable the base stations to monitor the network topology and performance. As a result, HWN is quite limited since both the source and destination must reside within the same cell.

Both MCN and HWN are further analyzed in [19] and enhancements are proposed. A single data and a single control channel are used in [19]. A contention-free protocol is used for beaconing to obtain neighborhood information. However, little detail has been mentioned with regard to obtaining a collision-free schedule. In addition, this neighborhood information must be sent to the base station to construct useful topological information. Although this can be done through flooding or unicasting via sending LSPs, the adopted method is not addressed in much detail. Similar to DSR, routes are acquired and maintained using RREQ, RREP and RERR packets and route cache maintenance. Dijkstra's shortest path algorithm is used to find the favorable path to the destination. The number of nodes in the capture area pertaining to the two stations constituting a wireless link is used as its cost. This metric may not always give a reasonable estimate of the neighbors' activity, and is vulnerable to underestimation of neighborhood traffic in case of irregular network topologies. The analytical model is simple and is based on the assumption that the closer the nodes are to the base the more the probability of error and the heavier their loads. The analytical model in [17], despite its shortcomings, is far more realistic.

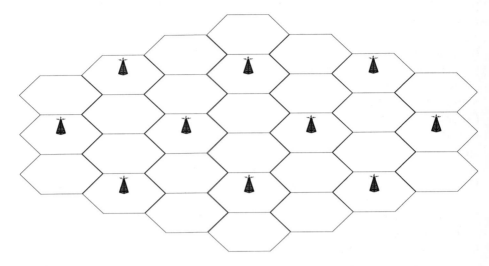

Figure 3. The MCN-b architecture [17]: fewer base stations are deployed.

Simulations have been performed to study the effect of relaying in hybrid wireless networks on the radio aspects [20]. The experiments assume perfect power control and take into account a single cell only, whereas the path loss exponent, number of wireless stations, and number of hops to the base station, have all been varied. Routing is significantly simplified by assuming that each node knows the location information of all the other nodes residing in the same cell beforehand. The location-based forwarding algorithm does not take shadowing into account since in some cases it is possible to use a neighbor that is slightly farther from the base station. Some experiments have been carried out for a CDMA system that uses a simplified time division protocol. However, time slot scheduling is suboptimal and the time slots are chosen according to the effect of the current network state on the assignment (in terms of the signal-to-noise ratio). Thus, multi-hop relaying reportedly degrades coverage. This is expected to change if a realistic power control algorithm for CDMA systems (which are interference-limited) is used along with efficient routing and resource allocation schemes.

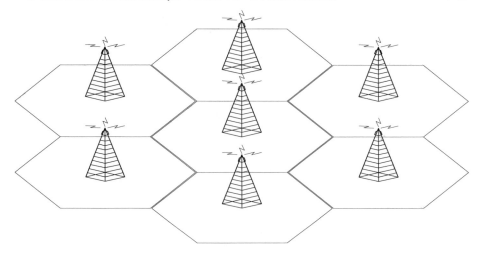

Figure 4. The MCN-p architecture [17]: the same set of base stations is used while the transmission power is reduced.

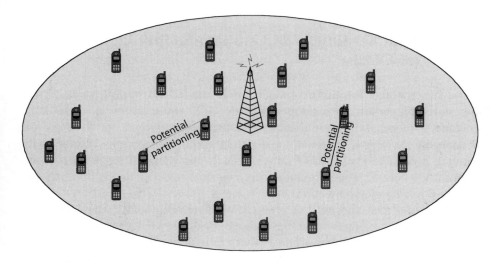

Figure 5. HWN: default mode of operation is ad hoc; base station dictates increasing the power to avoid network partitioning and achieve cellular-like throughput (at least).

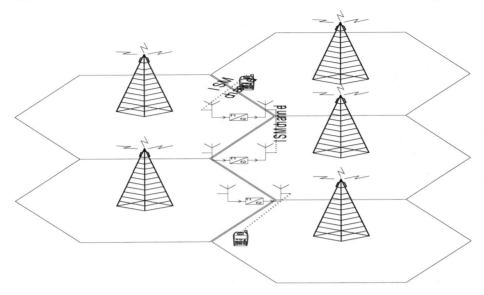

Figure 6. The ISM band is used in iCAR for inter-ARS, MT-ARS, ARS-MT communications; wireless stations may not communicate directly; placement of ARSs is critical to their success.

3. THE AD HOC-CELLULAR ARCHITECTURE (A-CELL)

In our scheme, the uplink capacity is enhanced via multi-hopping. All the wireless stations are equipped with GPS transceivers and are thus capable of computing their distances to the base station. For the sake of conserving the scarce bandwidth and reducing the overheads associated with routing, a node sends a RREQ only if it is no more than H hops away from the base[2], where H is the maximum number of distinct channels that may be assigned to a single connection.

We herein also address the problem of increasing spatial reuse through using directive antennas. To the author's best knowledge, this is the first time that directive antennas are used by mobile nodes to reduce interference, conserve energy and enhance spatial reuse in a multi-hop TDD W-CDMA wireless network. Each wireless node is equipped with an array of M directional antennas. The gain of the directive antenna is higher than that of an isotropic antenna, and is inversely proportional to its beamwidth. The antennas have perfectly non-overlapping beamforming directions. During

---

[2] A node is 0 hops away from the base if it can communicate directly with the base

non-unicast reception, selection diversity is used and the antenna that acquires the largest *SINR* is used by the subsequent, if any, unicast transmission. Only nodes whose instantaneous battery capacity exceeds the predefined energy threshold (γ) participate in routing. To maximize the probability of finding a path to the destination, the call initiator broadcasts the RREQ packet omni-directionally (or, as shown in Figure 7, using its forward antennas towards the base station) to its neighbors. The RREQ contains the location of the initiator. Upon receipt of the RREQ, a neighbor rebroadcasts the RREQ provided that its distance from the initiator is less than or equal to R and none of its neighbors had already broadcast the RREQ (see Figure 8). Otherwise, the RREQ packet is dropped from its MAC queue. The neighbor also stores its own location information in the RREQ. This will be used by the next hop to the base station to find out the most appropriate antenna (from the perspective of the previous node's location).

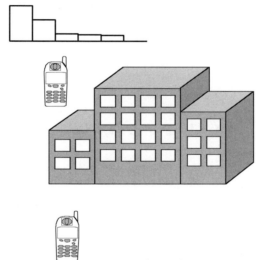

RREQ sent with *M/2* of its antenna elements to enforce forward path setup

Figure 7. A-Cell controlled route setup using *M/2* antenna elements.

When the route is acknowledged by the base station, every node records its location information in the RREP and forwards it towards the source (using the stored route record). Hence, every node on the route (including the initiator) is capable of adjusting its directional antenna. The base station is required to assign channels to the A-Cell network in an efficient manner that will minimize interference while increasing spatial reuse [21].

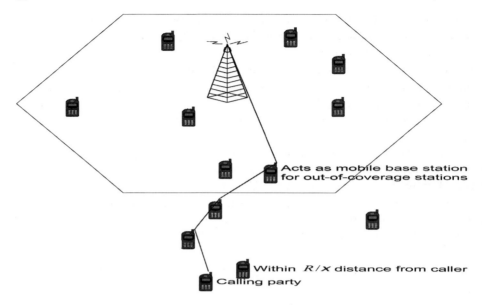

Figure 8. A-Cell utilizes directive antennas, GPS, and mobile base stations to enhance coverage.

4.   A-CELL MODELING

Let N_r be the remaining number of W-CDMA channels not presently in use (from a total of N channels) and K the number of hops that separate the base station from the call initiator. An incoming call is blocked if one of the following is true:

1. $K > H$
2. $K \leq H$ and $K > N_r$

Let A-Cell connection Ω consist of K_Ω hops, then j_Ω (A-Cell) channels are required:

$$j_\Omega = K_\Omega + I, I = \begin{cases} 1 : K_\Omega \neq 0 \\ 0 : K_\Omega = 0 \end{cases}, j = N - N_r - i = \sum_{\Omega=1}^{N-N_r-i} j_\Omega \,,$$

where i is the number of occupied non-A-Cell W-CDMA channels, and j is the total number of W-CDMA channels assigned to the A-Cell connections.

It can be shown that the A-Cell architecture can be modeled with a multidimensional continuous-time Markov chain whose state space is expressed as follows:

$$S = \{(i, j, j_1, j_2, ..., j_{\lfloor j/2 \rfloor}) : i + j = N - N_r, \sum_{\Omega=1}^{N-N_r-i} j_\Omega = j, j_\Omega \neq 0\}$$

Define for convenience S_a^f as the fth element in the tuple of state $S_a \in S$ and l the location (within the ordered tuple) of the newly added A-Cell connection. Let q be the number of freed channels when an A-Cell call terminates, and let $t \in \{1, ..., \lfloor j/2 \rfloor\}$ be the location of the first occurrence of q within the ordered portion of the present state. In the above Markov chain, the transition probabilities are as follows:

$$P_{S_a S_b} = \begin{cases} \dfrac{\psi(S_a, q).\mu}{\lambda + S_a^1.\mu + \sum_{\Omega=1}^{\lfloor S_a^2/2 \rfloor} I_\Omega.\mu}, S_a^1 = S_b^1, S_b^2 = S_a^2 - q, S_b^f = \begin{cases} S_a^f : 2 < f < t, \\ S_a^{f+1} : f \geq t \end{cases} \\[4ex] \dfrac{S_a^1.\mu}{\lambda + S_a^1.\mu + \sum_{\Omega=1}^{\lfloor S_a^2/2 \rfloor} I_\Omega.\mu}, S_b^1 = S_a^1 - 1, S_b^f = S_a^f : f > 1 \\[4ex] \dfrac{\delta(S_b^1 - 1).\lambda}{\lambda + S_a^1.\mu + \sum_{\Omega=1}^{\lfloor S_a^2/2 \rfloor} I_\Omega.\mu}, S_a^1 = S_b^1, S_b^2 = S_a^2 + S_b^l, S_b^f = \begin{cases} S_a^f : 2 < f < l \\ S_a^{f-1} : f > l \end{cases} \\[4ex] \dfrac{\delta(0).\lambda}{\lambda + S_a^1.\mu + \sum_{\Omega=1}^{\lfloor S_a^2/2 \rfloor} I_\Omega.\mu}, S_b^1 = S_a^1 + 1, S_b^f = S_a^f : f > 1 \end{cases}$$

In the above, $\psi(S_a, q)$ is the number of present A-Cell connections consisting of q channels, and $\delta(K)$ is the probability that exactly K hops are required to reach the base station.

5.   A-CELL CALL BLOCKING

Call blocking was briefly explained in the previous section. An incoming call is blocked if the required resources are not available at the time the connection is requested in which case Erlang's lost calls or, equivalently, Lost Calls Cleared (LCC) model is assumed. In this Section, we will formulate the blocking probability for our A-Cell network. This will give us insights with regard to enhancing A-Cell through Advanced Ad hoc Preemption (AAP), which is proposed and explained in [21].

Now, let $P_n = \lim_{t \to \infty} P\{s(t) = S_n\}$ be the stationary distribution of the chain. Thus, A-Cell's blocking probability, B, can be expressed as follows:

$$B = \sum_{n:S_n^1+S_n^2=N} P_n + \sum_{n:S_n^1+S_n^2<N} \sum_{K=N-S_n^1-S_n^2}^{H} \delta(K).P_n$$

6.   CONCLUSION AND FUTURE RESEARCH

In this Chapter, we proposed A-Cell, a novel architecture for supporting multi-hopping in UMTS TDD W-CDMA cellular networks. A-Cell enhances the capacity of a cellular network and conserves power by transmitting over short distances. Directive antennas are used to increase spatial reuse by simultaneously assigning channels within the same cell (and in low-rate and out-of-coverage areas) as well as in adjacent cells. In our future research, we will find a closed-form for the Markovian model and will analytically study the characteristics of A-Cell. In addition, we have been investigating the optimal allocation of resources in a multi-hop W-CDMA network with directive antennas [21], [24]. Load balancing in cellular networks using an ad hoc overlay network is also a topic of current and future research [21]. Finally, billing in an A-Cell wireless network remains an open area of research.

REFERENCES

[1] V. Sreng, H. Yanikomeroglu and D. Falconer, "Coverage Enhancement through Peer-to-Peer Relaying in Cellular Radio Networks," *www.sce.carleton.ca/faculty/yanikomeroglu/Pub/twireless_vs.pdf*.

[2] D. Johnson and D. Maltz, " dynamic Source Routing in Ad hoc Wireless Networks," *Mobile Computing*, Chapter 5, pp. 153-181, Kluwer Academic Publishers, 1996.

[3] T. Rouse, I. Band, and S. McLaughlin, "Capacity and Power Investigation of Opportunity Driven Multiple Access (ODMA) Networks in TDD-CDMA Based Systems," *IEEE International Conference on Communications (ICC) 2002*.

[4] H. Hsieh and R. Sivakumar, "Performance Comparison of Cellular and Multi-hop Wireless Networks: A Quantitative Study," *ACM SIGMETRICS 2001*.

[5] S. De, O. Tonguz, H. Wu, and C. Qiao, "Integrated Cellular and Ad hoc Relay (iCAR) Systems: Pushing the Performance Limits of Conventional Wireless Networks," *35th Annual Hawaii International Conference on System Sciences (HICSS 2002)*.

[6] A. Bria et al, "4th-Generation Wireless Infrastructures: Scenarios and Research Challenges," *IEEE Personal Communications*, Dec. 2001, pp. 25-31.

[7] A. Spilling, A. Nix, M. Beach and T. Harrold, "Self-Organization in Future Mobile Communications," *Electronics and Communication Engineering Journal*, June 2000, pp. 133-147.

[8] T. Harrold and A. Nix, "Intelligent Relaying for Future Personal Communication Systems," *IEE Colloquium on Capacity and Range Enhancement Techniques for Third Generation Mobile Communications and Beyond*, Feb. 2000, Digest No. 00/003, pp. 9/1-5.

[9] Y. Yamao et al, "Multi-hop Radio Access Cellular Concept for Fourth-Generation Mobile Communications Systems," *IEEE PIMRC 2002*, pp. 59-63.

[10] H. Xia, et al, "Radio Propagation Characteristics for Line-of-Sight Microcellular and Personal Communications," *IEEE Transactions on Antenna Propagation*, vol. 41, pp. 1439-1447, 1993.

[11] J. Walfisch and H. Bertoni, "A Theoretical Model of UHF Propagation in Urban Environments, *IEEE Transactions on Antenna Propagation*, vol. 36, pp. 1788- 1796, 1988.

[12] H. Xia, "A Simplified Analytical Model for Predicting Path Loss in Urban and Suburban Environments," *IEEE Transactions on Vehicular Technology*, vol. 46, pp. 1040-1046, 1997.

[13] 3rd Generation Partnership Project; Technical Specification Group Radio Access Network; Opportunity Driven Multiple Access (3G TR 25.924 version 1.0.0), 1999.

[14] J. Laneman and J. Wornell, "Energy-Efficient Antenna Sharing and Relaying for Wireless Networks," *IEEE WCNC 2000*, pp. 7-12.

[15] F. Balazs, G. Jeney, and L. Pap, "Capacity Expansion Capabilities in ODMA Systems," *WPMC 1999*.

[16] Y. Hsu and Y. Lin, "Base-Centric Routing Protocol for Multihop Cellular Networks," *IEEE GLOBECOM 2002*.

[17] Y. Lin and Y. Hsu, "Multihop Cellular: A New Architecture for Wireless Communications, *IEEE INFOCOM 2000*, pp. 1273-1282.

[18] Y. Lin et al, "Multihop Wireless IEEE 802.11 LANs: A Prototype Implementation," *IEEE International Conference on Communications (ICC) 1999*.

[19] K. Kumar, B. Manoj, and C. Murthy, "On the Use of Multiple Hops in Next Generation Cellular Architectures," *www.cs.berkeley.edu/~kjk/pubs/icon.ps.gz*.

[20] O. Mantel, N. Scully, and A. Mawora, "Radio Aspects of Hybrid Wireless Ad-hoc Networks," *IEEE Vehicular Technology Conference 2001*, pp. 1139-1143.

[21] A. Safwat, "Distributed Call Admission Control and Dynamic Channel Allocation in 4G+ TDD A-Cell Networks," *Technical Report*, Department of Electrical and Computer Engineering, Queen's University, in progress.

[22] A. Safwat, H. Hassanein, and H. Mouftah, "A Framework for Energy Efficiency in Wireless Multi-hop Ad hoc Networks," *Proceedings of IEEE ICC 2003*.

[23] A. Safwat, H. Hassanein, and H. Mouftah, "Optimal Cross-Layer Designs for Energy-Efficient Wireless Ad hoc and Sensor Networks," *Proceedings of IEEE IPCCC 2003*.

[24] A. Safwat, "ECCA for Future Generations of WLANs and 4G+ Wireless Networks with Directive Antennas," *Technical Report*, Department of Electrical and Computer Engineering, Queen's University, 2003.

Chapter 4

THIRD GENERATION
WCDMA RADIO EVOLUTION

Harri Holma and Antti Toskala
Nokia Networks
System Technologies
P.O.Box 301
FIN-00045 NOKIA GROUP
Finland
Tel: +358-40-513 2710
antti.toskala@nokia.com

Abstract   *3rd Generation Partner Ship Project (3GPP)* produced the first full version of the *Wideband CDMA (WCDMA)* standard at the end of 1999. Release'99, meets all the IMT-2000 requirements, including 2 Mbps data rate with variable bit rate capability, support of multi-service, quality of service differentiation and efficient packet data. The Release'5 specifications were created in March 2002 and they contain downlink packet data operation enhancement, under the title *High Speed Downlink Packet Access (HSDPA)*. HSDPA utilizes *Hybrid Automatic Repeat Request (HARQ)*, link adaptation and higher order modulation for improving data spectral efficiency and for pushing bit rates beyond 10 Mbps. The further 3GPP releases will study the enhancements of packet data performance in uplink. Other important features in future 3GPP releases include advanced antenna technologies and WCDMA standard for new spectrum allocations. The paper describes the main solutions of 3GPP WCDMA standard in more detail.

Keywords: Wideband CDMA (WCDMA), UMTS, High Speed Downlink Packet Access (HSDPA), beamforming, WCDMA radio evolution

1.   3GPP Release'99 – Commercial WCDMA Deployment

The first commercial WCDMA networks using 3GPP standard have been launched in Europe and in Japan during 2002-2003 with more operators to follow. The WCDMA bit rate capabilities reach 2 Mbps in the first version of the specifications, with later evolution reaching up to 10 Mbps downlink capability. In the first phase terminals up to 384 kbps downlink data transmission capability is available with the support for the new data services, e.g. video conferencing and fast email and internet access. A single physical connection can support more than one service, even if the services have different quality requirements. This avoids multi-code transmission in case there are two services running simultaneously as they can be dynamically multiplexed on a single physical resource. This allows also easy terminal support for service multiplexing, such as simultaneous speech and multimedia messaging.

WCDMA Release'99 supports variable user data rates, Bandwidth on Demand. Each user is allocated frames of 10 ms duration, during which the user data rate is kept constant. However, the data capacity among the users can change from frame to frame. This fast radio capacity allocation (or the limits for variation in the uplink) is controlled and co-ordinated by the radio resource management functions in the network to achieve optimum throughput for packet data services and to ensure sufficient *Quality of Service (QoS)* for circuit switched users.

WCDMA Release'99 supports QoS mechanisms to provide services from low delay conversational class to flexible background class. These QoS mechanisms improve the efficiency of the air interface as the radio network can optimize the resource allocations according to the QoS requirements of each service, thus avoiding over-dimensioning of the network.

The variable bit rate and service multiplexing capability is illustrated in Figure 4.1, resource allocation in Figure 4.2 and QoS mechanism in Figure 4.3.

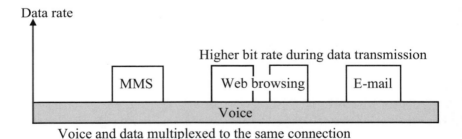

Voice and data multiplexed to the same connection

Figure 4.1. Variable bit rate and service multiplexing in WCDMA Release'99

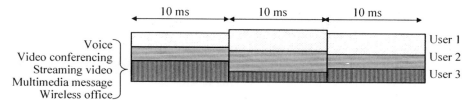

Figure 4.2. Resource allocation in WCDMA Release'99

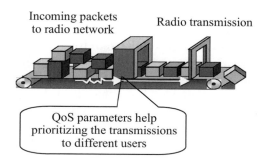

Figure 4.3. QoS mechanisms provides more efficient utilization of radio resources

WCDMA Release'99 supports several transport channels for flexible packet data operation. The following alternatives exist in WCDMA downlink for packet data:

- *Dedicated Channel (DCH)*
- *Downlink Shared Channel (DSCH)*
- *Forward Access Channel (FACH)*

The DCH can be used for any type of the service up to 2 Mbps, and it has a fixed spreading factor in the downlink. The DCH is fast power controlled and may be operated in macro diversity as well.

The DSCH has been developed to operate always together with an associated DCH. This allows defining the DSCH channel properties to be best suited for non-real time packet data while leaving the data with tight delay budget to be carried by DCH. The DSCH in contrast to DCH has dynamically varying spreading factor informed to the terminal on 10 ms frame-by-frame basis with physical layer signaling carried on the DCH. This allows dynamic multiplexing of several users to share the DSCH code resource and thus optimizing the orthogonal code resource and base station hardware usage in the downlink. DSCH can utilize power control based on the associated DCH. DSCH is not operated in soft handover. DSCH is not expected to be available in the first commercial networks, but DCH will be used for providing data services.

The FACH can be used for downlink packet data as well. The FACH is operated on its own, and it is sent with a fixed spreading factor and typically with rather high power level since it does not support fast power control, as power control feedback in the uplink is not available. FACH does not use soft handover.

The uplink counterpart for FACH is *Random Access Channel (RACH)*, which is intended for short duration 10/20 ms packet data transmission in the uplink. There exists also another uplink option, named *Common Packet Channel (CPCH)* to enable longer packet bursts up to 640 ms with power control, but that is not foreseen to be neither part of the first phase WCDMA network deployment nor the terminal implementations.

The end user performance is optimized by low network latency in addition to the flexible data rates. The latency is represented by round trip time. The short round trip time is beneficial also to enhance the performance of packet data protocols. Short round trip time is preferred in order to take full benefit of the WCDMA high bit rate capability and avoid TCP slow start effects. The end-to-end round trip times will be clearly below 200 ms in the first WCDMA systems [1]. Such short round trip times allow to support such real time packet switched applications like Voice over IP or packet switched 2-way video. The round trip time components are shown in Figure 4.4. The round trip time will be further reduced with the introduction of High Speed Downlink Packet Access.

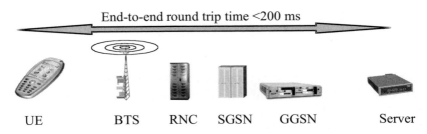

End-to-end round trip time <200 ms

UE   BTS   RNC   SGSN   GGSN   Server

Figure 4.4. End-to-end round trip times in WCDMA

The WCDMA downlink spectral efficiency can be improved with transmit diversity, where base station transmission utilizes two transmit antennas while terminal reception is done with a single antenna to keep the terminal complexity reasonable. Release'99 contains two modes of operation:
- Open loop transmit diversity, intended to be used e.g. on common channels but also on dedicated channels
- Closed loop transmit diversity, which relies on the uplink feedback to adjust phase or phase and amplitude between the two transmit

antennas. This is applicable only when there is feedback existing in the uplink, which means in Release'99 DCH/DSCH.

Typical WCDMA Release'99 spectral efficiency is approx. 0.2 bits/s/Hz/cell in macro cells and higher in micro cells. The gain of transmit diversity in spectral efficiency can be typically up to 25-40%. [1]

WCDMA Release'99 provides smooth interworking with the GSM/EDGE networks including common core network elements, multimode terminals, inter-system handovers and harmonized QoS parameters. The interworking is illustrated in Figure 4.5. GSM and WCDMA are two complementary radio access systems, one optimized for existing GSM spectrum and the other for new UMTS spectrum. The standardization of both WCDMA and GSM/EDGE takes place in 3GPP. Further alignment of the WCDMA and GSM/EDGE radio access network architecture has been completed in Release 5 as the Iu-mode was introduced to GSM/EDGE radio access network standards. This enables to connect both networks using the common Iu interface to the core network.

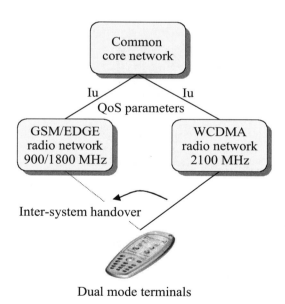

Figure 4.5. Harmonized GSM/EDGE and WCDMA standards

2.   3GPP Release 5 – High Speed Downlink Packet Access for Enhanced data performance

For Release 5 specifications, released March 2002, the *High Speed Downlink Packet Access (HSPDA)* was completed, which is the most significant radio related update since the release of the first version of 3GPP WCDMA specifications. HSPDA is based on distributed architecture where the processing is closer to the air interface at the base station (Node B) for low delay link adaptation. The key technologies used with HSDPA are:

 – Node B based scheduling for the downlink packet data operation
 – Higher order modulation
 – Adaptive modulation and coding
 – *Hybrid Automatic Repeat Request (HARQ)*
 – Physical layer feedback of the momentary channel condition

The HSDPA principle with Node B based scheduling is illustrated in Figure 4.6.

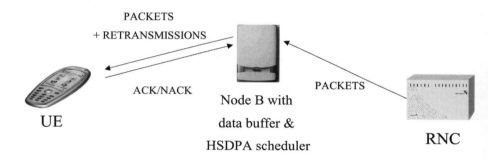

Figure 4.6. WCDMA HSDPA operation.

The HSDPA operation is carried on the *High Speed Downlink Shared Channel (HS-DSCH)*, which has fundamental differences when compared to DSCH in Rel'99/Rel'4. The key differences to the DSCH are:

 – 2 ms frame length, while with DSCH the frame length is 10, 20, 40 or 80 ms.
 – Fixed spreading factor 16 with maximum of 15 codes, while with DSCH the spreading factor may vary between 256 and 4.
 – HS-DSCH supports also *16 Quadrature Amplitude Modulation (16QAM)* modulation, in addition to *Quadrature Phase Shift Keying (QPSK)* of DSCH.
 – Link adaptation, while DSCH is power controlled with the DCH.
 – Physical layer combining of retransmissions with HARQ

The HS-DSCH is accompanied by the *(High Speed Shared Control Channel (HS-SCCH)*, which carries the necessary information for demodulation of the HS-DSCH such as what codes to despread and what is the modulation and necessary information for HARQ whether the transmission is a new one or a retransmission. Different retransmission may also have different redundancy versions, thus in case retransmissions are not identical (incremental redundancy) also the redundancy version needs to be provided to the terminal.

HSDPA extends the WCDMA bit rates up to 10 Mbps. The higher peak bit rates are obtained with higher order modulation, 16-QAM, and with adaptive coding and modulation schemes. The theoretical bit rates bit rates are shown in Table 4.1. The maximum bit rate with QPSK modulation is 5.3 Mbps and with 16-QAM 10.7 Mbps. Without any channel coding, up to 14.4 Mbps could be achieved. The terminal capability classes start from 900 kbps and 1.8 Mbps with QPSK only modulation and 3.6 Mbps with 16-QAM modulation. The highest capability class supports the maximum theoretical bit rate of 14.4 Mbps.

Table 4.1. Theoretical bit rates with HSDPA

Modulation	Coding rate	Max. bit rate
	¼	1.8 Mbps
QPSK	2/4	3.6 Mbps
	¾	5.3 Mbps
16-QAM	2/4	7.2 Mbps
	¾	10.7 Mbps

The HSDPA concept offers over 100% higher peak user bit rates than Release'99 in practical deployments. HS-DSCH bit rates are comparable to *Digital Subscriber Line (DSL)* modem bit rates. The mean user bit rates in large macro cell environment can exceed 1 Mbps and in small micro cells 5 Mbps. The HSDPA concept is able to support not only non real time UMTS QoS classes but also real time UMTS QoS classes with guaranteed bit rates.

Short 2 ms frame length in HSDPA allows also to minimise the round trip time which enables shorter network latency and better response times. Fast Layer 1 retransmissions can guarantee that the delay variations are minimized even with retransmissions.

The cell throughput refers here to the total number bits per second transmitted to the users through one cell. The cell throughput increases with HSDPA compared to the Release'99 because Hybrid ARQ combines packet retransmission with the earlier transmission, and no transmissions are wasted. 16-QAM modulation provides higher bit rates than QPSK of Release'99 with the same usage of orthogonal codes.

Typical throughput values are shown in Table 4.2. HSDPA increases the cell throughput 100% compared to the Release'99. The cell throughput with HSDPA depends on the interference environment: the inter-path interference and the inter-cell interference. The dependency is higher than with Release'99 WCDMA. The main reason is that the higher order modulation of 16-QAM requires high *C/I* and is therefore sensitive to amount of interference. The highest throughput is obtained with low inter-path interference and low inter-cell interference. For micro cell, the HS-DSCH can support up to 5 Mbps per sector per carrier, i.e. 1 bit/s/Hz/cell.

Table 4.2. Cell throughput with Release'99 DSCH and R5 HSDPA in macro cells [2]

	DSCH	HSDPA
Cell throughput	1.4 Mbps	2.7 Mbps

3.   WCDMA Performance Enhancement – smart antenna beamforming for higher capacity and coverage

The WCDMA air interface spectral efficiency and user bit rates are interference limited, in particular with HSDPA. If the interference levels could be reduced, the bit rates could be further improved. Base station beamforming allows reducing the interference levels by transmitting the signal via a narrow beam to the desired user, thus causing less interference to the other users. From the terminal point of view the beamforming is supported in WCDMA Release'99 air interface while necessary information elements on WCDMA radio access network internal interfaces are covered in Release 5 specifications. The support of radio resource management functionalities for beamforming in Iub interface between base station and radio network controller, RNC, is part of Release 6. This approach ensures that once the beamforming is taken into use, it can rely on the support of all terminals, as the later additions in the network side are not visible for the terminals of earlier Releases.

Beamforming is illustrated in Figure 4.7 where one scrambling code is shared between several beams. That approach brings the advantage that the spreading codes under one scrambling code are orthogonal reducing the interference levels further. The capacity of the WCDMA sector can be typically improved 100-150% with 4 beams compared to single antenna transmission. [3]

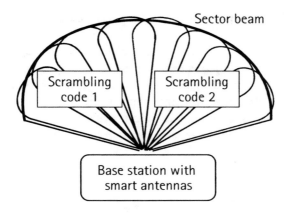

Figure 4.7. Beamforming smart antennas in WCDMA

4. WCDMA evolution beyond Release 5

In 3GPP standardization further work is being done on the WCDMA evolution for Release 6 due at first half of 2004 and for further releases as well. From the radio point of view the area getting most attention in 3GPP is the enhanced uplink DCH feasibility study. Enhanced uplink DCH basically studies similar methods as are included in HSDPA in downlink direction. The studied uplink methods are not expected to increase the peak data rates, but the focus is an improved coverage, reduced delay and higher system capacity of the packet based uplink services. Also the delay related to initiating uplink packet transmission is an area that is being studied for improvements. The proposed concept under study is illustrated in Figure 4.8. The specific issues that are under study include HARQ in the uplink direction as well as Node B controlled fast uplink scheduling.

In the uplink obviously there are fundamental differences, with the key issue being handling of the total transmission power resource. In the downlink direction the power resource is centralized while in the uplink the power resource available for an individual user is limited by the terminal power amplifier capabilities. This clearly indicates that aiming time division approach would not be preferred in CDMA based uplink. Another difference between uplink and downlink is the need for higher order modulation, which is useful in downlink with limited number of codes, while uplink could utilize more codes and lower order modulation.

Other area requiring specific attention is the operation in soft handover, where the receiving base station in the uplink will vary as a function of the terminal movement and changes in radio conditions. In this case use of e.g.

non-self decodable retransmissions does not lead to the best possible outcome.

The work on the details is on-going regarding the issues like interleaving lengths, with the currently considered cases being 10 ms and 2 ms as well as what new signaling need needs to be added. How to ensure reliable transmission and efficient transmission of the new control information requires also close attention. The latest description of the issues being investigated and the resulting conclusions can be found in [4].

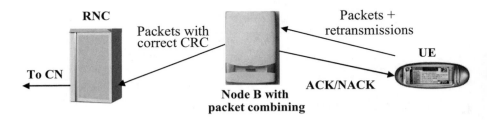

Figure 4.8. Enhanced uplink DCH operation with HARQ in the uplink.

The other areas relevant for WCDMA evolution in 3GPP include:
- New frequency variants of WCDMA, such as the utilization of 2.5 GHz spectrum and 1.7/2.1 GHz. The latter is relevant in USA. The main new frequency bands are shown in Figure 4.9. There are also other existing cellular bands, currently used by 2G technologies, being considered for WCDMA, such as the 850 MHz band.
- Advanced antenna technologies, including enhancements for the beamforming capabilities in the network side as well as transmit diversity technologies with single terminal receiver antenna or with two receiver antennas in the terminal, aiming to increased capacity or even higher peak bit rates than possible currently.
- Further topics under study include UTRAN architecture evolution to investigate whether there would be need to do any improvements for the UTRAN internal architecture related standards.
- The support of multicast services is also being worked on to provide efficient means for sharing a common content with several users.

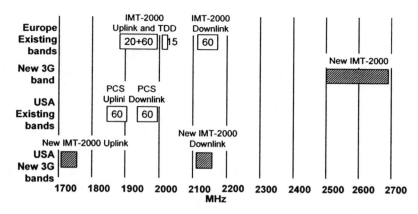

Figure 4.9. Example new frequency bands for WCDMA

In addition to the beamforming antenna solutions, the HSDPA bit rates could be increased by using several transmitter antennas in the base station and several receiver antennas in the mobile. This approach is called *Multiple Input Multiple Output (MIMO)*. The higher data rates are obtained since the same spreading code can be reused in different antennas. To distinguish the several substreams sharing the same code, the mobile uses multiple antennas and spatial signal processing. An example MIMO receiver with 2 antennas is shown in Figure 4.10. The space-time Rake combiner is the multiple antenna generalization of the conventional rake combiner.

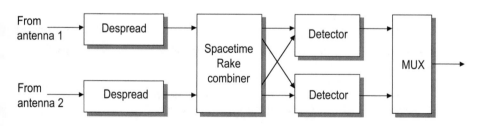

Figure 4.10. MIMO receiver

The performance of MIMO techniques is evaluated in [2]. The results from the simplified scenario show that using two antennas in the base station and in the mobile, the bit rate can be increased from 10.8 Mbps to 14.4 Mbps. With four antennas the bit rate can be increased to 21.6 Mbps with the same C/I requirement. During Release 6/7 further work on the more realistic performance impact and practical complexity aspects of MIMO technologies will be carried out in 3GPP standardization.

5. Conclusions

WCDMA commercial deployment has started with 3GPP Release'99 specifications, which supports flexible data services together with voice service. The practical data rates with first terminal extend beyond 300 kbps. Release'99 is optimized for varying QoS requirements from real time video, interactive web browsing and gaming to delay tolerant data transmission. WCDMA Release'99 allows the introduction of a large variety of new attractive wireless data services.

WCDMA Release'5 makes the data services more attractive and improves the efficiency of the service delivery with the introduction of High Speed Downlink Packet Access, HSDPA, which boosts the practical user bit rates beyond 1 Mbps and the theoretical peak bit rates beyond 10 Mbps. HSDPA improves also significantly the spectral efficiency of the non real time packet data services.

WCDMA standardization work beyond Release 5 is going on with targets to improve the uplink packet data performance and to further increase the practical bit rates of HSDPA with advanced antenna technologies. Further on the radio access network side the aim is to study whether the UTRAN architecture would need modifications to meet the need for increased data services. All this is aiming to ensure that WCDMA can provide low enough cost per bit for emerging market of mobile data applications with increasing end user performance requirements.

REFERENCES

[1]   Holma, H. and Toskala, A. "WCDMA for UMTS", 2nd edition, John Wiley & Sons, 2002.

[2]   3GPP Technical Report 25.848, "Physical layer aspects of UTRA High Speed Downlink Packet Access", version 4.0.0, March 2001.

[3]   Ramiro-Moreno, J., Pedersen, K.I. and Mogensen, P.E., "Capacity Gain of Beamforming Techniques in a WCDMA System under Channelization Code Constraints", submitted to IEEE Journal on Selected Areas of Communications

[4]   3GPP Technical Report 25.896, "Feasibility study for enhanced uplink for UTRA FDD", version 1.0.0, September 2003.

Chapter 5

EVOLUTION OF CDMA FROM INTERFERENCE-LIMITED TO NOISE-LIMITED

Hsiao-Hwa Chen

Institute of Communications Engineering
National Sun Yat-Sen University
70 Lien Hai Road, Kaohsiung, Taiwan, ROC
hshwchen@mail.nsysu.edu.tw

Abstract   This chapter will use descriptive language to address the issues on CDMA technological evolution, starting with the very basics of CDMA technologies, including various primary CDMA technologies and their core: spreading codes or sequences, such as M-sequences, Gold codes, Kasami codes, Walsh-Hadamard Sequences, OVSF Codes and complementary codes and their correlation statistical information, reflected in their auto-correlation functions (ACFs) and cross-correlation functions (CCFs), whose characteristics play a pivotal role in determining the overall performance of a CDMA system. Then, the cause of multiple access interference (MAI) and multipath interference (MI) in a CDMA-based wireless system will be addressed from the perspectives of spreading codes and system operation modes, followed by discussion on different methods to mitigate MAI and MI in a conventional CDMA system, such as open-loop and closed-loop power control, multi-user detection, RAKE receiver, antenna-array techniques, pilot-aided CDMA signal detection and up-link synchronization control, etc. The discussions on various features of state-of-the-arts CDMA technologies lead to the introduction of a new concept on *isotropic CDMA air-link technology*, which consists of two related sectors: isotropic spreading codes and isotropic spreading modulation, which can offer an homogenous performance in both synchronous and asynchronous CDMA channels. A novel CDMA architecture based on complete complementary codes, namely CC-CDMA, is presented as an example of the CDMA system using the isotropic spreading technology, characterized by its attractive isotropic MAI-free property. The lack of isotropic MI-free property in the CC-CDMA system motivates us to work out an even more desirable CDMA system, CC/DS-CDMA, which is designed based on combinational use of complementary codes and traditional direct-sequence spreading technique and can offer a truly interference-free operation in both up-link and down-link channels. With the help of such a superior isotropic MAI-free and isotropic MI-free property, the CC/DS-CDMA can operate virtually in an interference-free environment, ensuring a performance limited only by AWGN, no longer by interferences. A list of complementary codes with different PG values (from 8 to 512) is also given in Appendix at the end of this chapter. Based on the discussion on conventional CDMA technologies and introduction of two novel CDMA architectures, it is concluded in this chapter that the evolution of CDMA technologies will eventually make it happen that a CDMA system could offer a noise-limited performance.

Apart from the inclusion of two innovative CDMA architectures, this chapter covers also various new CDMA based wireless standards seldom reported elsewhere in the open literature, such as TD-SCDMA and LAS-CDMA, etc., together with a wide collection of reference sources associated with them, as well as those on other commonly referred CDMA based standards, such as IS-95A/B, cdma2000, UMTS-UTRA, W-CDMA, etc., providing a rather comprehensive coverage about the evolution of CDMA technologies from interference-limited to noise-limited.

Keywords: CDMA, spreading code, multiple access interference, multipath interference

1.   Introduction

Code division multiple access (CDMA) has emerged as the most important multiple access technology for the second and third generations (2G-3G) wireless communication systems, exemplified by its applications in many important standards, such as IS-95 [1][2], cdma2000 [3], UMTS-UTRA [4], W-CDMA [5] and TD-SCDMA [6], etc., which were proposed by TIA/EIA of the US, ETSI of Europe, ARIB of Japan and CATT of China, respectively. It will be likely that the CDMA technology will continue to be a primary air-link architecture for the future or *beyond* 3G (B3G) wireless communications, although some other new multiple access technologies have also gained attention in the community, such as ultra-wideband technology (UWB) [7]-[12] and even some renovated versions of TDMA. Basically, the UWB technology can also be considered as a special type of spread spectrum technique and it works on very narrow pulses based on pulse position modulation (PPM) or time-hopping technique. Although The UWB technique can provide a relatively higher transmission rate up to 600 Mb/s and a superior capability to mitigate multipath interference due to its high time resolution but the current study has indicated that it is only suitable for the applications covering a relatively short distance less than 50 meters [8][13]. Therefore, at least at the moment the UWB techniques are still far from mature for the applications in a wide-area mobile cellular system, which is still the most important sector of the B3G wireless.

CDMA is a multiple access technology that divides users based on orthogonality or quasi-orthogonality of their signature codes or simply CDMA codes. There are three primarily different types of pure CDMA technologies that have been extensively investigated in the recent two decades, direct sequence (DS) CDMA, frequency hopping (FH) CDMA and time hopping (TH) CDMA. Each user in a DS-CDMA system should use a code to spread its information bit stream directly by multiplication or modular-two addition operation, which is also the simplest and most popular CDMA scheme of the three. The FH-CDMA uses a multi-tone oscillator to generate multiple discrete carrier frequencies and each user in the system will choose a particular frequency-hopping pattern among those carriers that is governed by a specific sequence, which should be orthogonal or quasi-orthogonal to the others. Depending on the hopping rate relative to the data rate, FH-CDMA can also be classified into two sub-categories: slow-hopping and fast-hopping techniques. The majority of currently available FH-CDMA systems are using slow-hopping scheme. The third type of CDMA or TH-CDMA is found to be much less widely used than the previous two due mainly to its implementation difficulty and hardware cost associated with a transmitter that should provide an extremely high dynamic range and very high switching speed. As mentioned earlier, the UWB technique can, in a way or other, be viewed as a type of TH-CDMA systems [14].

There are also many different types of hybrid CDMA schemes, which can be formed by various combinations of DS, FH and TH, together with multi-carrier (MC) and multi-tone (MT) techniques, as shown in Figure (5.1), where the family tree of various forms of CDMA technologies is depicted, where CC-CDMA and DS/CC-CDMA are two new CDMA schemes to be introduced and then discussed in the later part of this chapter and *offset stacking* is a new spreading modulation technique used in the CC-CDMA architecture. It is stressed that this chapter will only concerns the DS based CDMA systems and its evolution issues. However, the conclusions drawn here may also be found equally relevant to the other CDMA schemes.

One of the most important characteristics of a CDMA system is that it allows all users to send their information at the same frequency band and the same time duration simultaneously but at different codes. Therefore, it is obvious that the orthogonality or quasi-orthogonality among

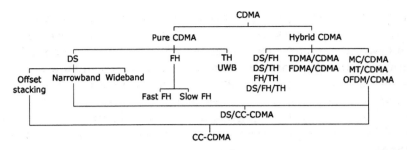

Figure 5.1. Family tree of various CDMA technologies. Offset-stacking is a bandwidth-efficient spreading technique used in CC-CDMA, which together with DS/CC-CDMA is the CDMA architecture to be introduced in the later part of this chapter.

the codes or sequences plays a critical role. In fact, we should define the two important roles of the codes or sequences involved in a CDMA system: one is to act as the signature codes (to accomplish code division multiple access) and the other is to spread the data bits (to spread signal bandwidth to achieve a processing gain). It should be emphasized that the roles of the former and the latter are not necessarily given to the same code in a particular CDMA air link architecture. For instance in the up-link channels of IS-95A/B [1][2], the signature codes are long M-sequences and the spreading codes are 64-ary Walsh-Hadamard functions. On the other hand, the down-link of the IS-95A/B standards uses 64-ary Walsh-Hadamard sequences as both the spreading codes and the signature codes [15].

Then, it comes to the question why CDMA has gained so much attention as the most popular air-interface technology for the current 2G and 3G, possibly also for B3G wireless communications. The main reasons can be summarized in the sequel. First, so far the CDMA is the only technology that can mitigate multipath interference (MI) problem in a very cost effective way, which otherwise should be tackled by using other relatively complicated sub-systems in FDMA and TDMA systems. Second, the current CDMA technology can offer on the average a far much better capacity than its counterparts, such as FDMA and TDMA systems, to meet the increasing demand for mobile cellular applications in the world. Third, the overall bandwidth efficiency of a CDMA system is considered to be much higher than that using conventional multiple access technologies, thus giving an operator much more incentives to adopt it due to extremely high price of spectra. Finally, relatively low peak emission power level of a CDMA transmitter offers a unique capability for CDMA based systems to overlay the existing radio services currently in operation without incurring noticeable interference with each other.

However, we have to admit that current CDMA systems are still far from perfect. It is a well-known fact that a CDMA system is always considered to be interference-limited due mainly to the existence of multiple access interference (MAI) and multipath interference (MI), which are the two major contributors to the limitation of capacity or performance in any CDMA based system currently in operation, including all mature 2G and 3G architectures [1]-[6]. Very likely, the following questions will always come to the mind of anybody who have learned the basic knowledge of CDMA:

- Do CDMA systems always deserve only interference-limited performance?

- Why a CDMA system has to work with so many complicated auxiliary sub-systems, such as close-loop and open-loop power control, RAKE receiver, rate-matching algorithms [4][5], up-link synchronization control [3][6], multi-user detection, etc., to just name a few as examples.

- Can we get rid of all of those complicated sub-systems to make a simple and yet effective CDMA?

Many people may think it is only a dream to make an ideal CDMA that never comes true, but we would like to offer our different views in this chapter through addressing the issues related to the evolution of CDMA technologies from currently available 2G and 3G systems to the new concept to develop an ideal CDMA for the future. Here we will also present some of our thoughts to engineer a new CDMA architecture with a greatly enhanced capability to mitigate MAI and MI, a critical issue associated with a noise-limited CDMA.

Several assumptions should be made to facilitate the description and discussion given in this chapter. Firstly, we should limit our discussions to DS-CDMA systems only and we will not address the issues related to other CDMA schemes, such as FH-CDMA and TH-CDMA (including the UWB techniques). Secondly, in such a DS-CDMA system of interest to us, data signal spreading will be fulfilled using short codes (with the chip width being T_c), whose length is exactly the same as the duration of one data bit (T_b), or the processing gain (PG) of such a DS-CDMA system is equal to $N = T_b/T_c$. On the other words, this chapter will not deal with the situation where a long spreading code, whose length is longer than the width of one data bit, is used to spread data bit stream. Thirdly, we will concern a wireless system with full-duplex operation, which consists of mobile terminals and a base station (BS). The transmission link from mobiles to a BS is referred to as up-link, and transmission link in the reverse direction is called down-link. The block diagram of a generic DS-CDMA system concerned in this chapter is shown in Figure (5.2), where we are interested only in a DS-CDMA system with K users, each of which is assigned *one* unique code for CDMA purpose, and the signal of concern is data source 1. Thus, any forms of M-ary CDMA schemes are not the system of interest here [16]-[18].

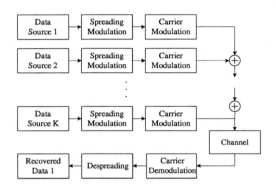

Figure 5.2.   A generic K-user DS-CDMA system model, where only up-link channels of the system are concerned.

2.   The Basics of CDMA Codes

As the name suggests, the CDMA codes, whose characteristics will govern the performance and limitations of a CDMA system, play an essential role in a CDMA system architecture. For instance, the use of orthogonal variable spreading factor (OVSF) codes in UMTS-UTRA [4] and W-CDMA [5] standards requires that a dedicated rate-matching algorithm has to be carried out in the transceivers involved whenever user data transmission rate changes to match a specific spreading factor or the system wants to admit as many users as possible in a cell. In addition, the rate change in UMTS-UTRA and W-CDMA can be made only at the multiples of two, meaning that the continuous rate change is impossible. This requirement is a direct consequence of the OVSF code generation tree structure, where the codes in the upper layers bear a lower spreading factor; whereas those in the lower layers offer a higher spreading factor. Therefore, occupancy of a node in the upper layers effectively blocks all nodes in the lower layers, meaning that less users will be accommodated in a cell. The rate-matching algorithms [19]-[23]

indeed consume a great amount of hardware and software resource and affect the overall performance, such as increased computation power and processing latency, etc. Therefore, the choice of CDMA codes is extremely important and should be exercised very carefully at very early stage of a CDMA system design; otherwise the short-comes of the system architecture due to the use of unsuitable CDMA codes will carry on for ever with a standard.

There are many different ways to characterize the CDMA codes, but no ones else can be more effective and intuitive than the auto-correlation function (ACF) and cross-correlation function (CCF), which should be discussed in detail as follows.

2.1   Auto-Correlation Function

The ACF is defined as the result of chip-wise convolution or simply correlation operation between two time-shifted versions of the same code, which can be further classified into two sub-categories: periodic ACF and aperiodic ACF, depending on the same and different bit patterns, respectively, in two consecutive bits of received bit stream during the correlation process at a CDMA receiver, as shown in the left-sided two branches in Figure (5.3). In a practical CDMA system, usually the periodic ACFs and aperiodic ACFs appear equal likely due to the fact that "+1" and "-1" appear equal probable in received binary bit stream.

The in-phase ACF or ACF peak, which is often equal to its length or PG value (N), affects detection efficiency of desirable signal in a CDMA receiver, where a correlator or matched filter is used. On the other hand, the out-of-phase ACFs of a CDMA code will be harmless if no multipath effect is present. However, they will contribute to MAI and will seriously affect system performance under MI.

2.2   Cross-Correlation Function

The CCF is defined as the result of chip-wise convolution operation between two different spreading codes in a family of the codes. Due to the similar reason as mentioned earlier for the ACF, there are also two different types of CCF: periodic CCF and aperiodic CCF; the former is mainly found in synchronous transmission channels, such as down-link channels in a wireless system, and the latter can appear in either synchronous or asynchronous channels. In contrast to out-of-phase ACF that will contribute to MAI only under multipath channels, the CCF always contributes to MAI, no matter whether or not multipath effect is present. On the other hand, the out-of-phase ACF will become harmful if and only if a multipath channel is concerned; otherwise it will never yield MAI at a correlator receiver. Obviously, the MAI is one of the most serious threats to jeopardize detection efficiency of a CDMA receiver using either correlator or RAKE and thus has to be kept under a sufficiently low level to ensure a satisfactory performance. Tables (5.1) and (5.2) list all correlation functions of a CDMA code and their merit behavior in a CDMA system.

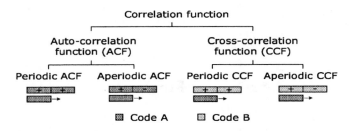

Figure 5.3.   Classification of correlation functions of CDMA codes.

Table 5.1.   Auto-correlation functions (ACFs) of CDMA codes and their merit behavior in a CDMA system

ACF	IPP[a]-ACF	OPP[b]-ACF	IPAP[c]-ACF	OPAP[d]-ACF
Cause	Correlator	AMP[e] channel	Correlator	AMP channel
Frequency	Once a bit	High	Once a bit	High
Behavior	Wanted signal	MAI under MI	Wanted signal	Self-interference
ITD[f]	Enhance	Impair	Enhance	Impair

[a] IPP: In-phase periodic.
[b] OPP: Out-of-phase periodic.
[c] IPAP: In-phase aperiodic.
[d] OPAP: Out-of-phase aperiodic.
[e] AMP: Asynchronous multipath.
[f] ITD: Impact to detection

2.3   Traditional CDMA Codes

The searching for promising CDMA codes or sequences used to be a very active research topic, which can be traced back to as early as the 60th. Numerous CDMA codes have been proposed [24]-[45] and their performance and possible applications in a CDMA system were investigated in the open literature [46]-[55]. In the following text, we will briefly discuss some of frequently referred CDMA codes, including both

Table 5.2. Cross-correlation functions (CCFs) of CDMA codes and their merit behavior in a CDMA system

CCF	IPP[a]-CCF	OPP[b]-CCF	IPAP[c]-CCF	OPAP[d]-CCF
Cause	Syn. channel	AMP[e] channel	Syn. channel	AMP channel
Frequency	Once a bit	High	Once a bit	High
Behavior	MAI	MAI	MAI	MAI
ITD[f]	Impair	Impair	Impair	Impair

[a] IPP: In-phase periodic.
[b] OPP: Out-of-phase periodic.
[c] IPAP: In-phase aperiodic.
[d] OPAP: Out-of-phase aperiodic.
[e] AMP: asynchronous multipath.
[f] ITD: Impact to detection

quasi-orthogonal codes (such as M-sequences, Gold codes, Kasami codes, etc.) and orthogonal codes (such as Walsh-Hadamard Sequences, OVSF codes, complementary codes, etc.), with their fundamental characteristics. In addition to those commonly referred CDMA codes listed above, there are many other less widely quoted ones, such as GMW codes [25], No codes [26], Bent codes [46], etc., which will not be discussed in this chapter. For more detail treatment of them, the readers can refer to the literature given in the references [24]-[55].

2.3.1   M-Sequences.   M-sequences [28]-[34] are also called maximum-length sequence, whose name reflects the fact that they can be generated by using a primitive polynomial in GF(2) with a specific degree n and have the longest possible length using any polynomial in GF(2) with the same degree. The simplicity of the M-sequences is also reflected in their sequence generator structure, where only a single shift-register is required together with a feedback logic that depends on the primitive polynomial concerned. Basically, M-sequences are not suitable for CDMA applications due to their uncontrollable CCFs. However, their out-of-phase ACFs are always "-1", making them suitable for some special applications such as radar and synchronization-control systems, where a high time resolution is of ultimate importance. Nevertheless, it should also be pointed out that it is still possible for us to find a *relatively small* group of M-sequences that do maintain a reasonable low CCFs among them, such that they can be used as CDMA codes.

2.3.2   Gold Codes.   One of the most popular quasi-orthogonal CDMA codes should be Gold code [24][34], which was first studied by R.

Gold in 1968 and has been extensively used in many commercial CDMA systems, including IS-93 and W-CDMA standards. The popularity of the Gold codes stems from the two main reasons listed below. First, any pair of Gold code families offer a uniquely controllable three-leveled CCFs, and so do their out-of-phase ACFs, the maximal value of which is equal to $2^{(n+1)/2} + 1$, where n is the degree of the polynomial generating the Gold codes. This maximal CCF value of the Gold codes is relatively low if compared to those of the others. This characteristic feature of the Gold codes make them very suitable for CDMA applications with a predictable performance in terms of MAI. Second, a family of Gold codes can be generated in a very simple logic with a pair of shift registers, each of which bears the same structure as that to generate an M-sequence. In other words, a pair of primitive polynomials are required to generate a complete family of Gold codes, whose size can be as large as $2n+1$, where n stands for the degree of the primitive polynomials used for generation of the Gold codes. Therefore, the relatively large family size makes Gold codes a popular choice as CDMA codes, being able to support many users in a cell. It should be pointed out here that the family size of CDMA codes should also be considered as an important resource, which could be utilized to increase effective transmission rate in a channel. For instance, if the family size is large enough, each user could be assigned multiple codes, instead of only one as the case in a conventional CDMA system, such that the use of distinct codes for the same user can effectively deliver more information than that in the case with single code assignment, given the same total bandwidth. This is just what has been done in M-ary CDMA systems [16]-[18]. In this sense, a large family size is definitely a plus to any CDMA codes.

2.3.3   Kasami Codes.   There are two different types of Kasami codes [34][38], small-set Kasami codes and large-set Kasami codes. The major difference between the two lies in their CCF, the former of which is lower than the latter. The maximal value of the CCF for the small-set Kasami codes is $2^{n/2} + 1$ (where n is the degree of the polynomials generating the small-set Kasami codes), which is already very close to the Welch bound [48]. In fact, a family of the large-set Kasami codes can be divided into two sub-sets, one of which are a family of small-set Kasami codes and the other are in fact a family of Gold codes at the same degree. Thus, in this sense, the Kasami codes bear many similar features as Gold codes discussed earlier. It should be pointed out that, although the small-set Kasami codes have a rather low peak CCF that is important for MAI reduction, they form a relatively small family size, which limits its wide applications as signature codes in a CDMA system.

2.3.4   Walsh-Hadamard Sequences.   Walsh-Hadamard sequences [39]-[41] can be obtained from a Walsh-Hadamard matrix, either rows or columns of which can be taken as spreading sequences that are perfectly orthogonal with one another if and only if they are used in a synchronous transmission mode. Walsh-Hadamard sequences have been applied to IS-95A/B [1]-[2] as well as cdma2000 [3] standards as either channelization codes for down-links channels or spreading sequences for up-link applications. It is necessary to emphasize that, although Walsh-Hadamard sequences are referred to as *ideally* or *perfect* orthogonal codes, their out-of-phase ACFs are very high, which will seriously affect the asynchronous up-link signal reception in a CDMA system under multipath effect, where received signal from a particular mobile consists of multiple replica with different delays, and thus out-of-phase ACFs will contribute substantially as a part of MAI at the receiver. In addition, the out-of-phase CCFs of Walsh-Hadamard sequences are also rather high, giving rise to a substantial increase of MAI. Therefore in this sense, we have a strong reason to doubt the suitability to use Walsh-Hadamard sequences as signature or spreading codes in any CDMA system, which ought to work inevitably in a multipath environment.

2.3.5   OVSF Codes.   Another important orthogonal codes extensively reported are orthogonal variable spreading factor (OVSF) codes, which have been made famous due to its application in three major 3G standards: UMTS-UTRA [4], W-CDMA [5] and TD-SCDMA [6], proposed by ETSI (Europe), ARIB (Japan) and CATT (China), respectively, as their IMT-2000 candidate proposals to ITU roughly at the same time in 1998. As its name has suggested, the OVSF codes are generated to fit variable spreading factors (SFs) or the code lengths under a special code generation tree, on which the codes with larger SFs form the lower layers and those with smaller SFs form the upper layers. The SFs can be made variable in the multiples of two from 4 to 256 based on a so called *rate-matching* algorithm, as specified in the standards. Thus, possible data rates can also be made variable only in the multiples of two. For instance, if a user requires a data rate of 5 units, the system has to assign the user a bandwidth associated with a data rate of 8 units, wasting about 37.5% bandwidth resource. The characteristics of ACFs and CCFs of OVSF codes are very similar to those of Walsh-Hadamard sequences with the identical length. Again similar to the Walsh-Hadamard codes, the OVSF codes perform badly under asynchronous up-link channels, where the orthogonality among the codes virtually does not exist. In fact, if all OVSF codes with a fixed SF are arranged into a matrix, we will readily find that it is exactly the same as a Walsh-Hadamard matrix. In this sense, the OVSF codes should not be considered as a new CDMA codes at all.

Due to the fact that both Walsh-Hadamard codes and OVSF codes possess very high out-of-phase ACFs and CCFs, they should not be considered as orthogonal codes, which are supposed to offer an ideal ACFs and CCFs with all possible time-shifts. However, there exist genuine orthogonal codes called complete complementary codes, which are of great interest to us in this chapter and will be discussed in much a great detail as follows.

2.3.6   Complete Complementary Codes.

The study on complementary codes (CC) can be traced back to the 60's, when Golay [42] and Turyn [43] first studied pairs of binary complementary codes whose autocorrelation function is zero for all even shifts except the zero shift. However, the main interest on complementary codes at that time was not for their possible applications in CDMA systems but in radar systems. Later, Suehiro [44][45] extended the concept to the generation of the CC code families whose autocorrelation function is zero for all even and odd shifts except the zero shift and whose cross-correlation function for any pair is zero for all possible shifts. The work carried out in [44][45] had paved the way for practical applications of the CC codes in modern CDMA systems, one possible architecture of which has been proposed and studied in [69].

There exist several fundamental distinctions between the traditional CDMA codes (such as Gold codes, M-sequences, Walsh-Hadamard codes, etc.) and the CC codes. Firstly, the orthogonality of the CC codes is based on a *flock* of element codes jointly, instead of a single code as the traditional CDMA codes. In other words, every user in a CC code based CDMA (or simply CC-CDMA) system will be assigned a flock of element codes as its signature code, which ought to be transmitted via different channels (either in FDM or TDM mode) and to arrive at a correlator receiver at the same time to produce an autocorrelation peak. Thus, all conventional spreading codes, either quasi-orthogonal or orthogonal codes, are also called *unitary* codes, because they work simply on an one-code-per-user basis. Secondly, the processing gain of the CC codes is equal to the *congregated length* of a *flock* of element codes. For the CC codes of lengths $L=4$ and $L=16$ as shown in Table (5.3), their processing gains are equal to $4 \times 2 = 8$ and $16 \times 4 = 64$, respectively. Thirdly, zero CCFs and zero out-of-phase ACFs are ensured for any relative shifts between two codes, which has made the CC codes different from the conventional orthogonal codes, such as Walsh-Hadamard sequences and OVSF codes, whose out-of-phase ACFs can never be zero.

Table 5.3. Two examples of complete complementary codes with element code lengths $L = 4$ and $L = 16$

Element code: $L=4$		Element code: $L=16$
Flock 1	A_0:+++−	Flock 1	A_0:+++++−+−++−−+−−+
			A_1:+−+−+++++−−+++−−
			A_2:++−−+−−+++++++−+−
			A_3:+−−+++−−+−+−++++
	A_1:+−++	Flock 2	B_0:++++−+−+++−−−++−
			B_1:+−+−−−−−+−−+−−++
			B_2:++−−−++−++++−+−+
			B_3:+−−+−−+++−+−−−−−
Flock 2	B_0:++−+	Flock 3	C_0:+++++−+−−−−++−++−
			C_1:+−+−+++++−++−−−++
			C_2:++−−+−−+−−−−−+−+
			C_3:+−−+++−−−+−+−−−−−
	B_1:+−−−	Flock 4	D_0:++++−+−+−−+++−−+
			D_1:+−+−−−−−−−++−++−−
			D_2:++−−−++−−−−−−+−+−
			D_3:+−−+−−++−+−+++++

Table 5.4. Family-sizes and flock-sizes for complete complementary codes with various element code lengths L

Element code length $(L = 4^n)^a$	4	16	64	256	1024	4096
PG $(L\sqrt{L})$	8	64	512	4096	32768	262144
Family size $(\sqrt{L})$	2	4	8	16	32	64
Flock size $(\sqrt{L})$	2	4	8	16	32	64

^{a} n can be any integer.

Also due to this fact, the CC codes could be referred to as truly perfect orthogonal codes. This property is extremely important to give us a great hope to further enhance the performance of a CDMA system.

Table (5.3) gives two examples of CC code of concern in the proposed CC-CDMA system model to be introduced in the following text. More examples of CC codes (whose PG values are from 8 to 512) can be found in Appendix given at the end of this chapter. Table (5.4) shows the *flock* and family sizes for various CC codes with different element code lengths (L). For more detail information on code generation procedure and other properties of the CC codes, the readers may refer to [42]-[46].

The reasons why we have paid so much attention to the CC codes lie on the fact that the correlation properties of the CC code are based on several (always an even number) element codes, instead of only one single code as the case in conventional spreading codes, such as Gold codes and OVSF codes, etc. This observation is significant to make the CC codes different from the conventional codes. While it is very difficult to ensure an ideal ACFs and CCFs of conventional or *unitary* spreading codes whose correlation function is based on a single code, it may be possible for us to formulate an ideal ACFs and CCFs that are based on the sum of individual ACFs and CCFs of a flock (always an even number) of element codes, as long as the non-zero values of out-of-phase ACFs and CCFs for different element codes within the same flock can be canceled in the sum operation. Therefore, this gives us a much greater degree of freedom in the code design process and it is no longer necessary for us to strictly require zero out-of-phase ACFs and CCFs for each individual element code.

The examples for both ACF and CCF of a particular CC code of PG equal to $16 \times 4 = 64$ are shown in Figures (5.4) and (5.5), respectively, from which it can be clearly seen that, although the auto-correlation functions of individual element codes are not ideal (there are many non-zero side lobes, as shown in Figure (5.4)), the sum of them yields, $R_A(\tau)$, an ideal ACF, which is just what we want. The same observation can also be made with regard to the CCF of some particular CC codes or $R_{A,B}(\tau)$, as shown in Figure (5.5). The same characteristic feature can also be found to any other CC codes, whose detail information has been given in Appendix of this chapter. This desirable features of CC codes can not be found from any other conventional or unitary CDMA codes available so far, including all *so called* orthogonal codes, such as Walsh-Hadamard Sequences and OVSF codes, etc.

One example for the application of the CC codes can already be found in TD-LAS system, which has been approved by 3GPP2 as an enhanced standard [56], in which a pair-wise CC codes, or called LS codes in the TD-LAS specification, have been used as spreading codes of the users. In other words, each user in the TD-LAS system is assigned two element codes (namely S section and C section), which are sent in the same time slot in a time-division-multiplex mode or TDM mode. Combined with LA codes, the use of LS codes in TD-LAS creates a unique interference-free window (IFW), which spans about a few to a few tens of chips beside ACF main lobe, making it possible for the system to employ some multi-level digital modems, such as 16-QAM, to further improve the bandwidth efficiency in its air-link sector. In April 2002, the TD-LAS system has undergone a successful field trial in Shanghai, demonstrating the great potential of the CC codes in the future wireless communications. Unfortunately, the TD-LAS system can only offer an

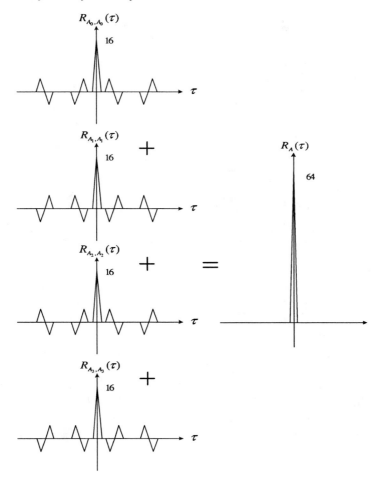

Figure 5.4. Auto-correlation function of a complete complementary code with its element code length being 16 and PG 16×4=64 and detail information of the CC code of concern is given in Table (5.3), where $R_{A_0,A_0}(\tau)$ is the ACF of element code A_0, $R_{A_1,A_1}(\tau)$ is the ACF of element code A_1, $R_{A_2,A_2}(\tau)$ is the ACF of element code A_2, $R_{A_3,A_3}(\tau)$ is the ACF of element code A_3 and $R_A(\tau)$ is the sum of $R_{A_0,A_0}(\tau)$, $R_{A_1,A_1}(\tau)$, $R_{A_2,A_2}(\tau)$ and $R_{A_3,A_3}(\tau)$.

IFW, which is still much smaller than the code-length. In Section 4 of this chapter, we will introduce two novel CDMA architectures based on the CC codes, which can offer an MAI-free window as large as the code-length. It will be further illustrated that the application of the CC codes can offer a new dimension to implement the B3G wireless.

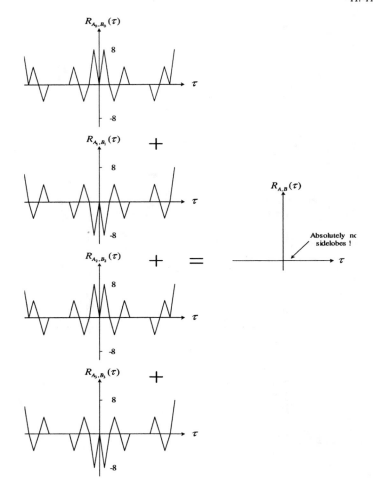

Figure 5.5. Cross-correlation function of a complete complementary code with its element code length being 16 and PG 16×4=64 and detail information of the CC code of concern is given in Table (5.3)where $R_{A_0,B_0}(\tau)$ is the CCF of element codes A_0 and B_0, $R_{A_1,B_1}(\tau)$ is the CCF of element codes A_1 and B_1, $R_{A_2,B_2}(\tau)$ is the CCF of element codes A_2 and B_2, $R_{A_3,B_3}(\tau)$ is the CCF of element codes A_3 and B_3, and $R_{A,B}(\tau)$ is the sum of $R_{A_0,B_0}(\tau)$, $R_{A_1,B_1}(\tau)$, $R_{A_2,B_2}(\tau)$ and $R_{A_3,B_3}(\tau)$.

3.   Multiple Access Interference and Multipath Interference

There are two major sources of interference in a CDMA system; one is MAI and the other is the MI. The former is because of imperfect CCFs of CDMA codes and the latter is caused by combined effect of out-of-phase ACFs and CCFs of spreading codes used by the system in a multipath channel. The different transmission modes in synchronous down-link and asynchronous up-link channels can further complicate their impact to the performance of a CDMA system.

3.1 Multiple Access Interference (MAI)

The down-link channels in a wireless system are referred to as the transmission direction from a BS to mobile terminals within the cell and usually are synchronous in a way that the bit-streams from different mobiles arrived at a particular mobile are aligned bit-by-bit in time. To reflect this characteristic feature of the down-link channels, the delays for the transmissions from a BS to a particular mobile should be considered constant. Without multipath effect, the MAI in the down-link channels will be caused simply by the periodic CCFs of all CDMA codes active in the system. However, due to the multipath propagation, the MAI will consist of two parts, one part being the CCFs between the code of interest and all other active codes as well as their multipath returns, and the other part being the out-of-phase ACFs of the code concerned at a receiver due to the fact that the receiver will receive several replicas of the code of concern with different delays. Because of the possibility that two consecutive bits have either same or different signs, both aperiodic out-of-phase ACFs and periodic out-of-phase ACFs will contribute to MAI with equal probability.

MAI in up-link channels will also consist of two different parts: the CCFs among different user codes and the out-of-phase ACFs due to multipath effect. In either an AWGN up-link channel or a multipath up-link channel, both periodic and aperiodic CCFs will be involved in MAI. Figure (5.6) lists all possible MAI contributions from non-ideal ACFs and CCFs of the CDMA codes, where the words in the parenthesis indicate the environment where the contributions will be effective.

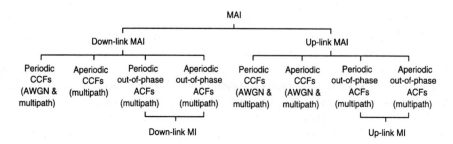

Figure 5.6. Illustration of all possible MAI contributions from non-ideal ACFs and CCFs of CDMA codes (MI has been viewed as a part of MAI in this figure).

3.2   Multipath Interference (MI)

Multipath interference is another major impairing factor in a CDMA system, which usually causes serious inter-symbol interference (ISI) at a conventional CDMA receiver based on matched filter or RAKE. In a multipath (either synchronous or asynchronous) channel, transmitted signals will undergo different propagation paths to reach a specific receiver, in which all those multipath returns will be involved either harmfully or usefully in signal detection process.

The multipath effect can be best illustrated in terms of both time-domain and frequency-domain. To do so, let us first define the relative delay between two consecutive paths as *inter-path delay*. If the inter-path delay is smaller than one chip width, a traditional matched-filter (or correlator) or even a RAKE (with only one sample per chip) will not be able to distinguish individual multipath returns. In this case, if the number of multipath returns is very small, such as only two or three rays, they usually pose little threat to the signal detection process at a receiver and they can be treated as a single multipath return if each chip will be sampled only once. However, if the number of multipath returns (or delay spread) is relatively large, all those closely located multipath returns will span a time duration longer than a chip or even a symbol, causing a serious ISI. On the other hand, if the inter-path delay of a multipath channel is larger than the chip width, all multipath returns will be considered as resolvable by using a correlator receiver and care should also be given not to let them interfere with one another, especially under near-far effect. The capability for a correlator to capture individual multipath returns is one of the most important reasons why CDMA technology has become a popular choice in 2-3G wireless communication systems.

The multipath effect can also be clearly illustrated in the frequency-domain. Since a multipath channel can be well modeled by a delayed-tap-line filter with the coefficient in each tap representing the path gain of a particular multipath return and its delay element being the inter-path delay, as defined earlier, without losing generality. For simplicity, let us look at a two-path channel model shown in Figure (5.7a), and thus only one delay element and one coefficient are involved in such a simple delayed-tap-line channel model, as shown in Figure (5.7b). The transfer function of the impulse response for this two-path channel model is shown in Figure (5.7c), where a comb-like frequency-domain transfer function can be observed, clearly showing the frequency-selectivity of the multipath channel. Similar observation can be made for a multipath channel with more than one delayed returns. Therefore, a multipath channel can also be called a frequency-selective channel.

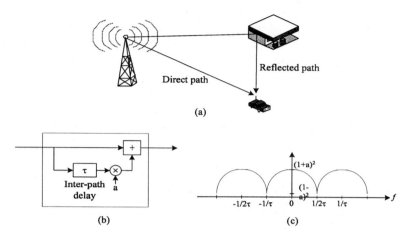

Figure 5.7. (a) A two-path channel model; (b) delayed-tap-line filter in the time-domain; and (c) frequency-selectivity in the frequency-domain.

Obviously, the cause of MI at a receiver is different from that of MAI from the viewpoint of CDMA codes. As mentioned earlier, MAI is mainly caused by non-ideal CCFs of the CDMA codes. However, the MI is due mainly to imperfect or non-zero out-of-phase ACFs of the CDMA codes. Assume that the inter-path delay is larger than the chip width. Then, the multipath returns will not be harmful to the signal detection if every CDMA code possesses an ideal ACF, meaning that all (both periodic and aperiodic) out-of-phase ACFs are zero. The output from a correlator will consist of clearly separated auto-correlation peaks, each of which represents an individual multipath return and can be detected one by one if symbol synchronization is achieved. However, if the out-of-phase ACFs of the CDMA codes of concern are not zero, the ACF side lobes following the earlier auto-correlation peak will seriously interfere the detection of the later auto-correlation peaks, especially if the earlier peak is much stronger than the later ones due to the near-far effect in a CDMA system. Both periodic and aperiodic out-of-phase ACFs will be involved in MI owing to the fact that data symbols could appear either +1 or -1 equal likely. In fact, Figure (5.6) has already taken the MI into account. Therefore, in this sense the MI can be effectively considered as a special form of MAI. Nevertheless, MI is still different from MAI in terms of their generation mechanism.

4.   Techniques to Combat MAI and MI

Combined effect of MAI and MI poses a great threat to successful signal detection at a traditional CDMA receiver. Therefore, many feasible techniques have been developed and studied in open literature to combat them [54-62], some of which will be discussed in the following text and have been found wide application in many commercial CDMA systems.

4.1   Power Control vs. Near-Far Effect

In early days of CDMA system development, such as Qualcomm's effort to develop the IS-95 standard, a great amount of attention has been paid to solve near-far effect in a CDMA system. The near-far effect stems from the fact that some near-by unwanted transmissions (appearing as CCFs between the desirable and unwanted codes) overwhelm a far-away desirable transmission (appearing as an ACF peak of desirable code), jeopardizing user separation process at a CDMA receiver. Therefore, the near-far effect is because of non-ideal CCFs of a CDMA code family of concern and can be considered as one of the direct consequences of MAI. In particular, it should be noted that the near-far effect usually has a stronger influence on up-link channels than down-link channels due to asynchronous transmissions in the up-link channels, where mobile users are distributed in different places in a cell and their received power levels looked at a BS receiver can vary greatly from one another. On the other hand, the synchronous transmission from the same source (or a BS) in the down-link makes it possible for a mobile to receive all signals from the BS at almost equal power. Therefore, there is no near-far effect in the down-link channels if only intra-cell transmissions are concerned. In other words, the near-far effect in the down-link channels is mainly caused by the transmissions from different cells or base stations, if they are not separated by different frequency bands.

The common practice to combat the near-far effect is to use precision power control techniques, which usually consist of two sectors: one is open-loop power control and the other is closed-loop power control. The former always proceeds before the latter and they should work jointly to ensure that all signals from different mobiles in a cell will be received by a BS at almost equal power. With the help of the power control techniques, the transmission power level of every mobile in a cell is under constant and effective control by the BS. Another advantage to use the power control is to limit unnecessary power emission in the whole cellular system to reduce the inter-cell as well as intra-cell interference, contributing to the overall capacity improvement of a CDMA system.

4.2   Multi-User Joint Detection against MAI

Another very effective way to combat the MAI is the use of multi-user detection (MUD), which has been an extremely active research topic in recent ten years [57]-[65]. The basic ideal of the MUD was motivated by the fact that a single-user based receiver, such as a matched filter correlator or a RAKE, always treats other transmissions as unwanted interference in a form of MAI that should be suppressed as much as possible in the detection process therein. Such detection methodologies simply ignore the correlation information among the CDMA codes (or MAI) appeared in the whole system and all of that correlation among the users have not been utilized as a useful information to assist the detection of all signals. On the other hand, the MUD algorithms take the correlation among the users (or MAI) into account in a positive manner and user signal detection proceeds one-by-one in a certain order as an effort to maximize the detection efficiency as a whole. Some MUD schemes (not all of them), such as decorrelating detector, have an ideal near-far resistance property in a non-multipath channel, and thus they can be also used an a counter-measure against the near-far problem in a CDMA system to replace or save the complex power control system that has to be used otherwise. However, it has to be pointed out that in the presence of multipath effect almost none of MUD schemes could offer perfect near-far resistance.

There are two major categories of MUDs: one is linear and the other is non-linear schemes. It has been widely acknowledged in the literature that the linear MUDs have a relatively simple structure than the non-linear schemes and thus they have been given much more attention for their potential applications in a practical CDMA system. In most current 3G standards, such as cdma2000 [3], UMTS-UTRA [4], W-CDMA [5] and TD-SCDMA [6], the MUD has been specified as an option. However, due to the concern of complexity, this option will remain as an option only in a real system implementation and most mobile network operators are still reluctant to activate them at this moment.

Two important linear MUD schemes have to be addressed briefly in this chapter; one is the decorrelating detector [64] and the other is MMSE detector [65]. Decorrelating detector, as its name suggests, performs MUD via decorrelating correlation among user signals by using a simple correlation matrix inversion operation. Some major properties of the scheme can be summarized as follows. First, it can eliminate MAI completely and thus offers a perfect near-far resistance in AWGN channel, which is important for the applications in particular in up-link channels. Secondly, it needs correlation matrix inversion operation, which may produce some undesirable side-effects, one of which is the noise-enhancement problem due partly to the ill-conditioned correlation matrix and partly to the fact that it never takes noise term into account in its decorrelating process. On the other hand, a MMSE detector takes both MAI and noise into account in its objective function to minimize the mean square error of detection and thus it offers a better performance than decorrelating detector especially when signal-to-noise ratio is relatively low in the channel. It should be pointed out that a MUD in the multipath channel behaves very much differently if compared with that in AWGN channel. Usually successful operation of a MUD in a multipath channel requires full information of the channel, such as the impulse response of the channel, etc. Therefore, a MUD working in multipath channels can be very complex. To overcome this problem, many adaptive MUD schemes [59,60] have been proposed such that they can perform the joint signal detection with only very a little or even no channel information.

The analysis of a MUD scheme in a down-link channel is much simpler than that in an up-link channel, where all user transmissions are asynchronous. However, with the help of extended correlation matrix, an asynchronous system can be treated as an enlarged equivalent synchronous one with only adding more *virtual user signals* in its dimension-extended correlation matrix. Thus, theoretically speaking, any asynchronous MUD can always be solved by this method without losing generality.

4.3   RAKE Receiver to Overcome MI

A salient feature of a CDMA system is its capability to offer a high time resolution in distinguish different multipath waves and thus makes it possible to re-combine them in a constructive way after separating them, which effectively yields an attractive multipath-diversity with the help of a RAKE receiver. While it is still possible for TDMA and FDMA systems to combat MI using one way or other, the great complexity to implement similar capability in a TDMA or FDMA system is a major concern.

A RAKE receiver works with several parallel correlators or *fingers*, each of which is to capture different multipath returns that might undergo different propagation delays to make a constructive re-combination, yielding an enhanced decision variable at the output. There are two major re-combining schemes in a RAKE; one is called equal gain combining (EGC) and the other maximal ratio combining (MRC), though there is the third option called *selective combining*, which works simply to choose the best signal as the output and is obviously not as popular as the EGC and MRC. In a EGC-RAKE, all captured multipath components are equally weighted before delay and summation operations and thus it does not require any information about the multipath channel, which is important to reduce the complexity of the receiver and to make it adaptive to any type of multipath channels. On the other hand, an MRC-RAKE offers an optimum performance in terms of maximal SNR achievable at the output by giving different weights to different captured multipath returns according to the known delay profile of the channel. Therefore, successful operation of a MRC-RAKE virtually requires full information of the multipath channel of concern, which can be obtained only by resorting other complicated channel estimation techniques, such as a dedicated pilot signaling, etc. In this sense, an MRC-RAKE is very difficult to work adaptively, especially in a fast fading multipath channel with a high Doppler spread due to great mobility of terminals, without the assistance of an independent channel sounding system. Figure (5.8) shows a generic MRC-RAKE receiver model, where only three fingers are presented for descriptive simplicity.

In addition to the complexity of a RAKE receiver, its performance can also be sensitively affected by MAI characteristics of a CDMA system. As mentioned earlier, each finger in a RAKE receiver should capture a particular multipath signal encoded by a specific code of interest, in addition to which all other unwanted transmissions together with their multipath components (contributing as MAI due to non-ideal CCFs) as well as the other multipath returns of the wanted signal (contributing as MI due to non-zero out-of-phase ACFs) will also come into the finger to act jointly as interference to the wanted particular multipath return signal. Obviously, if the characteristics of CCFs and ACFs of the CDMA codes of concern are not good enough, effectiveness of a RAKE using either EGC or MRC combining will be severely affected and sometimes it can perform very badly especially if there exist many strong later-coming multipath returns followed by the first path. For the mathematics treatment of this subject, the readers may refer to our previous work in the reference [66].

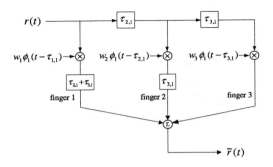

Figure 5.8. A generic model for a three-finger RAKE receiver, where $\tau_{i,j}$ stands for the inter-path delay between the i-th and j-th paths, w_i is the path weight for the i-th multipath return, $\phi_1(t)$ is the spreading code of interest.

4.4   Beam-Forming Techniques against Co-Channel Interference

Beam-forming technique is another effective way to combat MAI or commonly called *co-channel interference* under the context of antenna-array research area. What we refer here with respect to the antenna-array techniques is either smart antenna or switched beam system, both of which have been used in some CDMA based systems to improve the system performance under co-channel interference. The principle of antenna-array techniques is based on various beam-forming algorithms, whose conceptual block diagram is shown in Figure (5.9). It should be clarified that what we concern in antenna-array techniques here is only to achieve some suitable beam-pattern at either a transmitter or a receiver and thus is different from what is called space-time coded MIMO systems.

An antenna-array system consists of several antenna elements, whose space should be made large enough (usually at least 0.5 to 1 wavelength is required), in order to obtain decorrelation among the signals received at different elements. An antenna-array can be used by either a transmitter or a receiver. By using various beam-forming algorithms (such as MVDR, MUSIC, LMS, RLS algorithms, etc.), an antenna-array system will generate a directional beam pattern, whose width is dependent of the number of elements used; the more and then the narrower. With such a very narrow directional beam pattern, a transceiver using an antenna-array system can effectively suppress the co-channel interference generated outside the beam width. If an antenna-array is used as a transmitter antenna in a BS, it can help to project or direct BS signals to some specific mobiles to reduce interference to other mobiles. If an

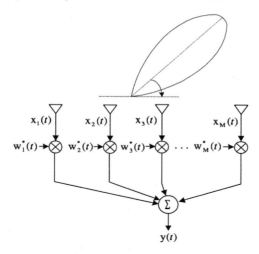

Figure 5.9. A conceptual block diagram of an antenna-array system with M weighted antenna elements.

antenna-array is used as a receiver antenna at a BS, it can assist the BS to focus on the mobiles it wants to receive their signals to suppress other possible interference outside the beam width. Similar to the aforementioned MUD algorithms, a beam forming algorithm can also be made adaptive to follow the change of the direction of arrival (DOA) of target signal with the help of a specific training signal sent by the target, which should be repeated within a time duration shorter than that a substantial change in DOA of the target signal may happen.

However, the usefulness of antenna-array systems is because of the existence of MAI due to imperfect CCFs among the spreading codes. Otherwise, the co-channel interference will no longer be present and then the antenna-array may not be needed.

4.5   Pilot-Aided CDMA Signal Detection

As mentioned earlier, sometimes CDMA signal detection needs the knowledge of multipath channel, such as the MRC-RAKE and MUD, etc. Therefore, the channel estimation becomes a necessity in those CDMA applications. Either a dedicated pilot signal or interleaving a pilot signal with data signal can be used for channel estimation purpose. Usually, a dedicated pilot channel is feasible only for down-link channel signal detection because of the motivation to ease signal detection process at mobiles and the relatively easy allocation of the resource (such as available power and bandwidth, etc.) in a BS.

In order to improve the accuracy of channel estimation, an ideal pilot signaling design should preferably bear the following three important characteristics features (or so called 3-*same conditions* [63]):

- The pilot signal should be sent at the *same frequency* as that for data signal to ensure reliable channel estimation in frequency-selective channels;

- The pilot channel in a CDMA system should share the *same code* as that for data channels to ensure the availability of identical MAI statistics;

- The channel estimation should be carried out at the *same time* as (or as close as possible to) that of data detection to combat fast fading of the channel due to the mobility of the terminals. For similar reason, time duration of the pilot signal should be made as short as possible to facilitate the real-time channel estimation due to the concern on latency in processing the pilot signal.

It should be stressed here that the needs for a pilot signal is due to the requirement for channel estimation, which originally pertains to the requirements of some specific CDMA signal detection schemes, such as MRC-RAKE and MUD, to mitigate MAI and MI. Obviously, if there is neither MAI nor MI, the complicated pilot signaling is not necessary either in a CDMA system.

4.6   Up-Link Synchronization Control

To pave a way for successful application of orthogonal codes in asynchronous up-link channels, up-link synchronization control is necessary, which has been considered as an option in UMTS-UTRA [4] and W-CDMA [5] standards. However, real workable scheme has been implemented solely in TD-SCDMA [6] standard as an important part of the system architecture. Similar to the power control algorithm, there are two sectors of up-link synchronization control: open-loop sector and closed-loop sector, which ought to work jointly to achieve an accurate synchronization, up to 1/8 chip, as specified in the TD-SCDMA standard [67][68]. With the help of such an accurate up-link synchronization control algorithm, the transmission channels in the up-link have been converted into quasi-synchronous ones, effectively enhancing the detection efficiency in up-link channel of a CDMA system, which is often a bottleneck to the whole air-link section. The procedure of the up-link synchronization control algorithm can briefly be explained as follows.

During the cell-search procedure in a TD-SCDMA system, a mobile will capture the information in down-link broadcasting slots to know the power level of transmitted signal from a BS, based on which the mobile can roughly estimate the distance from the BS using a simple free-space propagation law to complete the open-loop up-link synchronous control stage. With this knowledge, the mobile will send a testing burst in a special slot dedicated only for up-link testing bursts, called UpPTS slot. If this testing burst has fell within the *search-window* at the BS receiver, the testing burst will be successfully received and the BS will know if the timing for the mobile to send its burst correct or not. If not, the BS should send *SS* instruction in the next down-link slots to ask the mobile adjust its transmission timing to complete the closed-loop up-link synchronization control cycle. It is specified in the TD-SCDMA standard that the initial up-link synchronization procedure has to be finished within four sub-frames, followed by the up-link synchronization tracking process. A detail illustration of both open-loop and closed-loop up-link synchronization control algorithm implemented by TD-SCDMA is shown in Figure (5.10), where a scenario with three mobiles communicating with a BS is illustrated with UE3 is the mobile of interest, which wants to proceed up-link synchronization with the BS, and UE1 and UE2 are the mobiles that have already established communication links with the BS.

Also with the similar argument, the need of up-link synchronization control in TD-SCDMA system is because of its use of OVSF codes, which are orthogonal codes and perform poorly in asynchronous up-link channels due to the fact that the characteristics of their ACFs and CCFs in an asynchronous channel are very bad. However, it is still natural for us to question the justification to introduce such a complicated up-link synchronization control system only for the application of orthogonal OVSF codes in up-link channels. Why do not we think about the other better solutions, such as using some new spreading codes with inherent isotropic or symmetrical performance? This indeed opens an interesting issue, which should be discussed extensively in the next section.

5.   Isotropic CDMA Air-Link Technologies

The previous section has addressed the issues on various techniques used by a traditional CDMA system to mitigate the problems associated with MAI and MI. Those discussions have revealed the complex nature

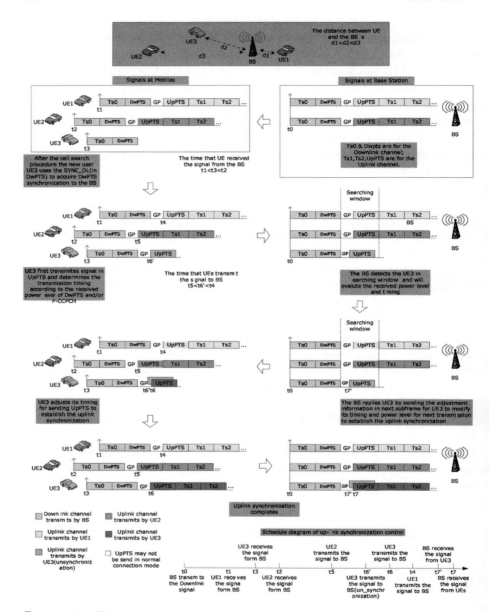

Figure 5.10. Illustration of open-loop and closed-loop up-link synchronization control algorithm specified by TD-SCDMA standard.

of a traditional CDMA system. Those techniques, such as power control, MUD, RAKE receiver, antenna-array beamforming, pilot signaling and up-link synchronization control, etc., neither of which can be considered simple, have to be used to combat the two major impairing factors, MAI and MI, arisen mainly from inherent defects in the CDMA codes.

It has also been shown in the previous section that, the operational performance in up-link and down-link channels of current 2-3G systems remains to be quiet different due to their distinct transmission modes, which have caused a lot of troubles. Various techniques have been introduced to deal with the problems stemmed from the asymmetrical operation modes in a full-duplex wireless system, as discussed in a great detail earlier in the previous sections. It has been seen that the up-link synchronization control technique offers an alternative way to convert an asynchronous up-link channel into a quasi-synchronous one with an accuracy no larger than 1/8 chip, such that the orthogonal OVSF codes could be used as channelization codes therein. Nevertheless, the up-link synchronization control should not be considered as a cost-effective way and it requires a set of very complicated open-loop and closed-loop algorithms to fulfill the synchronization control, as shown in Figure (5.10). Honestly speaking, the efficiency of the up-link synchronous control algorithm suggested in TD-SCDMA standard has not been openly demonstrated in a real system yet, as the standard itself still remains to be the paper work only and no fully-functional practical system is working today. Even if the algorithm does work well in a real system, our question is: can we find a better solution to make up-link and down-link identical in terms of their performance? Probably the most cost-effective way is to work out some new spreading codes and spreading modulation technologies, which could inherently offer symmetrical performance for both up-link and down-link channels.

5.1   Why Isotropic Spreading Techniques?

In a traditional CDMA based system (such as IS-95 [1][2], cdma2000 [3] or W-CDMA [4][5]), little effort has been made so far to ensure an *isotropic* or homogenous air-link performance in both synchronous down-link and asynchronous up-link. The up-link (also called *reverse link*) usually suffers much more impairing effects than its opposite direction, down-link (also called *forward link*), due to the reasons explained in the sequel. First, the transmitting power in up-link is always limited owing to the battery capacity constraint in a mobile handset, whose effective range is further reduced due to the fact that users often pre-

Table 5.5. Different operational environments for up-link and down-link transmissions in a mobile cellular scenario.

	Down-Link (Forward Link)	Up-Link (Reverse Link)
Transmission mode	Synchronous	Asynchronous
Power source	AC Power	Battery
RF Tx power	$\geq$10W	$\leq$0.5W
Antenna beam pattern	Directional\Omni-directional	Omni-directional
Antenna elevation	High (roof-mounted)	Low (hand-held)
Propagation loss	Medium	High
SNIR[a] at Rx	medium~high	Very low~low
Possible Rx diversity	Time\Frequency	Space\Time\Frequency
Pilot-signal multiplexing	TIDP[b]\DPCC[c]	TIDP[b]
Rx joint detection	Difficult	Possible

[a] SNIR: Signal-to-noise-and-interference ratio.
[b] TIDP: Time interleaved data and pilot.
[c] DPCC: Dedicated pilot code channel.

fer to place mobile phone calls indoors. Thus, most radiating energy is trapped inside a room and only a very small fraction of the energy could get through windows. Second, difference in antenna elevation heights between a mobile phone and a base station also determines that the up-link channels will more likely suffer from shadowing, local scattering, propagation or penetration loss, etc. than the down-link does. Third, the up-link is always asynchronous in a sense that data symbols from different mobiles in a cell arrive at the BS are not necessarily aligned in time and thus out-of-phase and aperiodic CCFs, rather than in-phase and periodic CCFs, of user signature codes govern the air-link performance. Unfortunately, the former is often much more difficult to control than the latter, causing a much greater MAI than that possible in down-link channels. Table (5.5) compares different operational environments for up-link and down-link channels in a mobile cellular system. It should be stressed that we have no way to alter all those existing impairing factors in up-link and down-link channels shown in Table (5.5), which is a reality we have to deal with. However, what we can do is to work out some new air-interface technologies that could hopefully counteract some of those disadvantages, and it motivates us to propose isotropic spreading techniques (ISTs), which is a focal point of this chapter, to overcome the problems.

The necessity for the isotropic spreading techniques is also driven by the needs for high-speed up-link transmission in the future wireless systems. Obviously, a slower up-link in a 2G system might be tolerable simply because the demand for high-speed up-load in most 2G applications is much less likely than that for download. However, the situation in the upcoming 3G applications may change and high-speed up-load applications will become commonplace when a mobile terminal with a CCD camera wants to send digital pictures or even video clips to someone else via the wireless networks. The future B3G systems need to support *mobile server* applications in a mobile terminal. A typical scenario could be that a news journalist would use his or her B3G mobile terminals with a high-resolution video camera to capture the scenes of interest and transmit the video signals simultaneously through broadband up-link to all-IP backbone network, from where the scenes can be viewed on a real-time basis by worldwide Internet subscribers. In this case, the journalist's mobile unit will act as a *mobile server* of the information source. Therefore, there is an imperative need to improve the performance of vulnerable up-link channels.

5.2   What is Isotropic Spreading Technique?

The *isotropic spreading techniques* (IST's) discussed in this section consist of two fundamental elements: isotropic spreading code (ISC) and isotropic spreading modulation (ISM), which ought to work jointly to yield satisfactory performance in both synchronous and asynchronous channels. In this sense, the spreading techniques employed in all current CDMA based 2-3G systems [1]-[6] do not belong to the category of IST. For instance, the orthogonal codes that have already been applied to the 2-3G systems, such as Walsh-Hadamard codes and OVSF codes, are not the ISC's, because their performance in asynchronous up-link channels is hardly comparable to that in synchronous down-link channels. As a matter of fact, those orthogonal codes are not orthogonal at all when used in an asynchronous channel as their out-of-phase ACFs and out-of-phase CCFs become totally uncontrollable, causing an unacceptably high MAI.

There are two important characteristic features of the IST's, which are of great interest to us. One is *isotropic MAI-free property* and the other is *isotropic MI-free property*. The former requires that they should ensure zero CCFs between any pair of the signature codes at any relative time shift in either synchronous or asynchronous channel; and the later specifies that their out-of-phase ACFs for any non-zero relative time shift should be zero in either synchronous or asynchronous channel. At the moment when this chapter is written, I have not seen any report on similar techniques in the open literature. It should be stressed, however, that an ISC family not necessarily carries the both features, i.e. isotropic MAI-free and isotropic MI-free, at the same time. In other words, some may possess only isotropic MAI-free property but not the isotropic MI-free property or vise versa; and only a very few of them retain the both: isotropic MAI-free and isotropic MI-free properties, which of course are most desirable to us to improve the performance of a CDMA system.

5.3   MAI-Free IST and CC-CDMA

MAI poses as a great danger to successful signal detection in a CDMA-based wireless system. It is well known that the MAI is the result of non-zero CCFs between user signature codes that undergo direct sequence (DS) spreading. Even in a synchronous CDMA channel the MAI can be rampant if the signature codes are not selected properly, let alone apply them to an asynchronous channel. Several possible ways can be considered to improve on this. One is to combine synchronization control technique with orthogonal signature codes, as discussed earlier in this chapter; the other way is to replace the conventional spreading techniques with isotropic MAI-free IST's to eliminate the MAI. The example of the latter has been shown in [69], which makes use of complete complementary (CC) codes together with offset-stacked spreading modulation to enable isotropic MAI-free operation in both synchronous down-link and asynchronous up-link channels. In the text followed, we will explain why a CC-CDMA can achieve isotropic MAI-free operation.

5.3.1   Performance of CC-CDMA in MI-free Channels

The core of the CC-CDMA architecture [69] is the use of orthogonal complete complementary codes, whose fundamental properties have been discussed in section 2. The conceptual diagrams for the proposed CC-CDMA system in a multipath-free channel are shown in Figures (5.11) and (5.12), where up-link and down-link spreading modulated signals for a two-user system are illustrated. Each of the two users therein employs two $L=4$ element codes as its signature code, which is exactly the same as the one listed in Table (5.3). The information bits (b_{11}, b_{12}, ...) and (b_{21}, b_{22}, ...), which are assumed to be all +1's for illustration simplicity in the figures, are spreading modulated by element codes that are *offset stacked*, each bit been shifted by one chip relative to its previous one.

Figures (5.11) and (5.12) illustrate that the CC-CDMA architecture can offer MAI-free operation in both up-link (asynchronous channel) and

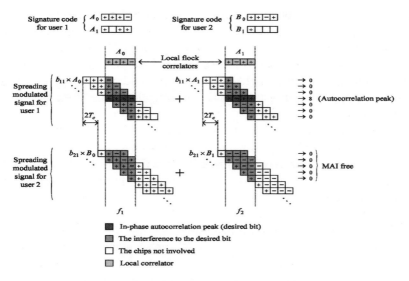

Figure 5.11. Down-link signal reception in a 2-user CC-CDMA system in MAI-AWGM channel with element code length being $L=4$, where the user 1 is the intended one.

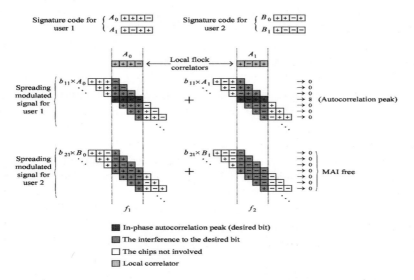

Figure 5.12. Up-link signal reception in a 2-user CC-CDMA system in MAI-AWGM channel with element code length being $L=4$, where the user 1 is the intended one.

down-link (synchronous channel) transmissions because of the unique properties of the CC codes. It has been assumed that the relative delay between the two users in Figure (5.12) takes the multiples of chips. If this assumption does not hold, it can be shown as well that the resultant MAI level is far less than that of a conventional CDMA system. It should also be pointed out that the rate change through adjusting the number of offset chips between neighboring two stacked bits, as to be explained in the later part of this subsection, will not affect the isotropic MAI-free operation of the CC-CDMA system.

The isotropic MAI-free property of the CC-CDMA architecture is significant in terms of its potential to greatly enhance system capacity in a wireless system. It is well known that a CDMA system is always an interference-limited system, whose capacity is dependent of the average co-channel interference contributed from all transmissions using different codes in the same band. The co-channel interference in a conventional CDMA system is caused in principle by non-ideal CCFs and out-of-phase ACFs of the codes concerned. In such a system, it is impossible to eliminate the co-channel interference, especially in the up-link channel, where bit streams from different mobiles are asynchronous such that the orthogonality among the codes virtually does not exist. On the contrary, the CC-CDMA system is unique due to the fact that an excellent orthogonality among transmitting codes is preserved even in asynchronous up-link channels, making a truly isotropic MAI-free operation for both up-link and down-link transmissions. The satisfactory performance of the system in multipath environment, as shown by the data given later, is also partly attributable to this property.

It should also be admitted that the two element codes for each of the user signature codes concerned in Figures (5.11) and (5.12) have to be sent separately through different channels, here being referred to different carriers, f_1 and f_2. Therefore, the CC-CDMA architecture is a multi-carrier CDMA system, bearing many similar characteristics of a multi-carrier CDMA system, except that it does not have the capability to mitigate frequency-selective fading through carrier frequency diversity because all element codes sent by different carriers bear indispensable information to form an ideal correlation function. Therefore, there is no redundancy in the information carried via different carriers. It is also possible for us to use orthogonal carriers, spaced by $1/T_c$ (where T_c denotes the chip width), to send all those element codes from the same user separately to further enhance the bandwidth efficiency of the system.

The bit error rate (BER) of the CC-CDMA system under MAI and AWGN is evaluated using computer simulations. The obtained BER performance of the CC-CDMA system is compared with that of conventional CDMA systems using Gold codes and Walsh-Hadamard sequences under identical operation environment. For each of the systems concerned here, a matched-filter (single-correlator) is used at a receiver. Both up-link and down-link are simulated considering various numbers of users and processing gains. Figures (5.13) and (5.14) typify the results we have obtained. The former shows the MAI sensitivity comparison of BER in up-link (asynchronous) channel with a processing gain of 64 for the CC codes, being comparable to that for Gold code of 63 and Walsh-Hadamard sequences of length 64. The latter compares the impact of link operation modes on BER in both up-link (asynchronous) and down-link (synchronous) channels. The both figures show the BER for the 1st user only as the intended one and similar results can be obtained for the BER of the others. It is observed from Figures (5.13) and (5.14) that at least 3 dB gain is obtainable from the CC-CDMA system if compared with the conventional CDMA systems using traditional CDMA codes. One of the most striking observations for the CC-CDMA system is its almost identical BER performance (shown in Figure (5.13)) regardless of the number of users, where two curves representing different numbers of users (one and four users respectively) in the system are virtually overlap with each other, exemplifying the MAI-free operation of the CC-CDMA system. On the other hand, the BER for a CDMA system using tradi-tional codes is MAI-dependent or interference-limited; the more active users are present, the worse it performs, as shown in Figure (5.13).

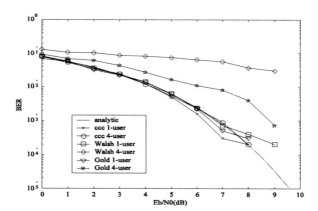

Figure 5.13. MAI sensitivity comparison of BER for CC-CDMA and conventional CDMA systems in MAI-AWGN channel using matched filter receiver. Lengths of Gold-code, CC code and Walsh-Hadamard code are 63, 4×16 and 64, respectively.

If compared with the spreading modulation technique used in tradi-tional CDMA systems, the offset stacking spreading modulation used in

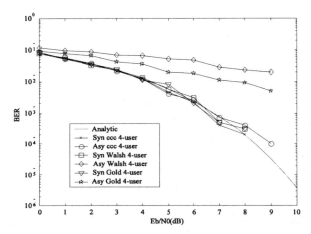

Figure 5.14. Up- and down-link BER comparison for CC-CDMA and conventional CDMA systems in MAI-AWGN channel using matched filter receiver. Lengths of Gold-code, CC code and Walsh-Hadamard code are 63 and 4×16 and 64, respectively.

the CC-CDMA has the following salient features. The most obvious one is that bit stream here is no longer aligned in time one bit after another, as shown in Figures (5.11) and (5.12). Instead, a new bit will start right after one chip delay relative to the previous bit, which is spread by an element code of length L. Another important characteristic attribute of the CC-CDMA system is that such an *offset stacked* spreading modulation method is in particular well suited for multi-rate data transmission in multi-media services, whose algorithm is termed as *rate-matching* in the current 3G mobile communication standards [20]-[23]. The unique *offset stacked* spreading method used by the proposed CC-CDMA system can easily slow down data transmission rate by simply shifting more than one chip (at most L chips) between neighboring two offset stacked bits. If L chips are shifted between two consecutive bits, the new system reduces to a conventional DS-CDMA system, yielding the lowest data rate. On the other hand, the highest data rate is achieved if only one chip is shifted between two neighboring offset stacked bits. In doing so, the highest spreading efficiency equal to one can be achieved, implying that every chip is capable to carry one bit information. As the bandwidth of a CDMA system is uniquely determined by the chip width of spreading codes used, a higher spreading efficiency simply means higher bandwidth efficiency. Thus, the proposed CC-CDMA architecture is capable to deliver much higher bandwidth efficiency than a conventional CDMA architecture with the same processing gain. Figure (5.15) shows the principle for a CC-CDMA system to offer an agile rate-matching algorithm for any possible data transmission rate.

It should be stressed that the *inherent* ability for the CC-CDMA system to facilitate multi-rate transmissions is based on its special spreading codes or complete complementary codes and innovative *offset stacked* spreading modulation scheme, which can not be applied to other traditional spreading codes, such as Gold codes, OVSF codes, Walsh-Hadamard codes, etc. The current 3G UMTS-UTRA [4] and W-CDMA [5] architectures have to rely on a complex and always inefficient rate-matching algorithm to adjust data transmission rate by selecting appropriate variable-length OVSF codes [20]-[23] according to a specific spreading factor and data rate requirement on the services. On the contrary, the CC-CDMA system is able to alter the data transmission rate on the fly: without the need of searching for suitable codes with a particular spreading factor. What to do is just to shift more or less chips between two neighboring offset stacked bits to slow down or speed up the data rate. That's it and no more overhead rate-matching algorithms!

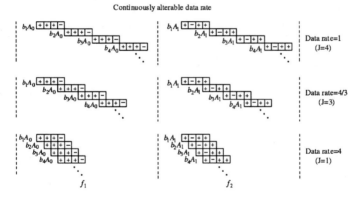

Figure 5.15. Agility to perform the rate-matching in CC-CDMA system without *stopping and searching* for the codes with specific spreading factors as required in UMTS-UTRA and W-CDMA standards. J is the number of chips to shift before the next bit is sent.

Another important feature of the rate-change scheme adopted by the CC-CDMA architecture is that the same processing gain will apply to different data transmission rates. However, the rate-matching algorithm proposed by the UMTS-UTRA [4] and W-CDMA [5] standards is processing gain dependent; the slower the transmission rate is, the higher the processing gain or spreading factor will be, if transmission bandwidth is kept constant. To maintain an even detection efficiency at a receiver, the transmitter has to adjust the transmitting power for different rate services, which surely further complicates both transmitter and receiver hardware.

The *offset stacked* spreading technique also helps to support asymmetrical transmissions in up-link and down-link channels, pertaining to the Internet services in future all-IP wireless networks. The data rates in a slow up-link and a fast down-link can be made truly scalable, such that the *rate-on-demand* is readily achievable by simply adjusting the offset chips between two neighboring spreading modulated bits.

5.3.2   Performance of CC-CDMA in Multipath Channels.

Next, let us examine the operation and performance of the CC-CDMA system under multipath channels.

It is well known that a conventional CDMA receiver always uses a RAKE to collect dispersed energy among different reflection paths to achieve a multipath diversity. Therefore, the RAKE receiver is considered as a must to all conventional CDMA systems, including all currently operational 2G (IS-95A/B) and 3G systems [1]-[6]. However, in the CC-CDMA architecture presented in this chapter the RAKE receiver becomes useless due to its unique property of offset stacked spreading modulation technique employed in the system. To illustrate how the CC-CDMA makes the RAKE receiver obsolete, let us refer to Figure (5.16), where a simple multipath channel consisting of three equally strong reflection rays is considered with its inter-path delay being one chip (in fact, any other inter-path delay can also be used to obtain the same result). The RAKE receiver has three fingers to capture these three paths, which should be combined coherently. Three columns in Figure (5.16) show the output signals from three fingers and the shaded parts are the chips involved in the RAKE combining algorithm. Due to the use of offset stacked spreading technique in the CC-CDMA system, there are in total five bits (b_1, b_2, b_3, b_4 and b_5) are relevant to the RAKE combining procedure, where it is assumed that $b_3=+1$ is the desirable bit. Therefore, b_1, b_2, b_4 and b_5 are all interfering terms, whose all three possible error-causing patterns are (b_1, $2b_2$, $2b_4$, b_5)=(1, -2, -2, -1), (-1, -2, -2, 1) and (-1, -2, -2, -1), respectively. Note that among total 16 possible combinations of the binary bits, b_1, b_2, b_4 and b_5, only three of them cause errors. Therefore, the error probability simple turns out to be 3/16=0.1875 (if each path carries the same strength).

From this example, we can see that the use of a RAKE receiver in the CC-CDMA system still causes an irreducible BER=0.1875 (with three identically strong paths), which is obviously not acceptable. Therefore, an adaptive recursive multipath signal reception filter is designed particularly for the CC-CDMA system, as shown in Figure 5.17, where the receiver consists of two key modules; the lower part is to estimate the channel impulse response and the upper part is to coherently combine signals in different paths to yield a boosted-up decision variable before a decision device.

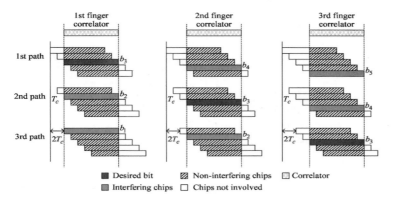

3 Error causing patterns out of 16 (resulting in an irreducible BER=3/16)

b_1	$2 \cdot b_2$	$2 \cdot b_4$	b_5
1	-2	-2	-1
-1	-2	-2	1
-1	-2	-2	-1

Figure 5.16. The illustration of incapability for a RAKE to offer a satisfactory performance under multipath interference in CC-CDMA system.

For this adaptive recursive filter to work properly, a dedicated pilot signal should be added to the CC-CDMA system, which should be spreading-coded by a signature code different from those used for data channels in the down-link transmission, and should be time-interleaved with user data frames in the up-link channel transmission, as shown in Figure (5.18). The rationale behind the difference in the pilot signals for up-link and down-link channels is explained in the sequel. The down-link transmission is a synchronous channel from the same source (a BS) and thus one dedicated pilot signaling channel is justified, considering that a relatively strong pilot is helpful for mobiles to lock onto it for controlling information retrieval. On the other hand, the up-link transmissions are asynchronous from different mobiles. Therefore, it will consume a lot more signature codes if every mobile is assigned two codes: one for data traffic and the other for pilot signaling. Thus, the pilot signaling has to be time-interleaved with user data traffic in the up-link channels. The time-interleaved pilot signals in the up-link channels can also assist the BS to perform adaptive beam forming, required by a smart antenna

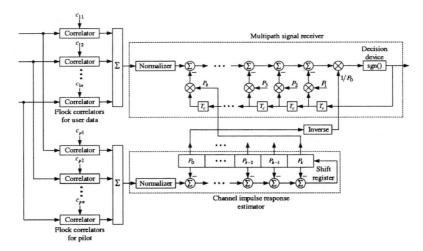

Figure 5.17. A recursive filter receiver for signal detection under multipath interference in CC-CDMA system.

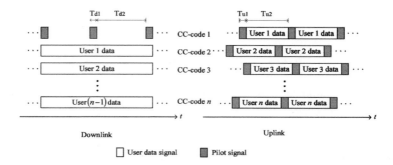

Figure 5.18. Pilot signaling for up-link and down-link channels in CC-CDMA system.

system. The signal reception in both up-link and down-link channels can use the same recursive filter (as shown in Figure (5.17)) for channel impulse response estimation as long as the receiver achieves frame synchronization with the incoming signal. In fact, the pilot signals in both up-link and down-link channels consist of a series of short pulses, whose durations (T_{d1} and T_{u1}) should be made longer than the delay spread of the channel and whose repetition periods (T_{d2} and T_{u2}) should be made shorter than the coherent time of the channel to adaptively follow the variation of the mobile channel.

The detail procedure for the recursive multipath signal reception filter to estimate the channel impulse response and to detect data signal is illustrated step by step in Figures (5.19) and (5.20), where it has been

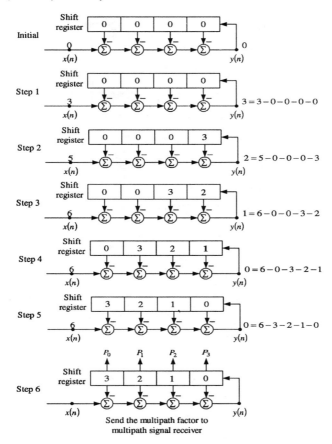

Figure 5.19. Step-by-step illustration for channel impulse response estimation using recursive multipath signal reception filter in CC-CDMA system.

assumed that a 3-ray multipath channel is concerned with its mean path strength being 3, 2 and 1 respectively. It is also assumed that exactly the same CC codes as concerned in Figures (5.11) and (5.12) are used, with one signature code ($c_{p1} = A_0$ and $c_{p2} = A_1$) being used for pilot channel and the other ($c_{11} = B_0$ and $c_{12} = B_1$) for the user data channel, if only the down-link transmission is considered in this example. In this illustration we presume that each pilot pulse consists of continuous five "+1", which is longer than the channel delay spread (three chips) in this case. The input sequence (3, 5, 6, 6, 6) to the left side of multipath channel estimator in Figure (5.19) is the received pilot signal after being convoluted with multipath channel impulse response and local flock correlators. It is seen from Figure (5.19) that the channel impulse response

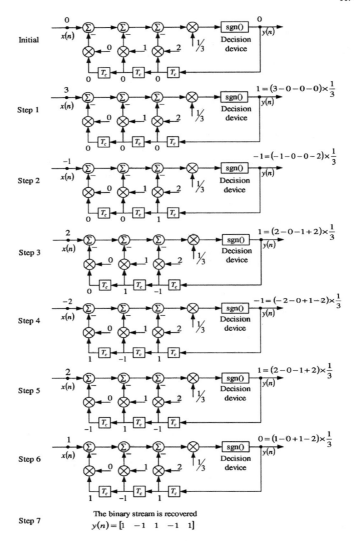

Figure 5.20. Signal detection procedure of the recursive multipath signal reception filter based on channel impulse response estimates with recovered bit stream being $y(n)=(1, -1, 1, -1, 1)$ in CC-CDMA system.

can be estimated accurately and saved in the output register at the end of the algorithm. The obtained channel estimates will then be passed on to the *multipath signal receiver* in the upper portion of Figure (5.17) to detect the signal contaminated with multipath interference, whose procedure is shown in Figure (5.20), where it is assumed that the originally transmitted binary bit stream is $(1, -1, 1, -1, 1)$. The input data to the multipath signal receiver, (3, -1, 2, -2, 2, 1), is the received signal of transmitted bit stream after going through multipath channel and local flock correlators.

The proposed recursive multipath signal reception filter possesses several advantages. Firstly, it offers a very agile structure, the core of which is made up of two transversal filters; one for channel impulse response estimation and the other for data detection. Secondly, working jointly with the pilot signaling it performs very well in terms of the accuracy in channel impulse response estimation, as shown in Figure (5.19) and the obtained BER results followed. The multipath channel equalization and signal coherent-combining are actually implemented at the same time in the proposed scheme under a relatively simple hardware structure. Thirdly, it can operate adaptively to the channel characteristics variation without needing the prior knowledge of the channel, such as inter-path delay and relative strength of different paths, etc. On the contrary, a RAKE receiver in a conventional CDMA system requires every individual path gain coefficient for maximal ratio combining, which themselves are usually unknown and thus have to be estimated by resorting to other complex algorithms.

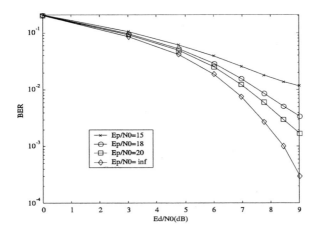

Figure 5.21. BER for CC-CDMA in down-link multipath channel using recursive filter, where pilot is used to estimate multipath parameters. Delay profile=[0.5774, 0.5774, 0.5774], inter path delay=1 chip, PG of CC code 4×16, user number=4.

The performance of the CC-CDMA architecture with the recursive filter for multipath signal reception is shown in Figures (5.21) to (5.24), where two typical scenarios are considered: one for down-link performance and the other for up-link performance, similar to the performance comparison made for the MAI-AWGN channel concerned in Fig-

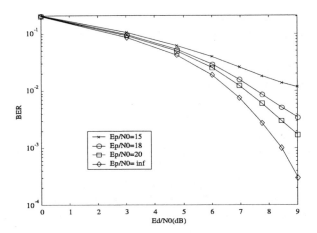

Figure 5.22. BER for CC-CDMA in up-link multipath channel using recursive filter, where pilot is used to estimate multipath parameters. Delay profile=[0.5774, 0.5774, 0.5774], inter path delay=1 chip, inter user delay=3 chip, PG of CC code 4×16, and user number=4.

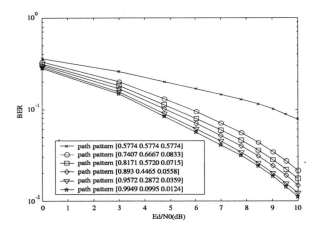

Figure 5.23. BER for CC-CDMA in down-link multipath channel using recursive filter, where pilot is used to estimate multipath parameters. Normalized multipath patterns are considered with inter path delay=1 chip, PG of CC code = 4×6, pilot power=signal power, and user number=4.

ures (5.13) and (5.14). It is observed from the figures that the almost identical BER performance for both up-link and down-link channels is presented. The transmitting power of the pilot channel has a very strong influence on the BER of the CC-CDMA as a whole, as shown in Figures (5.21) and (5.22). Thus, in order to ensure a satisfactory performance,

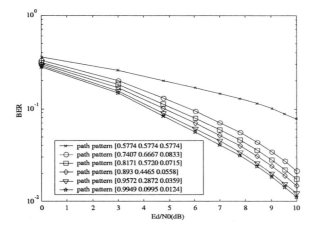

Figure 5.24. BER for CC-CDMA in up-link multipath channel using recursive filter, where pilot is used to estimate multipath parameters, Normalized multipath patterns are considered with inter path delay=1 chip, inter user delay=3 chip, PG of CC code = 4×6, pilot power=signal power, and user number=4.

E_p/N_o should be made at least 18 dB; otherwise the errors in channel estimation will seriously affect the data detection process followed. Figures (5.23) and (5.24) show the performance of a CC-CDMA system under various multipath channels, which consist of three multipath returns and whose delay profile are normalized to reflect the fact that the total transmitting power for different channels is fixed.

5.3.3 Pros and Cons for CC-CDMA.
The major characteristic features of the CC-CDMA system are summarized as follows.

- It can offer attractive isotropic MAI-free operation for wireless applications, which is hardly achievable by using any other CDMA technologies available at the moment.

- The use of the ISM (or offset stacking spreading modulation) can greatly increase the spreading efficiency (SE), which is defined as bit(s) per chip. It is noted that currently available 2-3G CDMA based systems [1]-[6] can offer an SE at merely about $1/N$ bit per chip, where N is the spreading factor of the systems; whereas the CC-CDMA can push the SE figure up to one bit per chip, about N times higher than that obtainable in the 2-3G systems. As the system bandwidth is determined by the chip-width of spreading-modulated signal, a greater SE means higher bandwidth efficiency.

Thus, the above example of the MAI-free IST has a great potential to increase the overall bandwidth efficiency of a wireless system.

- Some other advantages of the CC-CDMA system include its agility to perform rate-matching algorithm for multi-media services, the flexibility in supporting asymmetrical traffic in up-link and down-link channels, etc. Those who are interested in more detail information on the CC-CDMA and the generation of CC codes may refer to [69].

The CC-CDMA architecture has also its own technical limitations, which can be listed below.

- Although the CC-CDMA system does offer an isotropic MAI-free operation, it should use a relatively complex receiver (a dedicated recursive filter) to detect signal in multipath channel with the help of the pilot signaling, as shown in [69]. In order words, the CC-CDMA architecture proposed earlier [69] does not have isotropic MI-free property, without which the performance improvement under multipath channel is limited.

- Furthermore, the CC-CDMA needs some multi-level digital carrier modulation, such as QAM, to fulfill carrier transmission due to its multi-leveled baseband signal formulation after offset stacked spreading modulation, which further complicate transceiver hardware and motivates us to search for some other better solutions. In fact, the high bandwidth efficiency of the CC-CDMA relies on the high detection efficiency of the modulation scheme. Unfortunately, due to the need of multi-level digital modulation scheme, the CC-CDMA system will still face the same problem associated with the reduced detection efficiency as that in a multi-level modem under noise and interference.

- Another concern with the CC-CDMA system is that a relatively small number of users can be supported with a family of the CC codes. Take $L=64$ CC code family as an example. It is seen from Table (5.3) that such a family has only 4 flocks of codes, each of which can be assigned to one channel (for either pilot or data). If more users should be supported, long CC codes have to be used. On the other hand, the maximum length of the CC codes is in fact limited by maximal number of different baseband signal levels manageable in a digital modem, as mentioned previously. One possible solution to this problem is to introduce frequency-divisions on top of the code-divisions in each frequency band to create more transmission channels. Yet another possibility is to use long code scrambling together with the short CC code spreading to increase the number of users, which can be supported in the same frequency band. However, doing so might incur some other concerns about the system complexity and bandwidth efficiency, etc. Therefore, the most worthwhile work is to search for some other complementary codes, which retain all advantages of CC codes but with a larger code-set size.

Therefore, we have some strong reason to find some more preferable CDMA systems possessing both isotropic MAI-free and isotropic MI-free properties. One possible solution is the combination of CC codes and direct-sequence spreading modulation, or CC/DS-CDMA scheme, which provides very attractive properties, such as isotropic MAI-free and isotropic MI-free, near-far effect resistance, etc., and it works on a very simple receiver hardware, a correlator or matched-filter, to facilitate user portable terminal miniaturization. The detail about the CC/DS-CDMA scheme will be covered in the sub-section followed.

As the last remark before ending this sub-section, it should be emphasized that ordinary orthogonal codes can never guarantee isotropic MAI-free operation. For instance, the application of Walsh-Hadamard sequences as channelization codes in the IS-95 standard [1]-[2] is only limited to its down-link channels that are synchronous. On the other hand, its up-link channels do not use Walsh-Hadamard sequences for channelization due to the fact that they perform extremely poor with very high aperiodic CCFs in asynchronous channels. Therefore, it is a totally different story here to talk about an isotropic MAI-free ISC and an orthogonal code family. To ensure an isotropic MAI-free operation a proper joint-selection of ISC and ISM is very important, such as the example of CC-CDMA system that has been shown in this sub-suction and works on CC codes (as the ISC) and offset stacked spreading modulation (as the ISM).

5.4   MI-free IST and CC/DS-CDMA

Multipath interference (MI) is another serious impairing factor that the B3G wireless systems have to deal with, which is further complicated by the fact that extremely high data rate required in the B3G systems makes it more likely to happen that the transmitting signal bandwidth becomes much wider than channel coherent bandwidth, thus been suffered from severe frequency-selective fading. The traditional counter-measure against the frequency-selective fading is to use a RAKE receiver, consisting of several parallel correlator *fingers* to capture different multipath returns and then combine them in either a coherent or non-coherent way to achieve multipath diversity. We have to admit that the RAKE does work well in most circumstances in traditional 2-3G systems to overcome the MI. However, our question here is that whether it is the *best* solution for the 2-3G as well as future B3G systems. The answer should be negative due to the following two main arguments. First, the capability for each finger of a RAKE receiver to separate or isolate individual multipath return is limited because, even if a finger could track the instance that a particular multipath return appears, the output from that finger is inevitably the combination of all possible unwanted interference from other users' transmissions plus the side lobes of other multipath returns of its self-transmission, in addition to the useful autocorrelation peak. Thus, the signal-to-interference ratio at the output of a RAKE finger is sensitively affected by the other unwanted transmissions and the side-lobes of other multipath returns of its self-transmission. Second, for a RAKE to work satisfactorily, a receiver needs to acquire virtually all information about the channel, i.e., the delay profile or impulse response of the channel, before a maximal ratio combining (MRC) RAKE algorithm can be performed. Therefore, the channel estimation is a must for any MRC-RAKE, adding a great complexity burden to overall receiver hardware. Therefore, it sounds much more desirable for a CDMA system to overcome the frequency-selective fading by its own signaling merits, rather than depending on a complex receiver, such as MRC-RAKE and etc. In other words, it will be strongly recommended that the signaling of a CDMA system could be implanted some inherent resistance against frequency-selective fading, which is what we want to get from the isotropic MI-free IST's. The very core of the IST's is the use of isotropic spreading codes with an ideal property that ensures all nonzero relative shifted versions of autocorrelation function to be zero in either synchronous or asynchronous channels, except for the autocorrelation peak at the zero shift. In doing so, delayed multipath returns will never cause any harmful interference to the precedent main-path signal detection when sampled at the instances of the main-path arrivals. Our very recent study has revealed that some types of complementary codes do exhibit such ideal isotropic MI-free property and they can possibly be implemented by using simple digital technology. With the help of those IST's, an isotropic MI-free operation environment can be created with or without RAKE receiver, greatly improving signal reception quality and reducing the receiver hardware complexity.

As already demonstrated in the previous subsection [69], the combination of CC codes and offset stacked spreading modulation can offer an isotropic MAI-free operation. Unfortunately, it does not offer an isotropic MI-free operation due mainly to the use of the offset stacked spreading modulation, which causes strong mutual interference from different multipath returns. Nevertheless, it is very interesting to know from our recent study that the combination of the CC codes and conventional direct sequence spreading modulation can offer an isotropic MI-free operation, while still retaining its isotropic MAI-free property. With those desirable features, a CC/DS-CDMA architecture has been proposed by us in this chapter for the possible B3G wireless applications, which succeeds in getting rid of all those malefic MAI and MI interferences and transforming a CDMA system back to work in a noise-limited environment.

Figure (5.25) shows the conceptual block diagram of a K-user CC/DS-CDMA system with the MAI-free and MI-free IST using CC codes (as the ISC) and DS spreading modulation (as the ISM). It is noted that the CC/DS-CDMA system shown in Figure (5.25) differs substantially from the CC-CDMA system proposed in Figures (5.11) and (5.12) [66] in terms of complexity of the transmitters as well as the receivers. Especially, the spread modulated signal in the CC/DS-CDMA system is no longer multi-leveled and thus a simple BPSK modulation is sufficient here, resulting a much simplified transceiver hardware if compared with that of the CC-CDMA system [69], whose working principle is shown in Figures (5.11) and (5.12).

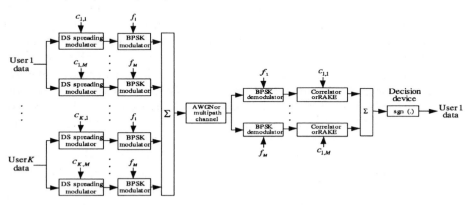

Figure 5.25. Block diagram of CC/DS-CDMA system with isotropic MAI-free and MI-free property, where the signal of interest to the receiver is user 1's transmission.

To illustrate the MI-free operation of the CC/DS-CDMA system shown in Figure (5.25), let us consider the following simple example of the complete complementary codes: $c_{11} = + + + -$, $c_{12} = + - + +$, $c_{21} = + + - +$, $c_{22} = + - - -$, where we have assumed $K=M=2$ in this exam-

ple, corresponding to the parameters concerned in Figure (5.25). The
users 1 and 2 are assigned c_{11}, c_{12} and c_{21}, c_{22} as their signature codes,
respectively. Presumably the receiver wants to detect the signal from
user 1, or c_{11} and c_{12} are the element codes of interest to the receiver.
Figures (5.26) and (5.27) show the decoding process of the MI-free IST
based on the complete complementary codes: c_{11}, c_{12}, c_{21} and c_{22}, as
defined earlier. Without losing generality, it is further assumed that
the user bit stream consists of three consecutive bits, whose values are
$(b_{11}, b_{12}, b_{13}) = (+1, -1, +1)$ for user 1 and $(b_{21}, b_{22}, b_{23}) = (+1, +1,$
-1) for user 2 respectively, and there are two equal-strength multipath
returns for each user's transmission, whose inter-path delay is one chip
only. Figures (5.26) and (5.27) illustrate the detection process from the
1st bit to the 3rd bit in the up-link and down-link channels. In Figure
(5.27) it is further assumed that the inter-user delay is also one chip
for illustration clarity. The receiver, whose structure is shown in Figure (5.25) (note that RAKE is not used in this example), uses a simple
matched filter or correlator to detect the signal from user 1. The decoding process proceeds bit by bit, given that the receiver has acquired bit
synchronization with incoming signal. From Figures (5.26) and (5.27) it
can be easily seen that the MI does not cause any impairing effect on
the data decoding process, even though the receiver only uses a simple
correlator. The values assumed for the inter-path delay, inter-user delay
and number of multipath returns in Figures (5.26) and (5.27) are only
for the illustration clarity and simplicity, and the same results will apply
even for any other settings of inter-path and inter-user delays (different
from one chip). The readers can verify it by drawing the similar graphs
by their own. This interesting result tells us the fact that the combinatory use of CC codes and traditional DS spreading can become the core
of a new type of CDMA system, or CC/DS-CDMA architecture, which
creates a truly isotropic MI-free operation environment with a simple
correlator, without using any complex receiver implementations, such as
MRC-RAKE, etc.

As pointed out earlier, the CC/DS-CDMA architecture shown in Figure (5.25) can also offer an isotropic MAI-free property, in addition to its
isotropic MI-free property, as shown in Figures (5.28) and (5.29), where
exactly the same CC codes, the DS spreading modulator, the information
bit stream pattern and receiver structure as shown in Figures (5.26) and
(5.27) are concerned. It is evident that the combined isotropic MAI-free
and isotropic MI-free properties of the proposed CC/DS-CDMA architecture can offer a superior air-link performance for the new generation
of wireless communications, which can finally make our dream come true:
turning an interference-limited CDMA into a noise-limited system, leav-

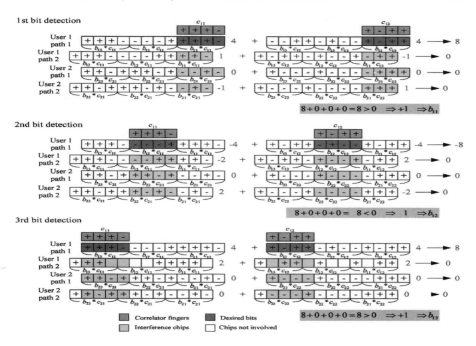

Figure 5.26. Illustration of MI-free operation for a 2-user CC/DS-CDMA system in synchronous down-link channels, where a two-path multipath channel is considered with the inter-path delay being one chip.

ing us a great room to push system capacity and data transmission rate into an even higher level than ever possible before.

It is not surprising that the proposed CC/DS-CDMA system can also employ a RAKE to further improve its detection efficiency, while retaining the isotropic MAI-free and isotropic MI-free properties. Figures (5.30) and (5.31), where a two-finger maximum ratio combining RAKE is considered in a 2-ray multipath channel with equal strength and only the detection procedure for the 2nd bit is shown for illustration clarity, are given to illustrate the detection process in up-link and down-link channels, respectively, in the same operational scenario as concerned in Figures (5.28) and (5.29). It is observed from Figures (5.30) and (5.31) that absolute value of the decision variable formed at the output of a MRC-RAKE in the receiver has been doubled (equal to 16) if compared to that (being 8) with a simple correlator in Figures (5.28) and (5.29). Such a result can be interpreted as an improved immunity against noise and other malicious external interference from the channel. Figures (5.32) to (5.34) are the BER performance to verify by sim-

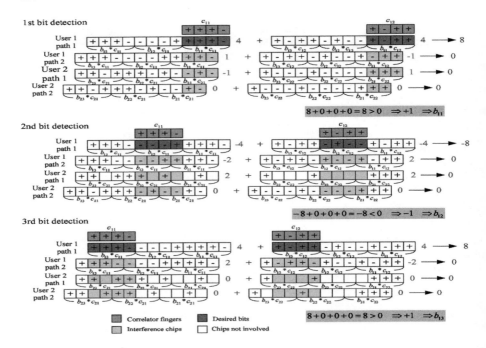

Figure 5.27. Illustration of MI-free operation for a 2-user CC/DS-CDMA system in asynchronous up-link channels, where a two-path multipath channel is considered with the inter-path delay being one chip and inter-user delay being also one chip.

ulations the three major characteristics features of the CC/DS-CDMA system: its MAI-free property, its ideal isotropic link performance and its inherent frequency-selective fading resistance or MI-free property, respectively. In particular, Figure (5.32) offers a comparison between the CC/DS-CDMA and conventional DS-CDMA with either Gold or Walsh-Hadamard codes in terms of their impairing effect of MAI on the BER performance. For each system, two scenarios are compared: one is a single user system and the other is the 4-user system. The former represents the operation without MAI and the latter studies the performance under MAI.

It is seen from Figure (5.32) that the CC/DS-CDMA system offers a perfect MAI-free operation, indicated by the almost overlapped two solid curves (with circle and cross markers) in the figures. Figure (5.33) compares the effect of link operation modes (either asynchronous up-link or synchronous down-link) on the BER performance of the CC/DS-CDMA and DS-CDMA systems. Again, the two solid curves (with circle and cross markers) are almost coincided, showing that the air-link opera-

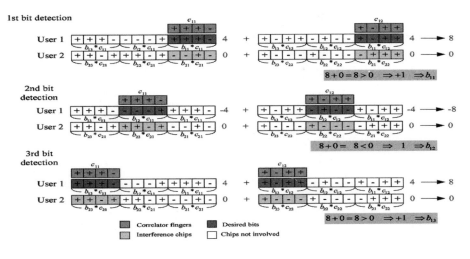

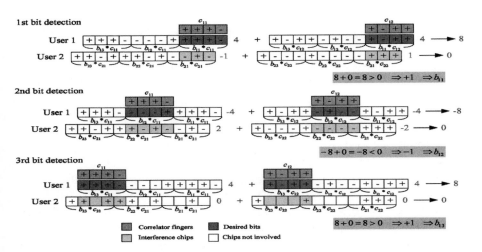

Figure 5.29. Illustration of MAI-free operation for a 2-user CC/DS-CDMA system in asynchronous up-link channels, where inter-user delay is assumed to be one chip.

tion mode exerts little impact on the overall link BER performance of a CC/DS-CDMA system. In other words, Figure (5.33) just verifies that a CC/DS-CDMA system indeed retains an isotropic operation property, which is independent of link operation modes. The frequency-selective fading resistance for different systems is shown in Figure (5.34), where the CC/DS-CDMA outperforms the other two counterparts with its al-

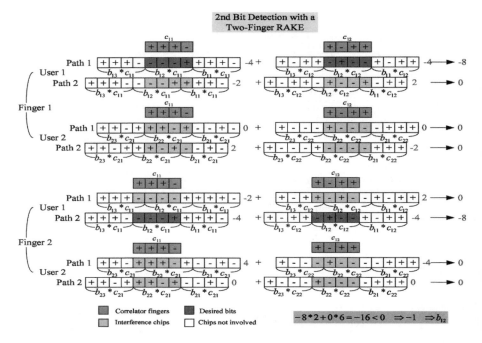

Figure 5.30. Illustration of RAKE receiver operation for a 2-user CC/DS-CDMA system in synchronous down-link channels where inter-path delay is assumed as one chip.

most identical performance under two very much different operational scenarios: the AWGN channel and 4-path multipath channel with each path having equal strength, representing a severe multipath interference or frequency-selective fading. We can also see from Figures (5.32) to (5.34) that the DS-CDMA system with 64-chip Walsh-Hadamard codes offers the worst performance, being extremely sensitive to MAI, link-operational modes and frequency-selective fading effects. The performance for the DS-CDMA system using 63-chip Gold code (indicated by two long-dash curves with circle and cross markers in Figures (5.32) to (5.34)) is in between the CC/DS-CDMA and Walsh-Hadamard DS-CDMA systems.

To show how well a system can perform relatively to a noise-limited system, the BER for a BPSK system in AWGN channel is also plotted in Figures (5.32) to (5.34), as given by a solid line curve (without marker). It can be observed from the figures that the CC/DS-CDMA system offers a performance very close to that for a BPSK system under AWGN channel, regardless of MAI, MI and the link operation modes (synchro-

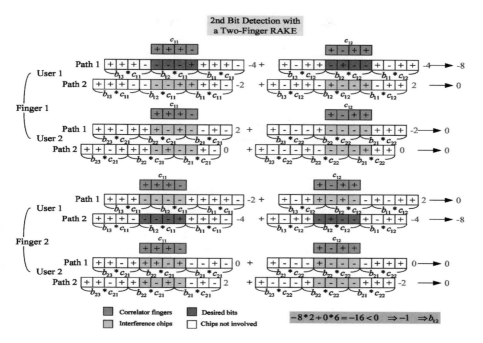

Figure 5.31. Illustration of RAKE receiver operation for a 2-user CC/DS-CDMA system in asynchronous up-link channels where both inter-path delay and inter-user delay are assumed to be one chip.

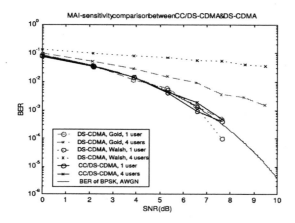

Figure 5.32. Comparison of conventional DS-CDMA and CC/DS-CDMA systems in terms of their MAI-resistance.

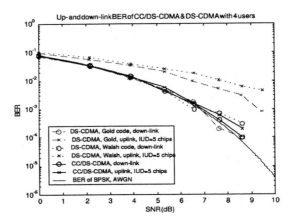

Figure 5.33. Comparison of conventional DS-CDMA and CC/DS-CDMA systems in terms of their sensitivity against operation modes (in either synchronous or asynchronous channels). IUD stands for *inter-user delay*.

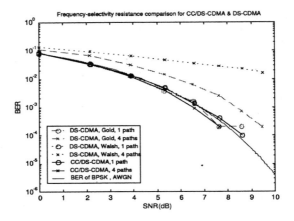

Figure 5.34. Comparison of conventional DS-CDMA and CC/DS-CDMA systems in terms of their ability to mitigate multipath interference or frequency-selective fading.

nous or asynchronous). This observation can be translated into that the performance of a CC/DS-CDMA system is indeed noise-limited. However, the performance of a conventional DS-CDMA system using either orthogonal codes (such as Walsh-Hadamard code) or quasi-orthogonal codes (such as Gold code) is clearly interference-limited.

5.5   Merits of CC/DS-CDMA

We have already demonstrated that the CC/DS-CDMA system architecture based on the IST can offer isotropic MAI-free and isotropic MI-free operation, which attribute to several operational advantages otherwise not possible with conventional CDMA technologies.

- First, it is not necessary for the CC/DS-CDMA system to use a RAKE receiver to mitigate frequency-selective fading in the channel. Only a simple correlator or matched filter is sufficient to yield satisfactory detection efficiency, thus greatly reducing the hardware cost of the receivers. Similarly, it also paves a way for a CC/DS-CDMA receiver to detect signals in a truly blind fashion without any prior channel information estimation. On the other hand, a RAKE has to acquire detailed channel information, such as delays and amplitudes of all multipath returns, for its successful operation of maximal-ratio-combining (MRC) algorithm, whose costly hardware in a mobile handset should never be underestimated.

- Second, as mentioned earlier, the CC/DS-CDMA receiver can also use a RAKE to separate different multipath returns to realize a genuine *multipath diversity* for further improved detection efficiency. It should be emphasized that the use of a RAKE in the CC/DS-CDMA system can yield a much better results than that possible in a conventional DS-CDMA system, because the output from each finger in the CC/DS-CDMA system consists of only autocorrelation peak at a particular time and nothing else. On the contrary, the output from each finger in a conventional CDMA system is a mixture of useful part (the autocorrelation peak at the instance) as well as a wealth of other unwanted components, including those nontrivial side lobes of other multipath returns in both self-transmission and other users' transmissions. Therefore, the CC/DS-CDMA system is in particular suitable for the operation in the channels with severe multipath interference due to its unique frequency-selective-fading resistance, as shown in Figure (5.34). Furthermore, if a RAKE is used in a CC/DS-CDMA receiver, the equal gain combining (EGC) can be considered to yield a satisfactory detection efficiency, due to the fact that the output from each finger in a CC/DS-CDMA receiver contains only the useful part: the autocorrelation peak at a particular sampling time. The advantage to use EGC instead of MRC in a CC/DS-CDMA RAKE is to save a complicated multipath amplitude estimation algorithm. Thus, an EGC-RAKE in a CC/DS-CDMA receiver requires only multipath return delays information, which is much easier to obtain than the multipath return amplitudes.

- Third, as shown in Figure (5.32), the isotropic MAI-free property in the CC/DS-CDMA architecture makes it unnecessary to use multi-user detection to de-correlate the transmission signals from different users, due to the fact that the transmissions from different users in the CC/DS-CDMA system are already pre-decorrelated at transmitter side because of its MAI-free signaling structure. In order words, the signaling design in the CC/DS-CDMA architecture has already taken into account multi-user signal decorrelation mechanism, such that the signal detection efficiency in a CC/DS-CDMA receiver can be substantially improved even without using a complex and error-prone multi-user detection algorithm.

- Fourth, owing to the isotropic MAI-free and MI-free properties, the near-far effect will virtually cause no harm to the signal detection process at a correlator receiver in a CC/DS-CDMA system, as long as bit-synchronization, which should not be too difficult to implement with the help of a pilot, can be achieved prior to the data detection process. In other words, the CC/DS-CDMA is a deliberated-designed system with an excellent near-far resistance. Therefore, a complicated open-loop and closed-loop power control system is no longer a necessity to CC/DS-CDMA, saving a great amount of hardware and software overhead in user terminals. More precisely, the power control in the CC/DS-CDMA system is used merely to reduce power consumption and unnecessary power emission at terminals, whose requirements on response time and power control accuracy can be made much more relaxed than necessary in a conventional CDMA system.

- Fifth, the CC/DS-CDMA system in principle does not care about the number of co-channel unwanted transmissions within a cell because of its isotropic MAI-free property. Therefore, the signal processing advantage from beam forming provided by smart-antenna technology has become less relevant to the improvement in signal detection efficiency of a CC/DS-CDMA system, as long as the system does not need users' particular spatial information, such as direction-of-arrivals of the incoming signals. More specifically, the beam-patterns adopted at a CC/DS-CDMA receiver have little direct impact on the signal-to-interference-cum-noise ratio if compared to that in a conventional CDMA system without isotropic

MAI-free and MI-free properties. Therefore, an omni-directional antenna is enough for CC/DS-CDMA system to perform satisfactorily, without needing costly antenna array system.

- Finally, the CC/DS-CDMA scheme is in particular well suited for detection of the data appearing at the edges of a message, as shown in the upper parts of Figures (5.26), (5.27) and (5.29), which illustrate the signal detection process for the first (or leftmost) bit of the message. In detecting those bits, partial CCFs of the codes will play an important role. It is seen from the figures that the CC/DS-CDMA system yields zero partial CCFs, which ensure an ideal performance for signal detection even on the edges of a message or packet. This observation is significant due to the concern of dominant bursty-type traffic in future wireless systems, where frequent on-and-off operating fashion in packet-switched applications will force a receiver to work constantly on detection of packet-edges. Therefore, in this sense, the CC/DS-CDMA is a perfect choice for the future all-IP wireless applications where bursty traffic will be dominant.

In summary, the CC/DS-CDMA architecture based on IST's can create an interference-free operational environment due to its superior isotropic MAI-free and isotropic MI-free properties, which can successfully transform an interference-limited operational environment into a noise-limited one. Thus, a great gain in the capacity and data throughput in a CDMA system can be expected. The major operational advantages of the CC/DS-CDMA architecture, if compared with a conventional DS-CDMA, are listed in Table (5.6).

6. Summary

CDMA technology is at a critical stage on its evolutional path. The current major CDMA-based standards all work on almost the same principle with direct-sequence spreading modulation and conventional spreading codes, suffering from serious interference problems stemmed from their undesirable spreading codes characteristics, and thus their performance is strictly limited by combined effect of extremely unpredictable MAI and MI. The conventional spreading codes applied to various CDMA-based 2-3G standards are never optimal in terms of their correlation properties as well as their asymmetrical feature when working in synchronous down-link and asynchronous up-link, causing many problems in rate-matching, channelization and co-channel inter-

Table 5.6. Comparison between CC/DS-CDMA and conventional DS-CDMA.

	CC/DS-CDMA	Conventional DS-CDMA
Detection Efficiency	Excellent	Medium~good (with RAKE)
Multipath Diversity	Excellent	Medium
RAKE Receiver	Optional	A Must
FSF[a] Resistance	Excellent	None (without RAKE)
Near-far Resistance	Excellent	None
Precision Power Control	Not Necessary	A Must
Multi-User Detection	Not Necessary	Recommended
Up-Link Sync. Control	Not Necessary	Recommended
Smart Antenna	Not Necessary	Recommended
Rx Hardware	Simple	Complex
System Capacity	High	Low~medium
Down-Link MAI	Low	Low
Up-Link MAI	Low	Medium~high
Down-Link MI	Low	Medium~high
Up-Link MI	Low	Medium~high
Bursty traffic	Well-suited	Poorly-performed

[a] FSF: Frequency-Selective Fading.

ference, etc. The concept of isotropic spreading technologies suggested in this chapter serves well as a direction finder for the future research on CDMA technological development, based on which two innovative CDMA architectures, CC-CDMA and CC/DS-CDMA, are introduced herein. In particular, the CC/DS-CDMA system can offer a superior isotropic interference-free operation in a wireless communication system, showing a great potential for CDMA systems to offer a real noise-limited performance. We have a strong belief that the B3G wireless should definitely work on a completely new CDMA platform based on interference-free architecture, rather than on the conventional CDMA technologies, which have suffered from irreducible interference. The bandwidth-efficiency of a B3G CDMA system should be achieved only based on its superior power-efficiency, which is the bottleneck to hinder any further improvement on overall performance of current 2-3G wireless. Isotropic interference-free CDMA technologies, which can greatly improve the power-efficiency and then the overall performance, should play an indispensable role in the future wireless air-link architecture. The proposed two novel CDMA schemes have shown the way leading to the development of a practical noise-limited CDMA architecture, which is our ultimate goal to achieve on CDMA evolution.   •

Acknowledgments

The author wishes to thank the Editor of this book, Prof. Mohsen Guizani, for his encouragement and support in preparing this chapter. He would like also to thank Mr. Jin-Xiao Lin, Mr. Hsin-Wei Chiu and Mr. Zhemin Lin, Institute of Communications Engineering, National Sun Yat-Sen University, Taiwan, for their help in generating some figures and plottings presented in this chapter. The author would like also to thank National Science Council (NSC), Taiwan, for the following research grants, NSC 89-2213-E-005-032, NSC 89-2213-E-005-029, NSC 89-2213-E-005-032, NSC 90-2213-E-110-054, NSC 91-2213-E-110-032, NSC 92-2213-E-110-015, NSC 93-2213-E-110-012, NSC 92-2213-E-110-047 and NSC 92-2213-E-110-048, which have partially supported the related research work.

Appendix: Complete Complementary Codes with PG being 8 to 512

The following list gives the complementary codes with their PG values changing from 8 to 512, which are commonly used in a practical CDMA system. Note that the commons "," are used to separate different element codes within a specific flock and blankets "()" are used to indicate flocks of a complementary code sets.

1 PG=8

Length of element codes=4, flock size=2

$(+ + + -, + - + +)$

$(+ + - +, + - - -)$

2 PG=16

Length of element codes=8, flock size=2

$(+ + + - + + - +, + - + + + - - -)$

$(+ + + - - - + -, + - + + - + + +)$

3 PG=32

Length of element codes=16, flock size=2

$(+ + + - + + - + + + + + - - - + -, + - + + + - - - + - + + - + + +)$

$(+ + + - + + - + - - - + + + - +, + - + + + - - - - + - - + - - -)$

4 PG=64

(1) Length of element codes=32, flock size=2

$(+ + + - + + - + + + + + - - - + - + + + - + + - + - - - + + + - +, + - + + + - - - + - + + - + + + + - + + + - - - + - - + - - -)$

$(+ + + - + + - + + + + + - - - + - - - - + - - + - + + + + - - - + -, + - + + + - - - + - + + - + + + - + - - - + + + + - + + - + + +)$

(2) Length of element codes=16, flock size=4

$(+ + + + + + - + - + + - - + - - +, + - + - + + + + + + - - + + + - -, + + - - + - - + + + + + + - + -, + - - + + + + - - + - + - + + + +)$

$(+ + + + + - + - + + + - - - + + -, + - + - - - - - + - - + - - + +, + + - - - + + - + + + + + - + - +, + - - + - - + + + - + - - - - -)$

$(+ + + + + + - + - - - + + - + + -, + - + - + + + + - + + - - - + +, + + - - + - - + - - - - + - + +, + - - + + + + - - - + - + - - - -)$

$(+ + + + - + - + - - + + + - - +, + - + - - - - - - + + - + + - -, + + + - - - - + + - + - + -, + - - + - - + + - + - + + + + + +)$

5 PG=128

Length of element codes=64, flock size=2

$(+ + + - + + - + + + + + - - - + - + + + - + + - + - - - + + + - + + + + - + + - + + + + - - - + - - - - + - - + - + + + + - - - + -, + - + + + + + - - - + - + + + - + + + + - - - + - + + - - - +)$

$(+ + + - + + - + + + + + - - - + - + + + - + + - + - - - + + + - + - - - + - - - + + + + + + + + - + + - + - - - + + + - +, + - + + + - - - + - + + - + + + + - + + + - - - + - - + - - - - + - - + - - - + - + + + + - - - + - - + - - -)$

6 PG=256

(1) Length of element codes=128, flock size=2

(+ + + − + + − + + + + − − − + − + + + − + + − + − − − + + + − + + +
+ − + + − + + + + − − − + − − − − + − − + − + + + + − − − + − + + + − +
+ − + + + + − − − + − + + + − + + − + − − − + + + − + − − − + − − + − −
− − + + + − + + + + − + + − + − − − + + + − +, + − + + + − − − + − +
+ − + + + + − + + + − − − − + − − + − − − + − + + + − − − + − + + + − +
+ + − + − − − + + + + − + + − + + + + − + + + − − − + − + + − + + +
+ − + + + − − − − + − − + − − − − + − − − + + + − + − − + − − − + − + +
+ − − − − + − − + − − −)

(+ + + − + + − + + + + − − − + − + + + − + + − + − − − + + + − + + +
+ − + + − + + + + − − − + − − − − + − − + − + + + + − − − + − − − − + − −
+ − − − − + + + − + − − − + − − + − + + + + − − − + − + + + − + + − + +
+ + − − − + − − − − + − − + − + + + − − − +, + − + + + + − − − + − + + +
+ + + + − + + + + − − − − + − − + − − − + − + + + + − − − + − + + + − +
− + − − − + + + + − + + − + + + + − + − − − + + + − + − − + − − − − + − −
− + + + + − + + − + + + + − + + + − − − + − + + − + + + − + − − − +
+ + + − + + − + + +)

(2) Length of element codes=64, flock size=4

(+ + + + + − + − + + − − + − − + + + + + − + − + + + − − − + + − + +
+ + + − + − − − + + − + + − + + + + − + − + − − + + + − − +, + − + −
+ + + + + − − + + + − − + − + − − − − − + − − + − − + + + − + − + + +
+ − + + − − − + + + − + − − − − − − + + − + + − −, + + − − + − − + + +
+ + + − + − + + − − − + + − + + + + − + − + + + − − + − − + − − − − −
+ − + + + − − − + + − − − − − + − + −, + − − + + + − − + − + − + + + +
+ − − + − − + + + − + − − − − − + − − + + + − − − + − + − − − − + − − + −
− + + − + − + + + + +)

(+ + + + + − + − + + − − + − − + − − + − − − − + − + − − + + + − − + + + +
+ + − + − − − + + − + + − + + − − − − − + − + − + + + − − + + −, + − + − + +
+ + + − − + + + − − − + − + + + + + − + + − + + − + + − − + − + − + + + +
− + + − − − + + − + − + + + + + + − − + − − + +, + + − − + − − + + +
+ + + − + − − − + + + − − + − − − − + − + − + + − − + − − + − − − − − + −
+ − − + + + − − + + + + + − + − +, + − − + + + − − + − + − + + + + −
+ + − + + − − − + − + + + + + + − − + + + − − − + − + − − − − − + + −
+ + − − + − + − − − − −)

(+ + + + + − + − + + − − + − − + + + + + − + − + + + − − − + + − − −
− − − + − + + + − − + − − + − − − − + − + − + + − − − + + −, + − + − + +
+ + + − − + + + − − + − + − − − − − + − − + − − + + − + − + − − − − + − −
+ + + − − − + − + + + + + + − − + − − + +, + + − − + − − + + + + + +
− + − + + − − − + + − + + + + − + − + − − + + − + + − + + + + + + + − +
− − − + + + − − + + + + + − + − +, + − − + + + − − + − + − + + + + +
− − + − − + + + − + − − − − − − + + − − − + + + − + − + + + + − + + − +
+ − − + − + − − − − −)

(+ + + + + − + − + + − − + − − + − − + − − − + − + − − − + + + − − + − − − −
− + − + + + − − + − − + + + + + − + − + − − + + + − − +, + − + − + +
+ + + − − + + + − − − + − + + + + + − + + − + + − + + − − + − − − − − +
− − + + + − − + − + − − − − − − + + − + + − −, + + − − + − − + + + + +
+ − + − − − + + + − − + − − − − + − + − − + + − + + − + + − + + + + + − + −
+ + − − − + + − − − − − + − + −, + − − + + + − − + − + − + + + + + − + +

```
- + + - - - + - + + + + + - + + - - - + + + - + - + + + + + - - + - -
+ + - + - + + + + + )
```

7 PG=512

(1) Length of element codes=256, flock size=2

```
(+ + + - + + - + + + + - - - + - + + + - + + - + - - - + + + - + + +
+ - + + - + + + + - - - + - - - - + - - + - + + + - - - + - + + + - +
+ - + + + + - - - + - + + + - + + - + - - - + + + - + - - - + - - + - -
- - + + + - + + + + - + + - + - - - + + + - + + + + - + + - + + + +
- - - + - + + + - + + - + - - - + + + - + + + + - + + - + + + + - - -
+ - - - - + - - + - + + + - - - + - - - - + - - + - - - + + + - + - - -
+ - - + - + + + - - - + - + + + - + + - + + + + - - - + - - - + - - +
- + + + - - - + -, + - + + + - - - + - + + - + + + + - + + + - - - - +
- - + - - - + - + + + - - - + - + + - + + + - + - - - + + + + - + + -
+ + + + - + + + - - - + - + + - + + + + - + + + - - - - + - - + - - -
- + - - - + + + - + - - + - - - + - + + + - - - - + - - + - - - + - + +
+ - - - + - + + - + + + + - + + + - - - + - - + - - - + - + + + - - -
+ - + + - + + + - + - - - + + + + - + + - + + + - + - - - + + + - +
- - + - - - - + - - - + + + + - + + - + + + + - + + + - - - + - + + + -
+ + + - + - - - + + + + - + + - + + + )
```

```
(+ + + - + + - + + + + - - - + - + + + - + + - + - - - + + + - + + +
+ - + + - + + + + - - - + - - - - + - - + - + + + - - - + - + + + - +
+ - + + + + - - - + - + + + - + + - + - - - + + + - + - - - + - - + - -
- - + + + - + + + + - + + - + - - - + + + - + - - - + - - + - - - - + +
+ - + - - - + - - + - + + + - - - + - - - - + - - + - - - + + + - + + +
+ + - + + - + - - - + + + - + + + + - + + - + + + + - - - + - + + +
- + + - + - - - + + + - + - - - + - - + - - - - + + + - + + + + - + + -
+ - - - + + + - +, + - + + + - - - + - + + - + + + + - + + + - - - -
+ - - + - - - + - + + + - - - + - + + - + + + - + - - - + + + + - + +
- + + + + - + + + - - - + - + + - + + + - + + + - - - - + - - + - - -
- - + - - - + + + - + - - + - - - + - + + + - - - + - - + - - - + - + +
- + + + - + - - + - - - - + - - - + + + + - + + - + + + - + - - - + +
+ - + - - + - - - + - + + + - - - + - - + - + + + - + - - + - +
+ - + + + + - + + + - - - - + - - + - - - + - - + + + - + - - + - -
- + - + + + - - - + - - + - - - )
```

(2) Length of element codes=64, flock size=8

```
(+ + + + + + + + + - + - + - + - + + - - + + - - + - - + + - - + + +
+ + - - - - + - + - - + - + + + - - - - + + + - - + - + + -, + - + - + -
+ - + + + + + + + + + - - + + - - + + + - - + + - - + - + - - + - +
+ + + + - - - - + - - + - + + - + + - - - - + +, + + - - + + - - + - -
+ + - - + + + + + + + + + + + + - + - + - + - + + - - - - + + + - - + -
+ + - + + + + - - - - + - + - - + - +, + - - + + - - + + + - - + + - -
+ - + - + - + - + + + + + + + + + + - - + - + + - + + - - - - + + + -
+ - - + - + + + + + + - - - -, + + + + - - - - + - + - - + - + + + - - - -
+ + + - - + - + + - + + + + + + + + + + + - + - + - + - + + - - + + - -
+ - - + + - - +, + - + - - + - + + + + + + - - - - + - - + - + + - + + -
- - - + + + - + - + - + - + + + + + + + + + + - - + + - - + + + - - +
+ - -, + + + - - - - + + + - - + - + + - + + + + - - - - + - + - - + - +
+ + - - + + - - + - - + + - - + + + + + + + + + + + + - + - + - + -, + -
```

$- + - + + - + + - - - - + + + - + - - + - + + + + + - - - - + - - + + -$
$- + + + - - + + - - + - + - + - + - + + + + + + + +)$

$(+ + + + + + + + - + - + - + - + + + - - + + - - - + + - - + + - + +$
$+ + - - - - - + - + + - + - + + - - - - + + - + + - + - - +, + - + - + -$
$+ - - - - - - - - - + - - + + - - + - - + + - - + + + - + - - + - + - - -$
$- + + + + + - - + - + + - - - + + + + - -, + + - - + + - - - + + - - +$
$+ - + + + + + + + + - + - + - + - + + + - - - - + + - + + - + - - +$
$+ + + + - - - - - + - + + - + -, + - - + + - - + - - + + - - + + + - +$
$- + - + - - - - - - - - + - - + - + + - - - + + + + - - + - + - - + - +$
$- - - - + + + +, + + + + - - - - - + - + + - + - + + - - - + + - + +$
$- + - - + + + + + + + + + - + - + - + - + + + - - + + + - - - + + - -$
$+ + -, + - + - - + - + - - - + + + + + + - - + - + + - - - + + + + - -$
$+ - + - + - + - - - - - - - - - + - - + + - - + - - + + - - + +, + + - - -$
$- + + - + + - + - + - - + + + + + - - - - - + - + + - + - + + - - + + - - -$
$+ + - - + + - + + + + + + + + + - + - + - + - +, + - - + - + + - - - +$
$+ + + - - + - + - - + - + - - - - + + + + + - - + + - - + - - + + - - +$
$+ + - + - + - + - - - - - - - - -)$

$(+ + + + + + + + + - + - + - + - - - + + - - + + - + + - - + + - + +$
$+ + - - - + - + - - + - + - - + + + + - - - + + - + - - +, + - + - + -$
$+ - + + + + + + + + - + + - - + + - - - + + - - + + + - + - - + - +$
$+ + + + - - - - - + + - + - - + - - + + + + - -, + + - - + + - - + - -$
$+ + - - + - - - - - - - - - + - + - + - + + + - - - - + + + - + - + + +$
$- - - - + + + + - + - + + - + -, + - - + + - - + + + - - + + - - - + -$
$+ - + - + - - - - - - - - - + - - + + - + + - + + - - - + + - + - + + - + -$
$- - - - + + + +, + + + + - - - - + - + - - + - + - - + + + + - - - + +$
$- + - - + + + + + + + + + + + - + - + - + - - - + + - - + + - + + - -$
$+ + -, + - + - + - - + - + + + + + + - - - - - + + - + - - + - - + + + + - -$
$+ - + - + - + - + + + + + + + + + - + + - - + + - - - + + - - + +, + +$
$- - - - + + + - - + - + + - - - - - + + + + - + - + + - + - + + - - + +$
$- - + - - + + - - + - - - - - - - - - + - + - + - +, + - - + - + + - + + -$
$- - - + + - + - + + + - + - - - - - + + + + + - - + + - - + + + - - + + -$
$- - + - + - + - + - - - - - - - -)$

$(+ + + + + + + + + - + - + - + - + - - + + - - + + + - - + + - - + + +$
$+ + - - - - - + - + + - + - - - + + + + - - + - - + - + + -, + - + - + -$
$+ - - - - - - - - - + + - - + + - + + - - + + - - + - + - - + - + - - -$
$- + + + + + - + + - + - - + + + - - - - + +, + + - - + + - - - + + - - +$
$+ - - - - - - - - - + - + - + - + - + + - - - - + + - + + - + - - + - - -$
$- + + + + + - + - - + - +, + - - + + - - + - - + + - - + + - + - + - +$
$- + + + + + + + + + + + - - + - + + - - - + + + + - - - + - + + - + -$
$+ + + + - - - -, + + + + - - - - - + - + + - + - - - + + + + - - + - -$
$+ - + + - + + + + + + + + + - + - + - + - + - - + + - - + + + - - + +$
$- - +, + - + - - + - + - - - + + + + - + + - + - - + + + - - - - + +$
$+ - + - + - + - - - - - - - - - + + - - + + - + + - - + + - -, + + + - - -$
$- + + - + + - + - - + - - - + + + + + - + - - + - + + + - - + + + - - -$
$+ + - - + + - - - - - - - - - + - + - + - + -, + - - + - + + - - - + + +$
$+ - - - + - + + - + - + + + + - - - + - - + + - - + - - + + - - + + -$
$+ - + - + - + + + + + + + + +)$

$(+ + + + + + + + + - + - + - + - + + - - + + - - + - - + + - - + - -$
$- - + + + + + - + - + + - + - - - + + + + - - - + + - + - - +, + - + - +$

```
 - + - + + + + + + + + - - + + - - + + + - - + + - - - + - + + - +
 - - - - + + + + - + + - + - - + - - + + + + - -, + + - - + + - - + - -
 + + - - + + + + + + + + + - + - + - + - - - + + + + - - - + + - +
 - - + - - - - + + + + - + - + + - + -, + - - + + - - + + + - - + + - -
 + - + - + - + - + + + + + + + + - + + - + - - + - - + + + + - - - +
 - + + - + - - - - + + + +, + + + + - - - - + - + - - + - + + + - - - -
 + + + - - + - + + - - - - - - - - - + - + - + - + - - + + - - + + - +
 + - - + + -, + - + - - + - + + + + + - - - - + - - + - + + - + + - - - -
 + + - + - + - + - + - - - - - - - - + + - - + + - - - + + - - + +, + +
 - - - + + + - - + - + + - + + + + - - - - + - + - - + - + - - + + - -
 + + - + + - - + + - - - - - - - - - + - + - + - +, + - - + - + + - + +
 - - - + + + - + - - + - + + + + + - - - - + + - - + + - - - + + - -
 + + - + - + - + - + - - - - - - - - )

(+ + + + + + + + - + - + - + - + + + - - + + - - - + + - - + + - - -
 - - + + + + + - + - - + - + - - + + + + - - + - - + - + + -, + - + - +
 - + - - - - - - - - - + - + + - - + - - + + - - + + - + - + - + - + +
 + + - - - - + + - + - + + + - - - - + +, + + - - + + - - - + + - - +
 + - + + + + + + + + - + - + - + - + - - + + + + - - + - - + - + + -
 - - - + + + + + - + - - + - +, + - - + + - - + - - + + - - + + + - +
 - + - + - - - - - - - - - + + - + - - + + + - - - - + + - + - + + - +
 + + + + - - - -, + + + + - - - -, - + - + + + - + - + + - - - - + + - + +
 - + - - + - - - - - - - - + - + - + - + - - + + - - + + + - - + + - - +,
 + - + - - + - + - - - - + + + + + - - + - + + - - - + + + + - - - + - +
 - + - + + + + + + + + + - + + - - + + - + + - - + + - -, + + + - - - -
 + + - + + - + - - + + + + + - - - - - + - + + - + - - - + + - - + + +
 - - + + - - + - - - - - - - - + - + - + - +, + - - + - + + - - - + + +
 + - - + - + - - + - + - - - + + + + - + + - - + + - + + - - + + - - -
 + - + - + - + + + + + + + + + + )

(+ + + + + + + + + - + - + - + - - - + + - - + + - + + - - + + - - -
 - - + + + + - + - + + - + - + + - - - - + + + - - + - + + -, + - + - +
 - + - + + + + + + + + + - + + - - + + - - - + + - - + + - + - + + - +
 - - - - - + + + + + - - + - + + - + + - - - - + +, + + - - + + - - + - -
 + + - - + - - - - - - - - - + - + - + - + - - + + + + - - - + + - + - - +
 + + + + - - - - + - + - - + - +, + - - + + - - + + + + - - + + - - - + -
 + - + - + - - - - - - - - - + + - + - - + - - + + + + - - + - + - - + - +
 + + + + - - - -, + + + + - - - - + - + - - + - + - - + + + + - - - + +
 - + - - + - - - - - - - - - + - + - + - + + + - - + + + - - + - - + + - - +,
 + - + - - + - + + + + + + - - - - - + + - + - - + - - + + + + - - - + - +
 - + - + - - - - - - - - + - - + + - - + + + - - + + -, + + + - - - - + + +
 + - - + - + + + - - - - + + + + - + - + + - + - - - + + - - + + - + + -
 - + + - + + + + + + + + + + + - + - + - + -, + - - + - + + - + + - - - -
 + + - + - + + + - + - - - - + + + + - + + - - + + - - - + + - - + + +
 - + - + - + - + + + + + + + + + )

(+ + + + + + + + + - + - + - + - + - - + + - - + + + - - + + - - + - -
 - - + + + + + - + - - + - + + + - - - - + + - + + - + - - +, + - + - +
 - + - - - - - - - - - + + - - + + - + + - - + + - - - + - + + + - + - + +
 + + - - - - + - - + - + + - - + + + + - -, + + - - + + - - - + + - - +
 + - - - - - - - - - + - + - + - + - - - + + + + - - + - - + - + + - + +
 + + - - - - - + - + + - + -, + - - + + - - + - - + + - - + + - + - + - +
```

− + + + + + + + + + − + + − + − − + + + − − − − + + + − + − − + − +
− − − − + + + +, + + + + − − − − − + − + + − + − − − + + + + − − + − −
+ − + + − − − − − − − − − + − + − + − + − + + − − + + − − − + + − − + +
−, + − + − − + − + − − − − + + + + − + + − + − − + + + − − − − + + − + −
+ − + − + + + + + + + + + + + − − + + − − + − − + + − − + +, + + − − −
− + + − + + − + − − + − − − − + + + + + − + − − + − + − − + + − − + +
+ − − + + − − + + + + + + + + + + − + − + − + − +, + − − + − + + − − −
+ + + + − − − + − + + − + − + + + + − − − − − + + − − + + − + + − − +
+ − − + − + − + − + − − − − − − − − −)

References

[1] TIA/EIA-95A, (1995). Mobile Station-Base Station Compatibility Standard for Dual-Mode Wideband Spread Spectrum Cellular Systems.

[2] TIA/EIA-95B, (1997). Mobile Station-Base Station Compatibility Standard for Dual-Mode Wideband Spread Spectrum Cellular Systems. Baseline Version, July 31, 1997.

[3] Telecommunications Industry Association (TIA), (1998). The cdma2000 ITU-R RTT Candidate Submission (0.18).

[4] ETSI, (1998). The ETSI UMTS Terrestrial Radio Access (UTRA) ITU-R RTT Candidate Submission, Jan. 29, 1998.

[5] Association of Radio Industries and Businesses (ARIB), (1998). Japan's Proposal for Candidate Radio Transmission Technology on IMT-2000: W-CDMA, June 26, 1998.

[6] CATT, (1998). TD-SCDMA Radio Transmission Technology For IMT-2000 Candidate submission, Draft V.0.4, Sept. 1998.

[7] (2002). First Report and Order in the Matter of Revision of Part 15 of the Commission's Rules Regarding Ultra-Wideband Transmission Systems. FCC, released, ET Docket 98-153, FCC 02-48, April 2002.

[8] Mitchell T. (2001). Broad is the way. *IEE Review*, pp. 35-39, Jan. 2001.

[9] Leeper D. (2001). A long-term view of short-range wireless. *IEEE Computer*, 34(6): pp. 39-44, 2001.

[10] Van't Hof JP, Stancil DD. (2002). Ultra-wideband high data rate short range wireless links, Pages: 85-89, Proc. IEEE VTC 2002.

[11] Win MZ. (2002). A unified spectral analysis of generalized time-hopping spread-spectrum signals in the presence of timing jitter. *IEEE J. Select. Areas Communications*; 20(9): 1664-1676, 2002.

[12] Win MZ, Scholtz RA. (2002). Characterization of ultra-wide bandwidth wireless indoor channels: a communication-theoretic view. *IEEE J. Select. Areas Communications*, 20(9): 1613-1627, 2002.

[13] United States Code of Federal Regulations, Title 47, Part 15, U.S. Government Printing Office, http://www.access.gpo.gov/nara.cfr/index.html.

[14] Conroy JT, LoCicero JL, Ucci DR. (1999). Communication techniques using monopulse waveforms. Proc. MILCOM 1999: 2:1185-1191.

[15] Ross, A. H. M., and K. L. Gilhousen, (1996). CDMA Technology and the IS-95 North American Standard, in The Mobile Communications Handpaper, J. D. Gipson (Ed.), CRC Press, pp. 430-448, 1996.

[16] N. Guo and L. B. Milstein, (1999). On rate-variable multidimensional DS/SSMA with dynamic sequence sharing, *IEEE J. Select. Areas Commun.*, vol. 17, May 1999.

[17] C.-L. I and R. D. Gitlin, (1995). Multi-code CDMA wireless personal communications networks, in Proc. IEEE Int. Conf. Commun. (ICC'95), Seattle, WA, vol. 2, pp. 1060-1064.

[18] S. Sasaki, H. Kikuchi, H. Watanabe, and J. Zhu, (1994). Performance evaluation of parallel combinatory SSMA systems in Rayleigh fading channel, in Proc. IEEE 3rd Int. Symp. Spread Spectrum Techniques and Applications (ISSSTA'94), Oulu, Finland, vol. 1, pp. 198-202.

[19] Baey, S.; Dumas, M.; Dumas, M.-C., (2002). QoS tuning and resource sharing for UMTS WCDMA multiservice mobile, *IEEE Trans. on Mobile Computing*, Vol. 1, Issue: 3 , Page(s): 221 -235, Jul-Sep 2002.

[20] Insoo Sohn; Seung Chan Bang; (2000). Performance studies of rate matching for WCDMA mobile receiver, Vehicular Technology Conference, 2000. IEEE VTS-Fall VTC 2000. 52nd , Vol. 6 , Page(s): 2661 -2665, 2000.

[21] Kam, A.C.; Minn, T.; Siu, K.-Y. (2001). Supporting rate guarantee and fair access for bursty data traffic in W-CDMA, *IEEE J. Select. Areas Communications*, Volume: 19, Issue: 11, Page(s): 2121-2130, Nov 2001.

[22] Thit Minn; Kai-Yeung Siu, (2000). Dynamic assignment of orthogonal variable-spreading-factor codes in W-CDMA, *IEEE J. Select. Areas Communications*, Vol. 18, Issue: 8, Page(s): 1429-1440, Aug 2000.

[23] Fantacci, R.; Nannicini, S., (2000). Multiple access protocol for integration of variable bit rate multimedia traffic in UMTS/IMT-2000 based on wideband CDMA, *IEEE J. Select. Areas Communications*, Vol. 18, Issue: 8, Page(s): 1441-1454, Aug. 2000.

[24] R. Gold, (1968). Maximal recursive sequences with 3-valued recursive cross-correlation functions, *IEEE Trans. on Information Theory*, vol. IT-14, pp. 154-156, January 1968.

[25] R. A Scholtz and L. R. Welch. (1984). GMW Sequences. *IEEE Trans. on Information Theory*, Vol. 30, No. 3, pp. 548-553, May 1984.

[26] J-S. No and P. V. Kumar. (1989). A New Family of Binary Pseudo-random Sequences Having Optimal Periodic Correlation Properties and Large Linear Span. *IEEE Trans. on Information Theory*, Vol. 35, No. 2, pp. 371-379, March 1989.

[27] X. H. Chen and J. Oksman, (1992). BER performance analysis of 4-CCL and 5-CCL codes in slotted indoor CDMA systems, *IEE Proceedings - I*, vol. 139, pp. 79-84, February 1992.

[28] X. H. Chen; Lang, T.; Oksman, J.; (1996). Searching for quasi-optimal subfamilies of m-sequences for CDMA systems, Seventh IEEE International Symposium on Personal, Indoor and Mobile Radio Communications (PIMRC'96), vol. 1, Page(s): 113 -117, 15-18 Oct. 1996.

[29] Tirkel, A.Z. (1996). Cross correlation of m-sequences-some unusual coincidences, Spread Spectrum Techniques and Applications Proceedings, 1996 IEEE 4th International Symposium on, Page(s): 969 -973 vol.3, 22-25 Sept. 1996.

[30] Ito, T.; Sampei, S.; Morinaga, N., (2000). M-sequence based M-ary/SS scheme for high bit rate transmission in DS/CDMA systems, *Electronics Letters*, vol. 36 Issue: 6, Page(s): 574 -576, 16 Mar 2000.

[31] Tirkel, A.Z., (1996). Cross correlation of m-sequences-some unusual coincidences, Spread Spectrum Techniques and Applications Proceedings, 1996 IEEE 4th International Symposium on, vol. 3, Page(s): 969 -973, 22-25 Sep 1996.

[32] Imamura, K.; Guo-Zhen Xiao, (1992). On periodic sequences of the maximum linear complexity and M-sequences, Singapore ICCS/ISITA '92. 'Communications on the Move' , Page(s): 1219 -1221 vol.3, 16-20 Nov. 1992.

[33] Uehara, S.; Imamura, K., (1992). Some properties of the partial correlation of M-sequences, Singapore ICCS/ISITA '92. 'Communications on the Move', Page(s): 1222 -1223, vol.3, 16-20 Nov. 1992.

[34] Turkmani, A.M.D.; Goni, U.S., (1993). Performance evaluation of maximal-length, Gold and Kasami codes as spreading sequences in CDMA systems, Universal Personal Communications, 1993. Personal Communications: Gateway to the 21st Century Conference Record, 2nd International Conference on, vol. 2, Page(s): 970 -974, 12-15 Oct. 1993.

[35] Lahtonen, J, (1995). On the odd and the aperiodic correlation properties of the Kasami sequences, *IEEE Trans. on Information Theory*, vol. 41 Issue: 5, Page(s): 1506-1508, Sept. 1995.

[36] Lebedev, O.N.; Poliakov, I.L., (2000). Properties of composite Kasami sequence sets for wideband signals, 10th International Microwave Conference, 2000. *Microwave and Telecommunication Technology*, Page(s): 234 -235, 2000.

[37] Barghouthi, R.T.; Stuber, G.L.; (1994). Rapid sequence acquisition for DS/CDMA systems employing Kasami sequences, *IEEE*

Trans. on Communications, vol. 42, Issue: 2, Page(s): 1957-1968, Feb/Mar/Apr 1994.

[38] Komo, J.J.; Liu, S.-C.; (1990). Modified Kasami sequences for CDMA System Theory, Twenty-Second Southeastern Symposium on , Page(s): 219 -222, 11-13 Mar. 1990.

[39] Dongwook Lee; Hun Lee; Milstein, K.B.; (1998). Direct sequence spread spectrum Walsh-QPSK modulation, *IEEE Trans. on Communications*, vol. 46 Issue: 9, Page(s):1227 -1232, Sept. 1998.

[40] Joonyoung Cho; Youhan Kim; Kyungwhoon Cheun; (2000). A novel FHSS multiple-access network using M-ary orthogonal Walsh modulation, VTC 2000. IEEE VTS-Fall VTC 2000. 52nd, vol. 3, Page(s): 1134 -1141, 2000.

[41] Tsai, S.; Khaleghi, F.; Seong-Jun Oh; Vanghi, V.; (2001). Allocation of Walsh codes and quasi-orthogonal functions in cdma2000 forward link, VTC 2001 Fall. IEEE VTS 54th, Page(s): 747 -751 vol.2, 2001.

[42] M. J. E. Golay, (1961). Complementary series, *IRE Trans. Inform. Theory*, vol. IT-7, pp. 82-87, 1961.

[43] R. Turyn, (1963). Ambiguity function of complementary sequences, *IEEE Trans. on Information Theory*, vol. IT-9, pp. 46-47, Jan, 1963.

[44] N. Suehiro, (1982). Complete complementary code composed of N-multiple-shift orthogonal sequences, *Trans. IECE of Japan* (in Japanese), vol. J65-A, pp. 1247-1253, Dec. 1982.

[45] N. Suehiro and M. Hatori, (1988). N-Shift Cross-Orthogonal Sequences, *IEEE Trans. on Information Theory*, vol. IT-34, no. 1, pp. 143-146, January 1988.

[46] P. V. Kumar and R. A. Scholtz, (1983). Bounds on the linear span of Bent sequences, *IEEE Trans. on Information Theory*, pp. 854-862.

[47] Tao Lang; Xiao-Hua Chen, (1994). Comparison of correlation parameters of binary codes for DS/CDMA systems, Singapore ICCS '94. Conference Proceedings, vol. 3, Page(s): 1059 -1063, 14-18 Nov. 1994.

[48] L. R. Welch, (1974). Lower bounds on the maximum cross-correlation of signals, *IEEE Trans. on Information Theory*, vol. IT-20, pp. 397-399, 1974.

[49] M. B. Pursley, (1977). Performance evaluation for phase-coded spread-spectrum multiple-access communications - Part I: System analysis, *IEEE Trans. on Communications*, vol. COM-25, no. 8, pp. 795-799, Aug. 1977.

[50] M. B. Pursley and D. V. Sarwate, (1977). Performance evaluation for phase-coded spread-spectrum multiple-access communications - Part II: code sequence analysis, *IEEE Trans. on Communications*, vol. COM-25, no. 8, pp. 800-803, Aug. 1977.

[51] K. Yao, (1977). Error probability of asynchronous spread spectrum multiple access communication systems, *IEEE Trans. on Communications*, vol. 25, no. 8, pp. 803-809, Aug. 1977.

[52] D. V. Sarwate and M. B. Pursley, (1980). Cross-correlation properties of pseudorandom and related sequences, *Proceedings of the IEEE*, vol. 68, no. 5, pp. 593-620, May 1980.

[53] M. B. Pursley, D. V. Sarwate and W. E. Stark, (1982). Error probability for direct-sequence spread spectrum multiple-access communications - Part I: Upper and lower bounds, *IEEE Trans. on Communications*, vol. COM-30, no. 5, pp. 975-984, May 1982.

[54] E. A. Geraniotis and M. B. Pursley, (1982). Error probability for direct-sequence spread spectrum multiple-access communications - Part II: Approximations, *IEEE Trans. on Communications*, vol. COM-30, no. 5, pp. 985-995, May 1982.

[55] J. M. Holtzman, (1992). On calculating DS/SSMA error probabilities, Proceedings of IEEE 2nd, International Symposium on Spread Spectrum techniques and Applications (ISSSTA' 92), Yokohama, Japan, pp. 23-26, Dec. 1992.

[56] CWTS WG1 LAS-CDMA, (2001). Physical channels and mapping of transport channels onto physical channels, LAS TS 25.221, V1.0.0, July 17-17, 2001.

[57] Verdu, (1986). Minimum Probability of Error for Asynchronous Gaussian Mul-tiple Access, *IEEE Trans. on Information Theory*, Vol. IT-32, No. 1, pp. 85-96, Jan. 1986.

[58] Verdu, S., (1994). Adaptive Multiuser Detection, in Code Division Multiple Access Communications, S. G. Glisic and P. A. Leppanen, Eds., Boston: *Kluwer Academic Publishers*, pp. 97-116, 1994.

[59] Moshavi, S., (1996). Multiuser Detection for DS-CDMA Communications, *IEEE Communications Magazine*, pp. 12-36, Oct. 1996.

[60] Duel-Hallen, A., J. Holtzman, and Z. Zvonar, (1995). Multiuser Detection for CDMA Systems, *IEEE Personal Commun.*, pp. 4-8, April 1995.

[61] Hsiao-Hwa Chen & H. K. Sim, (2001). A new CDMA multiuser detector - orthogonal decision-feedback detector for asynchronous CDMA systems, *IEEE Trans. on Communications*, vol. 49, no. 9, pp.1649-1658, Sept. 2001.

[62] Hsiao-Hwa & Zhe-Qiang Liu, (2001). Zero-insertion adaptive minimum mean square error (MMSE) receiver for asynchronous CDMA multiuser detection, *IEEE Trans. on Vehicular Technology*, vol. 50, no. 2, pp. 557-569, March 2001.

[63] Hsiao-Hwa Chen, Yi-Ning Chang & Yu-Bing Wu, (2003). Single Code Cyclic Shift Detection - A Pilot Aided CDMA Multiuser De-

tector Without Using Explicit Knowledge of Signature Codes, *IE-ICE Trans. On Communications*, vol. E86-B, No. 4, pp. 1286-1296, April 2003.

[64] R. Lupas, S. Verdu, (1989). Linear Multi-user Detectors for Synchronous Code-Division Multiple-Access Channels, *IEEE Trans. On information Theory*, vol. 35, pp. 123-136, Jan 1989.

[65] R. Lupas, S. Verdu, (1990). Near-far Resistance of Multi-user Detectors in Asynchronous Channels, *IEEE Trans. on Communications*, vol. 38, pp. 496-508, April 1990.

[66] Hsiao-Hwa Chen, T. Lang & J. Oksman, (1996). Correlation statistics distribution convolution (CSDC) algorithm for studying CDMA indoor wireless systems with RAKE receiver, power control & multipath fading," *IEICE Trans. on Communications*, vol. E79-B, No. 10, Oct. 1996.

[67] CWTS TSM 01.00 V.2.0.0 (2001-07), China wireless Telecommunication Standard, 3G digital cellular telecommunications, TD-SCDMA System for Mobile (TSM), General Description, 2001.

[68] Hsiao-Hwa Chen, CX Fan & Willie W. Lu, (2002) China's Perspectives on 3G Mobile Communications & Beyond:TD-SCDMA Technology, *IEEE Wireless Communications*, pp. 48-59, April, 2002.

[69] Hsiao-Hwa Chen, J. F. Yeh & N. Seuhiro, (2001). A Multi-Carrier CDMA Architecture Based on Orthogonal Complementary Codes for New Generations of Wideband Wireless Communications, *IEEE Communications Magazine*, vol. 39, no. 10, pp. 126-135, October 2001.

Chapter 6

POWER CONTROL IMPLEMENTATION IN 3RD GENERATION CDMA SYSTEMS

Bassam Hashem

Research and Development Department
Saudi Telecom Company
Riyadh 11342
P.O.Box 261323
Saudi Arabia

Abstract   Power control is one of the main features in 3rd generation cellular CDMA systems. Power control reduces the interference in the system and hence increases the capacity. In this paper, we describe the basic operation of power control implementation in 3G CDMA systems. The inner loop and the outer loop algorithms are explained where the power commands are sent at a high rate and uncoded resulting in high error rate in receiving them. The problem of errors in the power control commands becomes worse during soft handoff where the terminal communicates with more than one base station simultaneously. Few algorithms are introduced during soft handoff to increase the system capacity like power control rate reduction and the adjustment loop . Adjusting the transmission rate is discussed also as another scheme to achieve the required quality.

Keywords: CDMA, Power control, Soft handoff, adaptive modulation, rate control, UMTS, CDMA2000

6.1   Introduction

CDMA systems are characterized as being interference limited. Reducing the interference results in a direct increase in the system capacity. The uplink of CDMA cellular systems suffers from the near-far problem. If all users send a fixed power level, then the signal received from a nearby user will be stronger than a signal received from a far user and may mask it. This is referred to as the near-far problem, Typically, the down link transmissions from the base station (BTS) to the mobiles employ orthogonal codes and hence the interference is reduced. Some of this orthogonality, however, is lost due the multipath fading. This is in addition to the interference from other base stations in the downlink. Power control hence is employed on both the down and the up inks.

Third generation systems are designed to serve terminals or UEs (users equipments) with different quality of service (QoS) requirements. Different QoS results in different block error rate (BLER) requirements which is governed by the energy per bit to noise ratio (*Eb/No*) where No of course is both the interference and the white noise. Hence the objective of the power control algorithm is to make sure that each terminal is achieving its required *Eb/No*. A lower *Eb/No* results in higher BLER while a higher *Eb/No* results in excessive interference which reduces the system capacity. It should be noted that it is not only the transmitted power that controls the achieved *Eb/No*. The transmission rate and the interference level   also affect the achieved *Eb/No*. The rest of this chapter is organized as follows. Sections 6.2 to 6.6 covers the inner loop algorithm while section 6.7 covers the outer loop. Then we discuss the operation of power control during soft handoff (SHO) where section 6.8 covers the uplink and section 6.9 covers the downlink. Section 6.12 covers the operation of power control algorithms during compressed mode. Finally we discuss how the rate can be changed to achieve the required block error rate (BLER) and explain how adaptive modulation is investigated for future systems as a mean to replace power control.

6.2   Inner Loop Power Control

The received power varies depending on the distance between the BTS and the mobile, the shadowing and the multipath fading. Variations due to distance and shadowing are location dependent and hence are reciprocal on both the downlink and the uplink. Thus, the mobile can determine these variations by measuring the average received power on the downlink. This is called open loop power control. Full duplex is usually achieved by providing different frequency channels on the downlink and the uplink. The frequency separation between these channels is much larger than the coherence bandwidth of the communication channel and thus both links tend to fade independently. To eliminate the variations due to multipath fading, the base station sends power control commands to the mobile asking the mobile to either increase or decrease its power by a fixed amount. This is called closed loop power control. Both the open loop and the closed loop constitute what is called the inner loop power control.

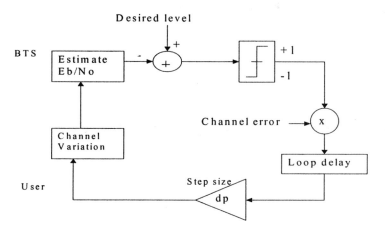

Figure 1 Inner loop power control

Figure 1 shows how the inner loop is implemented in 3G systems. This shows how the power control algorithm is implemented at the BTS to decide if the UE is to decrease or increase its power.

As was mentioned earlier, a similar algorithm is implemented at the UE to decide whether the BTS should increase or decrease its power. The base station will estimate the achieved *Eb/No* and compares it with the target *Eb/No*. The target *Eb/No* itself is controlled by the outer loop power control which will be explained later. If the achieved *Eb/No* is lower (higher) than the target *Eb/No*, the BTS sends an up (down) command asking the UE to increase (decrease) its power. This is done every slot or what is called sometimes power control group. The up command is sent as a 1 while the down command is sent as a 0 or vice versa (depending on the standard). The command it self may be received in error causing the UE to adjust its power in the wrong direction. Of course, the command will not reach the UE and be executed instantaneously but rather after some delay which is usually a one or two power control groups. After that, the UE adjusts the power by a certain step size. Standards support multiple step seizes. The received *Eb/No* can still not equal the target *Eb/No* due to variations in the channel conditions or the interference level. As we said earlier, in this paper, we assume *No* to include both the white noise and the interference.

6.3   Power control command rate

The variation in the communication channel is an essential factor in designing the power control algorithm. Such variation is a function of the

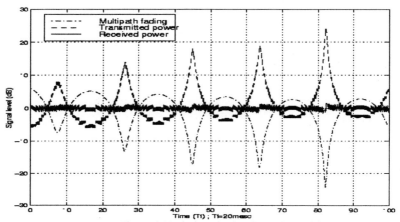
Figure 2 Variations in the channel condition

type of the communication channel (usually modeled as a Raleigh fading channel with multiple resolvable paths), the velocity of the UE and the carrier frequency. The rate at which the power commands is sent is determined by the rate at which the channel can change. Two common rates are 800 commands per second, which is adopted in the CDMA2000 standard, and 1500 commands per second, which is adopted in the UMTS (3GPP) standard. At low mobility, power control eliminates the variations in the received signal level and the received power can be made constant. This is shown in Figure 2 where 800 commands are sent to eliminate the variations in the channel condition for a UE moving at a speed of 5 Km/h in a two equal paths Raleigh fading channel. It is clear from the figure that the variations in the channel conditions can be eliminated at such a rate. This is usually referred to as perfect power control.

However, for fast users, power control will be slow to eliminate the variations in the received signal level completely and the residual error in the received signal power is known to follow a log-normal distribution with a standard deviation that is a function of the user velocity [3]. Many papers analyzing CDMA systems capacity do not include the impact of multipath fading on the system by assuming the power control algorithm to be perfect. This is, however, is only correct if the system has only one cell! A UE going into a deep fade transmits at high power level to compensate for the fade.

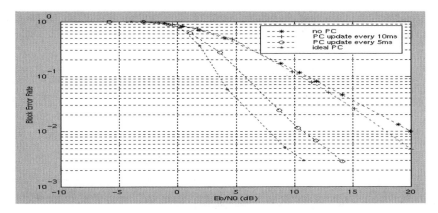

Figure 3: Performance comparison of raw BER vs. Eb/N0

This does not of course increases the intracell interference (interference from users communicating with the same BTS) and hence the impact of multipath fading in TDD can be neglected when estimating the intracell interference. However,transmitting at high power levels increases the intercell interference. Hence, variations in the channel conditions due to multipath fading may not be omitted in analyzing the multi cell system even if the power control is capable of tracking multipath fading. It is even shown that tracking deep fades reduces the system capacity due to the excessive intercell interference[3]. The ability to track deep fades is controlled by the dynamic range of the closed loop power control. A large dynamic range allows such tracking. A system that relies only on the closed loop (no open loop) power control to track variation in the channel due to distance, shadowing and multipath fading requires large dynamic range. The impact of such large dynamic range on increasing the intercell interference should be investigated carefully. Even though the open loop power control tracks changes due to distance and shadowing in an FDD system, the open loop suffers from a large estimation error and hence some designers may want to rely on the closed loop only [4].

The discussion on the difference capabilities of the open loop and the closed loop algorithm leads us to wonder if power control is needed in TDD mode. In TDD mode, both the transmitter and the receiver uses the same frequency band. Hence both downlink and uplink channels fade together. Closed loop power control still provides a gain in such a case but obviously the commands can be sent at a much slower rate than the FDD case. Figure 3 shows such a gain for a power control command rate of 100 and 200 commands per second. [5]

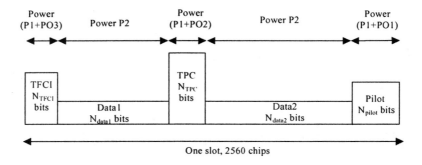

Figure 4: Down link slot format in the 3GPP standard (each frame is 15 slots)

6.4   Errors in the power control commands

Data bits are transmitted in frames. A frame is few msec long (10 msec or 20 msec). Data bits are coded and interleaved over one frame to give them more error protection. However, each frame has many slots (power control groups) where each of them has a unique transmit power control command (TPC command). These commands are sent uncoded to avoid the delay in the coding/decoding process, which increases their bit error rate (BER). To understand such issues more, Figure 4 shows the slot structure in the downlink of the 3GPP standard (UMTS). Few control fields are time multiplexed with the data bits. One of these fields is the transmit power control field (TPC) which can have N bits. The other filed that will have an impact in the power control algorithm is the TFCI field: transport format combination indicator. Data can be transmitted at different rates and the TFCI bits indicate at what rate the data is sent. The TFCI word is interleaved over the whole frame. One thing to notice here is the ability to send the down link control bits with extra power (offset from the data bits power) to reduce their error rate.

Two possible ways to decrease the TPC commands bit error rate (BER) are to increase the power of the TPC field or to increase the number of TPC bits to be transmitted in downlink. Figure 5 shows the BER probability for the TPC and the data bits (DPDCH) for different number of TPC bits and different power offsets. It is quiet clear from the figure how poor the TPC bits performance even when two bits are sent as a command (10%). The performance enhances when these commands are sent with extra powers. [6]

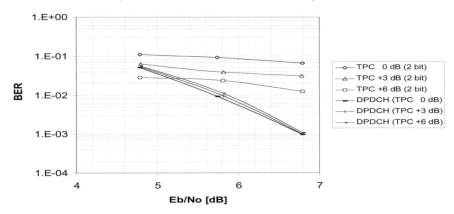

Figure 5: Downlink. TPC bit error rate as a function of the power offset in the TPC field

6.5 Power Control Step Size

One of the frequently asked questions is what is the optimum TPC step size? To answer such a question, we can look at Figure 6 which shows the required *Eb/No* for different step sizes. Of course, the lower the *Eb/No*, the better the step size. Conceptually, the small size should be small when the variations in the communications channel are small and vise versa. Looking at the figure, we see that the optimum step size is 0.5 dB when the channel is varying slowly (3 Km/h velocity). The optimum step size changes to 1 dB when the velocity changes to 10 Km/h. At a high velocity (100 Km/h), the step size is again 0.25 rather than being a big step size. This simply says that power control does not work at high speeds! As we have mentioned earlier, the power control does not adjust the power instantaneously but rather there is a delay in the loop. The assumed delay for the results in Figure 6 is one slot or one power control group. At high speeds, power control only creates more deviation in the received signal power. Hence, the larger the step size, the worse the situation becomes at high speeds. In other words, if the algorithm is not able to track the channel variations, it is better to turn it off.

Many papers are published investigating the gain from a variable step size and some suggest using also a delta modulation type of step size where a consecutive up (down) command results in larger step size. It might be hard to implement small step sizes in real life and hence these small step sizes can be emulated. When the step size is 0.25, for example, the UE can wait for two consecutive down steps before reducing its power by 0.5 dB. Thus even though the UE reduced its power by 0.5 dB over a two slots period, it is similar to reducing the power by 0.25 in each slot.

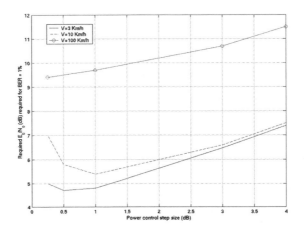

Figure 6: Received *Eb/No* performance of different step sizes at different velocities

6.6   Estimating *Eb/No*

Estimating the achieved *Eb/No* is one of the most tricky parts in the power control algorithm even though it looks like a straight down task The problem raises from the fact that 3G systems supports multiple rate services and to measure Eb, the transmission rate has to be known. As we have mentioned earlier, in the 3GPP standard, each frame can have a different transmission rate and the TFCI word indicates this rate. However, to get the TFCI word, the 15 slots have to be received. That is to say, the whole frame has to be received before we can know the transmission rate of the frame. During the frame, however, 15 TPC commands have to be issued. Hence, waiting till the end of the frame is not an option. Fortunately enough, the control bits are sent at a known rate and hence they can be used to estimate the *Eb/No* .The number of the control bits is much less than the number of the data bits and using them to estimate the achieved *Eb/No* increases the measurement error. Other schemes to be used include using the dedicated pilot channel. The pilot channel is, however, not power controlled which creates another type of problems.

6.7   Outer loop power control

The typical target for different services is the block error rate (BLER). Unfortunately, there is no one-to-one mapping between the BLER and the required *Eb/No*. The required *Eb/No* is a function of many factors that include channel type, mobile velocity, etc. Hence the target *Eb/No* is controlled by the outer loop. The idea behind the outer loop power control algorithm is very simple. If the receiver (UE in case of down link power control and BTS in case of uplink power control) is receiving error free frames, the target should be reduced and if it receives frames in error, the target should be increased. Assuming that the current achieved *Eb/No* equals the target *Eb/No*. Then decreasing the target will create an error in the next frame. That is to say, the system will oscillate between the error free and erroneous case. This does not achieve the target BLER. To achieve the target BLER the up step and the down step for adjusting the *Eb/No* should be governed by the following relation

$$Down\_Step = Up\_Step/(1/BLER - 1) \qquad (1)$$

That is to say, the *Eb/No* target is increased by a large step once a frame is found to be in error while the target is reduced slowly in case of error free transmission [8]. Of course, the above algorithm assumes that there are frames to be checked for erroneous! In case of DTX transmission (discontinuous transmission), the outer loop will not be able to do such checks and other means to estimate the link quality. On of such schemes is to use the bit error rate (BER) of the control bits. However, there is no one-to-one mapping between the BER of the control bits and the BLER of the data bits. Another issue here is if the UE should be allowed to control the target *Eb/No* for the down link power control? If the UE sets the target at the maximum allowed level, then the UE will be asking the BTS to send more and more power. This may make the link between the BTS and this particular UE error free but of course at the expense of excessive interference to the system.

6.8   Uplink Power control during soft handoff (SHO)

During SHO each BTS issues independently a TPC command to the UE. The UE can be receiving conflicting commands to adjust its power from the different BTSs. The "or of the downs" rule was used in the IS-95 standard to adjust the power in such a case. The rule is based on the idea of reducing the interference. If any base station is receiving extra power and asking the UE to decrease its power, the UE should decrease its power to reduce the interference. This is the optimum combining rule assuming that all the received TPC commands are error free. When the mobile is in soft handoff with N base stations, it will be receiving N signals, which are combined optimally. Hence, some of the down link connections (base station to UE) can be strong while others can be weak. The TPC commands are, however, not combined. This results in considerable increase in the TPC BER in the case of SHO. So, in the "or of the downs" combining method, all the base stations can be asking for an increase in power and due to an error in only one of the power commands bits, the mobile decreases its power. To overcome such a problem, the UE has to do a reliability check on the TPC commands before combining them and only combine reliable ones. This reliability check can be achieved by measuring the signal to interference ratio of the link. Another way would be to assign a weight to each TPC command before combing them or to have a soft decision on the TPC command rather than a hard one. To reduce the interference, if any reliable link asks for a decrease in the transmitted power, the power is reduced. Another way to reduce the variation in the UE transmitted power during SHO is to reduce the power control step size. Figure 7 shows a comparison between the different combining schemes. The target Eb/No in these simulations was 5 dB. Using the "or of the downs" rule we see that the achieved Eb/No is greater than 12 dB is 90%. This is a strange result since we said that the UE will tend to reduce its power when the 'or of the downs" rule is applied. To understand this we have to think of the outer loop function. In SHO , the BTSs start sending TPC commands to the UE which starts to reduce its power erroneously resulting in the frames being received in error. This triggers the outer loop that starts increasing the desired Eb/No target to high values resulting in the UE transmitting high power. That is to say, to avoid the impact of the erroneous behavior of the UE, the outer loop sets the Eb/No to high values. This is of course not helpful to the system capacity. We see from the figure that reducing the TPC step size to 0.25 dB (from 0.5 dB) still does not solve the problem. Ignoring the unreliable commands in the combining process clearly elevates the impact of the problem resulting in a performance close to the TPC error free case. [10]

6.9   Downlink power control during SHO

When the UE is in soft handoff (SHO), it combines the received signal from the different base stations that are in the active set and issues a single power command that all the base stations have to follow. The active set is the group of BTSs that are communicating with the UE simultaneously. The base stations in the active set demodulate the power command and adjust their powers independently. The mobile sends the power control command on the uplink. The uplinks between the mobile and the different base stations are independent. This results in different instantaneous power control error rates on the different reverse links. Thus even though the mobile sends a single power control command, the base stations in the active set can adjust their powers in opposite directions.

The deviation in the base stations transmitted powers results in a loss of the diversity gain that we would otherwise have from SHO. Also, due to the errors in the power control commands, one of the base stations can unnecessarily transmit at a high power level that increases the interference. It

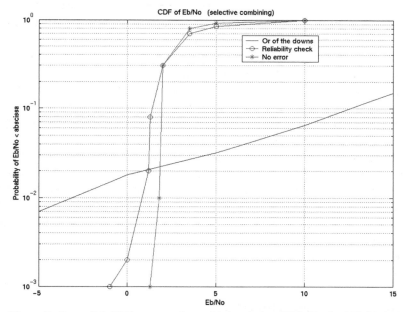

Figure 7: Downlink TPC commands combining during SHO (Desired *Eb/No* is 5dB)

could also erroneously transmit at a very low power and hence not contribute in the quality of the received signal.

To achieve the diversity gain, when in SHO, it is desirable to keep the base stations transmitted powers at comparable levels (not necessarily equal levels). One technique is to synchronize the base stations transmitted powers every few frames in order to make sure that their transmitted powers are as desired. This scheme however can be slow and requires signaling between the mobile and the base stations and a central controller. Another scheme for reducing the deviation in the base stations transmitted powers is to reduce the errors in the power control commands sent by the UE. By reducing the deviation, the requirement on synchronization frequency can be loosened.

The reduction in transmitted powers can be achieved by giving these commands extra power (power offset from the data bits). However, this scheme might not be a feasible option due to the power amplifier non-linearity problems at the UE. As an another option, the TPC commands error rate can be reduced by repeating the same power control command over more than one power control group.

6.9.1   Base Stations Transmitted Powers Synchronization

The base stations in the active set have to report their transmitted power levels to a central controller every M frames and the controller has to resynchronize their power levels. The central controller will send a new power level at which all the base stations in the active set have to transmit. This enforces all the base stations in the active set to transmit at equal power levels every M frames. There are different ways for the central controller to determine the new power level. The new power level can simply be the average of the reported base stations powers. A better algorithm will try to determine which reported power level is the closest to what the UE actually wants. This can be done by observing the quality of the uplink. The UE sends the TPC commands over the uplink and hence a better uplink results in fewer errors on the TPC commands. The fewer the errors on the TPC commands, the closer the transmitted power to what the UE requires. The reported uplink quality can be the BLER or the signal to interference ratio (SIR) or any other measurements that can indicate which link has lower TPC error rate.[11]

6.9.2   TPC rate reduction

The TPC rate reduction is applied to reduce the TPC error rate and hence the deviation in the BTSs transmitted powers is reduced. When in soft handover, the BTSs keep adjusting their power levels every slot. Assuming the frame to have 15 slots, this results in the BTSs adjusting their powers 15 times/frame. If the step size is 1dB and two BTSs adjust their powers in an opposite direction 4 times, this results in an 8 dB difference in their transmitted powers. This obviously will waste the diversity gain that we achieve by having the UE in soft handoff. When the UE is in soft handoff, it is receiving the signal from more than one BTS and probably via more than one fading path from each BTS. Thus, the high rate of 1500 commands may not be needed when in soft handoff. Reducing the power control rate during soft handoff results in the BTSs adjusting their transmitted powers less often which reduces the deviation in their transmitted powers.

Another advantage of reducing the power adjustment rate is giving more error protection to the TPC commands sent by the UE. Currently, the UE sends one power control command every slot. Another option would be that the UE sends one power control command every K slots. This means that the BTSs in soft handoff with the UE adjusts their transmitted powers only every K slots. The simplest way will be to repeat the same command over the K slots. However, this is obviously not the optimum way if the command is a multi step command in terms of error protection. Simple coding can be employed in such a case depending on the power control adjustment rate. This also provides some interleaving which will lower the power control command error rate. To achieve this the 3GPP standard supports two modes for uplink power control operation. The UE checks the downlink power control mode (DPC_MODE) before generating the TPC command:[2]

if DPC_MODE = 0 : the UE sends a unique TPC command in each slot and the TPC command generated is transmitted in the first available TPC field in the uplink DPCCH;

if DPC_MODE = 1 : the UE repeats the same TPC command over 3 slots and the new TPC command is transmitted such that there is a new command at the beginning of the frame.

Upon receiving the TPC commands, the BTS shall adjust its downlink power accordingly. For DPC_MODE = 0, BTS shall estimate the transmitted TPC command to be 0 or 1, and shall update the power every slot. If DPC_MODE = 1, BTS shall estimate the transmitted TPC command over three slots to be 0 or 1, and shall update the power every three slots.

After estimating the k:th TPC command, BTS shall adjust the current downlink power $P(k-1)$ [dB] to a new power $P(k)$ [dB] according to the following formula:

$$P(k) = P(k - 1) + P_{TPC}(k) + P_{bal}(k), \qquad (2)$$

where $P_{TPC}(k)$ is the kth power adjustment due to the inner loop power control, and $P_{bal}(k)$ [dB] is a correction according to the downlink power control procedure for balancing radio link powers towards a common reference power. The power balancing procedure is described later.

Simulations are shown in Figure 8 that show the impact of reducing the rate of the TPC commands from 1500 commands per second to 500 commands per second. The rate reduction is achieved by repeating the same TPC command over three slòts. [11] It can be seen that the rate reduction algorithm reduces the probability that the deviation in the transmitted powers exceeding 5 dB from 20% to 2%. This is obviously a big gain which can ensure that UEs in SHO are having the desired diversity gain. As was the case in the downlink, a small TPC step size is better for the uplink power control also when in SHO. The impact of the step size (delta) is shown in Figure 9 [14]

6.9.3   Adjustment loop

Adjustment loop is another scheme that is used to reduce the deviation in the BTSs transmitted powers during SHO. For adjustment loop, DL reference power P_{REF} and DL power convergence coefficient r ($0<r < 1$) are set in the active set BTSs during soft handover so that the two parameters are common to the BTSs. For simplicity, DL powers of two BTSs are considered in this explanation.

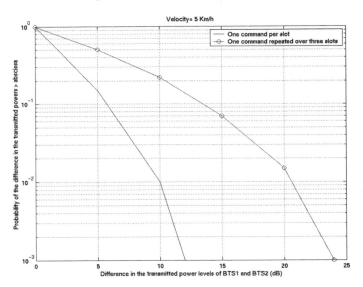

Figure 8: Deviation in the BTSs transmitted powers

Adjustment loop works in addition to inner loop power control, and DL power at slot i of two BTSs, P1(i), and P2(i), are updated at a certain interval (typically in every slot as in this explanation) as follows:

$$P1(i+1) = P1(i)+(1 - r)(P_{REF} - P1(i))+S_{INNERLOOP1}(i) \quad (3)$$
$$P2(i+1) = P2(i)+(1 - r)(P_{REF} - P2(i))+S_{INNERLOOP2}(i) \quad (4)$$

Where $S_{INNERLOOP1}$ and $S_{INNERLOOP2}$ are the inner loop power adjustments according to the received TPC commands. The difference in the BTSs transmitted powers is derived from equations (1) and (2) if TPC error does not occur i.e. SINNERLOOP1(i) and SINNERLOOP2(i) are equal.

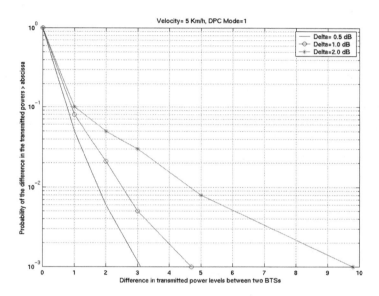

Figure 9: Impact of the step size (0.5 dB, 1dB and 2 dB) on the deviation between two BTSs transmitted powers

$$P1(i+1) - P2(i+1) = r(P1(i) - P2(i)) = r^i (P1(1) - P2(1)) \quad (5)$$

Equation (5) means that the difference converges at zero when r is smaller than one. The gain from this algorithm is limited by the fact that the BTS power can not be adjusted by tiny steps but still it is shown that such scheme can help in reducing the deviation in the BTSs transmission power [16].

6.10   Site Selection Diversity TPC (SSDT)

SSDT is an extreme case of downlink power control in a SHO region where the UE selects only one BTS to transmit at a time. SSDT operation is summarized as follows. The UE selects one of the BTSs from its active set to be 'primary', all other BTSs are classed as 'non primary'. The main objective is to transmit on the downlink from the primary BTS, thus reducing the interference caused by multiple transmissions in a soft handoff mode. A second objective is to achieve fast site selection without network intervention, thus maintaining the advantage of the soft handoff. In order to select a primary BTS, each BTS is assigned a temporary identification (ID) and UE periodically informs a primary BTS ID to the connecting BTSs. The non-primary BTSs selected by UE switch off the transmission power. The primary BTS ID is delivered by UE to the active BTSs via uplink feedback bits. SSDT activation, SSDT termination and ID assignment are all carried out by higher layer signaling. UE may also be commanded to use SSDT signaling in the uplink although BTSs would transmit the downlink without SSDT being active. In case SSDT is used in the uplink direction only, the processing in the UE for the radio links received in the downlink is as with macro diversity in non-SSDT case. Higher layers set the downlink operation mode for SSDT. [2]. Simulations in [17] show the gain from SSDT over conventional TPC. A sample of these results is shown in Table 1 to get a feeling of how large this gain could be. The gain from SSDT is achieved at low speeds where the UE is capable of exploiting the multipath fading diversity between multiple cells.

Table 1: Capacity [kb/s/MHz/Sector] of SSDT and conventional TPC for pedestrian UE in terms of number of RAKE fingers

	Infinite RAKE fingers	6 RAKE fingers	5 RAKE fingers
Conventional TPC	377	233	217
SSDT	479	361	341

6.11   Reduced active set

As we have seen in the previous section, site selection might be an alternative to the typical implementation of soft handoff in CDMA systems. One question here is if the data should be available at all the BTSs in the active set and be transmitted immediately when the UE selects the primary BTS or not. If the answer is yes, this will require more buffering at the BTSs. If the answer is no, the data should be sent from a central controller to the selected (primary) BTS. This of course causes more delay. This is one of the reasons why voice and data tolerating delay can be transmitted from a different group of cells. Voice clearly requires SHO more than data due to its delay intolerance. Data applications usually employ retransmission schemes

(ARQ) that allows receiving the data without errors even if the first transmission is erroneous. This lead to the concept of reduced active set. The concept simply says that there can be the regular active set where all the BTSs in the set transmits one service (voice for example) and a reduced number of these BTSs can transmit another service to the same UE. In the above case the communication channel will be different for the two services. This is because the two channels will be arriving through different paths (different BTSs). The power control algorithm has been modified in the CDMA2000 system to allow controlling the power of both channels independently. In IS200 the voice channel is called the fundamental channel while the data channels are referred to as supplemental channels. The power control bits are time multiplexed to allow controlling both fundamental and supplemental channels differently. In IS200, the TPC rate is 800 commands/sec. These commands can be split between the fundamental channel and the supplemental channel in three different ways: 800/0, 400/400 and 600/200.

6.12   Compressed mode

Third generation systems are expected to replace second generation systems in a gradual way. Operators may deploy 3G systems in urban areas first leaving the 2G systems covering the suburb areas. Hence, a UE should be able to make a hard handoff from 3G systems to 2G systems. To make such a handoff, the UE should be able to measure the signal strength of the 2G carriers. This is an example where inter-frequency measurements is required (a measurement on a carrier operating on a different frequency than the current carrier frequency).

There are two solutions for such an issue. First, the UE should have two receivers allowing it to do measurements on a separate carrier while receiving data on the current carrier. This is not a highly desirable solution due to the UE complexity increase. The other solution is to create a gap in the transmission allowing the UE to do the required measurements during this gap. This is referred to as the compressed mode or the slotted mode in the 3GPP standard. The interruption of the transmission creates a break in the operation of the power control algorithm. The inner loop requires receiving data to be able to estimate the achieved *Eb/No* and needs to issue TPC continuously to track the communication channel variations. Figure 10 shows how much a Raleigh fading channel can change during a gap of 8 slots. For a Doppler frequency of 40 Hz, the probability that the change being more than 5 dB is about 10%. Detailed procedures on how the power control algorithm should work after resuming the transmission (after the gap) can be found in [2]. The main principle in these procedures is to use a larger step size to adjust the *Eb/No* target faster after the gap to adjust the power in a faster way compared to the normal mode. Looking at these procedures, it is clear how important continuous transmission is for the operation of the power control algorithm. Power control algorithms and especially the closed loop were not designed for discontinuous transmission.

6.13 Power/Rate control

Even though the control of the rate in addition to the power is not a feature of early releases of 3G standards, it should be mentioned here since it is included in enhanced releases and it has large impact of the way we think of power control algorithms. Enhanced releases of 3G systems employ adaptive modulation/coding (ACM) for data transmission. As we have mentioned very early in this chapter, the objective of the TPC algorithm is to make sure that the received *Eb/No* is equal to the desired *Eb/No*, not less or more. Let us define *Eb/No*.

$$Eb/No= (S/R)/[(I+N)/W] \qquad (6)$$

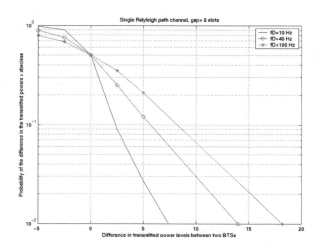

Figure 10: Complementary CDF of the change in the communication channel over 8 slots, the channel is a single path Rayleigh fading channel (fD is the Doppler frequency)

Where *S* is the signal power, *R* is the transmission rate, *I* is the interference level, *N* is the white noise and *W* is the system bandwidth. In typical power control algorithms, the power of the signal (*S*) is adjusted. This is obviously not the only parameter that can be adjusted to achieve the required *Eb/No*. The transmission rate (*R*) can be also adjusted where a lower transmission rate results in a higher *Eb/No*. This of course assumes that the service can tolerate the delay associated with reducing the transmission rate. Considering a UE in a deep fade. Typical power control algorithms increases the UE transmission powers which penalize all the other users by causing more interference. However, reducing the UE transmission rate penalize the affected UE by reducing its transmission rate.

Rather than changing the rate directly, ACM can be employed to change the required *Eb/No*. High speed data packet access channel (HSDPA) in 3GPP is an example where the channel is transmitted at a constant power. Served UEs measure the signal to interference ratio (SIR) of the downlink and report it back to the BTS. The BTS selects an ACM that allows the UE to receive the data at a desired BLER. The ACM levels include different modulation levels (QPSK, 16 QAM, 64 QAM, etc) and different coding rates (3/4, 1/2, 1/3, etc) This of course means that UEs closer to the BTS will have a higher data rate compared to users at the edge of the cell. To increase the system throughput, a scheduler is employed at the BTS that serves the users who have a good SIR and provides fairness between the different users by providing longer time for the disadvantaged users.

It might look like that the ACM concept is completely different from the power control algorithm. However, there are lots of similarities between the two schemes; after all both algorithms try to achieve the *Eb/No* that satisfies a certain BLER. Take as an example the outer loop. There is still no one to one mapping between the reported SIR and the required ACM level to achieve a certain BLER. An outer loop is still required for the ACM operation. Hence, many of the lessons learned from power control algorithms can be extended to ACM design.[21]-[23].

REFERENCES

1. Hashem B. and Yanikomeroglu H. ``Power Control for Third Generation Cellular Systems ''. The International Congress on Dynamics and Control of Systems, DYCONS99, Ottawa, Canada

2. 3GPP standard documents: TS 25.214 V4.1.0, "Physical layer procedures (FDD)"

3. Hashem B. and Sousa E., ``Reverse Link Capacity and Interference Statistics of Fixed Step Power Controlled DS/CDMA System Under Slow Multipath Fading", IEEE Transactions on Communications, December 1999

4. Jiangzhou Wang,"Open-loop power control error in cellular CDMA overlay system", IEEE Journal on Selected Areas in Communications, Volume: 19 Issue: 7 , July 2001 Page(s): 1246 –1254

5. 3GPP standard documents: TSGR1(00)0934, " The performance improvement from power control"\

6. 3GPP standard documents: TSGW1#3(99)171, "BER on power control bits"

7. 3GPP standard documents: TSGR1#6(99)821, "Optimum Power Control Step Size in Normal Mode"

8. Sampath, A.; Sarath Kumar, P.; Holtzman, J.M. On setting reverse link target SIR in a CDMA system Vehicular Technology Conference, 1997, IEEE 47th , Volume: 2 , 1997 Page(s): 929 -933 vol.2

9. 3GPP standard documents: TSGR1#3(99)213, "Potential downlink power control instabilities during soft handover"

10. Hashem B. and Secord N., ``Combining of Power Commands Received from Different Base Stations during Soft Handoff in Cellular CDMA Systems", International Symposium on Personal, Indoor and Mobile Radio Communications PIMRC '99 Osaka, Japan

11. Hashem B. and Le Strat E. ``On the Balancing of the Base Stations Transmitted Powers During Soft Handoff in Cellular CDMA Systems", ICCC2000, New Orleans, Louisiana, USA June 18-22, 2000

12. 3GPP standard documents: TSGR1#7 99b15, "Downlink Power Control Rate Reduction during Soft Handover"

13. Furukawa, H.; Harnage, K.; Ushirokawa, A." SSDT-site selection diversity transmission power control for CDMA forward link" IEEE Journal on Selected Areas in Communications, Volume:18 Issue: 8 , Aug. 2000 , Page(s): 1546 –1554

14. 3GPP standard documents: TSGR1#9(99)k02, "Downlink Power Control Step Size During Soft Handover"

15. 3GPP standard documents: TSGR1#7bis(99)E69, "Adjustment Loop in downlink power control during soft handover"

16. Hamabe, K , Adjustment loop transmit power control during soft handover in CDMA cellular systems, Vehicular Technology Conference, 2000. IEEE VTS Fall VTC 2000. 52nd , Volume: 4 , 2000 Page(s): 1519 -1523 vol.4

17. Takano, N.; Hamabe, K, Enhancement of site selection diversity transmit power control in CDMA cellular systems, Fall. IEEE VTS 54th Vehicular Technology Conference, 2001. VTC 2001, Volume: 2 , 2001, Page(s): 635 –6393

18. GPP standard documents: TSGR1#4(99)459, "Effect of slotted mode on the power control algorithm"

19. 3GPP standard documents: TR25.848 (Technical Report on Physical Layer Aspects of UTRA High Speed Downlink Packet Access Channel)

20. Lei Song; Mandayam, N.B.; Gajic, Z, Analysis of an up/down power control algorithm for the CDMA reverse link under fading , IEEE Journal on Selected Areas in Communications, Volume: 19 Issue: 2 , Feb 2001, Page(s):277 –286

21. Chi Wan Sung; Wing Shing Wong, Power control and rate management for wireless multimedia CDMA systems, IEEE Transactions on Communications, Volume: 49 Issue: 7 , July 2001 , Page(s): 1215 –1226

22. Mandyam, G.D.; Yi-Chyun Tseng, Packet scheduling in CDMA systems based on power control feedback Communications, IEEE International Conference on 2001. ICC 2001. Volume: 9 , 2001 Page(s): 2877 -2881 vol.9

23. Bender, P.; Black, P.; Grob, M.; Padovani, R.; Sindhushyana, N.; Viterbi, S. CDMA/HDR: a bandwidth efficient high speed wireless data service for nomadic users IEEE Communications Magazine , Volume: 38 Issue: 7 , July 2000, Page(s): 70 -77

Chapter 7

POWER CONTROL IN WIRELESS NETWORKS
Characteristics and Fundamentals

Fredrik Gunnarsson
Dept. of Electrical Engineering
Linköpings universitet
SE-581 83 Linköping, Sweden
fred@isy.liu.se

Abstract   The global communications systems critically rely on control algorithms of various kinds. In UMTS (universal mobile telephony system) – the third generation mobile telephony system just being launched, power control algorithms play an important role for efficient resource utilization. This chapter describes power control fundamentals including both theoretical and practical limitations. The relations to session management such as admission and congestion control is also addressed. Concepts and algorithms are illustrated by simple examples and simulations.

Keywords: power control, wireless networks, distributed control, stability, time delays, limitations, disturbance rejection, measurement errors, estimation errors.

Introduction

While the demand for access to services in wireless communications systems is exponentially growing, an increased interest in utilizing the available resources efficiently can be observed. A consequence of the limited availability of radio resources is that the users have to share these resources. Power control is seen as an important means to reduce mutual interference between the users, while compensating for time-varying propagation conditions. The objective with this chapter is to introduce power control algorithms and to point at some central and critical issues partly using a control theory framework.

Power control has been an area subject to extensive research in recent years. Some surveys include (Rosberg and Zander, 1998; Hanly and Tse, 1999; Gunnarsson and Gustafsson, 2002). A simplified radio link model is typically adopted to emphasize the network dynamics of power control. The transmitter is using the power $\bar{p}(t)$ to transmit data scrambled by a user-specific code. The channel is characterized by the power gain $\bar{g}(t)$ (< 1), where the "bar" notation indicates linear scale, while $g(t) = 10 \log_{10}(\bar{g}(t))$ is in logarithmic scale (dB). Correlating the received signal with the code, the receiver extracts the desired signal, which has the power $\bar{C}(t) = \bar{p}(t)\bar{g}(t)$, and is also subject to an interfering power $\bar{I}(t)$ from other connections. The perceived quality is related to the *signal-to-interference ratio (SIR)* $\bar{\gamma}(t) = \bar{p}(t)\bar{g}(t)/\bar{I}(t)$. For error-free transmission (and if the interference can be assumed Gaussian), the achievable data rate $R(t)$ is limited from above by (Shannon, 1956)

$$R(t) = W \log_2 (1 + \bar{\gamma}(t)) \ [\text{bits/s}], \tag{7.1}$$

where W is the bandwidth in Hertz. From a link perspective, the objective with power control can for example be

- to use constant power and variable coding and modulation to benefit from times of a favorable channel

- to maintain constant SIR and thereby constant data rate in liu of (7.1)

The former is information theoretically sound, and is further explored in e.g. (Goldsmith, 1997). In this chapter, the focus is on networks with relatively slow rate adaption mechanisms. Therefore the aim is the latter objective with power control to mitigate power gain and interference variations to maintain constant SIR and thereby constant rate. Figure 7.1 illustrates the two fundamentally different objectives.

Necessary system models and the notation are introduced in Section 1. Power control is also described, starting from an integrating controller, which is related to important proposals in the literature. Section 2 provides fundamental limitations of this controller on link level. Further network limitations are addressed in the subsequent section together with power control convergence results. Since power control is instrumental to other related resource management algorithms, a smorgasboard of different aspects of power control is brought up in Section 4.

1.   System Model

1.1   Power Gain

By neglecting data symbol level effects, the communication channel can be seen as a time varying power gain made up of three components $g(t) = g_p(t) + g_s(t) + g_m(t)$ as illustrated by Figure 7.2. The signal power decreases with distance $\bar{d}$ to the transmitter, and the path loss is modeled as $g_p = K_p - a_p \log_{10}(\bar{d})$ (Okamura et al., 1968). Terrain variations cause diffraction phenomenons and this shadow fading g_s is modeled as $ARMA(n_a, n_b)$-filtered Gaussian white noise (n_a is typically 1-2, $n_b = n_a - 1$, (Sørensen,

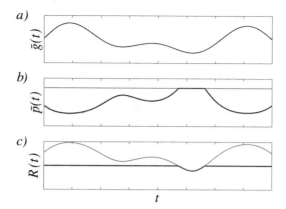

Figure 7.1. Illustration of power control objectives when subject to a varying power gain (a). b) Using a varying power (thick line) to mitigate channel variations or using maximum power (thin line) result in c) a constant data rate as long as less than maximum power is required (thick line) and a varying data rate depending on the channel (thin line) respectively.

1998)). The multipath model considers scattering of radio waves, yielding a rapidly varying gain g_m (Sklar, 1997).

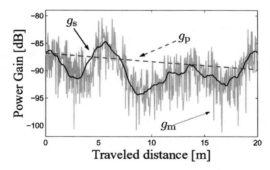

Figure 7.2. The power gain $g(t)$ is modeled as the sum of three components: path loss $g_p(t)$, shadow fading $g_s(t)$ and multipath fading $g_m(t)$. Here this is illustrated when moving from a reference point and away from the transmitter.

A *3GPP Typical Urban* (3GPP, 2003a) fast fading channel model will be utilized throughout the chapter. It models a multipath channel and a receiver unable to utilize all paths. Power gain values over 20 m are depicted in Fig. 7.3a. The frequency content of the power gain can be described independent of mobile velocity by expressing it with respect to the spatial frequency ξ [m^{-1}]. As seen in Fig. 7.3b, most of the frequency content is concentrated below 60 m^{-1}. For example the velocity v m/s means that the disturbance energy is concentrated below $60v$ Hz. Note that this is much higher than the Doppler frequency of each individual path (Jakes, 1974). This is due to the combination of several paths in the receiver.

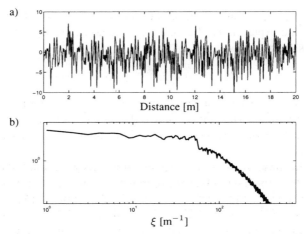

Figure 7.3. Realization of 3GPP Typical Urban power gain in dB a) with respect to traveled distance and b) in the frequency domain w.r.t. spatial frequency.

1.2   Wireless Networks

Consider a general network with m transmitters using the powers $\bar{p}_i(t)$ and m connected receivers. For generality, the base stations are seen as multiple transmitters (downlink) and multiple receivers (uplink). The signal between transmitter j and receiver i is attenuated by the power gain $\bar{g}_{ij}$. Thus receiver i connected to transmitter i will experience a desired signal power $\bar{C}_i(t) = \bar{p}_i(t)\bar{g}_{ii}(t)$ and an interference from other connections plus noise $\bar{I}_i(t)$. The *signal-to-interference ratio (SIR)*[1] at receiver i can be defined by

$$\bar{\gamma}_i(t) = \frac{\bar{C}_i(t)}{\bar{I}_i(t)} = \frac{\bar{g}_{ii}(t)\bar{p}_i(t)}{\sum_{j \neq i} \bar{g}_{ij}(t)\bar{p}_j(t) + \bar{\nu}_i(t)}, \qquad (7.2)$$

where $\bar{\nu}_i(t)$ is thermal noise at receiver i.

Depending on the receiver design, propagation conditions and the distance to the transmitter, the receiver is differently successful in utilizing the available desired signal power $\bar{p}_i\bar{g}_{ii}$. Assume that receiver i can utilize the fraction $\bar{\delta}_i(t)$ of the desired signal power. Then the remainder $\left(1 - \bar{\delta}_i(t)\right)\bar{p}_i\bar{g}_{ii}$ acts as interference, denoted *auto-interference* (Godlewski and Nuaymi, 1999). We will assume that the receiver efficiency changes slowly, and therefore can be considered constant. Hence, the SIR expression in Equation (7.2) transforms to

$$\bar{\gamma}_i(t) = \frac{\bar{\delta}_i \bar{g}_{ii}(t) \bar{p}_i(t)}{\sum_{j \neq i} \bar{g}_{ij}(t) \bar{p}_j(t) + \left(1 - \bar{\delta}_i\right) \bar{p}_i(t) \bar{g}_{ii}(t) + \bar{\nu}_i(t)}. \tag{7.3}$$

From now on, this quantity will be referred to as SIR. For efficient receivers, $\bar{\delta}_i = 1$, and the expressions (7.2) and (7.3) are equal. In logarithmic scale, the SIR expression becomes

$$\gamma_i(t) = p_i(t) + \delta_i + g_{ii}(t) - I_i(t). \tag{7.4}$$

It is also customary to use an alternative notation, where index i identifies the mobile, index j a base station, and j_i the base station mobile i is connected to. In addition, the distinction between uplink and downlink can be made more clear.

In the downlink, many of the interfering signals go through the same channels. It can therefore be motivated to denote the power gain from base station j to mobile i by $\bar{g}_{ij}^{\mathrm{DL}}(t)$.. Let $\bar{P}_j(t), j = 1, \ldots, B$ denote the total powers of the B base stations. Furthermore, the interfering signal powers from the same base station are attenuated by $\bar{\alpha}_i$ due to the effect of the channelization codes, which are orthogonal on the transmission side. Then, the alternative downlink SIR expression is given by

$$\bar{\gamma}_i^{\mathrm{DL}}(t) = \frac{\bar{\delta}_i \bar{g}_{ij_i}^{\mathrm{DL}}(t) \bar{p}_i(t)}{\bar{\alpha}_i (\bar{P}_{j_i}(t) - \bar{\delta}_i \bar{p}_i(t)) \bar{g}_{ij_i}^{\mathrm{DL}}(t) + \sum_{j \neq j_i} \bar{P}_j(t) \bar{g}_{ij}^{\mathrm{DL}}(t) + \bar{\nu}_i(t)}. \tag{7.5}$$

This expression can for example be used together with a downlink load assumption in terms of base station powers $\bar{P}_j(t) = \hat{\bar{P}}_j$ to analyze the downlink situation, perhaps conducted in a cell planning phase.

The uplink signals go through independent channels. However, it can still be instructive to use an alternative notation, where the power gain from mobile i to base station j is denoted $\bar{g}_{ij}^{\mathrm{UL}}$. If the uplink and downlink channel were reciprocal, then $\bar{g}_{ij}^{\mathrm{UL}} = \bar{g}_{ij}^{\mathrm{DL}}$. This uplink power gain notation requires an alternative uplink SIR definition

$$\bar{\gamma}_i^{\mathrm{UL}}(t) = \frac{\bar{\delta}_i \bar{g}_{ij_i}^{\mathrm{UL}}(t) \bar{p}_i(t)}{\sum_{k \neq i} \bar{g}_{kj_i}^{\mathrm{UL}}(t) \bar{p}_k(t) + \left(1 - \bar{\delta}_i\right) \bar{p}_i(t) \bar{g}_{ij_i}^{\mathrm{UL}}(t) + \bar{\nu}_{j_i}(t)}. \tag{7.6}$$

Note that the power gain matrix $\bar{G} = [\bar{g}_{ij}]$ is square $M \times M$, unlike the matrices $\bar{G}_{\mathrm{DL}} = [\bar{g}_{ij}^{\mathrm{DL}}]$ and $\bar{G}_{\mathrm{UL}} = [\bar{g}_{ij}^{\mathrm{UL}}]$, which are $M \times B$. The generic SIR expression (7.3) can be used through

$$\bar{g}_{ik} = \bar{g}_{ij_k}^{\mathrm{DL}}, \quad \bar{g}_{ik} = \bar{g}_{kj_i}^{\mathrm{DL}}$$

respectively.

1.3 Fundamental Power Control Algorithm

We adopt the log-linear power control model in (Dietrich et al., 1996; Blom et al., 1998) and consider all quantities in logarithmic scale (dB). With the power control objective to maintain constant SIR irrespective of time-varying power gain and interference power, it is natural to estimate SIR $\gamma_i(t)$ in the receiver and compare to the SIR reference $\gamma_i^t(t)$. The following integrating controller adjusts the power as long as the control error $e_i(t) = \gamma_i^t(t) - \gamma_i(t)$ is non-zero

$$p_i(t+1) = p_i(t) + \beta \left(\gamma_i^t(t) - \gamma_i(t) \right) = p_i(t) + \beta e_i(t), \qquad (7.7)$$

where the time index t refers to instants of power level updates in the transmitters. The linear control algorithm together with the SIR expression in (7.4) form a log-linear local control loop with the power gain, interference power and receiver inaccuracy as an additive disturbance. The local dynamical behavior can be conveniently described using a control theory framework. Introduce the *time-shift operator* [2] q as

$$q^{-n}p(t) = p(t-n), \ q^n p(t) = p(t+n) \qquad (7.8)$$

The integrating control algorithm in (7.7) can thereby be rewritten as

$$p_i(t) = \frac{\beta}{q-1} \hat{e}_i(t) = R(q)\hat{e}_i(t) \qquad (7.9)$$

Hence, the power control local loop is described by the block diagram in Figure 7.4.

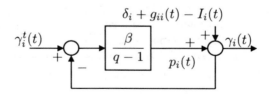

Figure 7.4. The local loop dynamics when employing the integrating controller in (7.9).

1.4 Relation to the Literature

The foundations for the distributed power control algorithms were laid in (Meyerhoff, 1974). The motivation in this article was not to obtain a distributed algorithm, but rather to make the centralized power control algorithms in Section 3 more computationally efficient. The proposed algorithms were based on results for iterative computations of eigenvectors.

These results are interpreted in (Grandhi et al., 1994; Foschini and Miljanic, 1993) as distributed power control utilizing only local information, and provided convergence results for the linear scale algorithm Distributed Power Control (DPC)

$$\bar{p}_i(t+1) = \bar{p}_i(t)\frac{\bar{\gamma}^t}{\bar{\gamma}_i(t)}. \tag{7.10}$$

The algorithms proposed in (Yates, 1995; Lee and Lin, 1996) aims at more careful control action compared to the DPC algorithm in (7.10). They can be rewritten as

$$\bar{p}_i(t+1) = \bar{p}_i(t)\left(\frac{\bar{\gamma}^t}{\bar{\gamma}_i(t)}\right)^\beta. \tag{7.11}$$

In logarithmic scale, this is (apart from individual target SIR's, $\bar{\gamma}_i^t(t)$) analogous to the integrating controller in (7.7), and DPC represents the specific case $\beta = 1$.

2. Fundamental Local Limitations

The power control loop in Figure 7.4 is essentially a block diagram describing the integrating controller together with theoretical and simplistic models. In practice, performance is fundamentally limited by several things as illustrated by Figure 7.5 (Gunnarsson, 2001).

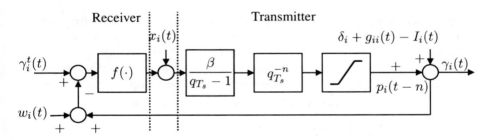

Figure 7.5. The local loop dynamics considering fundamental limitations and a realistic separation of the control loop into one receiver part and one transmitter part.

The limitations include that

- The **update rate** has a direct relation to the controller's ability of mitigating time-varying disturbances. It is indicated by the sample interval T_s together with the time shift operator q_{T_s}.

- The time-varying **measurement errors** $w_i(t)$ and **disturbances** $g_{ii}(t)$, $I_i(t)$ and δ_i make it difficult to equalize SIR.

- The **feedback bandwidth** limits the control information that is fed back from the receiver. It can be seen as quantization of the power control error $f(e_i(t))$.

- The **time delays** $q_{T_s}^{-n}$ effect the **local stability** of the control algorithm.

- The **transmitter power** is limited both from above and below.

- The feedback channel is subject to **feedback errors** $x_i(t)$, which further reduces the information about the current receiver situation.

If neglecting non-linear componentsdue to quantization and saturation, the remaining linear loop can be analyzed using linear systems theory. Clearly, the
power control objective is to maintain $\gamma_i(t) = \gamma_i^t(t)$ or equivalently $e_i(t) = 0$
despite limitations. From the block diagram in Fig. 7.5, we obtain

$$\gamma_i(t) = G(q)\gamma_i^t(t) + S(q)(g_{ii}(t) - I_i(t)) + G(q)w_i(t) \qquad (7.12a)$$

$$e_i(t) = S(q)\left(\gamma_i^t(t) - g_{ii}(t) + I_i(t)\right) + G(q)w_i(t), \qquad (7.12b)$$

where the dynamics is described by

$$G(q) = \frac{\beta}{q^n(q-1)+\beta}, \; S(q) = \frac{q^n(q-1)}{q^n(q-1)+\beta} \qquad (7.13)$$

$G(q)$ is referred to as the *closed-loop system*, and $S(q)$ as the *sensitivity function*. Note that $S(q) = 1 - G(q)$. The corresponding relations in the frequency
domain are obtained by replacing q by $e^{i\omega T_s}$. The closed-loop system describes
the tracking capability of the control algorithm, while the sensitivity function
relates to the disturbance suppression performance. The effects of measurement
(sensor) errors, however, is also captured by the closed-loop system. Furthermore, local loop stability is related to properties of $G(q)$.

2.1   Limited Update Rate

The actual time between consecutive power updates, the *sample interval*
T_s, varies from systems to system. For example $T_s = 0.48$ s in GSM and
$T_s = 1/1500$ s in WCDMA. To avoid confusion, we let the time index t
represent instants of power level updates in the transmitters. Seemingly, this
notation is equal to the assumption of synchronous updates, but the only needed
assumption is that all transmitters update their power levels within the time
frame of one sample interval.

Consider the ideal integrating controller in (7.7) with $\beta = 1$, update rate of
1500 Hz as in WCDMA and two mobiles with velocities 2.5 m/s and 12 m/s.
Performance is evaluated in terms of control error $\gamma_i^t(t) - \gamma_i(t)$ over time.
The power gain has a bandwidth of $60v$ Hz as described in Section 1.1, which

yields that the disturbance energy for the two mobiles is concentrated to up to 150 Hz and 720 Hz respectively. Both these frequencies are below the Nyquist frequency (750 Hz) and can thus be represented without alias. However, as seen in Figures 7.6a,b only the disturbance of the first mobile is compensated for. The answer lies in the sensitivity function in Figure 7.6c, where we note that only frequencies up to $\approx$ 200 Hz are suppressed.

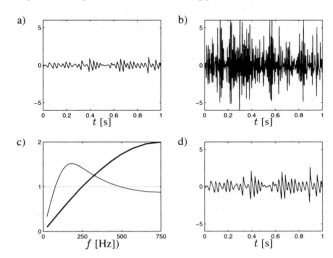

Figure 7.6. Performance of the integrating controller for $\beta = 1$ and no delay (a-c), and $\beta = 0.34$ and one slot delay (c-d). a) $e_i(t)$, slow mobile (v=2 m/s), b) $e_i(t)$, fast mobile (v=12 m/s), c) $|S(e^{i\omega T_s})|$ of the ideal controller (thick), with delay and $\beta = 0.34$ (thin) d) $e_i(t)$, slow mobile.

2.2 Measurement Errors and Disturbances

When no measurement noise is present, an intuitive design objective that stems from (7.12) is $S(q) = 0$ and $G(q) = 1$. This would result in perfect disturbance rejection and perfect tracking. However, even if this would have been possible (it is not according to the Bode integral[3]), it would not be interesting anyway when subject to noise. According to (7.12) this is equivalent to being maximally sensitive to measurement noise. Therefore, it is vital to consider measurement noise in the design and apply measurement filtering if necessary.

Measuring is not an instantaneous procedure, even though the measurements often are considered as samples of a continuous process. This is a relevant approximation in most power control cases. However, some related issues are brought up here.

In WCDMA, measurements are obtained from the fraction δ_s of the slot, which in turn corresponds to $T_s = 1/1500$ s. Typical values (Adachi et al., 1998) of δ_s include $\delta_s = 0.1$ (considering only the four pilot symbols out of

40 symbols) and $\delta_s = 0.25$ (considering the ten first symbols). These values depend on the data rate and channel configuration assigned to the user. A comparison of the filtering effects when considering a full slot average compared to a fractional slot average is found in Fig. 7.7. Aliasing is avoided if the frequency components over the Nyquist frequency are filtered out. This is almost the case when using the local average of the full slot (or measurement period). Conversely, aliasing effects are most likely when adopting local average over fractional slots.

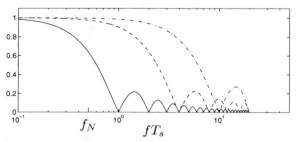

Figure 7.7. Filtering effect with respect to normalized frequency of a local average filter applied to the full slot (solid) and fractional slots $\delta_s = 0.25$ (dashed) and $\delta_s = 0.1$ (dash-dotted). The normalization is with respect to the frequency after down-sampling, which in this case is 40 times smaller.

Approaches for SIR estimation in different systems are discussed in (Ramakrishna et al., 1997; Andersin et al., 1998; Blom et al., 1999; Freris et al., 2001; Kurniawan et al., 2001). It is important to note that especially at low SIR levels, there is a significant bias in the SIR estimation with practically tractable schemes.

2.3   Limited Feedback Bandwidth

To implement a power control algorithm such as the integrating controller in (7.7), information about the error $e_i(t)$ has to be fed back. The feedback signaling bandwidth is limited in real systems. Typically, the communication is restricted to a fixed number of bits per second k. The evident trade-off is between error representation accuracy and feedback command rate. A single bit error representation allows k feedback commands per second, while m_e bits error representation allows k/m_e commands per second. This comparison is further explored in (Gunnarsson, 2001).

Different control error representations are proposed, for example: single bit (the sign of the error) (Salmasi and Gilhousen, 1991), k-bit linear quantizer (Sim et al., 1998) and k-bit logarithmic quantizer (Li et al., 2001). The single bit quantizer optimizes the update rate when the feedback bandwidth is fixed.

$$p_i(t+1) = p_i(t) + \beta\text{sign}\left(\gamma_i^t(t) - \gamma_i(t)\right) = p_i(t) + \beta\text{sign}(e_i(t)). \quad (7.14)$$

This is the standardized feedback communication for WCDMA (3GPP, 2003b).

The coded control error is transmitted over the feedback channel, and is therefore subject to possible bit errors. In WCDMA reference cases (Holma and Toskala, 2000) 4% of the single bit commands are assumed erroneous. In one alternative feedback communication scheme for WCDMA (3GPP, 2003b), commands are repeated three consecutive slots. This lowers the update rate to 500 Hz, but improves the command bit error probability from P_b to $P_b^2(3 - 2P_b) << P_b$ (for example from 4 % to 0.47 %).

2.4   Time Delays and Local Stability

Time delays affect stability as with any feedback controlled system, and therefore more careful control actions have to be imposed. However, time delays do not only affect stability. The sensitivity to disturbances is also degraded.

The dynamical behavior of a discrete-time linear system is closely related to the locations of the *closed-loop poles*. These are defined as the roots to the equation of the closed-loop system denominator equal to zero. In case of the integrating controller with closed loop system $G(q)$ as in (7.13), the closed-loop poles are the solutions to

$$q^n(q-1) + \beta = 0$$

Textbooks on linear systems theory, e.g. (Åström and Wittenmark, 1997), provide the following result for asymptotic stability (i.e. transients decay asymptotically).

Theorem 1 (Asymptotic Stability of Discrete-time Linear Systems) *A time-discrete linear feedback system is uniformly asymptotically stable, if and only if the closed-loop poles are strictly within the unit disc.*

With the integrating controller and in the typical time delay situation $n = 1$, it is easy to verify that this is equivalent to $\beta < 1$. Hence, DPC ($\beta = 1$) is not asymptotically stable when subject to time delays.

Furthermore, the negative effects of the time delay can be reduced by carefully selecting β. It can e.g. be optimized to provide fast reactions ($\beta = 0.34$ for $n = 1$ according to (Gunnarsson et al., 1999)). The time delay also effects the disturbance rejection capabilities of the controller. Figures 7.6c,d illustrate the increased sensitivity to disturbances, and the control error variance for the slow mobile traveling at 2 m/s.

With single-bit error representation as in (7.14) the local loop never converges to fully eliminate the control error. The sign function together with the linear components combine to result in a triangular wave oscillation of the control error with the period $4n+2$, where n is the time delay (Gunnarsson et al., 2001). This is clearly seen in Figure 7.8c, where we for example identify five oscillation periods between iteration 90 and iteration 120.

Time delays can be handled with careful linear design as discussed above. An alternative is to compensate for the delay in the measurement

$$\gamma_i(t) = p_i(t - n) + \delta_i + g_{ii}(t) - I_i(t). \tag{7.15}$$

If $p_i(t)$ is known in the receiver, the following compensation is plausible

$$\tilde{\gamma}_i(t) = \gamma_i(t) + p_i(t) - p_i(t - n) \tag{7.16}$$

which is referred to as Time Delay Compensation (TDC) (Gunnarsson and Gustafsson, 2001). This internal feedback alters the closed-loop system and is known in control theory as the *Smith predictor* (Åström and Wittenmark, 1997). With TDC, the sensitivity of the all-linear system becomes

$$G(q) = \frac{R(q)}{q^{n_p+n_m}(1 + R(q))}, \quad S(q) = \frac{q^{n_p+n_m}}{q^{n_p+n_m}(1 + R(q))} \tag{7.17}$$

TDC is particularly interesting when linear design is not an option, perhaps due to standardization as in the WCDMA case. Using (7.14), TDC can be rewritten as

$$\tilde{\gamma}_i(t) = \gamma_i(t) + \beta \sum_{k=t-n}^{t-1} \text{sign}(\gamma_i^t(t) - \gamma_i(t)) \tag{7.18}$$

i.e. adjusting the measurement with issued power control commands that not yet are reflected in the measurement. The benefit of TDC with single-bit error representation is illustrated in Figure 7.8, which also indicates the convergence dependency on β, and the fact that TDC and careful linear design provide roughly the same performance in the linear case.

2.5   Transmitter Power Limitations

One obvious limitation is availability of transmission power. These are partly due to amplifier and battery limitations, but the system may also further restrict the maximum power to limit the mutual interference between connections. It is also important to operate not too close to the limits on average, since the margin is needed to mitigate the fast fading. Furthermore, there is a direct relation between maximum power and coverage of the channel.

2.6   Power Control Error

Some analyses of the CDMA capacity are based on assumptions of the power control error statistics. For example, (Romero-Jerez et al., 2000) discusses ca-

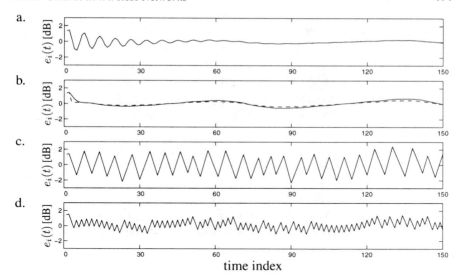

Figure 7.8. Local performance of the algorithms (7.7) in a-b and (7.14) in c-d subject to the typical delay situation $n = 1$. a. $\beta = 0.9$ b. $\beta = 1$ and TDC (dashed) and $\beta = 0.34$ (solid) c. $\beta = 1$ d. $\beta = 1$ and TDC.

pacity using Gaussian power control errors. The fundamental limitations described above in this section effects the error statistics, and simulations here for the integrating controller with single-bit control error representation (7.14) indicates that a Gaussian error model is relevant in most cases.

This algorithm has a superimposed stable oscillation due to time delays and the sign function. As seen in Figure 7.9, the power control error is not Gaussian. However, with both measurement errors and feedback errors as in realistic cases, a Gaussian power control error approximations is clearly plausible.

3. Global Stability

For practical reasons, power control algorithms in cellular radio systems are implemented in a distributed fashion. However, the local loops are interconnected via the interference between the loops, which affects the global dynamics as well as the capacity of the system. An important global issue is whether it is possible to accommodate all users with their service requirements. The power gains reflect the situation from the transmitters to the receivers, and the results are therefore applicable to both the uplink and the downlink.

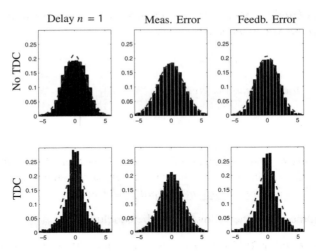

Figure 7.9. Power control error distribution without (top) and with (bottom) time delay compensation when subject to (left) one slot time delay, (middle) 3dB measurement error and (right) 4 % feedback error rate respectively.

3.1 Feasibility

The individual reference SIR:s and the power gains are considered constant in the global level analysis, where the latter is motivated by an assumption that the inner loops perfectly meet the provided SIR reference, and thereby mitigate the fast channel variations:

$$\frac{\bar{\delta}_i \bar{g}_{ii} \bar{p}_i}{\sum_{j \neq i} \bar{g}_{ij} \bar{p}_j + \left(1 - \bar{\delta}_i\right) \bar{p}_i \bar{g}_{ii} + \bar{\nu}_i} = \bar{\gamma}_i^t, \ \forall i \qquad (7.19)$$

Note that values in linear scale are used in this section. The aim is to characterize the system load, and a few definitions are needed. Introduce the matrices

$$\bar{\boldsymbol{\Gamma}}_t \stackrel{\triangle}{=} \operatorname{diag}(\bar{\gamma}_1^t, \ldots, \bar{\gamma}_m^t), \ \bar{\boldsymbol{Z}} = [\bar{z}_{ij}] \stackrel{\triangle}{=} \left[\frac{\bar{g}_{ij}}{\bar{g}_{ii}}\right], \ \bar{\boldsymbol{\Delta}} \stackrel{\triangle}{=} \operatorname{diag}\left(\bar{\delta}_1, \ldots, \bar{\delta}_m\right)$$

and vectors

$$\bar{\boldsymbol{p}} \stackrel{\triangle}{=} [\bar{p}_i], \ \bar{\boldsymbol{\eta}} = [\bar{\eta}_i] \stackrel{\triangle}{=} \left[\frac{\bar{\nu}_i}{\bar{g}_{ii}}\right].$$

The network itself puts restrictions on the achievable SIR's, and there exists an upper limit on the balanced SIR (same SIR to every connection). This is disclosed in the following theorem, neglecting auto-interference and assuming that the noise can be considered zero.

Theorem 2 (Zander, 1992) *With probability one, there exists a unique maximum achievable SIR in the noiseless case*

$$\bar{\gamma}^* = \max\{\bar{\gamma}_0 \mid \exists \bar{p} \geq 0 : \bar{\gamma}_i \geq \bar{\gamma}_0, \ \forall i\}.$$

Furthermore, the maximum is given by

$$\bar{\gamma}^* = \frac{1}{\bar{\lambda}^* - 1},$$

where $\bar{\lambda}^*$ is the largest real eigenvalue of $\bar{Z}$. Note that $\bar{\lambda}^* > 1$ implies that $\bar{\gamma}^* > 0$. Moreover, the optimal power vector $\bar{p}^*$ is the eigenvector of $\bar{\lambda}^*$ (i.e. $k\bar{p}^*$ for any $k \in \mathbb{R}^+$ constitutes an optimal power vector.).

Considering the noise, the following can be concluded:

Theorem 3 (Zander, 1993) *In the noisy case and with no power limitations, there exist power levels that meet the balanced SIR target $\bar{\gamma}_0^t$ if and only if $\bar{\gamma}_0^t < \bar{\gamma}^*$.*

Consider the following uplink example from Chapter 14[4]

$$\bar{G}_{\text{UL}} = \begin{pmatrix} 0.8 & 0.1 \\ 0.5 & 0.3 \\ 0.2 & 0.6 \end{pmatrix}, \ \bar{G} = \begin{pmatrix} 0.8 & 0.5 & 0.2 \\ 0.8 & 0.5 & 0.2 \\ 0.1 & 0.3 & 0.6 \end{pmatrix}, \ \bar{Z} = \begin{pmatrix} 1 & 0.63 & 0.25 \\ 1.6 & 1 & 0.4 \\ 0.17 & 0.5 & 1 \end{pmatrix}.$$

where the relation between $\bar{G}_{\text{UL}}$ and $\bar{G}$ is described in Section 1.2. The optimal balanced SIR from Theorem 2 is thus given by $\bar{\gamma}^* = 0.83$. Hence, only balanced SIR targets less than this imply a feasible power control problem, and it is possible o find power levels that meet the requirements of the users. This is generalized to multiple services below.

The effects of auto-interference are considered in (Godlewski and Nuaymi, 1999; Gunnarsson, 2000), and the requirements in Equation (7.19) can be vectorized to

$$\bar{p} = \bar{\Gamma}_t \left((\bar{\Delta}^{-1}\bar{Z} - E)\bar{p} + \bar{\Delta}^{-1}\bar{\eta} \right), \tag{7.20}$$

where E is the identity matrix. Solvability of the equation above is related to *feasibility* of the related *power control problem*, defined as:

Definition 1 (Feasibility) *A set of target SIR:s $\bar{\Gamma}_t$ is said to be feasible with respect to a network described by $\bar{Z}$, $\bar{\Delta}$ and $\bar{\eta}$, if it is possible to assign transmitter powers $\bar{p}$ so that the requirements in Equation (7.20) are met. Analogously, the power control problem $(\bar{Z}, \bar{\eta}, \bar{\Delta}, \bar{\Gamma}_t)$ is said to be feasible under the same condition. Otherwise, the target SIR:s and the power control problem are said to be infeasible.*

The degree of feasibility is described by the *feasibility relative load*, which is defined below.

Definition 2 (Feasibility Relative Load) *Given a power control problem* $(\bar{Z}, \bar{\eta}, \bar{\Delta}, \bar{\Gamma}_t)$, *the feasibility relative load* $\bar{L}_r \in \mathbb{R}^+$ *is defined by*

$$\bar{L}_r = \inf \left\{ \bar{x} \in \mathbb{R} \;:\; \frac{1}{\bar{x}} \bar{\Gamma}_t \text{ is feasible} \right\}$$

While Theorems 2 and 3 address the case when all users have the same service, the feasibility relative load describes the situation with multiple services. The following theorem captures the essentials regarding the feasibility relative load.

Theorem 4 (Gunnarsson, 2000) *Given a power control problem* $(\bar{Z}, \bar{\eta}, \bar{\Delta}, \bar{\Gamma}_t)$, *the feasibility relative load is obtained as*

$$\bar{L}_r = \max \operatorname{eig} \left\{ \bar{\Gamma}_t (\bar{\Delta}^{-1} \bar{Z} - E) \right\}.$$

Moreover, if $\bar{L}_r < 1$, *the power control problem is feasible, and there exists an optimal power assignment, given by*

$$\bar{p} = \left(E - \bar{\Gamma}_t (\bar{\Delta}^{-1} \bar{Z} - E) \right)^{-1} \bar{\Gamma}_t \bar{\Delta}^{-1} \bar{\eta}.$$

The power assignment above can of course be seen as a centralized strategy. However, since full information about the network is required to compile $\bar{Z}$ it is not plausible in practice. The result mainly serves as a performance bound.

Assume perfect receivers in the example, and that the three connections aims at the SIR targets $\bar{\Gamma}_t = \operatorname{diag}(1, 0.43, 0.67)$. According to Theorem 4, the feasibility relative load is $\bar{L}_r = 0.78 < 1$. The power control problem is thus feasible. Note that it is feasible despite the fact that one of the SIR targets is greater than the optimal balanced SIR[5], $\bar{\gamma}^* = 0.83$.

3.2   Convergence Results

3.2.1   Standard Interference Function.

The framework for addressing point-wise convergence of algorithms is based on (Bertsekas and Tsitsiklis, 1989), which has been refined and put into the context of power control algorithms by (Yates, 1995). An underlying assumption is that the G-matrix is constant. Let $\bar{p}$ denote the controllable resources (in a power control case, it is the power vector). Assume that the requirements to be met can be described by the inequality

$$\bar{p} \geq \bar{I}(\bar{p}), \tag{7.21}$$

where $\bar{I}(\bar{p})$ is referred to as the *interference function*. A vector $\bar{p} \geq 0$ is a feasible solution if it satisfies (7.21). Moreover, an interference function $\bar{I}(\bar{p})$

is feasible if (7.21) has a feasible solution. For an arbitrary interference function, the following definition applies:

Definition 3 (Standard Interference Function) *The interference function is* <u>*standard*</u> *if the following properties are satisfied for all* $\bar{p} \geq 0$.

- Positivity: $\bar{I}(\bar{p}) \geq 0$.

- Monotonicity: *if* $\bar{p} \geq \bar{p}'$ *then* $\bar{I}(\bar{p}) \geq \bar{I}(\bar{p}')$.

- Scalability: $\alpha \bar{I}(\bar{p}) > \bar{I}(\alpha \bar{p})$ *for all* $\alpha > 1$.

Given the requirements in (7.21), we want to study the convergence properties of the class of iterative algorithms described by

$$\bar{p}(t+1) = \bar{I}(\bar{p}(t)). \tag{7.22}$$

When $\bar{I}(\bar{p})$ is a standard interference function, the iteration (7.22) is denoted a *standard iterative algorithm*. The corresponding equilibrium is thus

$$\bar{p}(t+1) = \bar{I}(\bar{p}(t)) = \bar{p}(t).$$

The uniqueness of this equilibrium, is disclosed in the following theorem

Theorem 5 (Yates, 1995) *Let* $\bar{I}(\bar{p})$ *be standard and feasible. If the corresponding standard iterative algorithm (7.22) has an equilibrium, then it is* ̣*unique.*

Furthermore, the convergence of standard iterative algorithms is provided in Theorem 6

Theorem 6 (Yates, 1995) *If* $\bar{I}(\bar{p})$ *is standard and feasible, then the iterative algorithm (7.22) converges to a unique equilibrium* $\bar{p}_\infty$ *for any initial vector* $\bar{p}_0$.

This techniques can be used to analyze the stability and convergence of the integrating controller in the ideal situation without limitations described in Section 2.

Theorem 7 (Yates, 1995) *The iterative power control algorithm (7.7) converges to a unique equilibrium for any initial power vector if the power control problem is feasible.*

In the ideal case, but with time delays, the following can be concluded

Theorem 8 (Gunnarsson, 2000) *The iterative power control algorithm (7.7) with* $\beta = 1$, *time delay compensation (7.16) and subject to time delays* n *converges* $n+1$ *times slower than the same algorithm subject to no delays to a unique equilibrium for any initial power vector if the power control problem is feasible.*

3.2.2   Log-Linear Algorithms.   The analysis method in the previous section essentially applies to any controller. However, it is difficult to apply when considering algorithms based on log-linear design and when subject to time delays, other than in specific situations.

An alternative method is approximative and based on a linearization of the mutual cross-couplings between the connections via the interference and application of the *small gain theorem* (Zhou et al., 1995). The relative load is introduced in Definition 2. However, to support the stability results in this section, a different definition is also needed

Definition 4 (Cross-coupling Relative Load) *The cross-coupling relative load* $\bar{L}_{cc}$ *is defined as*

$$\bar{L}_{cc} \overset{\triangle}{=} \max_{1 \leq i \leq m} \left(1 - \frac{\bar{\nu}_i}{\bar{I}_i^t} \right)$$

where $\bar{I}_i^t$ is the interference at receiver i at the power control problem equilibrium (essentially using the power vector derived in Theorem 4).

This definition is considering the worst case situation over all receivers. Zero load means no mutual interference in the system, i.e. all the connections operate independently, and a unity load corresponds to an infinite interference, which constitute the upper limit. In the example in Section 3.1, the cross-coupling relative load is $\bar{L}_{cc} = 0.77$.

With this load definition, the following theorem can be concluded.

Theorem 9 (Gunnarsson, 2000) *Let $G(q)$ be the stable closed-loop transfer function of the local loop. Then the global system with linearized crosscouplings is stable if the power control problem is feasible and the following property is satisfied*

$$\sup_\omega |G(e^{i\omega})| < 1/\bar{L}_{cc},$$

Note that when cross-couplings can be neglected ($\bar{L}_{cc} = 0$), the power control problem is always feasible, and the only remaining stability requirement is stability of the local loop $G(q)$. Furthermore, a local power control design that meets

$$|G(e^{i\omega})| \leq 1$$

is always stable if locally stable and if the power control problem is feasible. The load level puts no further restrictions on global stability. From (7.17), we conclude that this is the case for the integrating controller (7.7) with TDC (7.16) for any time delay n.

If we consider the integrating controller in (7.7) and the typical time delay situation $n = 1$, the stability requirements in Theorem 9 can be simplified. The closed-loop system expression in (7.13) yields

$$\sup_{w} |G(e^{iw})| = \begin{cases} \dfrac{\beta}{\sqrt{(1-\beta)^2\left(1-\frac{1}{4\beta}\right)}} & \frac{1}{3} \le \beta < 1 \\ 1 & 0 < \beta < \frac{1}{3} \end{cases} \tag{7.23}$$

The stability requirements in terms of β are illustrated by Figure 7.10. Hence, in highly loaded systems the choice of β is more critical to avoid violating the global stability. In the example $\bar{L}_{cc} = 0.77$ and Theorem 9 can only guarantee

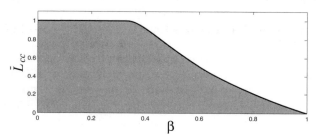

Figure 7.10. Requirements on cross-coupling relative load from Theorem 9 when using an integrating controller, subject to the typical time delay situation $n = 1$.

global stability for $\beta < 0.48$ according to (7.23).

3.2.3 Convergence of the Integrator with Single-Bit Error Representation.

The algorithm described in 7.14 with single-bit error representation never converges to a fixed point. As disclosed in Section 2.4, the relay feedback and the delays cause an oscillatory behavior around the reference signal. Instead, it converges to a region characterized by the following theorem.

Theorem 10 (Herdtner and Chong, 2000; Gunnarsson, 2000) *If the power control problem is feasible, the algorithm without and with the time delay compensation subject to a delay of totally n samples, $n = 0, 1, 2, \ldots$ converges to a region where the SIR error for every connection is bounded (in dB) by*

$$\begin{aligned} \textit{Without TDC:} \quad & |\gamma_i^t - \gamma_i(t)| \le 2(n+1)\beta \\ \textit{With TDC:} \quad & |\gamma_i^t - \gamma_i(t)| \le (n+2)\beta \end{aligned}$$

and β is the step size. The results also hold when subject to auto-interference.

Note that the error bound is tighter when using TDC, and also when using smaller β, which can be interpreted as the step-size. There is thus a trade-off between small tracking errors and fast responses to changes. Furthermore, the importance of well functioning admission and congestion control mechanisms is evident. Power control stability critically rely on their ability to preserve feasibility despite a time-varying network.

3.3   Global Stability Example

Some of the global stability results are illustrated in this section with network simulations using the example from Section 3.1. Figure 7.11 provides SIR error $e_i(t) = \gamma_i^t(t) - \gamma_i(t)$ plots for a number of different parameterizations. Even though the integrating controller (7.7) with perfect error representation and $\beta = 0.9$ yields a locally stable system according to Theorem 1, the global system is unstable and the mobile powers are oscillating between max and min powers. Clearly, the global stability requirements on the closed-loop system are not fulfilled at this load level. Theorem 9 and (7.23) guarantees stability for $\beta = 0.34$ which is illustrated in Figure 7.11b. Almost the same performance is obtained for $\beta = 1$ and time delay compensation in Figure 7.11c – a case which also is globally stable based on Theorem 9. The benefits of TDC are more evident when applied to the integrating controller with single-bit error representation. Comparative performances are provided in Figures 7.11d,e.

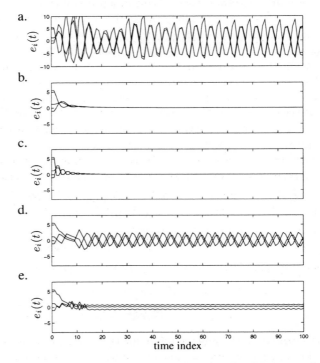

Figure 7.11.   Network simulations with the integrating controller (7.7) in a-c and with single-bit error representation (7.14) in d-e. a. $\beta = 0.9$, b. $\beta = 0.34$, c. $\beta = 1$ and TDC, d. $\beta = 1$, e. $\beta = 1$ and TDC.

4. Different Power Control Aspects

Some further power control aspects deserve extra attention, and the objective here is to guide the reader to a wider and more detailed set of references.

4.1 Soft Handover

Hitherto, the models have been focused on a situation, where all mobiles are connected solely to one base station. One core feature in DS-CDMA systems is soft handover, where the mobile can connect to several base stations simultaneously. For best performance, the power is adjusted to compensate for the power gain of the most favorable link. However, since one link is best at the time, the results above are still relevant.

In soft handover, power control becomes significantly different between uplink and downlink. In the uplink, information about the control error to the power control algorithm in the mobile is sent from all connected base stations. It is natural to control the power based on information from the base station with lowest control error, since this is the base station requesting the lowest transmission power. In the case of single-bit error representation, this means to only increase the power if all base stations report a positive control error. However, since the control error information is sent over a feedback channel, unfavorable links cause feedback errors. Therefore, it is important to not only consider the reported control error, but also the feedback channel quality (Grandell and Salonaho, 2001).

In the downlink, all the connected base stations receive control error information from the mobile. Thereby, the relations between the base station powers are maintained. Due to feedback errors, this information might be interpreted differently, and the base station powers might drift apart. To compensate for this drift, a centralized power balancing is proposed in the standards, see (3GPP, 2003c). The following should be seen as replacing equation (7.14)

$$p_i(t+1) = p_i(t) + \beta \text{sign}(e_i(t)) + p_{\text{bal},i} \tag{7.24}$$

The base station computes the average transmission power over a time frame $p_{\text{ave},i}$ and computes the balancing term as

$$p_{\text{bal},i} = (1 - r)(p_{\text{ref},i} - p_{\text{ave},i}), \tag{7.25}$$

where the adjustment ratio r and the reference power $p_{\text{ref},i}$ (possibly individual for each link) are signaled to the base station by the radio network controller (RNC), which controls the base stations.

The need for the central controlling node RNC is evident when considering soft handover. Since the uplink information is available in two different base stations, the actual service quality can only be evaluated in the RNC. Essentially, the RNC selects the data blocks from the base station with the best SIR (selection combining).

4.2   Outer Loop Power Control

Perceived quality of service might be well correlated to SIR, but it is not possible to set a SIR reference offline, that results in a specific service quality in terms of data rate and block error rate. Furthermore, the SIR estimate could be subject to errors as described in Section 2.2. Therefore, the inner control loop is operated in cascade with an outer loop, which adapts the SIR reference for a specific connection. Typically, the outer loop utilizes information about *block error rate (BLER)* and/or *bit error rate (BER)*.

Reliable communication can be seen as low block error rate requirements, which in turn means that it is very difficult to accurately estimate BLER. The errors appear rather seldom and it takes long time before the BLER estimate is stable. One approach increases the SIR reference significantly when an erroneous block is discovered, and decreases the SIR reference somewhat when an error-free block is received. (Niida et al., 2000) provides experimental results of outer loop power control using this method.

One block comprises many bits. Therefore, it is easier to obtain a good BER estimate. Then the relation between BER and BLER can be utilized to predict BLER based on BER measurements (Kawai et al., 1999).

To support soft handover, the uplink outer loop power control have to be implemented in the node controlling the base stations (RNC in WCDMA). This causes some additional round-trip delay of the control loop.

4.3   Up- and Downlink

The previous secions brought up aspects concerning implementationally differences between uplink and downlink related to soft handover and outer loop power control. Moreover, some other differences are also important.

Power control objectives are naturally different with a low power energy-efficiency focus in the uplink and a desire to utilize all available power in the downlink to use the available resuorces as much as possible to provide services.

Examples of the latter include High Speed Downlink Shared Channel (HS-DSCH) in the evolved WCDMA standard, where data service users are one by one allocated all base station power except what is needed to support conversational services. This means that the base station powers are always fully utilized, provided that there is always downlink data to transmit. Hence, downlink SIR in (7.5) is given by

$$\bar{\gamma}_i^{\mathrm{DL}}(t) = \frac{\bar{\delta}_i \bar{g}_{ij_i}^{\mathrm{DL}}(t) \bar{P}_{j_i}^{\mathrm{data}}(t)}{\bar{\alpha}_i (\bar{P}^{\max}(t) - \bar{\delta}_i \bar{P}_{j_i}^{\mathrm{data}}(t)) \bar{g}_{ij_i}^{\mathrm{DL}}(t) + \sum_{j \neq j_i} \bar{P}^{\max}(t) \bar{g}_{ij}^{\mathrm{DL}}(t) + \bar{\nu}_i(t)}.$$

where $\bar{P}^{\max}$ is the base station maximum power, and $\bar{P}^{\mathrm{data}}_{j_i}(t)$ the available power for data services at base station j_i. Naturally, the interference situation is more predictable than in the general case, since all base stations ideally are using maximum transmission power. Moreover, the base station resource utilization do not have to be coordinated over the base stations to the same extent for the same reason. This kind of time-sharing of resources and power is analyzed in (Berggren et al., 2001).

The uplink situation is different, since it is not likely that an individual user can utilize all available uplink resources. Instead, several users per base station have to be allocated power and resources for simultaneous use. Uplink power and resource allocation is for example addressed in (Villier et al., 2000). Another difficulty with the uplink is the challenge to characterize the uplink load. This is discussed in Chapter 14.

4.4 Power Control Algorithm Design

This chapter focuses on characteristics and fundamentals of power control algorithms. However, much performance can be gained, at least in theory, by adopting more advanced design methods. Some methods are briefly addressed here, with the objectives to improve link performance in terms of SIR target tracking in Section 4.4.1 and accelerate power control convergence in Section 4.4.2.

4.4.1 Local Power Control Design. The Smith predictor or TDC in (7.16) might compensate for some dynamical effects, but the controllers still show delayed reactions to changes in the power gains. One approach to improve the reactions is to predict the power gain. The following linear model structure is fitted to data

$$g_s(t) = \frac{C(q)}{A(q)} e_s(t), \quad \mathrm{Var}\left\{ e_s^2(t) \right\} = \sigma_e^2$$

Solve the Diophantine equation

$$q^{m-1} C(q) = A(q) F(q) + G(q)$$

for $F(q)$ and $G(q)$ yields the optimal m-step predictor (Åström and Wittenmark, 1997)

$$\hat{g}_s(t + m|t) = \frac{q G(q)}{C(q)} g_s(t) \tag{7.26}$$

Power gain prediction methods are further studied based on linear model structures (Ericsson and Millnert, 1996; Choel et al., 1999; Ekman et al., 2002) and nonlinear model structures (Ekman and Kubin, 1999; Tanskanen et al.,

1998; Zhang and Li, 1997). It is hard to give a general answer to whether linear or nonlinear models are most appropriate, or whether it is most suitable to predict $g(t)$ or $\bar{g}(t)$, since it depends on the fading situation and the controller objectives.

When a disturbance model and an optimal predictor is available as above, the step to employ minimum variance control is slightly short (Åström and Wittenmark, 1997; Gunnarsson, 2000):

$$R(q) = \frac{G(q)}{q(q-1)A(q)F(q)}, \qquad (7.27)$$

where $F(q)$ and $G(q)$ are obtained from the Diophantine equation above, and precautions are made in order to include integral action into the controller. Minimum-variance controllers are further explored in (Rintamäki and Koivo, 2001; Rintamäki et al., 2002) with linear models of the control error and real-time model estimation and adaptive controller parameters. Controllers like minimum variance control are more aggressive in the control action than the foundational integrating controller. This also means that the stability requirement in Theorem 9 critically limits the system load, which means that the gains in better SIR tracking can be lost due to lower system utilization.

As indicated in previous sections, it is not always justifiable to let a user disturb other connections significantly while aiming at a rather high SIR compared to the propagation conditions. In the proposed algorithm (Almgren et al., 1994), users aiming at using a high power are forced to use a lower SIR. The algorithm expressed in this framework is given by

$$R(q) = \frac{\beta}{q - \beta}, \qquad (7.28)$$

which does not include integral action. The algorithm and its convergence is further explored in (Yates et al., 1997).

In real systems, SIR is typically not readily available. One natural idea is to estimate SIR given available measurements. Approaches to do so in different systems are discussed in (Ramakrishna et al., 1997; Andersin et al., 1998; Blom et al., 1999; Freris et al., 2001; Kurniawan et al., 2001). A different approach is to base the power control on the available measurements directly (Ulukus and Yates, 1998).

4.4.2   Accelerating Power Control Convergence.   As established in Section 3.2, the foundational integrating power control algorithm converges provided that the power control problem is feasible. The convergence rate of the distributed algorithms from any initial power vector to the equilibrium is also related to the ability of the algorithms to compensate for fast power gain variations. It also depend on the system load (Hanly and Tse, 1999). Methods with the objective to improve the convergence rate, are likely improving the fading compensation capabilities as well. Such methods typically aims at efficiently and distributedly solving the power control matrix equation (7.20)

$$\bar{p} = \bar{\Gamma}_t \left((\bar{\Delta}^{-1}\bar{Z} - E)\bar{p} + \bar{\Delta}^{-1}\bar{\eta} \right).$$

This is essentially how the DPC algorithm (7.10) was motivated. Improvements include using successive overrelaxation techniques (Jäntti and Kim, 2000), Steffensson iterations (Li and Gajic, 2002), convex set projections (Rabee et al., 2001), and Gauss-Seidel iterations (Lelic and Gajic, 2002). Since such methods originates from numerical analysis, the robustness agains the fundamental limitations in Section 2 needs to be further investigated.

5.   Summary

This chapter surveys power control algorithms with a control theoretic perspective. The objective is to adjust the transmitter power $p_i(t)$ to obtain an acceptable signal-to-interference ratio $\gamma_i^t(t) - \gamma_i(t)$. Starting from the foundational integrating controller,

$$p_i(t+1) = p_i(t) + \beta \left(\gamma_i^t(t) - \gamma_i(t) \right) = p_i(t) + \beta e_i(t),$$

the practical situation is described via some fundamental limitations including

- limited power level update rate

- measurement errors and disturbances

- limited feedback bandwidth

- time delays and local stability

- transmitter power limitations.

Another natural limitation is global stability – i.e. stability of the entire network when operating several power control algorithms in parallel with mutual interference that needs to be compensated for. Moreover, in practical situations the power control algorithms have to consider connections to several base stations and the fact that it is difficult to address quality of service solely in terms of SIR.

Acknowledgments

This work has been in cooperation with Ericsson Research and supported by VINNOVA, which both are acknowledged.

Notes

1. In some situations it is more convenient to use the related *signal-to-total-interference ratio, (STIR)*

$$\bar{\beta}_i(t) = \frac{\bar{C}_i(t)}{\bar{I}_i(t) + \bar{C}_i(t)} = \frac{\bar{\gamma}_i(t)}{1 + \bar{\gamma}_i(t)}.$$

Furthermore, the term SIR/STIR is sometimes replaced by CIR/CTIR (*C - carrier*), essentially reflecting the radio frequency situation at the antenna, whereas SIR reflects the base band situation. It can also be important to distinguish between SIR before and after base band processing. Since the absolute values are of no importance here, the choice of terms is irrelevant.

2. For a more rigid discussion on a *q*-operator algebra, the reader is referred to (Åström and Wittenmark, 1997). The intuitive relations to the complex variable *z* of the *z*-transform are also addressed.

3. A fundamental constraint on the linear control performance and error suppression can be expressed in terms of the Bode integral constraint on S.

$$\int_0^\pi \log \left| S(e^{i\omega}) \right| d\omega = 0$$

This means that it is not possible to obtain $S(e^{i\omega T_s}) = 0$ for all frequencies.

4. Note that the matrices $\bar{Z}$ and $\bar{G}$ are defined differently in Chapter 14.

5. The corresponding example with signal-to-total-interference is essentially analogous. The optimal balanced STIR is then $\bar{\beta}^* = 0.45$, and the feasible example is given by $\bar{\beta}^t = (0.5, 0.3, 0.4)$.

References

3GPP (2003a). Deployment aspects. Technical Specification TSG RAN 25.943.

3GPP (2003b). Physical radio procedures. Technical Specification TSG RAN 25.214.

3GPP (2003c). UTRAN Iub interface NBAP signalling. Technical Specification TSG RAN 25.433.

Adachi, F., Sawahashi, M., and Suda, H. (1998). Wideband DS-CDMA for next-generation mobile communications systems. *IEEE Communications Magazine*, 36(9).

Almgren, M., Andersson, H., and Wallstedt, K. (1994). Power control in a cellular system. In *Proc. IEEE Vehicular Technology Conference*, Stockholm, Sweden.

Andersin, M., Mandayam, N., and Yates, D. (1998). Subspace based estimation of the signal to interference ratio for TDMA cellular systems. *Wireless Networks*, 4(3).

Åström, K. and Wittenmark, B. (1997). *Computer Controlled Systems – Theory and Design*. Prentice-Hall, Englewood Cliffs, NJ, USA, third edition.

Berggren, F., Kim, S.-L., Jæntti, R., and Zander, J. (2001). Joint power control and intracell scheduling of DS-CDMA nonreal time data. *IEEE Journal on Selected Areas in Communications*, 19(10).

Bertsekas, D. P. and Tsitsiklis, J. N. (1989). *Parallel and Distributed Computation*. Prentice Hall, Englewood Cliffs, NJ, USA.

Blom, J., Gunnarsson, F., and Gustafsson, F. (1998). Constrained power control subject to time delays. In *Proc. International Conference on Telecommunications*, Chalkidiki, Greece.

Blom, J., Gunnarsson, F., and Gustafsson, F. (1999). Estimation in cellular radio systems. In *Proc. IEEE International Conference on Acoustics, Speech, and Signal Processing.*, Phoenix, AZ, USA.

Choel, S., Chulajata, T., Kwon, H. M., Koh, B.-J., and Hong, S.-C. (1999). Linear prediction at base station for closed loop power control. In *Proc. IEEE Vehicular Technology Conference*, Houston, TX, USA.

Dietrich, P., Rao, R., Chockalingam, A., and Milstein, L. (1996). A log-linear closed loop power control model. In *Proc. IEEE Vehicular Technology Conference*, Atlanta, GA, USA.

Ekman, T., Ahlén, A., and Sternad, M. (2002). Unbiased power prediction of Rayleigh fading channels. In *Proc. IEEE Vehicular Technology Conference*, Vancouver, Canada.

Ekman, T. and Kubin, G. (1999). Nonlinear prediction of mobile radio channels: measurements and MARS model designs. In *Proc. IEEE International Conference on Acoustics, Speech, and Signal Processing*, Phoenix, AZ, USA.

Ericsson, A. and Millnert, M. (1996). Fast power control to counteract rayleigh fading in cellular radio systems. In *Proc. RVK*, Luleå, Sweden.

Foschini, G. and Miljanic, Z. (1993). A simple distributed autonomus power control algorithm and its convergence. *IEEE Transactions on Vehicular Technology*, 42(4).

Freris, N., Jeans, T., and Taaghol, P. (2001). Adaptive SIR estimation in DS-CDMA cellular systems using Kalman filtering. *IEE Electronics Letters*, 37(5).

Godlewski, P. and Nuaymi, L. (1999). Auto-interference analysis in cellular systems. In *Proc. IEEE Vehicular Technology Conference*, Houston, TX, USA.

Goldsmith, A. (1997). The capacity of downlink fading channels with variable rate and power. *IEEE Transactions on Vehicular Technology*, 46(3).

Grandell, J. and Salonaho, O. (2001). Closed-loop power control algorithms in soft handover for WCDMA systems. In *Proc. IEEE International Conference on Communications*, Helsinki, Finland.

Grandhi, S., Vijayan, R., and Goodman, D. (1994). Distributed power control in cellular radio systems. *IEEE Transactions on Communications*, 42(2).

Gunnarsson, F. (2000). *Power Control in Cellular Radio System: Analysis, Design and Estimation*. PhD thesis, Linköpings universitet, Linköping, Sweden.

Gunnarsson, F. (2001). Fundamental limitations of power control in WCDMA. In *Proc. IEEE Vehicular Technology Conference*, Atlantic City, NJ, USA.

Gunnarsson, F. and Gustafsson, F. (2001). Time delay compensation in power controlled cellular radio systems. *IEEE Communications Letters*, 5(7).

Gunnarsson, F. and Gustafsson, F. (2002). Power control in cellular radio systems – from a control theory perspective. In *Proc. IFAC World Congress*, Barcelona, Spain.

Gunnarsson, F., Gustafsson, F., and Blom, J. (1999). Pole placement design of power control algorithms. In *Proc. IEEE Vehicular Technology Conference*, Houston, TX, USA.

Gunnarsson, F., Gustafsson, F., and Blom, J. (2001). Dynamical effects of time delays and time delay compensation in power controlled DS-CDMA. *IEEE Journal on Selected Areas in Communications*, 19(1).

Hanly, S. and Tse, D.-N. (1999). Power control and capacity of spread spectrum wireless networks. *Automatica*, 35(12).

Holma, H. and Toskala, A., editors (2000). *WCDMA for UMTS. Radio Access for Third Generation Mobile Communications*. Wiley, New York, NY, USA.

Jakes, W. C. (1974). *Microwave mobile communications*. John Wiley & Sons, New York, NY, USA.

Jäntti, R. and Kim, S.-L. (2000). Second-order power control with asymptotically fast convergence. *IEEE Journal on Selected Areas in Communications*, 18(3).

Kawai, H., Suda, H., and Adachi, F. (1999). Outer-loop control of target SIR for fast transmit power control in turbo-coded W-CDMA mobile radio. *IEE Electronics Letters*, 35(9).

Kurniawan, A., Perreau, S., Choi, J., and Lever, K. (2001). SIR-based power control in third generation CDMA systems. In *Proc. International Conference on Information Communications & Signal Processing*, Singapore.

Lee, T. and Lin, J. (1996). A study on the distributed power control for cellular mobile systems. In *Proc. IEEE Vehicular Technology Conference*, Atlanta, GA, USA.

Lelic, D. and Gajic, Z. (2002). Gauss-Seidel iterations for SIR-based power updates for 3G wireless CDMA communications networks. In *Proc. Annual Allerton Conference on Communication, Control, and Computing*, Monticello, IL, USA.

Li, W., Dubey, V., and Law, C. (2001). A new generic multistep power control algorithm for the LEO satellite channel with high dynamics. To appear in *IEEE Communications Letters*.

Li, X. and Gajic, Z. (2002). An improved SIR-based power control for CDMA systems using Steffensen iterations. In *Proc. Information Science and Systems Conference*, Princeton, NJ, USA.

Meyerhoff, H. (1974). Method for computing the optimum power balance in multibeam satellites. *COMSAT Technical Review*, 4(1).

Niida, S., Suzuki, T., and Takeuchi, Y. (2000). Experimental results of outer-loop transmission power control using wideband-CDMA for IMT-2000. In *Proc. IEEE Vehicular Technology Conference*, Tokyo, Japan.

Okamura, Y., Ohmori, E., Kawano, T., and Fukuda, K. (1968). Field strength and its variability in VHF and UHF land-mobile radio service. *Review of the Electrical Communication Laboratory*, 16(9-10).

Rabee, S. A., Sharif, B. S., and Sali, S. (2001). Distributed power control algorithm in cellular radio systems using projection onto convex sets. In *Proc. IEEE Vehicular Technology Conference*, Atlantic City, NJ, USA.

Ramakrishna, D., Mandayam, N., and Yates, R. (1997). Subspace based estimation of the signal to interference ratio for CDMA cellular systems. In *Proc. IEEE Vehicular Technology Conference*, Phoenix, AZ, USA.

Rintamäki, M. and Koivo, H. (2001). Adaptive robust power contol for WCDMA systems. In *Proc. IEEE Vehicular Technology Conference*, Atlantic City, NJ, USA.

Rintamäki, M., Zenger, K., and Koivo, H. (2002). Self-tuning adaptive algorithms in the power control of WCDMA systems. In *Proc. Nordic Signal Processing Symposium*, Hurtigruten, Norway.

Romero-Jerez, J. M., Ruiz-Garcia, M., and Diaz-Estrella, A. (2000). Effects of power control errors and multipath fading on BER in a cellular CDMA system. In *Proc. IEEE Vehicular Technology Conference*, Boston, MA, USA.

Rosberg, Z. and Zander, J. (1998). Toward a framework for power control in cellular systems. *Wireless Networks*, 4(3).

Salmasi, A. and Gilhousen, S. (1991). On the system design aspects of code division multiple access (CDMA) applied to digital cellular and personal communications networks. In *Proc. IEEE Vehicular Technology Conference*, New York, NY, USA.

Shannon, C. (1956). The zero error capacity of a noisy channel. *IRE Transactions On Information Theory*, 2.

Sim, M., Gunawan, E., Soh, C., and Soong, B. (1998). Characteristics of closed loop power control algorithms for a cellular DS/CDMA system. *IEE Proceedings - Communications*, 147(5).

Sklar, B. (1997). Rayleigh fading channels in mobile digital communication systems. *IEEE Communications Magazine*, 35(7).

Sørensen, T. (1998). Correlation model for slow fading in a small urban macro cell. In *Proc. IEEE Personal, Indoor and Mobile Radio Communications*, Boston, MA, USA.

Tanskanen, J., Mattila, J., Hall, M., Korhonen, T., and Ovaska, S. (1998). Predictive estimators in CDMA closed loop power control. In *Proc. IEEE Vehicular Technology Conference*, Ottawa, Canada.

Ulukus, S. and Yates, R. (1998). Stochastic power control for cellular radio systems. *IEEE Transactions on Communications*, 46(6).

Villier, E., Legg, P., and Barrett, S. (2000). Packet data transmissions in a W-CDMA network-examples of uplink scheduling and performance. In *Proc. IEEE Vehicular Technology Conference*, Tokyo, Japan.

Yates, R. (1995). A framework for uplink power control in cellular radio systems. *IEEE Journal on Selected Areas in Communications*, 13(7).

Yates, R., Gupta, S., Rose, C., and Sohn, S. (1997). Soft dropping power control. In *Proc. IEEE Vehicular Technology Conference*, Phoenix, AZ, USA.

Zhang, Y. and Li, D. (1997). Power control based on adaptive prediction in the CDMA/TDD system. In *Proc. IEEE International Conference on Universal Personal Communications*, San Diego, CA, USA.

Zhou, K., Doyle, J., and K., G. (1995). *Robust and Optimal Control*. Prentice-Hall, Upper Saddle River, NJ, USA.

Chapter 8

AVERAGE OUTAGE DURATION OF WIRELESS COMMUNICATION SYSTEMS*

Lin Yang and Mohamed-Slim Alouini

Department of Electrical and Computer Engineering
University of Minnesota
Minneapolis, MN 55455, USA.
E-mail: <lyang,alouini@ece.umn.edu>

Abstract   Outage probability has been traditionally the most commonly used performance measure of wireless communication systems. However, in certain communication system applications such as adaptive transmission schemes, the outage probability does not provide enough information for the overall system design and configuration. In that case, in addition to the outage probability, the frequency of outages (or equivalently the level crossing rate (LCR)) and the average outage duration (AOD) are important performance measures for the proper selection of the transmission symbol rate, interleaver depth, packet length, and/or time slot duration. In this chapter, we present closed-form expressions for the LCR and AOD of some diversity combining schemes of interest with and without multiple co-channel interferences (CCI) and over independent, correlated, and/or unbalanced channels considering minimum signal-to-interference ratio (SIR) and/or desired signal power constraints. More specifically, the chapter presents generic results for the LCR and AOD (i) of maximal ratio combining systems subject to CCI operating over independent identically distributed (i.i.d.) Rician and/or Nakagami fading environments when a minimum desired signal power requirement is specified for satisfactory reception, (ii) of various selection combining diversity scheme in presence of multiple CCI and with both minimum SIR and desired signal power constraints over independent, correlated, and/or unbalanced channels. Corresponding numerical examples and plots illustrating the mathematical formalism are also provided and discussed.

*This work was supported in part by the National Science Foundation Grant No. CCR-9983462.

1. Introduction

In multiuser wireless communication systems which are both interference and power limited, an outage is declared whenever either the output signal-to-interference ratio (SIR) or signal-to-noise ratio falls below its pre-determined threshold. Outage probability has been traditionally the most commonly used important performance measure of cellular mobile radio communication systems (e.g, [1, 2, 3, 4, 5, 6, 7, 8, 9]). In particular, the overall system spectrum efficiency is usually determined by the required downlink outage probability. However, in certain communication system applications such as adaptive transmission schemes, the outage probability does not provide enough information for the overall system design and configuration. In that case, in addition to the outage probability, the frequency of outages (or equivalently the level crossing rate (LCR)) and the average outage duration (AOD) [10, 11, 12, 13] and more generally distribution of the outage duration [14] are important performance measures for the proper selection of the transmission symbol rate, interleaver depth, packet length, and/or time slot duration. LCR and AOD have been introduced and used to reflect the correlation properties (thus the second-order statistics) of fading channels and provide a dynamic representation of the system outage performance.

The performance of wireless systems is severely affected by fading and diversity-combining techniques are often used to mitigate fading and improve the reception quality. These techniques have also been found to be useful in combatting the effects of co-channel interference (CCI) in mobile radio environment. Especially, maximal ratio combining (MRC), which is optimal combining scheme in the absence of CCI but comes at the expense of complexity since MRC requires knowledge of all channel fading parameters. Another extreme case is selection combining (SC), which is one of the simplest micro-diversity combining methods [15], and which always picks the best branch as current operating branch to smooth out the deep fades and severe interferences. Previous related work on the performance analysis of MRC and SC systems has included the following. Outage probability of various diversity combining scheme with/without multiple CCI and over independent, correlated, and/or unbalanced channels considering either/both minimum SIR or/and desired signal power constraints has/have been extensively studied over the past two decades (see for example [1, 2, 3, 4, 5, 6, 16, 17, 18, 19, 7, 8, 9]). On the other hand, recent studies focusing on the analysis of the AOD and the LCR of MRC and SC diversity systems can be found in [20, 21, 22, 23] and in [22, 24, 23], respectively. Previous work on the effect of correlated fading on the AOD and LCR of SC diversity systems was done by

Adachi *et al.* [25]. More specifically, [25] presented a general method for calculating the LCR of correlated dual-branch SC system without CCI but ended up with approximate expressions for both the LCR and AOD.

All above published analytical work on the LCR and AOD of cellular mobile radio systems is still limited to purely power-limited systems [25, 20, 21, 22, 24, 23] or interference-limited systems subject to Rayleigh type of fading [10, 12]. Although the pure scattering (Rayleigh) fading model is reasonable for medium or large macrocellular systems, a line-of-sight or specular paths are more likely to exist in microcellular or picocellular indoor environments because cell size is smaller. In such kind of environments, the fading tends to rather follow a Rician or Nakagami type of distribution. Furthermore, in previous work on AOD evaluation, it was assumed that an outage is declared if either the desired signal power or the SIR falls below a predetermined protection ratio. While, this criterion is acceptable for purely power-limited or interference-limited systems, it tends to under-estimate the outage in particular for sectorized or lightly loaded systems with CCI in which an outage can occur not only because of low SIR but also because deep fades will result in a desired signal power below a minimum required level for satisfactory detection. To fill the remaining void in the area of performance analysis of LCR and AOD of diversity systems, the authors recently studied the LCR and AOD of some diversity combining schemes of interest with and without multiple CCI and over independent, correlated, and/or unbalanced channels considering minimum SIR and/or desired signal power constraints [13, 26, 27]. This chapter presents the main results of this study without going into all the details of the derivations.

2. Impact of CCI on the LCR and AOD of MRC

While special attention has been paid to the impact of the fading statistics of the desired and interfering users on the system outage probability (see for example [1, 3, 5, 28, 29, 7, 30] and references therein), published analytical work on the LCR and AOD of MRC-like cellular mobile radio systems is still limited to interference-limited systems subject to Rayleigh type of fading [10, 12]. In this section, we present closed-form expressions for the LCR and AOD of MRC in generalized fading environments and with the consideration of both minimum SIR and minimum desired signal power requirements for satisfactory reception.

2.1   Desired and Interference Signals Model

We consider a cellular mobile radio system in which the desired and interfering signals are independent and may have different fading statistics [31, 5, 32, 33, 7]. The desired signal is received over L independent identically distributed (i.i.d.) diversity (time, frequency, or space) paths with the same average fading power Ω_D and subject to flat Rayleigh, Nakagami (with fading parameter m_D), or Rician (with Rician factor K_D) type of fading. The total N_I active co-channel interfering signals (over the L diversity paths) are assumed to be i.i.d. with the same average fading power Ω_I and subject to flat Rayleigh, Nakagami (with fading parameter m_I), or Rician (with Rician factor K_I) type of fading.

The output SIR λ for an interference-limited system with MRC is given by [28, 34]

$$\lambda = \frac{\alpha_D^2}{\alpha_I^2} = \frac{\sum_{l=1}^{L} \alpha_{D,l}^2}{\sum_{n=1}^{N_I} \alpha_{I,n}^2}, \tag{8.1}$$

where α_D and α_I are the composite amplitudes of the desired user and co-channel interferers at the combiner output, $\alpha_{D,l}$ is the fading amplitude of the desired user over the lth diversity path and $\alpha_{I,n}$ is the fading amplitude of the nth co-channel interferer.

2.2   Method of Analysis

We consider wireless communication systems in which multiple CCI are present. Depending on propagation and fading conditions affecting the desired and interfering signals, the system is alternatively interference-limited or power-limited. For such kind of systems, an overall outage is declared whenever either the output SIR λ falls below a pre-determined threshold λ_{th}, i.e., $\lambda = \frac{\alpha_D^2}{\alpha_I^2} < \lambda_{\text{th}}$, or the received signal power of the desired user $s_D = \alpha_D^2$ falls below another pre-determined threshold s_{th}, i.e., $s_D < s_{\text{th}}$. In this section, we seek to evaluate the AOD, $T(\lambda_{\text{th}}, s_{\text{th}})$ [seconds] which is a measure to describe how long in average the system remains in the outage status.

Mathematically speaking, it can be shown that $T(\lambda_{\text{th}}, s_{\text{th}})$ is given by [13, Eqn. 8]

$$T(\lambda_{\text{th}}, s_{\text{th}}) = \frac{\text{P}_{\text{out}}}{N(\lambda_{\text{th}}, s_{\text{th}})}, \tag{8.2}$$

where $N(\lambda_{\text{th}}, s_{\text{th}})$ is the joint LCR that the SIR process crosses λ_{th} or the desired signal power process crosses s_{th} and $\text{P}_{\text{out}} = P[\lambda < \lambda_{\text{th}} \text{ or } s_D <$

s_{th}] is the system outage probability when a minimum desired signal power constraint is present.

The outage probability for various desired/interfering signal fading scenarios with a minimal signal power requirement has been extensively studied and closed-form expressions are available for several fading scenario combinations of interest (see for example [1, 3, 5, 7] for Nakagami/Rician[1], Rician/Nakagami, and Nakagami/Nakagami fading combinations and no diversity reception). Therefore, only the LCR $N(\lambda_{\text{th}}, s_{\text{th}})$ is needed to obtain the AOD in all above cases. However, it is worthwhile to point out that our approach and resulting new expressions for the LCR are also applicable to the diversity systems employing MRC.

Conditioning on α_I and using the total probability theorem, it can be shown that the LCR with a minimum desired signal power requirement can be written as

$$N(\lambda_{\text{th}}, s_{\text{th}}) = N\left(\alpha_D = \sqrt{\lambda_{\text{th}}}\alpha_I | \alpha_I \geq \sqrt{s_{\text{th}}/\lambda_{\text{th}}}\right) P\left(\alpha_I \geq \sqrt{s_{\text{th}}/\lambda_{\text{th}}}\right)$$
$$+ N(\alpha_D = \sqrt{s_{\text{th}}}) P\left(\alpha_I < \sqrt{s_{\text{th}}/\lambda_{\text{th}}}\right), \tag{8.3}$$

where $N\left(\alpha_D = \sqrt{s_{\text{th}}}\right)$ is the LCR of the signal power α_D^2 at level s_{th} and $N\left(\alpha_D = \sqrt{\lambda_{\text{th}}}\alpha_I \mid \alpha_I \geq \sqrt{s_{\text{th}}/\lambda_{\text{th}}}\right)$ is the LCR of the SIR λ at level λ_{th} given $\alpha_I \geq \sqrt{s_{\text{th}}/\lambda_{\text{th}}}$.

Closed-form expressions for $P\left(\alpha_I < \sqrt{s_{\text{th}}/\lambda_{\text{th}}}\right)$, $N\left(\alpha_D = \sqrt{s_{\text{th}}}\right)$, $P\left(\alpha_I \geq \sqrt{s_{\text{th}}/\lambda_{\text{th}}}\right)$ are available in the literature (e.g [35, 20, 36]), and $N\left(\alpha_D = \sqrt{\lambda_{\text{th}}}\alpha_I | \alpha_I \geq \sqrt{s_{\text{th}}/\lambda_{\text{th}}}\right)$ can be obtained by modifying the formula provided in [36, Eqn. (2.90)] as

$$N\left(\alpha_D = \sqrt{\lambda_{\text{th}}}\alpha_I | \alpha_I \geq \sqrt{s_{\text{th}}/\lambda_{\text{th}}}\right) = \int_0^\infty \dot{r}\, f_{r,\dot{r}|A}\left(\sqrt{\lambda_{\text{th}}}, \dot{r}\right) d\dot{r}, \tag{8.4}$$

where $r = \alpha_D/\alpha_I$, A denotes the event $\{\alpha_I \geq \sqrt{s_{\text{th}}/\lambda_{\text{th}}}\}$, and $f_{r,\dot{r}|A}(r, \dot{r})$ is the conditional joint probability density function (JPDF) of r and $\dot{r}$ given $\alpha_I \geq \sqrt{s_{\text{th}}/\lambda_{\text{th}}}$. As an extension of the formula [37, Eqn. (9)] (or equivalently [10, Eqn. (15)]), it can be shown that $f_{r,\dot{r}|A}(r, \dot{r})$ is given by

$$f_{r,\dot{r}|A}(r, \dot{r}) = \int_{\sqrt{\frac{s_{\text{th}}}{\lambda_{\text{th}}}}}^\infty \int_{-\infty}^\infty \alpha_I^2\, f_{\alpha_D}(\alpha_I r)\, f_{\dot{\alpha}_D}(\dot{r}\alpha_I + \alpha_I r)$$
$$\times\ f_{\alpha_I|A}(\alpha_I)\, f_{\dot{\alpha}_I}(\dot{\alpha}_I)\, d\dot{\alpha}_I d\alpha_I, \tag{8.5}$$

where $f_{\alpha_I|A}(\alpha_I)$ denote the truncated probability density function (PDF) of α_I given $\alpha_I \geq \sqrt{s_{\mathrm{th}}/\lambda_{\mathrm{th}}}$ and which can be obtained as

$$f_{\alpha_I|A}(\alpha_I) = \frac{f_{\alpha_I}(\alpha_I)}{P\left(\alpha_I \geq \sqrt{s_{\mathrm{th}}/\lambda_{\mathrm{th}}}\right)}, \quad \alpha_I \geq \sqrt{\frac{s_{\mathrm{th}}}{\lambda_{\mathrm{th}}}}. \tag{8.6}$$

2.3   Application to Common Fading Scenarios

In this section, we provide closed-form expressions for the LCR corresponding to typical fading combinations of interest: Nakagami/Nakagami, Rician/Nakagami, and Nakagami/Rician.

2.3.1   Nakagami/Nakagami Scenario.

We consider the case where both desired user and co-channel interferers are subject to Nakagami fading. It can be shown after considerable manipulations that the LCR with a minimum power requirement $N(\lambda_{\mathrm{th}}, s_{\mathrm{th}})$ is given by

$$
\begin{aligned}
N(\lambda_{\mathrm{th}}, s_{\mathrm{th}}) ={}& \sqrt{2\pi} f_D \frac{S_n^{\frac{2Lm_D-1}{2}}}{\Gamma(Lm_D)} \exp(-S_n) \left[1 - \frac{\Gamma(N_I m_I, \lambda_{nn} S_n)}{\Gamma(N_I m_I)}\right] \\
&+ \frac{\sqrt{2\pi}\,\Gamma\left(Lm_D + N_I m_I - \frac{1}{2}, S_n(1 + \lambda_{nn})\right)}{\Gamma(N_I m_I)\Gamma(Lm_D)} \\
&\times \left(f_D^2 + \frac{f_I^2}{\lambda_{nn}}\right)^{\frac{1}{2}} \frac{\lambda_{nn}^{N_I m_I}}{(1 + \lambda_{nn})^{Lm_D + N_I m_I - \frac{1}{2}}},
\end{aligned} \tag{8.7}
$$

where $S_n = \frac{s_{\mathrm{th}} m_D}{\Omega_D}$, $\lambda_{nn} = \frac{\Omega_D m_I}{\lambda_{\mathrm{th}} \Omega_I m_D}$, f_D and f_I are the maximum Doppler frequency shifts for the desired and interfering signals, respectively, and $\Gamma(\cdot, \cdot)$ is the incomplete Gamma function [38].

Special Cases

- Interference-limited case
  The LCR $N(\lambda_{\mathrm{th}})$ for interference-limited systems can be obtained by setting $s_{\mathrm{th}} = 0$ in (8.7) which results into

$$
\begin{aligned}
N(\lambda_{\mathrm{th}}) ={}& \frac{\sqrt{2\pi}\Gamma\left(Lm_D + N_I m_I - \frac{1}{2}\right)}{\Gamma(N_I m_I)\Gamma(Lm_D)} \left(f_D^2 + \frac{f_I^2}{\lambda_{nn}}\right)^{\frac{1}{2}} \\
&\times \frac{\lambda_{nn}^{N_I m_I}}{(1 + \lambda_{nn})^{Lm_D + N_I m_I - \frac{1}{2}}}.
\end{aligned} \tag{8.8}
$$

If $f_D = f_I = f_m$ and $m_D = m_I = 1$, (8.8) further reduces to the LCR of the interference-limited system in the Rayleigh/Rayleigh fading environment, which is given by

$$N(\lambda_{\text{th}}) = \frac{\sqrt{2\pi} f_m \Gamma\left(N_I + \frac{1}{2}\right)}{\Gamma(N_I)} \left(\frac{1}{1 + \frac{\lambda_{\text{th}} \Omega_I}{\Omega_D}}\right)^{N_I} \sqrt{\frac{\lambda_{\text{th}} \Omega_I}{\Omega_D}}, \quad (8.9)$$

in agreement with [10, Eqn. (17)] when $L = 1$, as expected.

- Power-limited case
  Setting $\lambda_{\text{th}} = 0$ in (8.7), we obtain the LCR $N(s_{\text{th}})$ of the signal power s_D at level s_{th} (purely power-limited case) which is given by

$$N(s_{\text{th}}) = \sqrt{2\pi} f_D \frac{S_n^{\frac{2Lm_D-1}{2}}}{\Gamma(Lm_D)} \exp(-S_n), \quad (8.10)$$

in agreement with Yacoub *et al.* L-fold MRC diversity result over i.i.d. Nakagami paths [22, Eqn. (36)] (or equivalently their result for no diversity case given in [35, Eqn. (17)]).

2.3.2   Rician/Nakagami Scenario.   When the desired user is subject to Rician type of fading whereas the co-channel interferers are subject to i.i.d. Nakagami fading, it can be shown that the LCR with a minimum power requirement is given by

$$
\begin{aligned}
N(\lambda_{\text{th}}, s_{\text{th}}) =\ & \sqrt{2\pi} f_D \frac{S_r^{\frac{L}{2}}}{(LK_D)^{\frac{L-1}{2}}} \exp(-LK_D - S_r) \\
& \times I_{L-1}\left(2\sqrt{LK_D S_r}\right) \left[1 - \frac{\Gamma(N_I m_I, \lambda_{rn} S_r)}{\Gamma(N_I m_I)}\right] \\
& + \frac{2^{\frac{3-L-2N_I m_I}{2}} \sqrt{\pi}}{\Gamma(N_I m_I)(LK_D)^{\frac{L-1}{2}}} \frac{\lambda_{rn}^{N_I m_I}}{(1 + \lambda_{rn})^{\frac{L}{2} + N_I m_I}} \\
& \times Q_{L+2N_I m_I - 1, L-1}\left(\sqrt{\frac{2LK_D}{1 + \lambda_{rn}}}, \sqrt{2S_r(1 + \lambda_{rn})}\right) \\
& \times \exp\left(-\frac{LK_D \lambda_{rn}}{1 + \lambda_{rn}}\right) \left(f_D^2 + \frac{f_I^2}{\lambda_{rn}}\right)^{\frac{1}{2}}, \quad (8.11)
\end{aligned}
$$

where $S_r = \frac{S_{\text{th}}(K_D+1)}{\Omega_D}$, $\lambda_{rn} = \frac{\Omega_D m_I}{\lambda_{\text{th}} \Omega_I (K_D+1)}$, and $Q_{m,n}(\cdot, \cdot)$ is the Nuttall Q-function defined by [39]. Similarly, one can get the closed-form results of the LCR for purely interference-limited/power-limited system in this scenario by setting $s_{\text{th}} = 0/\lambda_{\text{th}} = 0$ in (8.11) [13].

2.3.3   Nakagami/Rician Scenario.   The LCR can be shown
to be given in this scenario by

$$N\left(\lambda_{\text{th}}, s_{\text{th}}\right) = \sqrt{2\pi} f_D \frac{S_n^{\frac{2Lm_D-1}{2}}}{\Gamma(Lm_D)} \exp\left(-S_n\right)\left[1 - Q_{N_I}\left(\sqrt{2N_I K_I}, \sqrt{2S_n \lambda_{nr}}\right)\right]$$

$$+ \frac{2^{\frac{3-N_I-2Lm_D}{2}}\sqrt{\pi}}{\Gamma(Lm_D)(N_I K_I)^{\frac{N_I-1}{2}}} \frac{\lambda_{nr}^{\frac{N_I+1}{2}}}{(1+\lambda_{nr})^{\frac{N_I}{2}+Lm_D}}$$

$$\times Q_{N_I+2Lm_D-1,N_I-1}\left(\sqrt{\frac{2N_I K_I \lambda_{nr}}{1+\lambda_{nr}}}, \sqrt{2S_n(1+\lambda_{nr})}\right)$$

$$\times \exp\left(-\frac{N_I K_I}{1+\lambda_{nr}}\right)\left(f_D^2 + \frac{f_I^2}{\lambda_{nr}}\right)^{\frac{1}{2}}, \qquad (8.12)$$

where $\lambda_{nr} = \frac{\Omega_D(K_I+1)}{\lambda_{\text{th}}\Omega_I m_D}$ and $Q_n(a,b) = Q_{n,n-1}(a,b)/a^{n-1}$ is the generalized Marcum Q-function [39]. Similarly, the closed-form results of the
LCR for purely interference-limited/power-limited system in this sce-
nario can be obtained by by setting $s_{\text{th}} = 0/\lambda_{\text{th}} = 0$ in (8.11) [13].

2.4   Numerical Examples

Figure 8.1 plots the AOD versus the normalized SIR threshold $\frac{\Omega_D}{\Omega_I \lambda_{\text{th}}}$
when $L = 1$, $N_I = 6$, $m_D = 2$, $m_I = 2$, and for various values of the normalized signal power thresholds $\frac{s_{th}}{\Omega_D}$. From this figure, we can see that
the minimum desired signal power requirement causes a floor on the
AOD, since for $s_{\text{th}} > 0$ increasing the normalized SIR threshold above
a particular value does not reduce the AOD and the system becomes
power limited. This floor and the value of the normalized SIR threshold
at which this floor occurs depends on s_{th}/Ω_D for a fixed fading parameters. This can be seen from Fig. 8.2 which shows the AOD versus
the normalized desired signal power threshold for various values of the
normalized SIR threshold. On the other hand, Figs. 8.3 and 8.4 show
the effects of the speeds of desired and interfering users on the AOD.
Figure 8.3 plots the AOD versus the normalized SIR threshold $\frac{\Omega_D}{\Omega_I \lambda_{\text{th}}}$
when $L = 1$, $N_I = 6$, $f_D = f_I = 40$ Hz, $\frac{s_{th}}{\Omega_D} = -20$ dB, $m_D = 2$, and
$m_I = 2$, and for various values of the desired user maximum Doppler
frequencies. It is clear that the desired user Doppler frequency (or equivalently speed) has a significant impact on the system AOD. And from
Fig. 8.4, we can also conclude that the interfering users speed has a
negligible effect of the system AOD especially in the medium to high
normalized SIR threshold range. Furthermore, Figure 8.5 shows that for
interference-limited system a relatively significant decrease in the AOD

is obtained as the number of diversity paths L increases and on the other hand the increase in the number of interferers N_I results in performance degradation in AOD, especially when normalized SIR threshold $\frac{\Omega_D}{\Omega_I \lambda_{th}}$ is small. Obviously diminishing diversity gain is obtained as the number of diversity paths L increases.

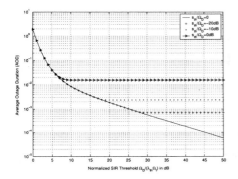

Figure 8.1. AOD versus $\frac{\Omega_D}{\Omega_I \lambda_{th}}$ for various values of $\frac{s_{th}}{\Omega_D}$ when $L = 1$, $N_I = 6$, $f_D = f_I = 40$ Hz, $m_D = 2$, and $m_I = 2$.

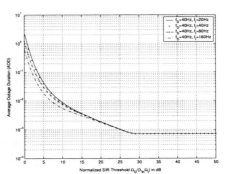

Figure 8.2. AOD versus $\frac{s_{th}}{\Omega_D}$ for various values of $\frac{\Omega_D}{\Omega_I \lambda_{th}}$ when $L = 1$, $N_I = 6$, $f_D = f_I = 40$ Hz, $m_D = 2$, and $m_I = 2$.

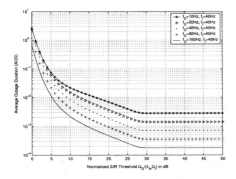

Figure 8.3. AOD versus $\frac{\Omega_D}{\Omega_I \lambda_{th}}$ for various values of the f_D when $L = 1$, $N_I = 6$, $f_I = 40$ Hz, $\frac{s_{th}}{\Omega_D} = -20$ dB, $m_D = 2$, and $m_I = 2$.

Figure 8.4. AOD versus $\frac{\Omega_D}{\Omega_I \lambda_{th}}$ for various values of f_I when $L = 1$, $N_I = 6$, $f_D = 40$ Hz, $\frac{s_{th}}{\Omega_D} = -20$ dB, $m_D = 2$, and $m_I = 2$.

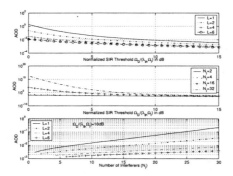

Figure 8.5. Effects of N_I and L on aod when $L = 1$, $N_I = 6$, $m_D = m_I = 2$, and $f_D = f_I = 40$Hz.

3.   Impact of CCI and Fading Correlation on the LCR and AOD of SC

3.1   System Model

For generality purpose, we consider a cellular mobile radio system where the mobile station has an L-branch antenna SC diversity combiner. The received signal at the ith antenna, operating in presence of N_I co-channel interferers, may be written as

$$r_i(t) = \alpha_{D,i}\cos(\omega_c t + \theta_i(t)) + \sum_{j=1}^{N_I} \alpha_{I,ij}\cos(\omega_c t + \psi_{ij}(t)) + n_i(t),$$

(8.13)

where ω_c is the carrier frequency, $\theta_i(t)$ and $\psi_{ij}(t)$ are the random phases of the desired signal and jth interfering signal, respectively, $n_i(t)$ is the additive Gaussian noise, $\alpha_{D,i}$ and $\alpha_{I,ij}$ are the received random envelopes at any time t of desired signal and jth interfering signal, respectively.

We consider the system in which the desired and interfering signals are independent and may have different fading statistics [31, 5, 32, 33, 7]. The received desired signal envelopes $\alpha_{D,i}, i = 1, \cdots, L$, are assumed to be independent but not necessarily identically distributed with the average fading power $\Omega_{D,i}, i = 1, \cdots, L$ and subject to flat Rayleigh, Nakagami, or Rician type of fading. The total $N_I L$ received co-channel interfering signal envelopes $\alpha_{I,ij}, i = 1, \cdots, L, j = 1, \cdots, N_I$, over the L diversity paths are assumed to be i.i.d. with the same average fading power Ω_I and subject to flat Rayleigh, Nakagami, or Rician type of fading.

SC is one of the simplest micro-diversity combining methods [15], which always picks the best branch as current operating branch to smooth out the deep fades and severe interferences. When a SC combiner is subject to CCI, the selection algorithm is not unique [40]. More specifically, the selection of branches for SC can be based on total (desired plus interference) power (i.e. best S+I algorithm), desired signal power (best signal power algorithm or best S algorithm), or SIR of the received signal at each branch output (best SIR algorithm).

Of the three algorithms, the best SIR algorithm usually provides the best performance for interference-limited systems, but is the most complex to implement, the best signal power algorithm requires identification and separation of desired signal and interferers (which can be achieved using different pilots [40]), and the best S+I algorithm is the most practical to implement. When SC is employed at the antenna array, the output SIR λ depends on the selection algorithm and as such is given by

$$\lambda = \frac{\alpha_D^2}{\alpha_I^2} = \begin{cases} \frac{\max_l\{\alpha_{D,l}^2\}}{\sum_{j=1}^{N_I}\alpha_{I,ij}^2}, & \text{For Best S Alg.} \\[3mm] \max_l\{\lambda_l\} = \max_l\{\frac{\alpha_{D,l}^2}{\sum_{j=1}^{N_I}\alpha_{I,lj}^2}\}, & \text{For Best SIR Alg.} \\[3mm] \frac{\alpha_{D,i}^2}{\sum_{j=1}^{N_I}\alpha_{I,ij}^2}\Big|_{\alpha_{D,i}^2+\sum_{j=1}^{N_I}\alpha_{I,ij}^2\max.}, & \text{For Best S+I Alg.} \end{cases} \quad , (8.14)$$

where α_D and α_I are the amplitude of the desired user and the composite amplitude of the co-channel interferers both at the SC combiner output.

3.2 Performance Comparison of Different SC Algorithms in Presence of CCI

Previous related work on the performance analysis of SC has included the following. Outage probability of SC in presence of CCI but without a minimum signal power requirement has been extensively studied over the past two decades (see for example [1, 2, 3, 4, 5, 6, 7, 8, 9]). Specifically, Jakes [40] derived the outage probability for the interference-limited case of SC in a single interferer Rayleigh fading scenario using the best signal power algorithm. Based on the best S+I algorithm, Schiff [41] presented the PDF of output SIR of SC in a single interferer Rayleigh fading scenario. Sowerby and Williamson [2] considered the multiple interferers case with SC over Rayleigh fading using both best SIR and best S+I algorithms. Abu-Dayya and Beaulieu [6] generalized the analysis of the interference-limited case of SC to the Nakagami fading environment for all three algorithm mentioned above (but obtained no closed-form for best S+I algorithm case). Finally, Okui considered dual-branch SC in correlated Nakagami fading using best signal power algorithm in [4] and using best SIR algorithm in [8]. Yang and Alouini in [9] recently studied the outage probability of SC in single interferer Rayleigh fading environment for best signal power algorithm with the consideration of both minimum SIR and signal power constraints. On the other hand, the LCR and AOD of SC over various fading channels but in absence of interference has been recently investigated in [24, 22, 23]. In [26], we complement these previous studies and present uniform analytical approaches for the outage probability, LCR, and AOD computation of SC diversity schemes in presence of multiple CCI. In our analysis we consider both minimum SIR and desired signal power constraints for all three algorithms. We then specialize our analysis to a low-complexity dual-branch SC receiver subject to multiple interferers over Rayleigh fading channels. This particular interference/fading scenario applies to small-size mobile terminals operating in heavily loaded macrocellular networks. Although more general fading scenarios (such as Rice and Nakagami) are also of interest, the relatively simple closed-form results obtained for the special case of interest provide useful insight for the design of SC diversity systems in presence of CCI.

3.2.1   General Method for SC Performance Analysis.

For communication systems which are both interference and power limited, the outage is defined in section 2.2. Extending the method used by Schiff in [41], outage probability P_{out} can be written for all three selection algorithms of SC as

$$P_{out} = 1 - \Pr[\text{System is not in outage}]$$

$$= 1 - \sum_{i=1}^{L} \Pr\left(\lambda_i > \lambda_{th}, s_{D,i} > s_{th}, \rho_i \text{ is the max of } \{\rho_k\}_{k=1}^{L}\right)$$

$$= 1 - \sum_{i=1}^{L} \underbrace{\iiint}_{s_D, \lambda, \rho \, \in \, G} f_i(s_D, \lambda, \rho) \prod_{\substack{k=1 \\ k \neq i}}^{L} F_k(\rho) \, d\rho \, d\lambda \, ds_D, \qquad (8.15)$$

where $f_i(s_D, \lambda, \rho)$ is the JPDF of $s_{D,i}$, λ_i, and ρ_i at the ith branch, $F_k(\rho)$ is the CDF of ρ_k at the kth branch, G stands for the non-outage area with respect to s_D, λ, and ρ, the symbol ρ represents the statistics used for selection algorithm. When the best signal power algorithm is employed, $\rho = s_D$ and $G = \{\lambda \geq \lambda_{th} \cap s_D \geq s_{th}\}$. If the selection is based on the best SIR algorithm, then $\rho = \lambda$ and still $G = \{\lambda \geq \lambda_{th} \cap s_D \geq s_{th}\}$. When the best S+I algorithm is employed, we have $\rho = s_D + s_I = \alpha_D^2 + \alpha_I^2$ and $G = \{s_D \geq s_{th} \cap s_D \leq \rho \leq \frac{1 + \lambda_{th}}{\lambda_{th}} s_D\}$ since the non-outage condition

$\{\lambda = \frac{s_D}{s_I} = \frac{s_D}{\rho - s_D} \geq \lambda_{\text{th}}$ and $s_D \geq s_{\text{th}}\}$ is equivalent to $\{s_D \leq \rho \leq \frac{1 + \lambda_{\text{th}}}{\lambda_{\text{th}}} s_D$ and $s_D \geq s_{\text{th}}\}$.

To evaluate the AOD, $T(\lambda_{\text{th}}, s_{\text{th}})$ in (8.2) for SC system, we also need to calculate LCR $N(\lambda_{\text{th}}, s_{\text{th}})$ which can be obtain by using the method provided in section 2.2, except that the conditional JPDF $f_{r,\dot{r}|A}(r, \dot{r})$ for SC can be written as [26, Eqn. (20)]

$$f_{r,\dot{r}|A}(r, \dot{r}) = \int_{\sqrt{\frac{s_{\text{th}}}{\lambda_{\text{th}}}}}^{\infty} \int_{-\infty}^{\infty} \alpha_I^2 \, f_{\alpha_D, \dot{\alpha}_D}(\alpha_I r, \dot{r}\alpha_I + \dot{\alpha}_I r)$$
$$\times f_{\alpha_I|A}(\alpha_I) \, f_{\dot{\alpha}_I}(\dot{\alpha}_I) \, d\dot{\alpha}_I d\alpha_I, \qquad (8.16)$$

where $f_{\alpha_D, \dot{\alpha}_D}(\alpha, \dot{\alpha})$ is the JPDF of α_D and its time derivative $\dot{\alpha}_D$ which depends on the selection algorithm employed in SC system and can be obtained in a uniform way for all three algorithms as [26, Eqn. (16)]

$$f_{\alpha_D, \dot{\alpha}_D}(\alpha, \dot{\alpha}) = \sum_{i=1}^{L} f_{\dot{\alpha}_{D,i}}(\dot{\alpha}) \underbrace{\int f_i(\alpha, \rho)}_{\rho \in Q'} \prod_{\substack{k=1 \\ k \neq i}}^{L} F_k(\rho) \, d\rho, \qquad (8.17)$$

where $f_i(\alpha, \rho)$ is the JPDF of $\alpha_{D,i}$ and ρ_i at the ith branch, $f_{\dot{\alpha}_{D,i}}(\dot{\alpha})$ is the PDF of $\dot{\alpha}_{D,i}$, the time derivative of $\alpha_{D,i}$, which is independent of $\alpha_{D,i}$ and $\alpha_{I,i}$ (therefore of ρ_i), and $Q' = \{\alpha_D = \alpha \cap 0 < \rho < \infty\}$.

3.2.2 Application to Different SC Algorithms.

Best Signal Power Algorithm. When the best signal power algorithm is employed at the SC combiner, the outage probability for dual-branch i.i.d. Rayleigh fading case can be shown by applying the method provided in last section to be given in closed-form as

$$\mathrm{P^S}_{\text{out}} = [1 - \exp(-S_{\text{r}})]^2 + 2 \sum_{n=0}^{N_I - 1} \frac{\lambda_{\text{rr}}^n}{n!}$$
$$\times \left\{ \frac{\Gamma[n + 1, S_{\text{r}}(1 + \lambda_{\text{rr}})]}{(1 + \lambda_{\text{rr}})^{n+1}} - \frac{\Gamma[n + 1, S_{\text{r}}(2 + \lambda_{\text{rr}})]}{(2 + \lambda_{\text{rr}})^{n+1}} \right\}. \qquad (8.18)$$

where $S_{\text{r}} = \frac{s_{\text{th}}}{\Omega_D}$ and $\lambda_{\text{rr}} = \frac{\Omega_D}{\lambda_{\text{th}}\Omega_I}$. Eqn. (8.18) can be shown to reduce to [9, Eqn. (5)]) when $N_I = 1$, as expected. The LCR with a minimum power requirement $N(\lambda_{\text{th}}, s_{\text{th}})$ for the best signal power algorithm can also be obtained in closed-form as

$$N_S\left(\lambda_{\text{th}},s_{\text{th}}\right)=2\sqrt{2\pi}f_D\sqrt{S_r}\left[\exp(-S_r)-\exp(-2S_r)\right]$$

$$\times\left[1-\frac{\Gamma(N_I,S_r\lambda_{\text{rr}})}{\Gamma(N_I)}\right]+\frac{2\sqrt{2\pi}\lambda_{\text{rr}}{}^{N_I}\left(f_D^2+\frac{f_I^2}{\lambda_{\text{rr}}}\right)^{\frac{1}{2}}}{\Gamma(N_I)}$$

$$\times\left\{\frac{\Gamma[N_I+\tfrac{1}{2},S_r(1+\lambda_{\text{rr}})]}{(1+\lambda_{\text{rr}})^{N_I+\frac{1}{2}}}-\frac{\Gamma[N_I+\tfrac{1}{2},S_r(2+\lambda_{\text{rr}})]}{(2+\lambda_{\text{rr}})^{N_I+\frac{1}{2}}}\right\}.$$

$$(8.19)$$

Then the corresponding AOD $T_S(\lambda_{\text{th}},s_{\text{th}})$ is just the ratio of (8.18) and (8.19) as per (8.2).

Special Cases

- Interference-limited Case
  The LCR $N(\lambda_{\text{th}})$ for interference-limited systems can be obtained by setting $s_{\text{th}}=0$ in (8.19) which results into the compact formula

$$N_S(\lambda_{\text{th}})=\frac{2\sqrt{2\pi}\lambda_{\text{rr}}{}^{N_I}\Gamma\left(N_I+\tfrac{1}{2}\right)\left(f_D^2+\frac{f_I^2}{\lambda_{\text{rr}}}\right)^{\frac{1}{2}}}{\Gamma(N_I)}$$

$$\times\left[\frac{1}{(1+\lambda_{\text{rr}})^{N_I+\frac{1}{2}}}-\frac{1}{(2+\lambda_{\text{rr}})^{N_I+\frac{1}{2}}}\right],\qquad(8.20)$$

and the corresponding P_{out} can be obtained by evaluating (8.18) at $s_{\text{th}}=0$ as

$$P^S{}_{\text{out}}=1-2\left(\frac{\lambda_{\text{rr}}}{1+\lambda_{\text{rr}}}\right)^{N_I}+\left(\frac{\lambda_{\text{rr}}}{2+\lambda_{\text{rr}}}\right)^{N_I}.\qquad(8.21)$$

Hence, as per (8.2), the resulting AOD is just the ratio of (8.21) and (8.20).

- Power-limited Case
  The outage probability of the power-limited case P_{out} can be obtained by evaluating (8.18) at $\lambda_{\text{th}}=0$ or [36, Eqn. (6.7)] as

$$P^S{}_{\text{out}}=[1-\exp(-S_r)]^2.\qquad(8.22)$$

Setting $\lambda_{\text{th}}=0$ in (8.19), we obtain the corresponding LCR $N(s_{\text{th}})$ of the signal power s_D at level s_{th} (purely power-limited case) which is given by

$$N_S(s_{\text{th}})=2\sqrt{2\pi}f_D\sqrt{S_r}\left[\exp(-S_r)-\exp(-2S_r)\right],\qquad(8.23)$$

and the resulting AOD is

$$T_{\mathrm{S}}\left(s_{\mathrm{th}}\right) = \frac{\exp(S_{\mathrm{r}}) - 1}{2\sqrt{2\pi} f_D \sqrt{S_{\mathrm{r}}}}, \tag{8.24}$$

in agreement with [24, Eqns. (18a) and (20a)] for the dual-branch case (or equivalently [22, Eqns. (15) and (16)] and [23, Eqn. (18)] with $L = 2$).

Best SIR Algorithm. When the selection decision is based on the best SIR algorithm, the diversity branch with the highest SIR is selected. The best SIR algorithm usually provides the best performance for interference-limited systems, but is the most complex to implement. In this section, we study the performance of SC system based on the best SIR algorithm for both interference-limited and power-limited systems.

When the best SIR algorithm is employed at the SC combiner, the outage probability for dual-branch i.i.d. Rayleigh fading case can be shown to be given by

$$
\begin{aligned}
P_{\mathrm{out}}^{\mathrm{SIR}} = {}& 1 - 2\exp(-S_{\mathrm{r}}) \left[1 - \frac{\Gamma(N_I, S_{\mathrm{r}}\lambda_{\mathrm{rr}})}{\Gamma(N_I)} \right] - 2\left(1 + \frac{1}{\lambda_{\mathrm{rr}}}\right)^{-N_I} \frac{\Gamma[N_I, S_{\mathrm{r}}(1+\lambda_{\mathrm{rr}})]}{\Gamma(N_I)} \\
& + 2N_I \sum_{k=0}^{N_I} \frac{1}{k!} \sum_{l=0}^{2N_I-1} \binom{2N_I-1}{l}(-1)^l S_{\mathrm{r}}^{l+1} \\
& \times \{\Gamma[k-l-1, S_{\mathrm{r}}] - \Gamma[k-l-1, S_{\mathrm{r}}(1+\lambda_{\mathrm{rr}})]\}.
\end{aligned} \tag{8.25}
$$

The corresponding LCR can be obtained as

$$
\begin{aligned}
N_{\mathrm{SIR}}\left(\lambda_{\mathrm{th}}, s_{\mathrm{th}}\right) = {}& 2\sqrt{2\pi} f_D \sqrt{S_{\mathrm{r}}} \left[\exp(-S_{\mathrm{r}}) \right. \\
& \left. - \frac{1}{\Gamma(N_I)} \sum_{l=0}^{2N_I-1} \binom{2N_I-1}{l}(-S_{\mathrm{r}})^l \Gamma(N_I-l, S_{\mathrm{r}}) \right] \\
& \times \left[1 - \frac{\Gamma(N_I, S_{\mathrm{r}}\lambda_{\mathrm{rr}})}{\Gamma(N_I)} \right] + \frac{2\sqrt{2\pi}\lambda_{\mathrm{rr}}^{N_I}\left(f_D^2 + \frac{f_I^2}{\lambda_{\mathrm{rr}}}\right)^{\frac{1}{2}}}{\Gamma(N_I)} \\
& \times \left\{ \frac{\Gamma[N_I+\frac{1}{2}, S_{\mathrm{r}}(1+\lambda_{\mathrm{rr}})]}{(1+\lambda_{\mathrm{rr}})^{N_I+\frac{1}{2}}} - \frac{2}{\lambda_{\mathrm{rr}}^{N_I+\frac{1}{2}}\Gamma(N_I)} \sum_{l=0}^{2N_I-1} \binom{2N_I-1}{l} \right. \\
& \left. \times (-\frac{1}{\lambda_{\mathrm{rr}}})^l \int_{\sqrt{s_{\mathrm{th}}\lambda_{\mathrm{rr}}}}^{\infty} y^{2l+2N_I} \exp\left(-y^2\right) \Gamma\left(N_I-l, \frac{y^2}{\lambda_{\mathrm{rr}}}\right) dy \right\}. \tag{8.26}
\end{aligned}
$$

Note that the difference between the LCR results for best signal power algorithm and best SIR algorithm given in (8.19) and (8.26) is only in the subtraction parts. Setting $s_{th} = 0/\lambda_{th} = 0$ in (8.25) and (8.26), one can get new results for the LCR and AOD for interference-limited case/power-limited case in closed-form [26].

Best S+I Algorithm.   In the best S+I algorithm, the diversity branch with the highest total (desired plus interference) power is selected. This algorithm is the most practical to implement among the three algorithms. In the following subsection, the performance of SC system based on the best S+I algorithm for both interference-limited and power-limited systems will be investigated.

When the best SIR algorithm is employed at the SC combiner, the outage probability for independent not necessarily identically distributed dual-branch Rayleigh fading case can be shown to be given by [26]

$$P_{out}^{SI} = 1 - \sum_{\substack{i=1 \\ k=3-i}}^{2} (I_1 - I_2 - I_3 + I_4), \qquad (8.27)$$

where

$$I_1 = \exp\left(-\frac{s_{th}}{\Omega_{D,i}}\right)\left[1 - \frac{\Gamma\left(N_I, \frac{s_{th}}{\lambda_{th}\Omega_I}\right)}{\Gamma(N_I)}\right]$$
$$+ \left(1 + \frac{\lambda_{th}\Omega_I}{\Omega_{D,i}}\right)^{-N_I} \frac{\Gamma\left(N_I, \frac{s_{th}}{\Omega_{D,i}} + \frac{s_{th}}{\lambda_{th}\Omega_I}\right)}{\Gamma(N_I)}, \qquad (8.28)$$

$$I_2 = \frac{1}{2^{N_I}\Omega_{D,i}\Gamma(N_I)} \sum_{t=0}^{N_I-1} \frac{\left(\frac{1}{2}\right)^t}{t!} \sum_{z=0}^{N_I-1} \binom{N_I-1}{z}\left(\frac{2}{\Omega_I}\right)^z$$
$$\times \Gamma(N_I+t-z) \sum_{m=0}^{N_I+t-z-1} \frac{\left(\frac{2}{\Omega_I}\right)^m}{m!}\left[\left(\frac{1}{\Omega_{D,i}} + \frac{1}{\Omega_I}\right)^{-m-z-1}\right.$$
$$\times \Gamma\left(m+z+1, \frac{s_{th}}{\Omega_{D,i}} + \frac{s_{th}}{\Omega_I}\right) - \left(\frac{1}{\Omega_{D,i}} + \frac{1}{\Omega_I} + \frac{2}{\Omega_I\lambda_{th}}\right)^{-m-z-1}$$
$$\times \left.\left(\frac{1+\lambda_{th}}{\lambda_{th}}\right)^m \Gamma\left(m+z+1, \frac{s_{th}}{\Omega_{D,i}} + \frac{s_{th}}{\Omega_I} + \frac{2s_{th}}{\lambda_{th}\Omega_I}\right)\right], \qquad (8.29)$$

$$I_3 = \frac{\left(1 - \frac{\Omega_I^2}{\Omega_{D,k}^2}\right)^{-N_I}}{\Omega_{D,i}\Gamma(N_I)} \sum_{z=0}^{N_I-1} \binom{N_I-1}{z} \left(\frac{1}{\Omega_I} - \frac{1}{\Omega_{D,k}}\right)^z$$

$$\times \Gamma(N_I - z) \sum_{m=0}^{N_I-z-1} \frac{\left(\frac{1}{\Omega_I} + \frac{1}{\Omega_{D,k}}\right)^m}{m!}$$

$$\times \left[\left(\frac{1}{\Omega_{D,i}} + \frac{1}{\Omega_{D,k}}\right)^{-m-z-1} \Gamma\left(m+z+1, \frac{s_{\text{th}}}{\Omega_{D,i}} + \frac{s_{\text{th}}}{\Omega_{D,k}}\right)\right.$$

$$- \left(\frac{1}{\Omega_{D,i}} + \frac{1+\lambda_{\text{th}}}{\Omega_{D,k}\lambda_{\text{th}}} + \frac{1}{\Omega_I\lambda_{\text{th}}}\right)^{-m-z-1} \left(\frac{1+\lambda_{\text{th}}}{\lambda_{\text{th}}}\right)^m$$

$$\left.\times \Gamma\left(m+z+1, \frac{s_{\text{th}}}{\Omega_{D,i}} + \frac{s_{\text{th}}(1+\lambda_{\text{th}})}{\lambda_{\text{th}}\Omega_{D,k}} + \frac{s_{\text{th}}}{\lambda_{\text{th}}\Omega_I}\right)\right], \tag{8.30}$$

$$I_4 = \frac{\left(1 - \frac{\Omega_I}{\Omega_{D,k}}\right)^{-N_I}}{\Omega_{D,i}\Omega_I^{N_I}\Gamma(N_I)} \sum_{z=0}^{N_I-1} \frac{\left(\frac{1}{\Omega_I} - \frac{1}{\Omega_{D,k}}\right)^z}{z!} \sum_{t=0}^{N_I-1} \binom{N_I-1}{t} (-1)^t$$

$$\times \left(\frac{2}{\Omega_I}\right)^{N_I+z-t} \Gamma(N_I+z-t) \sum_{m=0}^{N_I+z-t-1} \frac{\left(\frac{2}{\Omega_I}\right)^m}{m!} \left[\left(\frac{1}{\Omega_{D,i}} + \frac{1}{\Omega_I}\right)^{-m-t-1}\right.$$

$$\times \Gamma\left(m+t+1, \frac{s_{\text{th}}}{\Omega_{D,i}} + \frac{s_{\text{th}}}{\Omega_I}\right) - \left(\frac{1}{\Omega_{D,i}} + \frac{1}{\Omega_I} + \frac{2}{\Omega_I\lambda_{\text{th}}}\right)^{-m-t-1}$$

$$\left.\times \left(\frac{1+\lambda_{\text{th}}}{\lambda_{\text{th}}}\right)^m \Gamma\left(m+t+1, \frac{s_{\text{th}}}{\Omega_{D,i}} + \frac{s_{\text{th}}}{\Omega_I} + \frac{2s_{\text{th}}}{\lambda_{\text{th}}\Omega_I}\right)\right]. \tag{8.31}$$

The corresponding LCR for i.i.d Rayleigh case can be obtained as

$$N_{\text{SI}}(\lambda_{\text{th}}, s_{\text{th}}) = \sqrt{2\pi} f_D \sqrt{\Omega_D}(R_1 - R_2 - R_3 + R_4)|_{\alpha = \sqrt{s_{\text{th}}}}$$

$$\times \left[1 - \frac{\Gamma(N_I, S_r\lambda_{\text{rr}})}{\Gamma(N_I)}\right] + \frac{4\sqrt{2\pi}\sqrt{\frac{\lambda_{\text{th}}}{\Omega_D}}}{\Gamma(N_I)\Omega_I^{N_I}} \left(f_D^2 + \frac{f_I^2\Omega_I\lambda_{\text{th}}}{\Omega_D}\right)^{\frac{1}{2}}$$

$$\times (A_1 - A_2 - A_3 + A_4), \tag{8.32}$$

where R_1-R_4 and A_1-A_4 are given by

$$R_1 = \frac{2\alpha}{\Omega_{D,i}}\exp\left(-\frac{\alpha^2}{\Omega_{D,i}}\right), \tag{8.33}$$

$$R_2 = \frac{\alpha}{2^{N_I-1}\Omega_{D,i}\Gamma(N_I)}\exp\left(-\frac{\alpha^2}{\Omega_{D,i}}+\frac{\alpha^2}{\Omega_I}\right)\sum_{t=0}^{N_I-1}\frac{\left(\frac{1}{2}\right)^t}{t!}$$
$$\times\sum_{z=0}^{N_I-1}\binom{N_I-1}{z}\left(-\frac{2\alpha^2}{\Omega_I}\right)^z\Gamma\left(N_I+t-z,\frac{2\alpha^2}{\Omega_I}\right), \tag{8.34}$$

$$R_3 = \frac{2\alpha\left(1-\frac{\Omega_I^2}{\Omega_{D,k}^2}\right)^{-N_I}}{\Omega_{D,i}\Gamma(N_I)}\exp\left(-\frac{\alpha^2}{\Omega_{D,i}}+\frac{\alpha^2}{\Omega_I}\right)\sum_{z=0}^{N_I-1}\binom{N_I-1}{z}$$
$$\times\left(\frac{\alpha^2}{\Omega_I}-\frac{\alpha^2}{\Omega_{D,k}}\right)^z\Gamma\left(N_I-z,\frac{\alpha^2}{\Omega_I}+\frac{\alpha^2}{\Omega_{D,k}}\right), \tag{8.35}$$

$$R_4 = \frac{\alpha\left(1-\frac{\Omega_I}{\Omega_{D,k}}\right)^{-N_I}}{\Omega_{D,i}2^{N_I-1}\Gamma(N_I)}\exp\left(-\frac{\alpha^2}{\Omega_{D,i}}+\frac{\alpha^2}{\Omega_I}\right)\sum_{z=0}^{N_I-1}\frac{\frac{1}{2^z}\left(1-\frac{\Omega_I}{\Omega_{D,k}}\right)^z}{z!}$$
$$\times\sum_{t=0}^{N_I-1}\binom{N_I-1}{t}\left(\frac{2\alpha^2}{\Omega_I}\right)^t\Gamma\left(N_I+z-t,\frac{2\alpha^2}{\Omega_I}\right), \tag{8.36}$$

$$A_1 = \frac{1}{2}\left(\frac{1}{\Omega_I}+\frac{\lambda_{\mathrm{th}}}{\Omega_D}\right)^{-N_I-\frac{1}{2}}\Gamma\left(N_I+\frac{1}{2},\frac{s_{\mathrm{th}}}{\Omega_I\lambda_{\mathrm{th}}}+\frac{s_{\mathrm{th}}}{\Omega_D}\right), \tag{8.37}$$

$$A_2 = \frac{1}{2^{N_I+1}\Gamma(N_I)}\sum_{t=0}^{N_I-1}\frac{\left(\frac{1}{2}\right)^t}{t!}\sum_{z=0}^{N_I-1}\binom{N_I-1}{z}\left(-\frac{2\lambda_{\mathrm{th}}}{\Omega_I}\right)^z$$
$$\times\Gamma(N_I+t-z)\sum_{m=0}^{N_I+t-z-1}\frac{\left(\frac{2\lambda_{\mathrm{th}}}{\Omega_I}\right)^m}{m!}\left(\frac{\lambda_{\mathrm{th}}}{\Omega_D}+\frac{\lambda_{\mathrm{th}}}{\Omega_I}+\frac{1}{\Omega_I}\right)^{-N_I-m-z-\frac{1}{2}}$$
$$\times\Gamma\left(N_I+m+z+\frac{1}{2},\frac{s_{\mathrm{th}}}{\Omega_D}+\frac{s_{\mathrm{th}}}{\lambda_{\mathrm{th}}\Omega_I}+\frac{s_{\mathrm{th}}}{\Omega_I}\right), \tag{8.38}$$

$$A_3 = \frac{\left(1-\frac{\Omega_I^2}{\Omega_D^2}\right)^{-N_I}}{2\Gamma(N_I)}\sum_{z=0}^{N_I-1}\binom{N_I-1}{z}\left(\frac{\lambda_{\mathrm{th}}}{\Omega_I}-\frac{\lambda_{\mathrm{th}}}{\Omega_D}\right)^z$$
$$\times\Gamma(N_I-z)\sum_{m=0}^{N_I-z-1}\frac{\left(\frac{\lambda_{\mathrm{th}}}{\Omega_I}+\frac{\lambda_{\mathrm{th}}}{\Omega_D}\right)^m}{m!}\left(\frac{1}{\Omega_I}+\frac{2\lambda_{\mathrm{th}}}{\Omega_D}\right)^{-N_I-m-z-\frac{1}{2}}$$
$$\times\Gamma\left(N_I+m+z+\frac{1}{2},\frac{2s_{\mathrm{th}}}{\Omega_D}+\frac{s_{\mathrm{th}}}{\lambda_{\mathrm{th}}\Omega_I}\right), \tag{8.39}$$

$$A_4 = \frac{\left(1 - \frac{\Omega_I}{\Omega_D}\right)^{-N_I}}{2^{N_I+1}\Gamma(N_I)} \sum_{z=0}^{N_I-1} \frac{1}{2^z} \frac{\left(1 - \frac{\Omega_I}{\Omega_D}\right)^z}{z!} \sum_{t=0}^{N_I-1} \binom{N_I-1}{t} \left(\frac{2\lambda_{\mathrm{th}}}{\Omega_I}\right)^t$$

$$\times \Gamma(N_I+z-t) \sum_{m=0}^{N_I+z-t-1} \frac{\left(\frac{2\lambda_{\mathrm{th}}}{\Omega_I}\right)^m}{m!} \left(\frac{\lambda_{\mathrm{th}}}{\Omega_D} + \frac{\lambda_{\mathrm{th}}}{\Omega_I} + \frac{1}{\Omega_I}\right)^{-N_I-m-t-\frac{1}{2}}$$

$$\times \Gamma\left(N_I+m+t+\frac{1}{2}, \frac{s_{\mathrm{th}}}{\Omega_D} + \frac{s_{\mathrm{th}}}{\Omega_I} + \frac{s_{\mathrm{th}}}{\lambda_{\mathrm{th}}\Omega_I}\right). \tag{8.40}$$

3.2.3   Numerical Examples.   Figures 8.6, 8.7, and 8.8 compare the system performance of dual-branch SC systems for all three algorithms with a minimum power requirement over i.i.d. Rayleigh diversity paths and in presence of multiple i.i.d. interfering Rayleigh faded signals. In particular, these figures show the comparisons of the outage probabilities, AODs, and average LCRs of SC as function of average SIR $\frac{\Omega_D}{\Omega_I}$ when $\frac{s_{\mathrm{th}}}{\Omega_D} = -10$dB, $\lambda_{\mathrm{th}} = 8$dB, and $N_I = 1$ and $N_I = 6$. They show that in the senses of outage probability and AOD, the best SIR algorithm provides the best performance over low average SIR range and the best signal power algorithm and best S+I algorithm, which have nearly the same performance for all three performance criteria over the entire range of average SIR shown in these figures, outperform the best SIR algorithm for high average SIR range. This is in agreement with the simulation results of outage probability reported in [9]. However, from Fig. 8.8, one can see that the best SIR algorithm almost always provides the worst performance for average LCR. Therefore in comparison to the best signal power algorithm and the best S+I, the best SIR algorithm yields error bursts with lower probability and shorter duration (in average). However the frequency of these bursts of errors is higher with the best SIR algorithm. It is also shown in these figures that the performance differences of SC for all three algorithms is greater when single interferer is present (i.e. $N_I = 1$).

3.3   Impact of Fading Correlation on the LCR and AOD of SC

Outage probability of SC over independent and correlated channels has been extensively studied over the past two decades (e.g., [4, 6, 18, 9] and references therein). On the other hand, the LCR and AOD of SC in the absence of CCI over various independent fading channels has been recently investigated in [24, 22, 23]. Pioneering work on the effect of correlated fading on the AOD and LCR of SC diversity systems was done by Adachi *et al.* [25]. More specifically, [25] presented a general method for

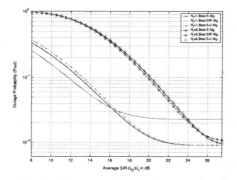

Figure 8.6. Comparison of the outage probabilities of SC for all three algorithms as function of average SIR $\frac{\Omega_D}{\Omega_I}$ when $\frac{s_{th}}{\Omega_D} = -10$dB, $\lambda_{th} = 8$dB, and $N_I = 1$ and $N_I = 6$.

Figure 8.7. Comparison of the AODs of SC for all three algorithms as function of average SIR $\frac{\Omega_D}{\Omega_I}$ when $\frac{s_{th}}{\Omega_D} = -10$dB, $\lambda_{th} = 8$dB, and $N_I = 1$ and $N_I = 6$.

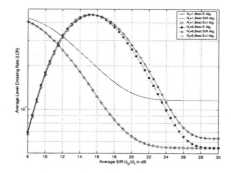

Figure 8.8. Comparison of the LCRs of SC for all three algorithms as function of average SIR $\frac{\Omega_D}{\Omega_I}$ when $\frac{s_{th}}{\Omega_D} = -10$dB, $\lambda_{th} = 8$dB, and $N_I = 1$ and $N_I = 6$.

calculating the LCR of correlated dual-branch SC system without CCI but ended up with approximate expressions for both the LCR and AOD. In [27], we complement these previous studies and present an exact and more general approach for the evaluation of the LCR and AOD of SC diversity systems with and without CCI operating over correlated and/or unbalanced multiple fading channels. The analysis is then specialized to a low-complexity dual-branch SC receiver over correlated Rayleigh and Nakagami fading channels. This particular scenario leads to relatively simple closed-form expressions and applies to small-size mobile terminals equipped with space antenna diversity where the antenna spacing is insufficient to provide independent fading among the various signal paths.

3.3.1   General Method.   We consider the SC systems employing the desired signal power algorithm. For communication systems which are both interference and power limited, the outage is defined in section 2.2. Outage probability of SC over correlated channels in presence of CCI with a minimal signal power requirement has been studied in [9] and closed-form expressions are available for the dual-branch i.i.d. diversity fading scenario combination of Rayleigh/Rayleigh in [9]. Therefore, to compute the AOD in (8.2), we just need to calculate the LCR $N(\lambda_{\text{th}}, s_{\text{th}})$ which can be obtained by using the method provided in section 2.2, except that the conditional JPDF $f_{r,\dot{r}|A}(r,\dot{r})$ for SC can be written as [26, Eqn. (20)][2]

$$
f_{r,\dot{r}|A}(r,\dot{r}) = \int_{\sqrt{\frac{s_{\text{th}}}{\lambda_{\text{th}}}}}^{\infty} \int_{-\infty}^{\infty} \alpha_I^2 \, f_{\alpha,\dot{\alpha}}(\alpha_I r, \dot{r}\alpha_I + \alpha_I r)
$$
$$
\times f_{\alpha_I|A}(\alpha_I) \, f_{\dot{\alpha}_I}(\dot{\alpha}_I) \, d\dot{\alpha}_I d\alpha_I, \tag{8.41}
$$

where the JPDF $f_{\alpha,\dot{\alpha}}(\alpha,\dot{\alpha})$ is given by

$$
f_{\alpha,\dot{\alpha}}(\alpha,\dot{\alpha})
$$
$$
= \sum_{l=1}^{L} f_{\dot{\alpha}_l}(\dot{\alpha}) \underbrace{\int_0^\alpha \cdots \int_0^\alpha}_{(L-1)-\text{fold}} f_{\alpha_1,\cdots\alpha_L}(\alpha_1,\cdots,\alpha_l=\alpha,\cdots,\alpha_L) \underbrace{d\alpha_1\cdots d\alpha_k\cdots d\alpha_L}_{\substack{(L-1)-\text{fold}\\ k\neq l}}
$$
$$
\tag{8.42}
$$

Furthermore, the LCR of purely power-limited SC system over correlated multiple channels can be obtained by applying Rice formula with (8.42) as

$$
N(\alpha_{\text{th}}) = \sum_{l=1}^{L} \frac{\sigma_l}{\sqrt{2\pi}} \underbrace{\int_0^{\alpha_{\text{th}}} \cdots \int_0^{\alpha_{\text{th}}}}_{(L-1)-\text{fold}} f_{\alpha_1,\cdots\alpha_L}(\alpha_1,\cdots,\alpha_l=\alpha_{\text{th}},\cdots,\alpha_L) \underbrace{d\alpha_1\cdots d\alpha_k\cdots d\alpha_L}_{\substack{(L-1)-\text{fold}\\ k\neq l}}
$$
$$
= \sum_{l=1}^{L} \frac{\sigma_l}{\sqrt{2\pi}} \frac{\partial F_{\alpha_1,\cdots\alpha_L}(\alpha_1,\cdots,\alpha_L)}{\partial \alpha_l}\Big|_{\{\alpha_k=\alpha_{\text{th}}\}_{k=1}^{L}}, \tag{8.43}
$$

where $F_{\alpha_1,\cdots\alpha_L}(\alpha_1,\cdots,\alpha_L)$ is the joint cumulative density function of $\alpha_1,\cdots,\alpha_L$ and $\sigma_l^2 = \Omega_x \pi^2 f_m^2$, in which f_m is the maximum Doppler frequency shift, and $\Omega_x = \Omega_l$ for Rayleigh fading and $\Omega_x = \Omega_l/m$ for Nakagami fading.

3.3.2   Application to the Bivariate Nakagami Scenario.

Let the desired signal be subject to bivariate Nakagami-m fading with PDF provided by [17, Eqn.(6.1)]. It can be shown that the LCR with a minimum power requirement $N\left(\lambda_{\text{th}}, s_{\text{th}}\right)$ is given by

$$
\begin{aligned}
&N\left(\lambda_{\text{th}}, s_{\text{th}}\right)\\
&= \Bigg\{ \sqrt{2\pi} f_m \frac{s_{\text{th}}^{m-0.5} m^{m-0.5}}{\Gamma(m)\Omega_1^{m-0.5}} \exp\left(-\frac{ms_{\text{th}}}{\Omega_1}\right) \left[1 - Q_m\left(\sqrt{\frac{2m\rho s_{\text{th}}}{(1-\rho)\Omega_1}}, \sqrt{\frac{2ms_{\text{th}}}{(1-\rho)\Omega_2}}\right)\right]\\
&\quad + \sqrt{2\pi} f_m \frac{s_{\text{th}}^{m-0.5} m^{m-0.5}}{\Gamma(m)\Omega_2^{m-0.5}} \exp\left(-\frac{ms_{\text{th}}}{\Omega_2}\right) \left[1 - Q_m\left(\sqrt{\frac{2m\rho s_{\text{th}}}{(1-\rho)\Omega_2}}, \sqrt{\frac{2ms_{\text{th}}}{(1-\rho)\Omega_1}}\right)\right] \Bigg\}\\
&\quad \times \left[1 - \frac{\Gamma\left(N_I m_I, \frac{m_I s_{\text{th}}}{\Omega_I \lambda_{\text{th}}}\right)}{\Gamma(N_I m_I)}\right] + \frac{\sqrt{2\pi} \left(\frac{m_I \Omega_1}{\lambda_{\text{th}} m\Omega_I}\right)^{N_I m_I}}{\Gamma(N_I m_I)\Gamma(m)} \left(f_m^2 + \frac{f_I^2 \lambda_{\text{th}} m\Omega_I}{m_I \Omega_1}\right)^{\frac{1}{2}}\\
&\quad \times \Bigg\{ \frac{\Gamma\left(m+N_I m_I - \frac{1}{2}, \frac{ms_{\text{th}}}{\Omega_1} + \frac{m_I s_{\text{th}}}{\Omega_I \lambda_{\text{th}}}\right)}{\left[1 + \left(\frac{m_I \Omega_1}{\lambda_{\text{th}} m\Omega_I}\right)\right]^{m+N_I m_I - \frac{1}{2}}} - 2\left(\frac{ms_{\text{th}}}{\Omega_1}\right)^{m+N_I m_I - \frac{1}{2}}\\
&\quad \times \int_1^{\infty} x^{2N_I m_I + 2m - 2} \exp\left[-\left(\frac{ms_{\text{th}}}{\Omega_1} + \frac{m_I s_{\text{th}}}{\Omega_I \lambda_{\text{th}}}\right)x^2\right]\\
&\quad \times Q_m\left(\sqrt{\frac{2ms_{\text{th}}\rho}{\Omega_1(1-\rho)}} x, \sqrt{\frac{2ms_{\text{th}}}{(1-\rho)\Omega_2}} x\right) dx \Bigg\}\\
&\quad + \frac{\sqrt{2\pi} \left(\frac{m_I \Omega_2}{\lambda_{\text{th}} m\Omega_I}\right)^{N_I m_I}}{\Gamma(N_I m_I)\Gamma(m)} \left(f_m^2 + \frac{f_I^2 \lambda_{\text{th}} m\Omega_I}{m_I \Omega_2}\right)^{\frac{1}{2}}\\
&\quad \times \Bigg\{ \frac{\Gamma\left(m+N_I m_I - \frac{1}{2}, \frac{ms_{\text{th}}}{\Omega_2} + \frac{m_I s_{\text{th}}}{\Omega_I \lambda_{\text{th}}}\right)}{\left[1 + \left(\frac{m_I \Omega_2}{\lambda_{\text{th}} m\Omega_I}\right)\right]^{m+N_I m_I - \frac{1}{2}}} - 2\left(\frac{ms_{\text{th}}}{\Omega_2}\right)^{m+N_I m_I - \frac{1}{2}}\\
&\quad \times \int_1^{\infty} x^{2N_I m_I + 2m - 2} \exp\left[-\left(\frac{ms_{\text{th}}}{\Omega_2} + \frac{m_I s_{\text{th}}}{\Omega_I \lambda_{\text{th}}}\right)x^2\right]\\
&\quad \times Q_m\left(\sqrt{\frac{2ms_{\text{th}}\rho}{\Omega_2(1-\rho)}} x, \sqrt{\frac{2ms_{\text{th}}}{(1-\rho)\Omega_1}} x\right) dx \Bigg\},
\end{aligned}
\tag{8.44}
$$

where $\rho = \frac{\text{cov}(\alpha_1^2, \alpha_2^2)}{\sqrt{\text{var}(\alpha_1^2)\text{var}(\alpha_2^2)}}$ is the correlation coefficient $(0 \le \rho < 1)$.

Special Cases

- Interference-limited Case
  The LCR $N(\lambda_{\text{th}})$ for interference-limited systems can be obtained

by setting $s_{\text{th}} = 0$ in (8.44) and using an appropriate change of variable which results into

$$
\begin{aligned}
N(\lambda_{\text{th}}) = & \frac{\sqrt{2\pi} \left(\frac{m_I \Omega_1}{\lambda_{\text{th}} m \Omega_I}\right)^{N_I m_I}}{\Gamma(N_I m_I) \Gamma(m)} \left(f_m^2 + \frac{f_I^2 \lambda_{\text{th}} m \Omega_I}{m_I \Omega_1}\right)^{\frac{1}{2}} \left\{ \frac{\Gamma(m + N_I m_I - \frac{1}{2})}{\left[1 + \left(\frac{m_I \Omega_1}{\lambda_{\text{th}} m \Omega_I}\right)\right]^{m + N_I m_I - \frac{1}{2}}} \right. \\
& - 2\left(\frac{m \lambda_{\text{th}}}{\Omega_1}\right)^{m + N_I m_I - \frac{1}{2}} \int_0^\infty y^{2 N_I m_I + 2m - 2} \exp\left[-\left(\frac{m \lambda_{\text{th}}}{\Omega_1} + \frac{m_I}{\Omega_I}\right) y^2\right] \\
& \times \left. Q_m\left(\sqrt{\frac{2 m \lambda_{\text{th}} \rho}{\Omega_1 (1 - \rho)}} y, \sqrt{\frac{2 m \lambda_{\text{th}}}{(1 - \rho) \Omega_2}} y\right) dy \right\} + \frac{\sqrt{2\pi} \left(\frac{m_I \Omega_2}{\lambda_{\text{th}} m \Omega_I}\right)^{N_I m_I}}{\Gamma(N_I m_I) \Gamma(m)} \\
& \times \left(f_m^2 + \frac{f_I^2 \lambda_{\text{th}} m \Omega_I}{m_I \Omega_2}\right)^{\frac{1}{2}} \left\{ \frac{\Gamma(m + N_I m_I - \frac{1}{2})}{\left[1 + \left(\frac{m_I \Omega_2}{\lambda_{\text{th}} m \Omega_I}\right)\right]^{m + N_I m_I - \frac{1}{2}}} \right. \\
& - 2\left(\frac{m \lambda_{\text{th}}}{\Omega_2}\right)^{m + N_I m_I - \frac{1}{2}} \int_0^\infty y^{2 N_I m_I + 2m - 2} \exp\left[-\left(\frac{m \lambda_{\text{th}}}{\Omega_2} + \frac{m_I}{\Omega_I}\right) y^2\right] \\
& \times \left. Q_m\left(\sqrt{\frac{2 m \lambda_{\text{th}} \rho}{\Omega_2 (1 - \rho)}} y, \sqrt{\frac{2 m \lambda_{\text{th}}}{(1 - \rho) \Omega_1}} y\right) dy \right\} .
\end{aligned}
\tag{8.45}
$$

- Power-limited Case

  Setting $\lambda_{\text{th}} = 0$ in (8.44), we obtain the corresponding LCR $N(s_{\text{th}})$ of the signal power s_D at level s_{th} (purely power-limited case) which is given by the nicely compact and symmetrical closed-form expression

$$
\begin{aligned}
N(s_{\text{th}}) = & \sqrt{2\pi} f_m \frac{s_{\text{th}}^{\frac{2m-1}{2}} m^{m - 0.5}}{\Gamma(m) \Omega_1^{m - 0.5}} \exp\left(-\frac{m s_{\text{th}}}{\Omega_1}\right) \\
& \times \left[1 - Q_m\left(\sqrt{\frac{2 m \rho s_{\text{th}}}{(1 - \rho) \Omega_1}}, \sqrt{\frac{2 m s_{\text{th}}}{(1 - \rho) \Omega_2}}\right)\right] \\
& + \sqrt{2\pi} f_m \frac{s_{\text{th}}^{\frac{2m-1}{2}} m^{m - 0.5}}{\Gamma(m) \Omega_2^{m - 0.5}} \exp\left(-\frac{m s_{\text{th}}}{\Omega_2}\right) \\
& \times \left[1 - Q_m\left(\sqrt{\frac{2 m \rho s_{\text{th}}}{(1 - \rho) \Omega_2}}, \sqrt{\frac{2 m s_{\text{th}}}{(1 - \rho) \Omega_1}}\right)\right] .
\end{aligned}
\tag{8.46}
$$

The corresponding outage probability for dual-branch SC over correlated Nakagami fading is available in the form of a single integral with finite limits in [17, Eqn. (6.26)]. As such taking a ratio of [17, Eqn. (6.26)] and (8.46) leads to the desired AOD result. Further setting $m = 1$ in (8.46) results in the LCR $N(\alpha_{\text{th}})$ for dual-branch SC over correlated Rayleigh fading.

- Independent and Balanced Rayleigh Channels Case
  As a double check, setting $m = m_I = 1$, $\Omega_1 = \Omega_2 = \Omega_D$, and $\rho = 0$ in (8.44), and with the help of the identity $Q_1(0, b) = \exp(-b^2/2)$ given in [39, Eqn. (2)] and the definition of the incomplete gamma function, we obtain the LCR $N_{\text{Ray}}(\lambda_{\text{th}}, s_{\text{th}})$ of dual-branch SC over independent and balanced Rayleigh branches as

$$
\begin{aligned}
&N_{\text{Ray}}(\lambda_{\text{th}}, s_{\text{th}}) \\
&= 2\sqrt{2\pi} f_m \sqrt{S_{\text{r}}} \left[\exp(-S_{\text{r}}) - \exp(-2S_{\text{r}}) \right] \\
&\times \left[1 - \frac{\Gamma(N_I, S_{\text{r}}\lambda_{\text{rr}})}{\Gamma(N_I)} \right] + \frac{2\sqrt{2\pi}\lambda_{\text{rr}}^{N_I}\left(f_m^2 + \frac{f_I^2}{\lambda_{\text{rr}}} \right)^{\frac{1}{2}}}{\Gamma(N_I)} \\
&\times \left\{ \frac{\Gamma[N_I + \frac{1}{2}, S_{\text{r}}(1+\lambda_{\text{rr}})]}{(1+\lambda_{\text{rr}})^{N_I+\frac{1}{2}}} - \frac{\Gamma[N_I + \frac{1}{2}, S_{\text{r}}(2+\lambda_{\text{rr}})]}{(2+\lambda_{\text{rr}})^{N_I+\frac{1}{2}}} \right\},
\end{aligned} \tag{8.47}
$$

where $S_{\text{r}} = \frac{s_{\text{th}}}{\Omega_D}$, $\lambda_{\text{rr}} = \frac{\Omega_D}{\lambda_{\text{th}}\Omega_I}$. Eqn.(8.47) is in perfect agreement with [26, Eqn. (36)].

3.3.3 Numerical Examples.

As numerical examples, Figs. 8.9 and 8.10 show the effects of average power unbalance and fading correlation on the AOD and LCR of power-limited SC system. We can see that, compared to a system with i.i.d. fading over the diversity paths, both the average power unbalance and the fading correlation induce a noneligible degradation on the AOD performance. Furthermore, the effect of the average fading power unbalance is more significant than that of fading correlation for the typical parameters used in these figures. Finally, the impact of fading correlation on the AOD is more important when the branches tend to be balanced.

Figs. 8.11 and 8.12 plot the AOD and LCR of power-limited SC system versus normalized outage threshold of first path $\alpha_{\text{th}}/\sqrt{\Omega_1/m}$ over correlated Nakagami channels for various values of the correlation coefficient and of the desired user Nakagami parameter when $\Omega_1 = \Omega_2$. One can see that, when the desired user Nakagami parameter m increases (i.e., the fading severity deceases), the AOD decreases significantly (i.e., better performance) and the LCR curve shifts to the right. Similar to the observation to Figs. 8.9 and 8.10, Figs. 8.11 and 8.12 also shown that the effect of the desired user Nakagami parameter is more significant than that of fading correlation. On the other hand, the impact of fading correlation on the AOD essentially remains the same as the Nakagami parameter is varied.

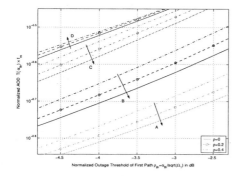

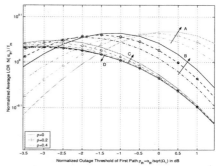

Figure 8.9. AOD of SC versus normalized outage threshold of first path $\alpha_{\text{th}}/\sqrt{\Omega_1}$ over correlated Rayleigh channels for various values of the correlation coefficient and of the average fading power unbalance. (A) $\Omega_1 = \Omega_2$, (B) $\Omega_1 = 2\,\Omega_2$, (C) $\Omega_1 = 5\,\Omega_2$, and (D) $\Omega_1 = 10\,\Omega_2$.

Figure 8.10. LCR of SC versus normalized outage threshold of first path $\alpha_{\text{th}}/\sqrt{\Omega_1}$ over correlated Rayleigh channels for various values of the correlation coefficient and of the average fading power unbalance. (A) $\Omega_1 = \Omega_2$, (B) $\Omega_1 = 2\,\Omega_2$, (C) $\Omega_1 = 5\,\Omega_2$, and (D) $\Omega_1 = 10\,\Omega_2$.

Figs. 8.13 and 8.14 show the effects of the number of interferers and fading correlation on the AOD and LCR of SC. Specifically, they plot the AOD and LCR as a function of normalized SIR threshold of first path $\frac{m_I \Omega_1}{\lambda_{\text{th}} \Omega_I m}$ over correlated Nakagami channels for various values of the correlation coefficient and of the number of interferers when $\Omega_1 = \Omega_2$, $m = m_I = 2$, $f_m = f_I = 40$Hz, and normalized desired signal power threshold of first path $\frac{s_{\text{th}} m}{\Omega_1} = -10$ dB. Similarly we can see that as the number of interferers N_I increases, LCR curve shifts to the right again and AOD performance degrades considerably. Furthermore, for the chosen parameters the effect of the number of interferers is also more significant than that of fading correlation.

As the last set of examples, Figs. 8.15, 8.16, and 8.17 show the effects of minimum signal power constraint on the outage probability, AOD, and LCR of SC. From these figures, we can see that the minimum desired signal power requirement causes a floor on the outage probability, AOD, and LCR, since for $s_{\text{th}} > 0$ increasing the normalized SIR threshold above a particular value does not reduce the outage probability, AOD, and LCR, and the system becomes noise-limited. This floor and the

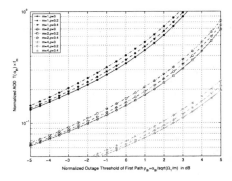

Figure 8.11. AOD of SC versus $\alpha_{\text{th}}/\sqrt{\Omega_1/m}$ over correlated Nakagami channels for various values of the correlation coefficient and of the desired user Nakagami parameter when $\Omega_1 = \Omega_2$.

Figure 8.12. LCR of SC versus $\alpha_{\text{th}}/\sqrt{\Omega_1/m}$ over correlated Nakagami channels for various values of the correlation coefficient and of the desired user Nakagami parameter when $\Omega_1 = \Omega_2$.

value of the average SIR at which this floor occurs depends on $s_{\text{th}}m/\Omega_1$ for the fixed fading parameters.

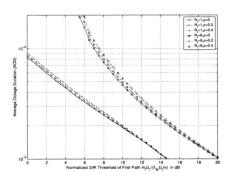

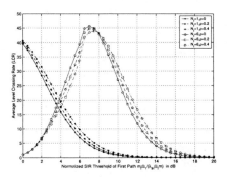

Figure 8.13. AOD of SC versus normalized SIR threshold of first path $\frac{m_I\Omega_1}{\lambda_{th}\Omega_I m}$ over correlated Nakagami channels for various values of the correlation coefficient and of the number of interferers when $\Omega_1 = \Omega_2$, $m = m_I = 2$, $f_m = f_I = 40$Hz, and normalized desired signal power threshold of first path $\frac{s_{th}m}{\Omega_1} = -10$ dB.

Figure 8.14. LCR of SC versus normalized SIR threshold of first path $\frac{m_I\Omega_1}{\lambda_{th}\Omega_I m}$ over correlated Nakagami channels for various values of the correlation coefficient and of the number of interferers when $\Omega_1 = \Omega_2$, $m = m_I = 2$, $f_m = f_I = 40$Hz, and normalized desired signal power threshold of first path $\frac{s_{th}m}{\Omega_1} = -10$ dB.

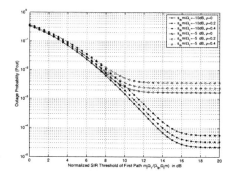

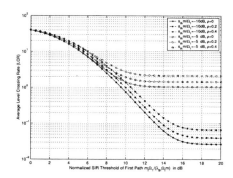

Figure 8.15. Outage Probability of SC $\frac{m_I \Omega_1}{\lambda_{th} \Omega_I m}$ over correlated Nakagami channels for various values of the correlation coefficient and of $\frac{s_{th} m}{\Omega_1}$ when $\Omega_1 = \Omega_2$, $m = m_I = 2$, $f_m = f_I = 40$Hz, and $N_I = 1$.

Figure 8.16. AOD of SC versus $\frac{m_I \Omega_1}{\lambda_{th} \Omega_I m}$ over correlated Nakagami channels for various values of the correlation coefficient and of $\frac{s_{th} m}{\Omega_1}$ when $\Omega_1 = \Omega_2$, $m = m_I = 2$, $f_m = f_I = 40$Hz, and $N_I = 1$.

Figure 8.17. LCR of SC versus $\frac{m_I \Omega_1}{\lambda_{th} \Omega_I m}$ over correlated Nakagami channels for various values of the correlation coefficient and of $\frac{s_{th} m}{\Omega_1}$ when $\Omega_1 = \Omega_2$, $m = m_I = 2$, $f_m = f_I = 40$Hz, and $N_I = 1$.

4. Conclusion

In this chapter, we presented closed-form expressions for the LCR and AOD of some diversity combining schemes of interest with and without multiple CCI and over independent, correlated, and/or unbalanced channels considering minimum SIR and/or desired signal power constraints. More specifically, it presented generic results for the LCR and AOD

(i) of MRC systems subject to CCI operating over i.i.d. Rician and/or Nakagami fading environments when a minimum desired signal power requirement is specified for satisfactory reception, (ii) of various SC diversity scheme in presence of multiple CCI and with both minimum SIR and desired signal power constraints over independent, correlated, and/or unbalanced channels. Our numerical examples showed that: (i) Specifying a certain minimum desired signal power requirement induces a floor on the AOD, (ii) The AOD is primarily affected by the the maximum Doppler frequencies (or equivalently the speed) and the amount of fading of the desired users, (iii) In the senses of outage probability and average outage duration, the best SIR algorithm provides the best performance in the low average SIR range while the best signal power algorithm and best S+I algorithm, which have nearly the same performance for all three performance criteria over the entire range of average SIR, outperform the best SIR algorithm for high average SIR range, and (iv) For typical parameters of interest the fading correlation has a noneligible impact on the system performance but its impact is less significant than that of the average fading power unbalance, desired user severity of fading, or the number of interferers.

Notes

1. By Nakagami/Rician, we mean that the desired user is subject to Nakagami fading, while the interferers are subject to Rician fading.

2. For simplicity, we use in this section α_i to stand for $\alpha_{D,i}$, Ω_i to represent $\Omega_{D,i}$, and α to be α_D.

References

[1] K. W. Sowerby and A. G. Williamson, "Outage probability calculations for multiple cochannel interferers in cellular mobile radio systems," *IEE Proc. (Pt. F)*, vol. 135, no. 3, pp. 208–215, June 1988.

[2] ——, "Selection diversity in multiple interferer mobile radio systems," *IEE Electron. Lett.*, vol. 24, pp. 1511–1513, November 1988.

[3] A. A. Abu-Dayya and N. C. Beaulieu, "Outage probabilities of cellular mobile radio systems with multiple Nakagami interferers," *IEEE Trans. Veh. Tech.*, vol. VT-40, no. 4, pp. 757–768, November 1991.

[4] S. Okui, "Probability of co-channel interference for selection diversity reception in the Nakagami m-fading channel," *IEE. Proc I*, vol. 139, no. 1, pp. 91–94, February 1992.

[5] Y. -D. Yao and A. U. H. Sheikh, "Investigation into cochannel interference in microcellular mobile radio systems," *IEEE Trans. Veh. Tech.*, vol. VT-41, no. 2, pp. 114–123, May 1992.

[6] A. A. Abu-Dayya and N. C. Beaulieu, "Outage probabilities of diversity cellular systems with cochannel inteference in Nakagami fading," *IEEE Trans. Veh. Tech.*, vol. VT-41, no. 4, pp. 343–355, November 1992.

[7] H. -C. Yang and M. -S. Alouini, "Outage probability of wireless communication systems with a minimum signal power constraint: Some closed-form expressions," *IEEE Trans. Veh. Tech.*, vol. 51, pp. 1689–1698, Nov. 2002.

[8] S. Okui, "Effects of CIR selection diversity with two correlated branches in the m-fading channel," *IEEE Trans. on Commun.*, vol. COM-48, pp. 1631–1633, October 2000.

[9] H. -C. Yang and M. -S. Alouini, "Outage probability of dual-branch diversity systems in presence of co-channel interference," in *Proc. of the 12th IEEE International Symposium on Personal, Indoor and Mobile Radio Communications (PIMRC'2001), San Diego, CA*, Sept. 2001, pp. 113–117, full journal paper in *IEEE Trans. Wireless Comm.*, vol. 2, pp. 310-319, Mar. 2003.

[10] J. -P. M. G. Linnartz and R. Prasad, "Threshold crossing rate and average non-fade duration in a Rayleigh-fading channel with multiple interferers," *Archiv Fur Elektronik und Ubertragungstechnik Electonics and Communication*, vol. 43, pp. 345–349, Nov./Dec. 1989.

[11] A. Abrardo and D. Sennati, "Outage statistics in CDMA mobile radio systems," *IEEE Commun. Letters*, vol. 4, no. 3, pp. 83–85, Mar. 2000.

[12] Z. Cao and Y. -D. Yao, "Definition and derivation of level crossing rate and average fade duration in an interference-limited environment," in *Proc. IEEE Veh. Tech. Conf. (VTC'Fall 01), Atlantic City, NJ*, Oct. 2001, pp. 1608–1611.

[13] L. Yang and M. -S. Alouini, "Average outage duration of multiuser wireless communication systems with a minimum signal power requirement," in *Proc. IEEE 55th Veh. Tech. Conf. (VTC'Spring 02), Birmingham, AL*, May 2002, pp. 1507–1511, full journal paper to appear in *IEEE Trans. Wireless Comm.*

[14] F. Graziosi and F. Santucci, "Distribution of outage intervals in macrodiversity cellular systems," *IEEE Jour. Sel. Areas Commun.*, vol. 17, no. 11, pp. 2011–2021, Nov. 1999.

[15] D. Brennan, "Linear diversity combining techniques," *Proc. IRE*, vol. 47, no. 6, pp. 1075–1102, June 1959.

[16] Y. Roy, J. -Y. Chouinard, and S. A. Mahmoud, "Selection diversity combining with multiple antennas for mm-wave indoor wireless channels," *IEEE J. Select. Areas Commun.*, vol. 14, no. 4, pp. 674–682, May 1996.

[17] M. K. Simon and M. -S. Alouini, *Digital Communications over Generalized Fading Channels: A Unified Approach to Performance Analysis.* Wiley, New York, 2000.

[18] ——, "A unified performance analysis of digital communications with dual selective combining diversity over correlated Rayleigh and Nakagami-m fading channels," *IEEE Trans. Commun.*, vol. COM-47, no. 1, pp. 33–43, January 1999, see also *Proc. Communication Theory Mini-Conference (CTMC-VII) in conjunction with IEEE Global Commun. Conf. (GLOBECOM'98)*, pp. 28-33, Sydney, Australia, November, 1998.

[19] M. -S. Alouini and M. K. Simon, "An MGF-based performance analysis of generalized selection combining over Rayleigh fading channels," *IEEE Trans. Commun.*, vol. 48, no. 3, pp. 401–415, March 2000.

[20] Y. -C. Ko, A. Abdi, M. -S. Alouini, and M. Kaveh, "Average outage duration of diversity systems over generalized fading channels," *IEEE Trans. Veh. Tech.*, vol. 51, pp. 1672–1680, Nov. 2002.

[21] X. Dong and N. C. Beaulieu, "Average level crossing rate and average fade duration of MRC and EGC diversity," in *Proc. IEEE Intern. Conf. Commun. (ICC'2001), Helsinki, Finland*, June 2001, pp. 1078–1083.

[22] M. D. Yacoub, C. R. C. M. da Silva, and J. E. V. Bautista, "Second-order statistics for diversity-combining techniques in Nakagami-fading channels," *IEEE Trans. Veh. Tech.*, vol. 50, no. 6, pp. 1464–1470, Nov. 2001.

[23] C. Iskander and P. Mathiopoulos, "Analytical level crossing rates and average fade durations for diversity techniques in Nakagami fading channels," *IEEE Trans. on Commun.*, vol. COM-50, no. 8, pp. 1301–1309, August 2002.

[24] X. Dong and N. C. Beaulieu, "Average level crossing rate and average fade duration of selection diversity," *IEEE Commun. Letters*, vol. 5, no. 10, pp. 396–398, Oct. 2001.

[25] F. Adachi, M. T. Feeney, and J. D. Parsons, "Effects of correlated fading on level crossing rates and average fade durations with pre-detection diversity reception," *IEE Proc. (Pt. F)*, vol. 135, no. 1, pp. 11–17, February 1988.

[26] L. Yang and M. -S. Alouini, "Performance comparison of different selection combining algorithms in presence of co-channel interference," submitted to *IEEE Trans. Veh. Tech.*, Dec. 2002.

[27] L. Yang and M. -S. Alouini, "An exact analysis of the impact of fading correlation on the average level crossing rate and average outage duration of selection combining," in *Proc. IEEE 57th Veh. Tech. Conf. (VTC Spring 2003), Jeju, Korea*, April 2003, pp. 241–245, full journal paper submitted to *IEEE Trans. Wireless Comm.*, Nov. 2003.

[28] A. Shah and A. M. Haimovich, "Performance analysis of maximal ratio combining and comparision with optimum combining for mobile radio communications with cochannel interference," *IEEE Trans. Veh. Tech.*, vol. 49, no. 4, pp. 1454–1463, July 2000.

[29] M. K. Simon and M. -S. Alouini, "On the difference of two chi-square variates with application to outage probability computation," *IEEE Trans. on Commun.*, vol. 49, pp. 1946–1954, Nov. 2001.

[30] V.A. Aalo and J. Zhang, "Performance analysis of maximal ratio combining in the presence of multiple equal-power cochannel interferers in a Nakagami fading channel," *IEEE Trans. Veh. Tech.*, vol. 50, no. 2, pp. 479–503, March 2001.

[31] Y. D. Yao and A. U. H. Sheikh, "Outage probability analysis for microcell mobile radio systems with co-channel interferers in Rician/Rayleigh fading environment," *IEE Electron. Lett.*, vol. 26, pp. 864–866, June 1990.

[32] J. R. Haug and D. R. Ucci, "Outage probability for microcellular radio systems in a Rayleigh/Rician fading environment," in *Proc. IEEE Int. Conf. on Commun. (ICC'92)*, June 1992, pp. 312.4.1–312.4.5.

[33] J. -C. Lin, W. -C. Kao, Y. T. Su, and T. -H. Lee, "Outage and coverage considerations for microcellular mobile radio systems in a shadowed-Rician/shadowed-Nakagami environment," *IEEE Trans. Veh. Tech.*, vol. VT-48, no. 1, pp. 66–75, January 1999.

[34] T. T. Tjhung and C. C. Chai, "Distribution of SIR and performance of DS-CDMA systems in lognormally shadowed Rician channels," *IEEE Trans. Veh. Tech.*, vol. VT-49, no. 4, pp. 1110–1125, July 2000.

[35] M. D. Yacoub, J. E. V. Bautista, and L. G. R. Guedes, "On higher order statistics of the Nakagami-m distribution," *IEEE Trans. Veh. Tech.*, vol. 48, no. 3, pp. 790–794, May 1999.

[36] G. L. Stüber, *Principles of Mobile Communication, 2nd ed.* Kluwer Academic Publishers, Norwell, Massachusetts, 2000.

[37] G. W. Lank and L. S. Reed, "Average time to loss of lock for an automatic frequency control loop with two fading signals and a related probability density function," *IEEE Trans. Inform. Theory*, vol. 12, no. 1, pp. 73–75, Jan. 1966.

[38] I. S. Gradshteyn and I. M. Ryzhik, *Table of Integrals, Series, and Products*, corr. and enlarg. ed. Orlando, FL: Academic Press, 1980.

[39] A. H. Nuttall, "Some integrals involving the Q function," Research Report of the Naval Underwater Systems Center, 1972.

[40] W. C. Jakes, *Microwave Mobile Communication*, 2nd ed. Piscataway, NJ: IEEE Press, 1994.

[41] L. Schiff, "Statistical suppression of interference with multiple Nakagami fading channels," *IEEE Trans. Veh. Tech.*, vol. VT-21, pp. 121–128, November 1972.

Chapter 9

ENHANCING TCP PERFORMANCE IN WIDE-AREA CELLULAR WIRELESS NETWORKS

Transport Level Approaches

Ekram Hossain and Nadim Parvez

University of Manitoba
Department of Electrical and Computer Engineering
Winnipeg, MB, Canada R3T 5V6
Tel: (204) 474 8908, Fax: (204) 261 4639
{ekram, parvez}@ee.umanitoba.ca

Abstract

Internet technology-based architectures and protocols for supporting multimedia traffic over wireless networks are evolving. Since TCP (Transmission Control Protocol)/IP (Internet Protocol) is the standard network protocol stack on the Internet, its use over the next-generation wireless mobile networks is a certainty. Performance of TCP would be one of the most critical issues in IP-based data networking over wireless links. This article presents a comprehensive study on the performances of the basic TCP variants (e.g., TCP Tahoe, TCP Reno, TCP New-Reno, SACK TCP, FACK TCP) in wide-area cellular wireless networks. For the basic TCP variants, an in-depth analysis of the transport-level system dynamics is presented based on computer simulations using *ns-2*. Impacts of variations in wireless channel error characteristics, number of concurrent TCP flows and wireless link bandwidth on the average TCP throughput and fairness performances are investigated. The maximum achievable throughput under window-based end-to-end transmission control is also evaluated and the throughput performances of TCP New-Reno, SACK TCP and FACK TCP are compared against this ideal TCP throughput performance. To this end, an overview of the major modifications to the basic TCP variants based on transport-level approaches to enhance TCP performance in wired-cum-wireless networks is presented.

Keywords:
>   Wired-cum-wireless networks, TCP, transport level approach, random and correlated error.

Introduction

Transport layer protocol such as TCP operates on an end-to-end basis and it's performance is one of the most critical issues in data networking over wireless links. A transport layer protocol is responsible for managing end-to-end flow and congestion control, providing reliability, security and QoS (Quality of Service) [1]. In OSI model, application protocol data unit (APDU) is passed down to the transport layer which is then called the transport service data unit (TSDU). A transport protocol header is added to this TSDU to form a transport protocol data unit (TPDU) and then it is passed to the network layer as a network service data unit (NSDU)[1].

TCP is the transport layer protocol used along with the unreliable network layer protocol IP (Internet Protocol) in todays Internet for non-real-time applications. The use of TCP/IP as the network protocol stack in the future generation wireless networks will leverage the rapidly evolving Internet technology in the wireless domain and enable to provide seamless wide-area Internet service to mobile users [2].

Transport protocol segments are transmitted in a wireless network as radio link level frames over the air-interface. For example, let us consider a TCP connection in a wide-area cellular wireless network, such as a GPRS (General Packet Radio Service) network. In case of transmission to a mobile from a fixed host in the Internet, after the SNDCP (Subnetwork Dependent Convergence Protocol) module in the SGSN (Serving GPRS Support Node) receives a TCP segment, it passes it to the LLC (Link Level Control) protocol (Fig. 9.1) [3]. The LLC protocol segments it into radio frames for transmission over the wireless link. Therefore, the performance of TCP will depend on the performance and service provided by the underlying radio link layer. In this article, however, we do not explicitly consider the impact of radio link protocols, rather we emphasize on the end-to-end (or transport-level) approaches while evaluating the performance of TCP in a wide-area wireless environment.

TCP is a connection-oriented reliable transport protocol consisting of three phases of operations: connection setup, data transfer, and connection termination. The connection set up procedure uses a three-way handshake where the connection is established after both ends of the connection inform each other of the set up process. The connection termination procedure uses a four-way handshake where both the active and the passive terminators send their own set of two-way handshake

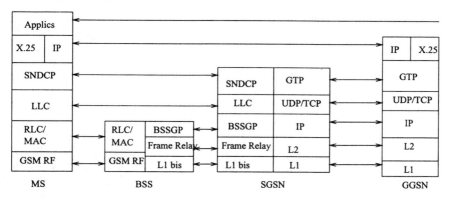

Figure 9.1. GPRS transmission protocol stack.

messages. Again, TCP is a bi-directional transport protocol that can transmit and receive at the same time (i.e., within the same connection set up). TCP is a byte-oriented transport protocol since it passes data to its peer in a byte by byte manner. For error control, TCP uses sequence number, cumulative positive acknowledgements with piggyback option for successfully transmitted packets and retransmission-based end-to-end error recovery. TCP uses a congestion control mechanism (to ensure network stability and prevent congestion collapse) which is interwined with a window-based flow control mechanism. The TCP sender detects a packet-loss either by the arrival of several duplicate cumulative ACKs (Acknowledgements) or by the absence of an acknowledgement during a timeout interval and it attributes the packet loss to network congestion. Upon detection of a packet loss at the TCP sender, the congestion control/avoidance mechanism is triggered which reduces the transmission window size multiplicatively and/or increases the retransmission timer exponentially, and consequently, the throughput is reduced.

The implementation of the TCP flow/congestion control mechanism is based on two sender-side state variables, namely, *congestion window (cwnd)*, $W(t)$ and the *slow-start threshold (ssthresh)*, $W_{th}(t)$ ([4]-[6]). During connection set up, the receiver advertises a maximum window size W_{max} and the TCP sender is not allowed to have more than $min(W_{max}, W(t))$ unacknowledged data packets outstanding at any given time t. For a new connection, $W(t)$ is generally initialized to 1. The basic window adaptation mechanism (which is triggered by ACKs) in all currently available TCP implementations is as follows:

i. If $W(t) < W_{th}(t)$ each ACK causes $W(t)$ to be incremented by 1. This is the *slow-start* phase.

ii. If $W(t) \geq W_{th}(t)$ each ACK causes $W(t)$ to be incremented by $\frac{1}{(int)W(t)}$. This is the *congestion avoidance* phase.

In case of timeout, TCP source updates $W(t)$ and $W_{th}(t)$ as follows: $W_{th}(t+) = \lceil \frac{W(t)}{2} \rceil$ and $W(t+) = W_{th}(t+)$ and then starts retransmitting from the first lost packet. However, when the TCP source receives a cumulative acknowledgement from the TCP sink acknowledging the already transmitted packets, it immediately starts transmission after the highest ACKed packet. After a timeout the sender also updates the RTO (Retransmission Time Out) value using exponential backoff. The backoff continues until an ACK is received for a packet transmitted exactly once.

TCP sinks can accept packets out of order but deliver them only in sequence to the user and they generate immediate ACKs. If a sink receives a packet out of order, it issues an ACK immediately for the last in order packet that was received. If a packet is lost (after a stream of correctly received packets), then the receiver keeps sending ACKs (called duplicate ACKs) with the sequence number of the first lost packet even if packets transmitted after the lost packet are correctly received.

1.   TCP in Wide-Area Cellular Wireless Environment

While the "traditional" TCP has been tuned for the last two decades to optimize its performance in wired networks where congestion is the main cause of any packet loss, it performs poorly in wireless environments due to the following factors:

- *Transmission losses on wireless link*: While in the wired transmission media the bit error rates are of $O(10^{-6}) - O(10^{-8})$, wireless media suffer significantly higher bit error rates such as $O(10^{-3}) - O(10^{-1})$ [7]. Therefore, the packet loss rate in a wireless link is an order of magnitude higher than that in a wired link. In a typical wireless scenario packet loss rate can be as high as 10% depending on channel fading and interference conditions and user mobility. When a packet is lost, TCP performs end-to-recovery by retransmitting the lost packet and frequent end-to-end retransmissions may reduce end-to-end throughput significantly. Again, any packet loss in the wireless link is misinterpreted by the TCP sender as a congestion loss and it triggers the congestion control mechanism which reduces the sender's transmission window size resulting in reduced end-to-end throughput. Moreover, some of the mechanisms like *fast retransmit* proposed for enhancing the basic TCP algorithm may even fail in a wireless network due to the low bandwidth-delay product (and hence small TCP window size) and correlated channel fading (and hence loss of back-to-back packets and ACKs). Therefore, in a wired-cum-wireless environment, random packet losses in the wireless link cause the TCP sender to underestimate the available network bandwidth and thereby reduces the application layer performance.

Also, the location and time-dependent channel errors may impact the achieved throughput fairness among multiple concurrent TCP flows[2] and this would presumably be different for the different TCP variants. Again, frequent channel errors and subsequent retransmissions may result in inefficient use of the limited battery power in the mobile devices.

- *Wireless link delays*: Presence of limited bandwidth wireless links in an end-to-end path generally results in increased end-to-end transmission delays (and hence longer round-trip times (RTTs)). Although 2.5G and and 3G cellular wireless networks (e.g., EG-PRS, UMTS, IMT-2000) will have increased bandwidth, factors such as link asymmetry due to different uplink and downlink bandwidth, bandwidth oscillation due to dynamic resource allocation, stronger FEC (Forward Error Correction) coding, interleaving, radio link level recovery will result in large BDP (Bandwidth-Delay Product)[3] wide-area cellular wireless networks [8].

Since the rate of increase in the congestion window size at the TCP sender is proportional to the rate of incmoing ACKs, the congestion window may increase at a much lower rate in the presence of wireless links and thus reduce the end-to-end throughput. In fact, the throughput of a TCP connection has been shown to vary as the inverse of the conmections RTT. Given a packet loss rate p, the maximum sending rate for a TCP sender is γ *bytes/sec*, for $\gamma \leq \frac{1.5*\sqrt{2/3}*B}{T*\sqrt{p}}$ where B is the packet size and T is the round trip time [9]. The inherent TCP bias against flows with longer RTT results in throughput unfairness among flows traversing the same number of hops but having different number of wireless links in the end-to-end path.

- *User mobility*: During a TCP session when a mobile user in a wireless cellular network moves from one cell to another, all necessary information must be transferred from the previous base station (BS) to the new base station which might cause a short duration of disconnection (typically of the order of several hunderds of milliseconds) during which no transmission takes place. Similar to that in the case of link errors, delays and packet losses during this "handoff" scenario trigger TCP congestion control mechanisms which results in reduced end-to-end throughput [10].

- *Short-lived TCP flows*: Data services which are more transactional than streaming in nature (e.g., web browsing on a smart phone, email) usually involve transmission of a rather small amount of data, and therefore, tend to require short TCP connection duration. It is possible that the entire transmission is completed while the TCP sender is in the slow-start phase and thus resulting in the under utilization of the network capacity.

There has been a flurry of recent works ([10]-[38]) on improving TCP performance over wireless networks. A taxonomy of the different proposed mechanisms to improve TCP performance in wireless networks is shown in Fig. 9.2. This article provides a unified study of the different transport-layer approaches to TCP design for the wired-cum-wireless networks. The performances of the different TCP variants including the newer variants such as TCP SACK (Selective Acknowledgement), TCP Vegas, TCP FACK (Forward Acknowledgement) are evaluated in a wide-area cellular network in terms of TCP throughput, fairness (in the case of multiple competing TCP flows). We consider wireless link outage due to channel impairments only and we consider both the random and the correlated wireless channel error conditions.

To gain insight into the performance behaviors of the different TCP variants, system dynamics characterized by the *slow-start*, the *fast retransmit/fast recovery* and the *timeout* events is analyzed. Note that, while the fairness of TCP congestion avoidance among competing flows in a wide-area cellular wireless network can be maintained through improved traffic management techniques (e.g., in [9]) and/or modified congestion avoidance algorithms (e.g., in [39]) in the wired Internet, it may deteriorate in the wireless last hop due to time and location-dependent channel errors.

Also, we evaluate the maximum achievable throughput under a window-based end-to-end transmission control by simulation and compare the throughput performances of several basic TCP variants against this ideal TCP throughput performance. To provide a unified study, several approaches to wireless TCP design based on transport level modifications to the basic variants are also discussed in the related context.

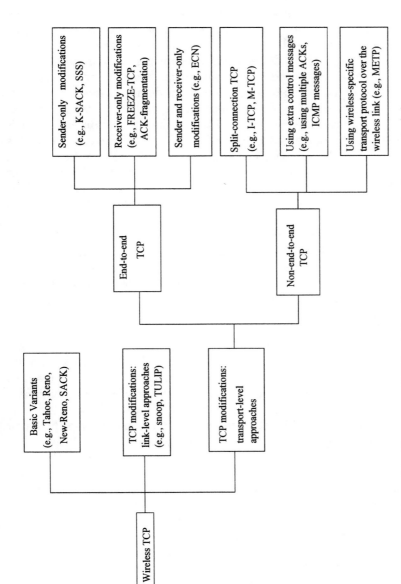

Figure 9.2. Taxonomy of TCP solutions (based on transport-level approaches) for wireless networks.

The motivation of this work is to investigate the effectiveness of different end-to-end TCP strategies in a wired-cum-wireless network under different wireless link conditions (e.g., random error, correlated error), and determine the limitations of TCP-like window-based flow control, which would be required to design more efficient end-to-end TCP. Note that, when IPSEC or other security mechanisms are employed to encrypt IP payloads, true end-to-end techniques based on transport level modifications of TCP would be required for wide-area wireless networks. In addition, it is feasible to implement end-to-end transport layer solutions in a wide-area wireless scenario where the connection end-point is a mobile host or a dedicated mobility/wireless-aware proxy since in this case the TCP implementations on stationary hosts need not be changed. Another concern is, when the data and the ACKs traverse different paths (e.g., in satellite networks), non-end-to-end solutions may cause serious problems, while end-to-end solutions are applicable over any type of networks and links.

For wide-area cellular wireless networks, end-to-end TCP solutions based on transport-level modifications which use some intermediary (such as a base station) may not maintain true end-to-end semantics (e.g., I-TCP [27]) and the intermediary may become a bottleneck due to the overhead involved in processing traffic associated with each connection and handing over the 'state' of each connection to the new intermediary during handoff.

Non-end-to-end TCP solutions such as those which use TCP-aware smarts in the base station (e.g., snoop protocol [13]) have scalability problems and may even degrade the performance of TCP in a wide-area cellular environment when the latency over the wireless link dominates the round trip time [22]. In addition, such a solution requires the base station to maintain significant state and is often tuned to specific flavor of TCP (e.g., snoop protocol does not work well with TCP Vegas).

Prior research closest to this work is that reported in [40], where the throughput performances of TCP Tahoe, OldTahoe, Reno and New-Reno in a localized wireless network were investigated under random packet losses using a stochastic model. The effects of coarse TCP timeout and the number of duplicate acknowledgements on TCP throughput performance were particularly emphasized. More general findings were reported in [41], where performance of TCP was investigated for a wide-area wireless network, particularly, for a network with high bandwidth-delay product and random packet losses. It was observed that random loss could lead to significant throughput degradation when the product of the packet loss (p) probability and the square of the bandwidth-delay product (i.e., $(\mu T)^2$, μ is the bandwidth and T is the delay) is larger than one. Also, TCP's unfairness towards connections with higher round trip delays was reported. A packet level performance comparison among TCP Tahoe, Reno, New-Reno and SACK in a wide area wired-Internet environment was presented in [42] considering scenarios where the number of packet drops in a transmission window varies from one to four.

2.   Basic TCP Variants

2.1   TCP Tahoe

TCP Tahoe uses *slow-start, congestion avoidance* and and *fast retransmit* mechanisms [5]. During *slow-start*, the congestion window increases exponentially (by one for each acknowledgement received) until it reaches the slow-start threshold (*ssthresh*), and during *congestion avoidance* the congestion window increases linearly by one per round trip time (RTT). The TCP sender goes into the *fast retransmit* mode when it receives *tcp_rexmt_thresh* (which is usually set to 3) number of duplicate acknowledgements. During *fast retransmit*, the sender retransmits the lost segment and enters into the *slow-start* phase by setting the congestion window to 1 and *ssthresh* to the half of congestion window. In addition, it forgets all outstanding data transmitted earlier [42]. When the loss is due to sporadic channel error, switching to *slow-start* mode causes the throughput to fall.

2.2   TCP Reno

TCP Reno is similar to TCP Tahoe except that in addition to *fast retransmit*, it also includes the *fast recovery* mechanism for a single segment loss. When the TCP sender receives *tcp_rexmt_thresh* number of duplicate acknowledgements, instead of switching to *slow-start* after *fast retransmit*, TCP Reno enters into *fast recovery*. During *fast recovery*, the sender sets *ssthresh* to the half of the congestion window and the new congestion window to the new *ssthresh* plus the number of duplicate acknowledgements received. TCP Reno remains in *fast recovery* until the lost segment which triggered the *fast retransmit* has been acknowledged. When the sender receives new acknowledgement(s), it exits *fast recovery* and resets the congestion window to *ssthresh* and thereby moves into congestion avoidance.

In case of congestion loss, the *fast recovery* mechanism keeps the average congestion window size high resulting in better throughput performance compared to TCP Tahoe. During *fast recovery*, each new duplicate acknowledgement increases the congestion window size by one. Although TCP Reno works fine for single loss, in case of multiple losses from the same transmission window the performance suffers since it exits *fast recovery* and enters into it again in a repeated fashion or goes to timeout.

2.3   TCP New-Reno

TCP New-Reno uses an augmented *fast recovery* mechanism where, unlike TCP Reno, *fast recovery* continues until all the segments which were outstanding during the start of the *fast recovery*, have been acknowledged [44]. This strategy helps to combat multiple losses without entering into *fast recovery* multiple times or causing timeout. In this case, a *partial acknowledgement*[4] is considered as an indication that the segment following the acknowledged one has been dropped from the same transmission window (or flight), and therefore, TCP New-Reno immediately retransmits the other lost segment indicated by the *partial acknowledgement* and remains in *fast recovery*. It takes one round trip time to detect each lost segment and to retransmit it.

2.4   SACK TCP

In SACK TCP, the receiver sends acknowledgements with SACK (Selective Acknowledgement) option when it receives out of order segments due to loss or out of order delivery [45]. The SACK option field contains a number of SACK blocks, where each SACK block reports a non-contiguous set of data that has been received and queued. The first block in SACK option reports the most recently received block.

The SACK TCP sender is an intelligent extension of that in TCP Reno. It only modifies the *fast recovery* mechanism of TCP Reno keeping the other mechanisms unchanged [46]. Similar to New-Reno, it can handle multiple packet losses from the same flight. It has a better estimation capability for the number of outstanding segments. The acknowledgement with SACK option enables the sender to determine explicitly which segments have been received or have been lost. To keep track of the acknowledged and lost segments, it maintains a data structure called *scoreboard*. Whenever the sender is allowed to transmit based on the congestion window size and the number of outstanding segments, it consults the *scoreboard* and transmits the missing segments. If there is no missing segment to retransmit, it transmits new segments. When a retransmitted segment is dropped, the sender detects it by a retransmit timeout. In case of timeout it retransmits the segment and enters into the *slow-start* phase.

SACK TCP maintains a variable called *pipe* to keep track of the number of outstanding segments. For each retransmission or new transmission, the sender increases *pipe* by one and for each received acknowledgement, it decreases *pipe* by one. For each received partial acknowledgement, *pipe* is decreased by two with the assumption that the original packet and the retransmitted packet have left the network. Since *pipe* represents the amount of outstanding data, the sender transmits only when *pipe* is less than the congestion window.

2.5 FACK TCP

FACK (Forward Acknowledgement) TCP is a variant of SACK TCP with a modified *fast recovery* mechanism [46]. It uses a better technique for estimating the number of outstanding segments. For this, it introduces two new variables *snd.fack* and *retran_data*, where *snd.fack* represents the forward most data held by the receiver and *retran_data* represents the size of the retransmitted data outstanding in the network.

In non-recovery states *snd.fack* is updated using the acknowledgement number in the TCP header and is equivalent to *snd.una* [4]. But when a SACK block is received during error reocovery, it is updated to the highest sequence number received by the receiver plus one regardless of the number of the intermediate dropped segments. Therefore, the outstanding data is estimated as *(snd.nxt - snd.fack) + retran_data*.

For each retransmission, the size of the *retran_data* is incremented by the corresponding segment size. When retransmitted segment leaves the network it is decreased by the same size. FACK TCP compares it's estimate for outstanding data against the congestion window size and decides on the number of transmissions. A timeout is forced in the case of loss of a retransmitted segment on the assumption that the congestion is persistent. It enters into the *fast recovery* mode when the sender receives *tcp_rexmt_thresh* number of duplicate acknowledgements or *(snd.fack - snd.una > tcp_rexmt_thresh * MSS)*[5]. In the case that several segments are lost but fewer than *tcp_rexmt_thresh* number of duplicate acknowledements have been received, the latter condition triggers the *fast recovery* phase sooner. Similar to New-Reno and SACK, FACK TCP terminates it's recovery phase upon receiving acknowledgement for all the segments that were outstanding during transition to *fast recovery*.

2.6 TCP Vegas

TCP Vegas [47] comes with a proactive congestion control mechanism in which network congestion is predicted based on the estimated *expected throughput* and the *actual throughput*. Expected throughput is estimated as *WindowSize/BaseRTT*, where *BaseRTT* is the minimum of all the measured RTTs and *WindowSize* is the size of the current congestion window. The actual throughput is calculated from the RTT for a tagged segment and the number of segments transmitted within that RTT. TCP Vegas compares the difference between the expected and the actual throughput against two thresholds α and β (where $\alpha < \beta$) and adjusts the transmission window accordingly. If the difference is smaller than α, the congestion window is increased linearly (by one per RTT) under the assumption that there is unutilized bandwidth available in the network. On the other hand, if the difference is larger than β, the congestion window is decreased, while if the difference lies between α and β the congestion window remains unchanged.

TCP Vegas has a fine grained timer expiry calculation mechanism to support early switching to *fast retransmit*. For this, the sender reads and records the system clock each time a segment is transmitted. When an acknowledgement arrives, it reads the clock again and calculates the fine grained RTT. TCP Vegas uses this fine-grained RTT estimate to calculate RTO. For each duplicate acknowledgement received, the sender checks the *Vegas expiry*[6] and if Vegas expiry occurs, the sender switches to *fast retransmit*. Similar to the other TCP variants, it switches to *fast retransmit* when it receives *tcp_rexmt_thresh* number of duplicate acknowledgements. Also, the sender switches to *slow-start* whenever the usual timeout occurs.

2.7   Comparison Among the Basic TCP Variants

Although the TCP senders for all the basic TCP variants rely on using binary positive acknowledgement, similar RTO estimation and timeout mechanisms, they use different congestion avoidance mechanisms. The major differences among the basic TCP variants are shown in Table 9.1.

The main difference between TCP Tahoe and TCP Reno is that the latter uses the *fast recovery* mechanism while the former does not. Although TCP Reno can handle single drop in a flight efficiently, it cannot handle multiple packet drops in a single flight very well. Tahoe does not memorize outstanding data when it switches to slow-start, but Reno does. Both TCP New-Reno and SACK TCP can handle multiple losses in a single flight more efficiently.

SACK TCP is different from TCP New-Reno in that SACK TCP can retransmit selectively. While TCP New-Reno retransmits starting from the segment corresponding to the duplicate acknowledgement or partial acknowledgement, SACK TCP retransmits this segment along with other missing segments from it's *scoreboard*. That is, the retransmission pool for a New-Reno sender can have at most one segment regardless of the number of packet drops. Therefore, it takes one RTT to recover

Table 9.1.   Differences among the basic TCP variants.

TCP Variant	Sender mechanism(s)
Tahoe	No *fast recovery*
Reno	Uses *fast recovery* to recover from single packet drops
New–Reno	Uses *fast recovery* to recover from multiple packet drops
SACK	Uses *fast recovery* to recover from multiple packet drops, better outstanding data estimation mechanism and larger retransmission pool
FACK	Similar to SACK, but uses a better outstanding data estimation mechanism and switches to *fast recovery* faster
Vegas	Uses proactive mechanism to control congestion window, switches to *fast retransmit* faster due to the fine-grained timer

each packet loss. On the other hand, since the retransmission pool for a SACK sender can have many lost segments, the recovery process becomes faster. Also, due to the *pipe* variable, SACK TCP has a better estimation of the number of outstanding segments.

The main advantage of FACK TCP over SACK TCP is that, by using *retran_data* and *snd.fack* the former performs a better estimation of the number of outstanding segments than the latter. Generally, the *pipe* estimation for FACK TCP becomes smaller compared to that for SACK TCP which allows the former to transmit more segments. Also, using the information from SACK block, sometimes it switches to *fast recovery* before receiving *tcp_rexmt_thresh* number of duplicate acknowledgements which is helpful when the transmission window size is small or the number of duplicate acknowledgements is too few to trigger *fast retransmit*.

By using *expected throughput*, *actual throughput* and the threshold parameters α, β, TCP Vegas can have a better estimation of the available bandwidth compared to the other TCP variants. Due to its own fine-grained timer management, TCP Vegas can switch to *fast retransmit* earlier which contributes to it's performance improvement. Also, since it reduces the congestion window to 3/4 instead of 1/2 during *fast retransmit* (when the segment that triggered *fast retransmit* has not been transmitted more than once), it helps to combat losses due to the sporadic wireless channel errors more efficiently.

3. Analysis of System Dynamics and Performance Evaluation

3.1 Simulation Model and Performance Measures

System dynamics under the different basic TCP variants, namely, TCP Tahoe, Reno, New-Reno, SACK, FACK and Vegas are investigated using *ns-2* [48] under varying wireless channel bandwidth, round trip delay, number of concurrent TCP connections and different wireless channel error characteristics in a wired-cum-wireless scenario where the mobile nodes act as TCP sinks. Also, two variants of New-Reno – one with *TCP-aware link level retransmission* (TLLR) and the other with *delayed acknowledgement* (DACK)-capable TCP sink, are considered.

It is assumed that the slowly varying shadow and path losses in a wireless link are perfectly compensated and that the channel quality variations due to multipath fading remain uncompensated. For a broad range of parameters, the packet error process in a mobile radio channel, where the multipath fading is considered to follow a Rayleigh distribution, can be modeled using a two-state Markov chain with transition probability matrix M_c [49]: $M_c = \begin{bmatrix} p & 1-p \\ 1-q & q \end{bmatrix}$ where p and $1-q$ are the probabilities that the jth packet transmission is successful, given that the $(j$-1)th packet transmission was successful or unsuccessful, respectively.

Also, the fading envelope is assumed to not change significantly during transmission of a TCP packet (which is of duration, say, t_{tcp}). The channel variation for each of the mobiles is assumed to be independent and determined by the two parameters p and q which, in turn, are determined by $f_d t_{tcp}$ [7], the normalized Doppler bandwidth and P_E, the average packet error probability. P_E describes the channel quality in terms of *fading margin F* [8]. Different values of P_E and $f_d t_{tcp}$ result in different degree of correlation in the fading process. When $f_d t_{tcp}$ is small, the fading process is highly correlated; on the other hand, for large values of $f_d t_{tcp}$, the fading process is almost independent. For a certain value of the average packet error rate P_E, the error burst-length increases as user mobility (and hence $f_d t_{tcp}$) decreases.

The network topology used in the simulation is shown in Fig. 9.3. The different performance measures are: *average per-flow throughput (γ)* (i.e., average amount of data successfully transmitted to a TCP sink per unit time), *energy consumption (e)* (average amount of energy dissipated in a TCP sink for successful reception of 1 *MB* of data) and *throughput fairness (f)*. For n concurrent TCP flows, the fairness index

f is calculated as follows[9]:

$$f = \frac{(\sum_{i=1}^n \gamma_i)^2}{n \times \sum_{i=1}^n \gamma_i^2}. \tag{9.1}$$

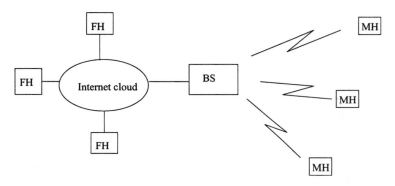

Figure 9.3. Simulation topology for wide-area cellular wireless network (FH $\equiv$ Fixed host, MH $\equiv$ mobile host, BS $\equiv$ base station/wireless router).

To calculate the energy consumption at the mobile nodes (primarily due to transmission of acknowledgements and reception of TCP packets), the ratio of transmisson power and receive power wattage is assumed to be 3 for all the simulation scenarios described in this article.

3.2   Simulation Results and Analysis

3.2.1   Scenario 1: Wireless link bandwidth $= 2$ *Mbps*, Internet delay $= 20$ *ms*, wireless link delay $= 10$ *ms*, single flow, maximum window size $= 15$, segment size $= 1000$ bytes, uniform wireless channel error.

In this scenario, for the *delayed acknowledgemnt* case, the interval for delaying the acknowledgement is assumed to be 10 *ms*. The simulation run time is taken to be 1000 *seconds*.

Under random error conditions, TCP New-Reno and SACK TCP provide remarkable performance improvement over TCP Tahoe and TCP Reno (Fig. 9.4). Although TCP Tahoe and TCP Reno exhibit similar performance, TCP Reno performs slightly better than TCP Tahoe when the error rate is smaller than 1.5%. Analysis of the system events such as the total number of timeouts (TO), total number of *fast retransmit/fast recovery* (FR), number of *fast retransmits* due to multiple packet drop in a transmission window (MD), number of segments newly transmitted during *fast recovery* (SND), and average of the transmission window sizes measured at the epoches of transitions to *fast retransmit* (FLT) reveal that under random error conditions the system dynamics for both TCP New-Reno and TCP SACK are more or less similar (Table 9.2), and therefore, their long-term performances are observed to be fairly close. FACK TCP is observed to perform better than each of TCP New-Reno and SACK TCP. TCP Vegas shows significant performance improvement over FACK TCP. However, the throughput performance of each of the above basic TCP variants is inferior to the performance of TCP New-Reno with TLLR (Fig. 9.4).

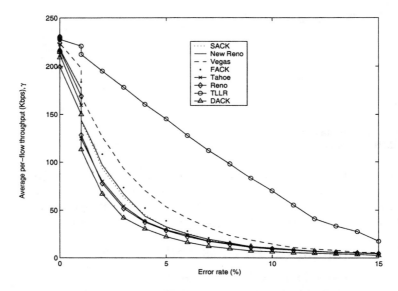

Figure 9.4. Throughput performances of different TCP variants (scenario 1).

For both TCP Tahoe and TCP Reno, the number of timeouts increases with increasing error rate, and the difference in the number of timeouts for these two cases during a certain observation interval is observed to be small. As the error rate increases, Reno suffers due to multiple packet drops in the same flight and also the number of timeouts increases compared to the number of *fast retransmits*. This offsets Reno's gain due to *fast recovery* and makes the performance closer to that of Tahoe.

As is observable from Table 9.2, at 1% error rate, for SACK TCP there are 115 events of *fast recovery* due to multiple packet drops (within a transmission window) among the total of 1252 *fast recovery* events, while for TCP New-Reno, among the total of 1237 *fast recovery* events 129 *fast recovery* events are due to multiple packet drops. Therefore, the ratio of *multiple drop fast recovery* to total *fast recovery* is smaller for SACK

Table 9.2. Fast retransmit/recovery analysis under random error loss (for 1000 *second* simulation run)

TCP variant	TO	FR	MD	SND	FLT
	Error rate = 1%				
Tahoe	53	1056			
Reno	100	1181	110	1909	11.59
New-Reno	27	1237	129	2013	11.92
SACK	25	1252	115	2435	11.88
FACK	17	1365	137	6668	12.27
TLLR	0	14	1	2	15.00
Vegas	22	1552 1072E			13.04
	Error rate = 10%				
Tahoe	691	295			
Reno	692	277	65	359	5.55
New-Reno	705	315	88	442	5.55
SACK	685	298	83	341	5.51
FACK	630	392	125	426	5.28
TLLR	20	95	58	134	13.94
Vegas	546	550 334E			5.89

TCP. This observation holds for other error rates as well. In this case, SACK TCP cannot exploit the advantage of selective retransmission more effectively (compared to TCP New-Reno).

Again, under random error conditions, SACK TCP may not be able to exploit the advantage of better *pipe* estimation. Referring to Table 9.2, for 1% error rate SACK TCP transmits on the average 1.94 (= 2435/1252) new segments during *fast recovery*, while TCP New-Reno transmits on the average 1.63 (= 2013/1237) new segments. This small difference may not result in significant performance improvement for SACK TCP as compared to TCP New-Reno. As the error rate increases, the difference may become even smaller.

The *delayed acknowledgement* (DACK) variant of TCP New-Reno is observed to perform the worst, which is primarily due to its slow response to transmission failure.

Due to the better estimation of the number of outstanding segments, FACK TCP performs better than SACK TCP under all error conditions. At 1% and 10% error rate, for FACK TCP the number of new segments transmitted during *fast recovery* is observed to be 6668 and 426, respectively, while for TCP SACK, the numbers are 2435 and 341 (Table 9.2). With increasing error rate, as the effect of timeout becomes more dominant, the number of new segments transmitted during *fast recovery* decreases.

The performance improvement in TCP Vegas is mainly due to its two unique features – Vegas expiry and congestion window reduction by a factor of 3/4. The Vegas expiry mechanism causes it to react very fast to segment loss. As can be observed in Table 9.2, for 1% error rate, during 1000 *second* of simulation Vegas experiences 1552 *fast retransmit*s among which 1072 were detected early by the Vegas fine-grained expiry mechanism (in Table 9.2, 'E' refers to fast retransmit detected by Vegas expiry). The *Vegas expiry* mechanism also reduces the number of timeouts and it is observed that among all the TCP variants it experiences the smallest number of timeouts.

The TCP New-Reno with TCP-aware link level retransmission (TLLR) scheme (e.g., snoop protocol [11]) offers the best performance (Fig. 9.4) due to the fact that the 'snoop agent' at the link level eliminates most of the fast retransmits and timeouts. For example, with TLLR, for 10% error rate, the number of fast retransmits and the number of timeouts are 20, and 95, respectively, while for New-Reno they are 705 and 315, respectively. As a result, the average flight size is always very high (e.g., 13.94 for 10% error rate) which is close to the assumed maximum transmission window size of 15, and consequently, better throughput performance is achieved.

Regarding energy consumption for transmission of a fixed amount of data (1 *MB*) to a mobile node using the different TCP variants, it is observed that TCP Vegas requires the highest amount of energy. This is due to the fact that, TCP Vegas results in the highest number of *fast retransmits*. For TCP New-Reno with the DACK option, the energy consumption is observed to be the lowest, which is due to it's conservative acknowledgement transmission policy. But as the error rate increases, energy consumption due to reception of the retransmitted segments becomes more dominant compared to that due to transmission of acknowledgements, and consequently, energy consumption for TCP New-Reno with DACK tends to be similar to that for the other TCP variants (Fig. 9.5).

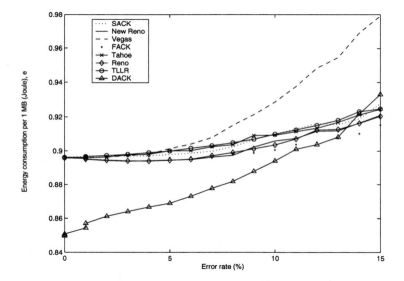

Figure 9.5.   Energy consumption for the different TCP variants (scenario 1).

In fact, in the high error rate regime none of the end-to-end TCP mechanisms performs well in a wide-area cellular wireless environment. For example, with 10% error rate, the achieved thoroughput is about 10 *Kbps* compared to 217 *Kbps* in the ideal case. This is due to the inability of TCP to differentiate between wireless loss and congestion loss.

3.2.2   Scenario 2: Wireless link bandwidth = 10 *Mbps*, Internet delay = 20 *ms*, wireless link delay = 10 *ms*, 5 concurrent flows, maximum window size = 15, segment size = 1000 bytes,

uniform wireless channel error. System dynamics (in terms of timeout, fast retransmit/fast recovery) and energy consumption per flow under multiple concurrent TCP connections are similar to those in case of a single connection (Figs. 9.6-9.7). But the average per-flow throughput is observed to be slightly higher than the single flow case (scenario 1) when the error rate is not too high (e.g., $\leq 5\%$).

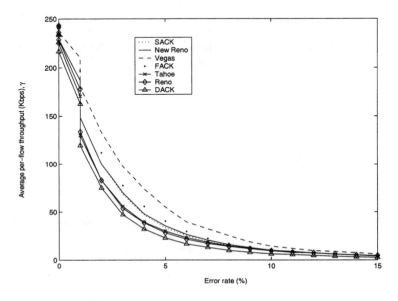

Figure 9.6. Throughput performances of different TCP variants (scenario 2).

In this case, we also observe the achieved throughput fairness among the different competing TCP connections. All of the TCP variants provide good long-term fairness and it is observed that even for error rate as large as 9%, the fairness index lies above 0.99. For error rate larger than 14%, the fairness index reduces to 98% (Fig. 9.8). Therefore, it can be concluded that the random wireless channel errors do not impact the TCP throughput fairness for the different TCP variants remarkably.

3.2.3   Scenario 3: Wireless link bandwidth = 10 *Mbps*, Internet delay = 20*ms*, wireless link delay = 10 *ms*, single flow, maximum window size = 75, segment size = 1000 bytes, uniform wireless channel error. This scenario is considered to compare the throughput performance in the case of multiple concurrent TCP flows with that in the case of a single TCP flow occupying the entire bandwidth. The average throughput of this case (which is

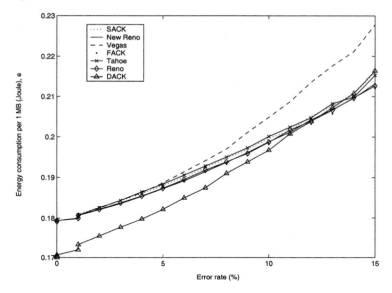

Figure 9.7. Energy consumption for the different TCP variants (scenario 2).

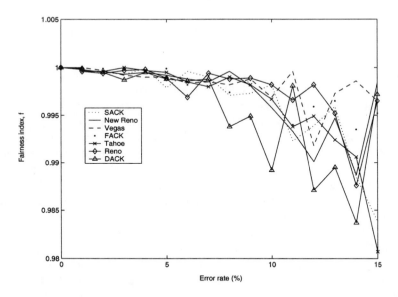

Figure 9.8. Throughput fairness for the different TCP variants (scenario 2).

expected to be five times that for a flow in scenario 2) becomes close
to the per-flow throughput in scenario 2 as the error rate increases be-

yond 1% (Fig. 9.9). The performance gain due to the high wireless link
bandwidth is observable only for very small error rates (e.g., 0%-0.05%).

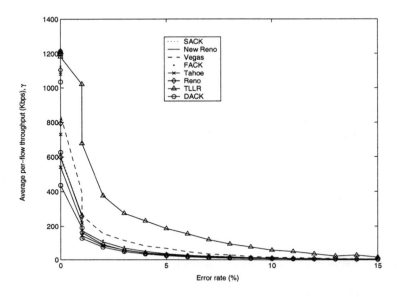

Figure 9.9.   Throughput performances of different TCP variants (scenario 3).

3.2.4   Scenario 4: Wireless link bandwidth = 2 *Mbps*, Internet delay = 20 *ms*, wireless link delay = 10 *ms*, single flow, maximum window size = 15, segment size = 1000 bytes, correlated wireless channel error, TCP New-Reno only.

In
this scenario, we choose to investigate only the performance of TCP
New-Reno (which is regarded as the de facto standard in the present Internet) under correlated channel error. For a particular average packet
error rate, the channel error correlation (as manifested in the lengths
of the error bursts) varies as the mobile speed varies [49]. For example, for the assumed wireless channel bandwidth and maximum segment
size, with 10% error rate, mobile speed of 1 *km/hr*, 3 *km/hr*, 5 *km/hr*,
10 *km/hr* and 100 *km/hr* correspond to burst-error size of 11.41, 3.85,
2.37, 1.42 and 1.11 segments, respectively. For 1% average error rate,
the corresponding burst-error sizes are 3.41, 1.39, 1.14, 1.04 and 1.01
segments, respectively.

The average throughput decreases with increased channel error correlation (Fig. 9.10). In the presence of highly correlated error, the acknowledgements from the mobile TCP sinks do not reach the TCP senders and
the TCP error recovery is triggered primarily by timeouts. For example,

Table 9.3. Fast retransmit/fast recovery analysis under correlated error loss (for 1000 *second* run)

Mobile speed	TO	FR	MD	SND	FLT
Error rate = 3%					
1 km/hr	83	11	11	8	14.82
3 km/hr	230	234	231	164	13.93
5 km/hr	259	328	319	172	13.64
10 km/hr	301	399	380	204	13.41
100 km/hr	300	404	384	191	13.45
Error rate = 5%					
1 km/hr	105	1	1	1	13.00
3 km/hr	189	16	14	20	11.62
5 km/hr	321	75	71	137	11.45
10 km/hr	451	172	161	294	11.59
100 km/hr	479	187	175	251	11.50
Error rate = 10%					
1 km/hr	76	0	0		
3 km/hr	140	0	0		
5 km/hr	298	0	0		
10 km/hr	532	0	0		
100 km/hr	507	0	0		

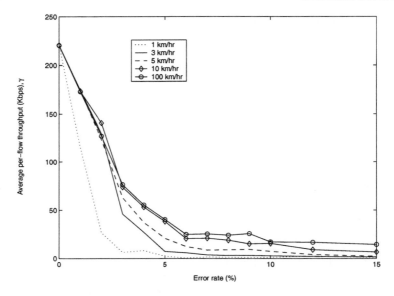

Figure 9.10. Throughput performance of TCP New-Reno (scenario 4).

with 10% error rate, it is observed that there is no *fast retransmit* at all
(Table 9.3).

One interesting observation is that, although the number of timeouts decreases/increases with decreasing/increasing mobile speed, the
average throughput decreases/increases. This is because, as user speed
decreases, the channel error correlation increases, as a consequence of
which the value of RTO generally becomes high. With an increase in
the value of RTO, the end-to-end error recovery after a timeout becomes
more sluggish, and consequently, the throughput deteriorates.

The amount of energy consumption is observed to increase with increasing channel error correlation.

3.2.5   Scenario 5: Wireless link bandwidth = 2 *Mbps*, In-
ternet delay = 20 *ms*, wireless link delay = 10 *ms*, single flow,
maximum window size = 15, segment size = 1000 bytes, correlated wireless channel error.   We assume mobile speed of 3
km/hr at which the burst-error size is 3.85 and 1.39 for error rate of
10% and 1%, respectively. The performance trends in this scenario are
observed to be radically different from those in the random error case
(Fig. 9.11). Throughput performance of TCP Vegas reduces by about
80% (compared to the uniform error case) when the error rate is 2%.

SACK TCP now performs better than TCP New-Reno over a range of error rates from 0% to 5%.

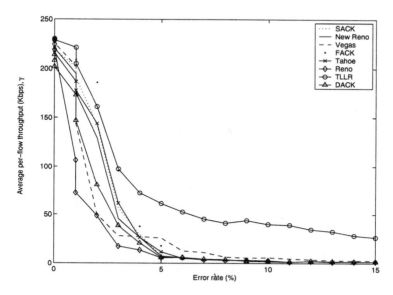

Figure 9.11. Throughput performances for the different TCP variants (scenario 5).

Since the system dynamics is now characterized by a large number of timeouts, high ratio of the number of multiple (three or more) packet drops to the number of single packet drops in a single window and loss of a large number of acknowledgements, performance of TCP Tahoe becomes comparable to the performance of SACK TCP. Also, we observe that the performance of TCP New-Reno with TCP-aware link level retransmission deteriorates considerably compared to that in the uniform error case.

TCP Reno is observed to perform the worst in this correlated error scenario. Although the number of fast retransmits is fewer than that in the uniform error case, most of the fast retransmits are due to multiple packet drops (three or more) in a single window. For example, with 1% error rate, there are 402 fast retransmits, among which 375 are due to multiple drops (Table 9.4). Again, 327 of the 375 packet drops are due to three or more packet drops in a single window. Most of these multiple (more than three) drop cases are observed to be followed by timeouts. In fact, when three or more packets are dropped from a window of data, the Reno sender is forced to wait for a timeout most of the time.

TCP Tahoe is observed to perform significantly better than Reno under correlated error scenarios. Since after switching to *slow-start*, Tahoe

Table 9.4.   Fast retransmit/fast recovery analysis under correlated error rate 1% and 10% in 1000 *second* run

TCP variant	TO	FR	MD	SND	FLT
Error rate = 1%					
Tahoe	8	539	538	0	14.98
Reno	431	402	375	226	13.01
New-Reno	7	544	541	1	14.99
SACK	4	553	547	1205	14.97
DACK	12	530	484	260	14.42
FACK	0	548	546	3284	14.93
TLLR	0	0	0		
Vegas	157	445			14.20
Error rate = 3%					
Tahoe	195	214	210	59	14.42
Reno	346	95	93	25	12.80
New-Rreno	230	234	231	164	13.93
SACK	330	251	246	870	14.02
DACK	254	215	210	103	11.82
FACK	159	349	348	1941	14.51
TLLR	0	0	0	0	0
Vegas	381 175E	164			10.82
Error rate = 10%					
Tahoe	140	0	0		
Reno	140	0	0		
New-Reno	140	0	0		
SACK	140	0	0		
DACK	145	0	0		
FACK	140	0	0		
TLLR	68	0	0		
Vegas	248 0E	0			

forgets all outstanding data that were transmitted earlier and increases its congestion window upon receipt of each acknowledgement, it results in multiple transmission attempts for same packets (some of which have presumably been lost already), and consequently, the throughput improves for the multiple drop cases.

Since the number of packet drops per transmission window increases in the correlated error case, SACK TCP performs better than New-Reno. TCP New-Reno requires $n \times RTT$ to recover from n packet drops in a single window whereas SACK recovers much faster. This is mainly due to its better pipe estimation method and usage of a larger retransmission pool. For the same reason FACK TCP performs even better than SACK TCP.

Since a large number of acknowledgements are lost due to correlated channel error, sluggishness in transmitting acknowledgements in the case of TCP New-Reno with DACK option does not significantly impact the throughput performance. For the same reason, sharp response to packet losses in the case of TCP Vegas is not very conducive to improving TCP throughput performance in the correlated error case. Under different channel error rates and channel error correlation, the fixed values of α and β do not work well. Also, TCP Vegas experiences the highest number of timeouts.

For error rate greater than 7%, the energy consumption is observed to be more or less same for all the TCP variants (except for Vegas and New-Reno with TLLR) (Fig. 9.12). This is due to the fact that for a large error rate all of the TCP variants experience almost the same number of timeouts and there may be no *fast retransmit* at all.

3.3   Summary of the Results

Throughput, fairness and energy performances of the different basic TCP variants have been investigated in a wide-area cellular wireless environment for both uniform and correlated wireless channel errors. The following provides a summary of the key observations:

- Implications of the fast retransmit and timeout events on the TCP performance largely depend on the wireless channel error characteristics (e.g., error rate and degree of channel error correlation). For example, in the case of TCP Vegas, sharp response of the sender due to its fine-grained expiry mechanism works well under uniform error case while the same strategy results in poor performance under correlated error scenarios. Therefore, although coarse timeout is undesirable under random losses, the ineffectiveness of the 'Vegas expiry' mechanism under correlated error cases

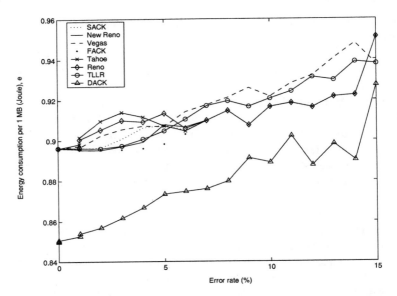

Figure 9.12. Energy consumption for the different TCP variants (scenario 5).

suggests that very sharp timeouts may also not be desirable under such error conditions.

- A method which provides better estimation for the number of out-standing segments is always conducive to the TCP throughput performance under both random and correlated error scenarios. Due to this reason, FACK TCP is observed to be consistently better than SACK TCP and TCP New-Reno under both random and correlated error scenarios.

- Since the impact of timeout becomes more dominant, the throughput performances of all of the basic end-to-end TCP variants (which primarily differ in the *fast recovery* mechanism) suffer under correlated error scenarios. Again, end-to-end protocols with link level retransmissions (e.g., snoop protocol) also suffer serious performance degradation in case of correlated channel errors.

- Lower degree of wireless channel error correlation is more conducive to energy saving at the mobile TCP sinks.

- In a wide-area wireless scenario, TCP connections with high bandwidth delay product may not get their fair share. Splitting a single TCP flow (with a high bandwidth-delay product) to multiple flows (say,

x flows) increases the throughput approximately x times for a certain range of low error rates.

3.4 Outlook

- TCP performance in a wired-cum-wireless environment can be improved significantly by using some transport level mechanism to differentiate between the congestion loss and wireless channel loss and then adjusting the window adaptation mechanism accordingly.

- The transmission window size at the TCP sender should be adapted differently depending on the degree of correlation in the wireless link errors. In addition, timer granularity may be adapted dynamically based on error correlation.

- The throughput performance of an end-to-end TCP designed for a wide-area cellular environment based on transport level modifications should be compared against the maximum achievable throughput envelope. This envelope, which can be characterized empirically based on simulation results, defines the maximum throughput achieveable under any TCP-like window-based end-to-end transmission control mechanism.

4. Maximum Achievable TCP Throughput Under Window-Based End-to-End Transmission Control

In this section, we obtain the upper bound on the TCP througput performance under window-based end-to-end transmission control (i.e., throughput of *ideal end-to-end TCP* (IE^2-TCP)) and compare it against the throughput performances of TCP New-Reno, SACK TCP and FACK TCP. For any window-based transmission control, the timeout mechanism is a must. Note that, all of the basic TCP variants use similar timeout mechanism along with exponential RTO backoff. Also, they use the *fast retransmit* mechanism (i.e., immediate retransmission of lost segment detected by *tcp_rexmit_thresh* number of duplicate ACKs). However, the basic end-to-end TCP variants mainly differ in the implementation of the *fast recovery* mechanism. Note that, for the maximum possible throughput performance in a wired-cum-wireless scenario, IE^2-TCP should be aggressive enough in retransmission and in measuring the outstanding data segments. The upper bound on the throughput performance (achievable by the IE^2-TCP) can be obtained using the simulation model described before based on the following assumptions:

- IE^2-TCP uses the timeout mechanism along with exponential RTO backoff and it uses *fast retransmit* mechanism to retransmit a lost segment after reception of *tcp_rexmit_thresh* number of duplicate ACKs.

- IE^2-TCP retransmits a lost segment after *tcp_rexmit_thresh* number of partial duplicate ACKs (in order to avoid some possible timeouts by retransmissions during *fast recovery*).

- Rather than using a go-back-N type of retransmission, IE^2-TCP uses selective acknowledgement (SACK)-based retransmission.

- IE^2-TCP uses the information in the SACK acknowledgement to estimate outstanding data.

- IE^2-TCP maintains the transmission window size equal to the receiver-advertized window size all the time.

Figs. 9.13-9.14 show some typical results on the long-term average throughput performance of IE^2-TCP when compared to those of SACK TCP, TCP New-Reno and FACK TCP under both random error and correlated error conditions. As is evident from Fig. 9.13, under random packet error in the wireless link IE^2-TCP can achieve a throughput which is higher by as much as 200% (e.g., for packet error rate of 4%) than that due to SACK TCP and FACK TCP. Improvement in throughput performance can be even higher under correlated error scenarios (e.g., 250% improvement for packet error rate of 4% with average error burst length of 3.85) (Fig. 9.14). Therefore, it can be concluded that the basic TCP variants perform significantly poor compared to the *ideal TCP* (i.e., IE^2-TCP) in a wired-cum-wireless scenario. Since the throughput performance of IE^2-TCP characterizes the envelope of the maximum possible throughput under a TCP-like window-based end-to-end transmission control, the TCP modifications based on end-to-end approaches should use this as the benchmark performance.

5.   Modifications to the Basic TCP Variants for Wireless Networks

5.1   Sender-Only Modifications

- *K-SACK* [15]: *K*-SACK is an efficient SACK-TCP variant of TCP New-Reno that requires modification at the TCP sender only. The novelty of this protocol is that it differentiates between wireless losses and congestion losses from anticipated loss pattern and behaves accordingly. During *fast recovery*, in case of wireless loss, it

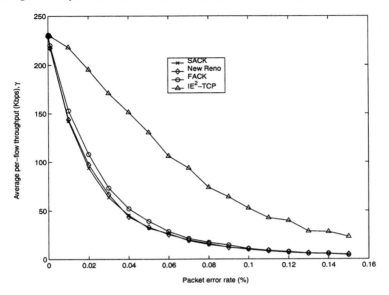

Figure 9.13. Throughput performances for SACK TCP, TCP New-Reno, FACK TCP and IE^2-TCP under random packet error in the wireless link.

does not halve the congestion window instead it stalls the growth of the window. The *fast recovery* phase is partitioned into two phases: *halt growth* phase and *K-recovery* phase. During *K*-recovery phase, the sender apprehends congestion only if it anticipates *K lookahead-loss* within a loss window of size *lwnd*, where *K* and *lwnd* are appropriately chosen parameters for the protocol.

The parameter *lwnd* can be chosen to be equal to the number of outstanding packets. The parameter *lookahead-loss* estimates the number of losses in the outstanding packets. For each such loss the sender has received either (i) at least *max_dupack* number of duplicate acknowledgements, or (ii) *max_dupsack* number of selective acknowledgements with higher sequence numbers. For lookahead-loss below *K* (e.g., $K = 2$), the sender assumes loss due to channel error.

Upon entering into the *fast recovery* phase, if lookahead-loss is less than *K*, the sender enters into the *halt growth* phase, otherwise to the *K-recovery* phase and transition between these phases occurs depending on the value of the lookahead-loss.

During the *halt growth* phase, congestion window remains frozen and the sender remains in this phase until the lookahead-loss be-

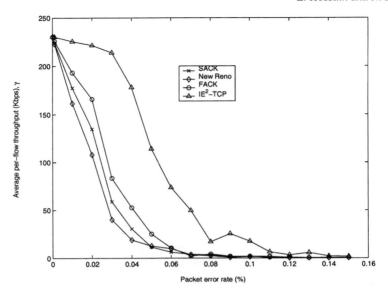

Figure 9.14. Throughput performances for SACK TCP, TCP New-Reno, FACK TCP and IE^2-TCP under correlated packet error in the wireless link (for mobile velocity of 3 km/hr).

comes at least K. When the sender enters into the K-recovery phase for the first time, *ssthresh* is set to $max(cwnd/2, 2 \times MSS)$ and *cwnd* is set to *ssthresh*. Upon re-entry, *ssthresh* remians unchanged, but congestion window is set to $max(cwnd/2, 1 \times MSS)$. This may cause the congestion window to change multiple times during the same *fast recovery* period. As a result, upon exiting *fast recovery*, depending on the value of *ssthresh* and *cwnd*, K-SACK TCP may find itself in the *slow start* phase instead of the *congestion avoidance* phase.

K-SACK performs a better *pipe* (i.e., number of packets in the link) estimation by measuring the *pipe* as the number of unacknowledged segments which are not marked *lost* or marked *lost* and retransmitted. This improved pipe estimation results in faster error recovery.

The main advantage of K-SACK is that it detects packet loss due to wireless channel error without explicit notification and it can be used seamlessly in wired or wired-cum-wireless environment without separate tuning. It was observed to perform very well under high *i.i.d.* error condition. Simulation results showed that, at 5% *i.i.d.* error rate the throughput was almost twice of that due to SACK-TCP over New Reno.

Although K-SACK performs well under random packet errors in the wireless channel, the performance of this scheme will suffer under correlated error conditions. This is due to the fact that, a moderate degree of correlated error will increase lookahead-loss to K and thus push K-SACK into K-recovery phase. Therefore, during *fast recovery* the sender may spend significant amount of time in *K-recovery* phase compared to that in the *halt growth* phase.

To combat the situation due to correlated packet loss in the wireless channel, higher value of K can be chosen. However, it will increase the chance of congestion collapse in the wired Internet. Since in a real Internet scenario, only a small percentage of packet losses (e.g., 15% [16]) is recovered by *fast recovery*, enhancement only in the *fast recovery* mechanism as in K-SACK is not likely to improve the TCP performance significantly compared to the other TCP variants in a wide-area wireless Internet. In additon, under pathological scenarios K-SACK would suffer from oscillation between the K-recovery and the *halt growth* phase which would reduce *cwnd* to small values, and consequently, the performance may fall below that of SACK TCP.

- *TCP Westwood (TCPW)* [17]: TCPW uses a sender-side modification of TCP congestion window adaptation algorithm which is based on estimation of the bandwidth used by the TCP connection via monitoring the rate of returning ACKs. The bandwidth estimate (BWE) is used to set the congestion window and and the slow-start threshold after a congestion episode (i.e., after three duplicate acknowledgements or after a timeout), and thereby, TCPW achieves a "faster" error recovery. The basic congestion window dynamics during *slow-start* and *congestion avoidance* are unchanged. For bandwidth estimation, TCPW uses a discrete-time exponential filter with variable coefficients as shown in (9.2) below

$$\hat{b}_k = \alpha_k \hat{b}_{k-1} + (1 - \alpha_k) \left(\frac{b_k + b_{k-1}}{2} \right) \tag{9.2}$$

where $\hat{b}_k$ is the filtered estimate of the available bandwidth at time $t = t_k$, $\alpha_k = (2\tau - \Delta_k)/(2\tau + \Delta_k)$, $\Delta_k = t_k - t_{k-1}$ and $1/\tau$ is the cutoff frequency of the filter. When the interarrival time Δ_k increases, the last value of $\hat{b}_{k-1}$ has less significance, while a recent $\hat{b}_{k-1}$ is given higher significance. The coefficient α_k decreases when the interarrival time increases, and thus the previous value of b_{k-1}

has less significance with respect to the last two samples which are weighted by $(1 - \alpha_k)/2$. Using the estimated bandwidth, the values of *ssthresh* and *cwnd* are updated as follows:

```
/* algo. for n dup acks */

if (n dupacks are received)
 ssthresh = (BWE * RTTmin)/seg_size;
 if (cwnd > ssthresh)
  cwnd =  ssthresh
 endif
endif

/* algo. for coarse timeout expiration */

if (coarse timeout expires)
  ssthresh = (BWE * RTTmin)/seg_size;
  if (ssthresh < 2)
    ssthresh = 2;
  endif;
  cwnd = 1;
endif
```

As an end-to-end solution to error and congestion control in mixed wired and wireless networks, TCPW was observed to provide significant throughput gain over TCP Reno and SACK TCP for both random and correlated error conditions in the wireless link. Also, the observed throughput fairness among TCPW flows across different round trip times was better compared to that for TCP Reno flows. This was due to the fact that, under TCPW long connections suffered less reduction in *cwnd* and *ssthresh*.

As the round trip time increases, the time for the feedback information to reach the TCP sender also increases which reduces the effectiveness of TCPW. Also, TCPW performs poorly when random packet loss rate exceeds a small threshold. During long correlated error burst, TCP timeouts occur which affect TCPW performance severely.

- *TCP Santa Cruz* [18]: TCP Santa Cruz (SC) uses new congestion control and error recovery strategies which are designed to work with path asymmetries, networks with lossy links, limited bandwidth, variable delay and out-of-order packet delivery. Since congestion losses are preceded by an increase in the network bottleneck queue (which may not be true for random losses in the wireless link), TCP-SC monitors the queue developing over a bottleneck link and determines whether congestion is incipient in the network and responds by increasing or decreasing the congestion window accordingly. The goals are to perform timely and efficient early retransmission of any lost packet, eliminate unnecessary retransmissions for correctly received packets when multiple losses occur within a window of data, and provide RTT estimates during periods of congestion and retransmission.

The congestion control algorithm in TCP-SC is based upon the measurement of *relative delay* that packets experience with respect to each other as packets are transmitted through the network. For each transmitted packet, TCP-SC sender maintains information about the transmission time of the packet, the arrival time of an ACK for that packet and arrival time of the data packet at the receiver (which is reported by the receiver in its ACK). Then for any two data packets i and j, the *relative forward delay* $D_{i,j}^F$ can be calculated as $D_{j,i}^F = R_{j,i} - S_{j,i}$, where $D_{j,i}^F$ represents the additional forward delay experienced by packet j with respect to packet i, $S_{j,i}$ is the time interval between the transmission of the packets, and $R_{j,i}$ is the interarrival time of data packets at the receiver. Based upon the values of $D_{j,i}^F$ over time, the change in the states of the queues (and specially the bottleneck queue) are determined. For example, if the sum of relative delays over an interval is 0, it implies that no additional congestion or queueing is present in the network at the end of the interval with respect to the beginning.

If n is the operating point of the congestion control algorithm (i.e., n is the desired number of packets in the bottleneck queue which is assumed to be one packet more than the BDP of the network), it attempts to maintain the total number of packets queued at the bottleneck link from the beginning of the connection until t_i (N_{t_i}) to n, where $N_{t_i} = N_{t_{i-1}} + M_{W_{i-1}}$, with $M_{W_{i-1}}$ being the additional amount of queueing introduced over the previous window W_{i-1}, and $N_{t_1} = M_{W_0}$. Now $M_{W_{i-1}}$ can be calculated based on the relative delay measurements as follows: $M_{W_{i-1}} = \frac{\sum D_k^F}{\bar{t}_{pkt}}$ where $\bar{t}_{pkt}$ is the average packet service time and k is the number of packet pairs within window W_{i-1}. Note that, the average packet service time can be calculated based on the timestamps returned by the receiver.

TCP-SC adjusts the congestion window once for each interval equal to the amount of time it takes to transmit one window of data. Over this interval, $M_{W_{i-1}}$ is calculated and at the end of the interval it is added to $N_{t_{i-1}}$. If $N_{t_i} < n - \delta$ (δ is some fraction of a packet), the congestion window is increased linearly, while if $N_{t_i} > n + \delta$, the congestion window is decreased linearly during the next interval, otherwise the congestion window is maintained at its current size. In this way, since adjustment of the congestion window does not depend on the arrival of ACKs, the congestion control algorithm becomes robust to ACK loss.

TCP-SC sender receives precise information on each packet correctly received and it uses a tighter estimate of the RTT per packet. TCP-SC recovers losses quickly without necessarily waiting for three duplicate acknowledgements from the receiver. Let packet i initially transmitted at time t_i is lost and marked as a hole in the ACK window. Packet i is retransmitted as soon as an ACK arrives for any packet transmitted at time t_x such that $t_x > t_i$, and $t_{current} - t_i > RTT_e$, where $t_{current}$ is the current time and RTT_e is the estimated round trip time of the connection.

TCP-SC was observed to provide high throughput and low end-to-end delay and delay variance over networks with a simple bottleneck link, networks with congestion in the reverse path of the connection, and networks which exhibit path asymmetry. TCP-SC can be implemented as a TCP option by utilizing the extra 40 bytes available in the options field of the TCP header.

TCP-SC does not differentiate between packet loss due to congestion and packet loss due to wireless link error. In fact, effectiveness of the proposed relative delay-based approach under correlated wireless channel losses is unclear.

- *Selective Slow Start (SSS)* [19]: Under this scheme, TCP attempts to distinguish between packet loss due to congestion and packet loss due to handoff based on the pattern of losses. For this, after a timeout the TCP sender counts the number of lost segments in the last S attempts including the current one and compares it with a pre-calculated *LIMIT*. If the number of lost segments is higher than *LIMIT*, the sender assumes that these losses are due to network congestion and initiates *slow start*, otherwise it continues to transmit at its current rate using the same timer values assuming that the losses are due to handoff. The value of the parameter *LIMIT* is set in a way that it can account for all segment losses during a handoff. For example, *LIMIT* can be set as *LIMIT* =

$\frac{K \times C \times T_h}{Seg}$, where C is the senders transmission rate (in *bps*), T_h is the average handoff duration (in *seconds*), Seg is the average segment size and K is a safety factor.

Although the *SSS* scheme was shown to perform better than standard TCP in wireless ATM environment for high handoff rate (e.g., $0.2/second$), the main limitation of this scheme is that for rate adaptation at the TCP sender it does not take wireless link conditions into account. Also, using a high value of *LIMIT* may cause congestion collapse in the wired Internet.

5.2   Receiver-Only Modifications

- *Freeze-TCP* [20]: This is a proactive mechanism in which the receiver notifies the sender of any impending 'blackout' situation (e.g., due to wireless channel fading or handoff) by 'zero window advertisement (ZWA)' and prevents the sender from entering into congestion avoidance phase. Upon receiving ZWA, the sender enters into the 'zero window probes (ZWP)' mode and freezes corresponding timers. While in ZWP mode, the sender transmits zero window probes and the interval between successive probes grows exponentially until it reaches 1 minute, where it remains constant. When the 'blackout' period is over, the receiver sends TR-ACKs (Triplicate Reconnection ACKs) for the last data segment successfully received to enable *fast retransmit* at the TCP sender.

  Ideally, the 'warning period' prior to disconnection (i.e., how much in advance should the receiver start ZWA) should be long enough to ensure that exactly one ZWA gets across the sender. If the warning period is longer than this, the sender will be forced into ZWP mode prematurely resulting in idle time prior to disconnection. If the warning period is too small, the receiver might not have enough time to send out a ZWA which will cause the sender's congestion window to drop.

  One drawback of this approach is that, the receiver needs to predict the impending disconnections. For this, some cross-layer information exchanges may be necessary.

- *ACK-Fragmentation Protocol* [21]: If the mobile host, after sending three duplicate acknowledgements, receives a new retransmitted packet from the sender, it sends N equally spaced cumulative acknowledgements for this new packet within one round-trip time. If for this received packet the sequence number of the first octet is S and last octet is F, the ith cumulative acknowledgement will acknowledge upto octet $S + i(F - S)/N$. The value of N can be computed from the random packet loss probability in the wireless link, the link capacity, round-trip time and buffer size at the bottleneck link [21].

The effectiveness of this protocol comes from the fact that increased number of acknowledgements within same round-trip time expedites the resumption of normal transmission window at the TCP sender. Moreover, it increases the chance that the sender receives at least one acknowledgement.

5.3   Modifications at the Sender and the Receiver

- *Wireless TCP (WTCP)* [22]: WTCP is an end-to-end reliable transport layer protocol for wired-cum-wireless networks which uses *rate-based* transmission control instead of window-based transmission control and the rate adaptation computations are performed by the receiver. Since the receiver performs the rate computations, the effects of delays variations and losses in the ACK path are eliminated. The goal of WTCP is to decrease the transmission rate aggressively in case of congestion (so that the congestion alleviates quickly) and decrease less aggressively in the case of incipient congestion to improve efficiency.

To distinguish between congestion-related and non-congestion-related packet loss, the WTCP receiver, using a history of packet losses, computes the mean and mean deviation in the number of packet losses when the network is predicted to be uncongested. If the loss is predicted to be due to congestion, the transmission rate is decreased aggressively by 50%. For non-congestion related packet loss WTCP uses the ratio of the average interpacket delay observed at the receiver to the interpacket delay at the sender (rather than using packet loss and retransmission timeouts) to control transmission rate. The WTCP receiver maintains two state variables *lavg_ratio* and *savg_ratio* for the long-term and short-term running averages of the ratio of the observed sending rate at the receiver to the actual sending rate at the sender. At any time, the receiver can be in one of the three states–*increase* (when *lavg_ratio* $> \alpha_+$ and *savg_ratio* $> \beta_+$), *decrease* (when *lavg_ratio* $> \alpha_-$ and *savg_ratio* $> \beta_-$), *maintain*.

WTCP uses SACK and no retransmission timer for loss recovery. The sender tunes the desired rate for ACK transmission by the receiver so that it receives at least one ACK in a threshold period of time and can react to the new transmission rate.

In summary, WTCP is significantly different from TCP and tuning of the different parameters would be required to optimize the performance of WTCP.

- *Explicit Congestion Notification (ECN) TCP* [23]: This proposal is based on the TCP with explicit congestion notification [24] mechanism where the routers should inform the TCP sender of incipient congestion so that the TCP senders can lower the transmission rate. For example, if RED (Random Early Detection) mechanism is used, a router signals incipient congestion to TCP by setting the Congestion Experienced (CE) bit in the IP header of the data packet when the average queue size lies between two thresholds min_{th} and max_{th}. In response, the receiver sets the 'ECN-Echo (ECE)' bit in the ACKs and when the sender receives an ACK packet with ECE bit set, it invokes the congestion control mechanism. In case of packet loss in the wireless channel, the receiver sends ACK packets without CE bit set and on receiving three duplicate ACKs without CE bit set the sender deduces that the packet loss is due to errors in the wireless channel. In this case, the sender does not invoke the *congestion avoidance* algorithm.

  In case of long outage in the wireless channel, this scheme may not work well. Again, this can be used only for ECN-capable TCP.

- *Eifel Algorithm* [25]: A large number of timeouts and fast retransmissions that occur at a TCP sender have been observed to be *spurious*. That is, the ACKs received at the sender after a timeout or fast retransmit are actually the ACKs for the original packets and not for the retransmitted packets. Spurious fast retransmits result when the packet reordering in the Internet results in a reordering length more than the duplicate ACK threshold. Due to spurious timeouts and spurious retransmissions, the TCP sender unnecessarily reduces the transmission rate and enters into a go-back-N type of retransmission mode. When the sender enters into the retransmission mode, it causes the receiver to send duplicate ACKs which may result in *fast retransmit* after a timeout resulting in further reduction in transmission rate. The *Eifel algorithm* was proposed to eliminate the problems caused by *spurious timeouts* and *spurious fast retransmits* in TCP.

  The Eifel algorithm uses the TCP timestamp option. Each packet transmitted by the sender is timestamped and the values of *ssthresh* and *cwnd* are also stored. The receiver returns the timestamp value of the packet in the corresponding ACK. After a timeout or a fast retransmit, the values of *ssthresh* and *cwnd* are updated according to the usual algorithms and the packet is retransmitted. After receiving the ACK, if the sender finds that it is for the original packet, it restores the values of *ssthresh* and *cwnd*.

This algorithm does not take packet loss due to wireless channel error into account, and therefore, does not solve the problems of TCP in a wired-cum-wireless network.

5.4 Modifications at Base Station: Non-End-to-End Transport Level Approaches

- *Indirect-TCP (I-TCP)* [27]: I-TCP splits a transport layer connection between the MH and the FH into two separate connections – one over the wireless link between the MH and its mobile support router (MSR) at the base station and the other between the MSR and the FH over the fixed network. After receiving the data packets destined to the MH, the MSR sends ACKs to the FH and forwards the packets to the MH using a separate transport protocol designed for better performance over wireless links. Such a transport protocol for the wireless link can support notification of events such as disconnections to the link-aware and mobility-aware applications. In this case, the mobile hosts can run simple wireless access protocol to communicate with the MSR and the MSR manages the communication overheads for communicating with the FHs. For example, when an MH wants to communicate with a FH, it sends a request to its current MSR, and the MSR then establishes a TCP connection with the FH. Whenever an MH moves to another cell, the entire connection needs to be moved to the new MSR which should be completely transparent to the fixed network.

  Performance of I-TCP was observed to be better than regular TCP in wide-area wireless networks. However, end-to-end TCP semantics are not maintained in this protocol and applications running on the mobile host have to be relinked with the I-TCP library. Also, I-TCP is not suitable for cases where the wireless link is not the last part of a connection path, because in such a case a particular connection may need to be split several times resulting in performance degradation.

- *Using ICMP Control Messages* [29]: A scheme for improving TCP performance in wireless networks using ICMP control messages was proposed in [29]. In this scheme, the BS transmits an ICMP-DEFER message to the FH when it fails in its first attempt to transmit the packet to the MH over the wireless link. On receiving the ICMP-DEFER message for a particular packet, the sender resets the corresponding timer. When all local retransmissions fail, the BS notifies the sender using ICMP-RETRANSMIT message so that the sender can retransmit the packet.

- *Using Multiple Acknowledgements* [30]: The scheme proposed in [30] maintains end-to-end semantics and uses two types of ACKs, namely, *partial ACK (ACK_p)* and *complete ACK (ACK_c)* to distinguish losses due to congestion from losses due to wireless link error. The sender uses the regular TCP mechanism with *slow-start, congestion avoidance, fast retransmit* and *fast recovery*. The complete acknowledgement ACK_c is sent by the receiver and the partial acknowledgement ACK_p is sent by the BS. ACK_p with sequence number N_a informs the sender that packet(s) with sequence numbers up to $N_a - 1$ have been received by the BS, and the BS is facing problems in forwarding the packets with sequence numbers from ACK_c up to $N_a - 1$ (i.e., it has not received the ACK from the wireless host even after waiting for the maximum possible delay time between the transmission of a packet from BS and reception of the ACK from the mobile host). Then the sender marks the corresponding packets and updates RTO to give the BS more time to perform retransmission. If timeout occurs, and the packets are still marked, the sender will backoff the timer only without invoking any congestion control method. On receipt of ACK_c, the sender works similar to a normal TCP sender.

  When the BS receives an in-sequence packet, or an out of sequence packet which is not in its buffer, the packet is buffered and transmitted to the mobile host. The BS starts a timer with the timeout value equal to the maximum time which can elapse between the reception of a packet at the BS and its acknowledgement from the mobile host. If the BS receives an out of sequence packet which is present in the buffer, it sends an ACK_p to the sender. If the local timer at the BS expires before receiving an acknowledgement from the mobile host, the BS retransmits the packets and sends an ACK_p to the sender.

- *Mobile-End Transport Protocol (METP)* [31]: METP hides the wireless link from the rest of the Internet in a wired-cum-wireless network by terminating the TCP connection at the BS on behalf of mobile host. TCP/IP between the BS and a mobile host is replaced by a low overhead protocol. All TCP connections are handled by METP at the BS which negotiates with another host in the Internet to open or close a TCP connection, and keeps the connection state and sending and receiving buffers. For data transfer from a mobile host to an Internet host through a TCP connection, the mobile host sends the data to the BS, and METP sends them out as TCP segments to the destination. When a TCP/IP packet destined to the mobile host arrives at the BS, METP sends an ACK and puts it in the receiving buffer from where a separate process transmits it to the mobile host. In case of temporary link failure, since METP at the BS continues to receive data from the fixed host, it sends out ACK with smaller advertised window. The advertized window is retained to its original level when the wireless link becomes good again.

METP uses link layer retransmissions and acknowledgements for reliable data delivery over the wireless link. If no acknowledgement is received immediately after a data frame transmission, the frame is retransmitted after a random backoff interval. To avoid buffer overflow, flow control is achieved by the METP at the receiver through periodically sending out a feedback packet to inform the sender of the available buffer space.

METP uses a header size of only 12 bytes for packets exchanged between the mobile host and the BS. The header size can be further reduced by using the idea of header compression. In case of handoff the old BS opens a separate TCP connection with the new BS and sends all data and state information.

The merits of this protocol are that by using small header it incurs less overhead for wireless transmission and it avoids activation of the TCP congestion control mechanisms in case of packet loss in the wireless channel. METP was observed to provide better throughput performance than split-connection TCP and split-connection TCP with selective ACK under varying handoff interval and BS buffer size. However, METP does not maintain the strict end-to-end semantics and suffers from the scalability and reliability problems (due to failure of the BS).

- *WTCP* [32]: WTCP uses a modified flow and error control protocol between the BS and the mobile host. After receiving a TCP/IP packet from a fixed host, WTCP at the BS buffers it (if this is the next packet expected from the fixed host or if the packet has a larger sequence number than expected) along with its arrival time. WTCP then transmits the buffered packet to the mobile host and when a packet is transmitted to the mobile host, the BS schedules a new timeout if there is no other timeout pending. Upon receiving the ACK from the mobile host, WTCP frees the corresponding buffer and sends the ACK to the fixed host. WTCP performs local error recovery based on duplicate acknowledgements or timeout. In case of timeout, WTCP reduces the transmission window size to one assuming subsequent bad channel condition. As soon as the BS receives an ACK, the transmission window size is set again to the receiver advertized window size. Also, upon reception of a duplicate ACK, WTCP opens the transmission window in full.

Another feature of WTCP is, it attempts to hide the time spent by the BS for local error recovery by adding that time to the timestamp value of the corresponding segment so that the TCP's round trip time estimation at the source is not affected. In this way, the TCP source's ability to effectively detect congestion in the wired network is not impacted.

Due to its aggressiveness in window adaptation, WTCP was observed to provide better/equal throughput performance compared to I-TCP and *snoop TCP*. However, a more optimized flow and congestion control scheme along with header compression (as in METP) can be used under similar scenario to provide even better performance.

A qualitative comparison among the different end-to-end TCP modifications is provided in Table 9.5.

6. Chapter Summary

Performances of the different basic TCP variants (e.g., TCP Tahoe, TCP Reno, TCP New-Reno, SACK TCP, FACK TCP, TCP Vegas) have been analyzed in a wide-area cellular network by using *ns-2*-based computer simulations. Impact of the wireless channel error characteristics on the slow-start, congestion avoidance and timeout mechanisms of TCP has been explored by analyzing the transport level system dynamics. The maximum achievable end-to-end throughput for a TCP-like window-based end-to-end transmission control mechanism has been evaluated which can serve as a benchmark for performance evaluation of any transport protocol designed (to enhance TCP performance in wide-area wireless networks) based on transport level modifications to the basic TCP mechanisms. A brief overview of the different proposed

Table 9.5. Comparison among the transport-level TCP modifications.

TCP Variant	Maintains end-to-end semantics	Can handle end-to-end encryption	Requires intermediaries' support	Can handle random packet loss	Can handle burst errors efficiently
K-SACK	Yes	Yes	No	Yes	No
Westwood	Yes	Yes	No	Yes	No
Santa Cruz	Yes	Yes	No	Yes	-
SSS	Yes	Yes	No	Yes	No
Freeze-TCP	Yes	Yes	No	Yes	Yes
ACK-fragmentation	Yes	Yes	No	Yes	No
WTCP (rate-based)	Yes	Yes	No	Yes	Yes
ECN	Yes	No	Yes	Yes	No
Eifel	Yes	Yes	No	No	No
I-TCP	No	No	Yes	Yes	No
Using ICMP	No	No	Yes	Yes	Yes
Multiple-ACK	Yes	No	Yes	Yes	Yes
METP	No	No	Yes	Yes	Yes
WTCP (window-based)	Yes	No	Yes	Yes	Yes

TCP mechanisms (based on transport level modifications) along with a qualitative performance comparison have also been presented.

Future research in end-to-end TCP design for wired-cum-wireless networks should address

- how to distinguish among packet losses due to wireless channel error and congestion error (e.g., by using some estimation/filtering mechanisms) and comparison among the different (estimation/filtering) mechanisms

- how to exploit the above information for adjusting the transmission rate at the sender in the case of packet loss to design end-to-end TCP based on modifications at the sender only

- fairness, energy-efficiency and TCP-friendliness of an end-to-end TCP customized for wired-cum-wireless networks.

Acknowledgments

The authors acknowledge the support from Telecommunications Research Labs (TR*Labs*), Winnipeg, Canada, for this research.

Notes

1. In the Internet community, both NPDU and TPDU are known as *packet* even though in OSI terminology a *segment* is referred to as a TPDU. The terms 'packet' and 'segment' are used interchangeably in this article.

2. The terms 'flow' and 'connection' are used interchangeably in this article.

3. The bandwidth-delay product for a TCP connnection refers to the product of the round trip delay (T) for the connection and the capacity of the bottleneck link (μ) in its path.

4. This refers to a new acknowledgement received during *fast recovery* which acknowledges some but not all of the packets that were outstanding at the start of the *fast recovery* phase.

5. MSS refers to the *Maximum Segment Size*.

6. Timeout caused by this fine-grained timer is referred to as 'Vegas expiry'.

7. Here, $f_d = v/\lambda$ = mobile speed/carrier wavelength. The value of t_{tcp} determines the minimum fade duration.

8. It refers to the maximum fading attenuation which still allows correct reception of a packet.

9. This is similar to the fairness function used in [50] to quantify the fairness in a shared resource system with n users: $F = \dfrac{\left(\sum_{i=1}^{n} x_i\right)^2}{n \sum_{i=1}^{n} x_i^2}$ (where x_i is the ith user's throughput).

References

[1] Iren, S., and P. D. Amer, "The transport layer: Tutorial and survey," *ACM Computer Surveys*, vol. 31, no. 4, pp. 360-405, Dec. 1999.

[2] Ayala, R., K. Basu, and S. Elliot, "Internet technology based infrastructure for mobile multimedia services," *Proc. IEEE WCNC'99*, pp. 109-113.

[3] Brasche, G., and B. Walke, "Concepts, services and protocols of the new GSM phase 2+ general packet radio service," *IEEE Communications Magazine*, Aug. 1997, pp. 94-104.

[4] Postel, J.,"Transmission control protocol," *Internet RFC 793*, 1981.

[5] Jacobson, V.,"Congestion avoidance and control," *ACM SIGCOMM Computer Communication Review*," vol. 18, no. 4, pp. 314-329, Aug. 1988.

[6] Allman, M., V. Paxson, and W. R. Stevens, "TCP congestion control," RFC 2581, Apr. 1999.

[7] Lee, W.C.Y. (1993). *Mobile Communication Design Fundamentals*. 2nd Edition, John Wiley and Sons.

[8] Inamura, H., G. Montenegro, R. Ludwig, A. Gurtov, and F. Khafizov, "TCP over second (2.5G) and third generation (3G) wireless networks," *Internet draft*, May 2002, URL: http://www.ietf.org/internet-drafts/draft-ietf-pilc-2.5g3g-08.

[9] Floyd, S., and K. Fall, "Promoting the use of end-to-end congestion control in the Internet," *IEEE/ACM Transactions on Networking*, May 1993, URL: http://www-nrg.ee.lbl.gov/floyd

[10] Cáceres, R., and L. Iftode, "Improving the performance of reliable transport protocols in mobile computing environments," *IEEE Journal on Selected Areas in Communications*, vol. 13, no. 5, pp. 850-857, June 1995.

[11] Balakrishnan, H., S. Seshan and R. Katz, "A comparison of mechanisms for improving TCP performance over wireless links," *IEEE/ACM Transactions on Networking*, vol. 5, pp. 756-769, Dec. 1997.

[12] Balakrishnan, H., S. Seshan, and R. H. Katz, "Improving reliable transport and handoff performance in cellular wireless networks," *ACM/Baltzer Wireless Networks*, vol. 1, no. 4, pp. 469-481, Dec. 1995.

[13] Balakrishnan, H., S. Seshan, E. Amir, and R. H. Katz, "Improving TCP/IP performance over wireless networks," *Proc. ACM MOBICOM'95*.

[14] Brown, K., and S. Singh, "M-TCP: TCP for mobile cellular networks," *Proc. IEEE INFOCOM'96*, 1996.

[15] Chrungoo, A., V. Gupta, H. Saran, and R. Shorey, "TCP K-SACK: A simple protocol to improve performance over lossy links," *Proc. IEEE GLOBECOM'01*, San Antonio, Texas, USA, Nov. 2001.

[16] Lin, D., and H. Hung, "TCP fast recovery strategies: Analysis and improvements," *Proc. IEEE INFOCOM'98*, Apr. 1998.

[17] Casetti, C., M. Gerla, S. Mascolo, M. Y. Sanadidi, and R. Wang, "TCP Westwood: End-to-end congestion control for wired/wireless networks," *Wireless Networks*, vol. , no. 8, pp. 467-479, 2002.

[18] Parsa, C., and J. J. Garcia-Luna-Aceves, "Improving TCP congestion control over Internets with heterogeneous transmission media," *Proc. IEEE Int. Conference on Network Protocols (ICNP'99)*, Toronto, Canada, Oct. 31-Nov. 3, 1999.

[19] Varshney, U., "Selective slow start: A simple algorithm for improving TCP performance in wireless ATM environment," *Proc. IEEE MILCOM'97*, pp. 465-469.

[20] Goff, T., J. Moronski, D. S. Phatak, and V. Gupta, "Freeze-TCP: A true end-to-end TCP enhancement mechanism for mobile environments," *Proc. IEEE INFOCOM'00*.

[21] Banerjee, D. N., "Improving wireless-wireline TCP interaction," submitted to *IEEE/ACM Transactions on Networking*.

[22] Sinha, P., N. Venkitaraman, T. Nandagopal, R. Sivakumar, and V. Bharghavan, "WTCP: A reliable transport protocol for wireless wide-area networks," *Proc. ACM MOBICOM'99*, Seattle, Washington, Aug. 1999.

[23] Ramani, R., and A. Karandikar, "Explicit congestion notification (ECN) in TCP over wireless networks," *Proc. IEEE Int. Conference on Personal Wireless Communications (ICPWC'00)*, pp. 495-499.

[24] Floyd, S., "TCP and explicit congestion notification," *ACM Computer Communication Review*, vol. 24, no. 5, pp. 10-23, Oct. 1994.

[25] Ludwig, R., and R. H. Katz, "The Eifel algorithm: Making TCP robust againt spurious retransmissions," *ACM Computer Communication Review*, vol. 30, no. 1, Jan. 2000.

[26] Vaidya, N., Overview of work in mobile-computing (Transparencies), http://www.cs.tamu.edu/faculty/vaidya/slides.ps

[27] Bakre, A., and B. R. Badrinath, "I-TCP: Indirect TCP for mobile hosts," *Proc. 15th IEEE Int. Conf. Distributed Computing Systems (ICDCS)*, pp. 136-143, May 1995.

[28] Balakrishnan, H., and R. Katz, "Explicit loss notification and wireless web performance," *Proc. IEEE GLOBECOM'98, Internet Mini Conference*, Sydney, Australia.

[29] Goel, S., and D. Sanghi, "Improving TCP performance over wireless links," *Proc. of IEEE Region Ten Conference on Global Connectivity in Energy, Computer Communication and Control (TENCON'98)*, Dec. 1998.

[30] Biaz, S., and N. Vaidya, "TCP over wireless networks using multiple acknowledgements," Texas A&M University, Technical Report 97-001, Jan. 1997.

[31] Wang, K.-Y, and S. K. Tripathi, "Mobile-end transport protocol: An alternative to TCP/IP over wireless links," *Proc. IEEE INFO-COM'98*.

[32] Ratnam, K., and I. Matta, "WTCP: An efficient mechanism for improving TCP performance over wireless links," *Proc. Third IEEE Symposium on Computers and Communications (ISCC'98)*, Athens, Greece, June 1998.

[33] Chan, M. C, and R. Ramjee, "TCP/IP performance over 3G wireless links with rate and delay variation," *Proc. ACM MOBICOM'02*, Sept. 2002.

[34] DeSimone, A., M. C. Chuah, and O. C. Yue, "Throughput performance of transport-layer protocol over wireless LANs," *Proc. IEEE GLOBECOM'93*, Dec. 1993.

[35] Parsa, C., and J. J. Garcia-Luna-Aceves, "Improving TCP Performance over wireless networks at the link layer," *ACM Mobile Networks and Applications*, Special Issue on Mobile Data Networks: Advanced Technologies and Services, vol. 5, no. 1, 2000, pp. 57-71.

[36] Vaidya, N., and M. Mehtha, "Delayed duplicate acknowledgements: A TCP-unaware approach to improve performance of TCP over wireless links," Texas A&M University, Technical Report 99-003, Feb. 1999.

[37] Chiasserini, C.-F., and M. Meo, "Improving TCP over wireless through adaptive link layer setting," *Proc. IEEE GLOBECOM'01*, San Antonio, TX, Nov. 2001.

[38] Wong, J. W. K., and V. C. M. Leung, "Improving end-to-end performance of TCP using link-layer retransmissions over mobile internetworks," *Proc. IEEE ICC'99*, pp. 324-328.

[39] Henderson, T.R., E. Sahouria, S. McCanne, and R. Katz, "On improving the fairness of TCP congestion avoidance," *Proc. IEEE GLOBECOM'98*, pp. 539-544.

[40] Kumar, A., "Comparative performance analysis of versions of TCP in a local network with a lossy link," *IEEE/ACM Transactions on Networking*, vol. 6, no. 4, pp. 485-498, Aug. 1998.

[41] Lakshman, T.V., and U. Madhow, "The performance of TCP/IP for networks with high bandwidth-delay products and random loss," *IEEE/ACM Transactions on Networking*, vol. 5, no. 3, pp. 336-350, June 1997.

[42] Fall, K., and S. Floyd, "Simulation-based comparisons of Tahoe, Reno, and Sack TCP," *ACM Computer Communication Review*, Jul. 1996.

[43] Mathis, M., J. Mahdavi, S. Floyd, and A. Romanow, "TCP selective acknowledgement options," *Internet RFC 2018*, 1996.

[44] Floyd, S., and T. Henderson, "The New-Reno modification to TCP's fast recovery algorithm," *RFC 2582*, Apr. 1999.

[45] Mathis, M., J. Mahdavi, S. Floyd, and A. Romanow, "TCP selective acknowledgement options," *RFC 2018*, Apr. 1996.

[46] Mathis, M., and J. Mahdavi, "Forward acknowledgement: Refining TCP congestion control," *Proc. ACM SIGCOMM'96*, pp. 281-291, 1996.

[47] Brakmo, L.S., and L. L. Peterson, "TCP Vegas: End to end congestion avoidance on a global Internet," *IEEE Journal on Selected Areas in Communications*, 1995.

[48] McCanne, S., and S. Floyd, "NS (Network Simulator)," 1995. URL http://www.isi.edu/nsnam/ns.

[49] Chockalingam, A., M. Zorzi, L. B. Milstein, and P. Venkataram, "Performance of a wireless access protocol on correlated Rayleigh fading channels with capture," *IEEE Transactions on Communications*, vol. 46, pp. 644-655, May 1998.

[50] Chiu, D., and R. Jain, "Analysis of increase and decrease algorithms for congestion avoidance in computer networks," *Computer Networks and ISDN Systems*, vol. 17, pp. 1-14, June 1989.

Chapter 10

MULTI-SERVICE WIRELESS INTERNET LINK ENHANCEMENTS

George Xylomenos and George C. Polyzos
Mobile Multimedia Laboratory
Department of Informatics
Athens University of Economics and Business
xgeorge@aueb.gr, polyzos@aueb.gr

Abstract   The deployment of several real-time multimedia applications over the Internet has motivated a considerable research effort on the provision of multiple services on the Internet. In order to extend this work over wireless links however, we must also take into account the performance limitations of wireless media. We survey various related approaches and conclude that link layer schemes provide a universal and localized solution. Based on simulations of application performance over many link layer schemes we show that different approaches work best for different applications. We present a multi-service link layer architecture which enhances the performance of diverse applications by concurrently supporting multiple link layer schemes. Simulations of multiple applications executing simultaneously show that this approach dramatically improves performance for all of them. We finally consider embedding this approach into a Quality of Service oriented Internet, discussing the traditional best-effort architecture, the Differentiated Services architecture and an advanced dynamic service discovery architecture.

Keywords: Wireless link layer protocols, quality of service, differentiated services.

1.   Introduction

The most important Internet related developments in the past few years were arguably the emergence of mechanisms for providing *Quality of Service* (QoS) guarantees to applications and the wide deployment of wireless access networks using technologies such as *Digital Cellular Communications* and *Wireless Local Area Networks* (WLANs). The

QoS effort is largely driven by the desire to migrate applications with
real-time delay constraints, such as voice telephony and video conferencing, from the circuit-switched telephony networks to the packet-switched
Internet. Many approaches have been proposed for the provision of adequate QoS for such applications, without requiring dramatic changes to
the Internet [1, 2]. These schemes provide differentiated treatment to
different traffic classes, in terms of packet priorities and drop probabilities, so as to combine the economies of statistical multiplexing with the
delay guarantees needed for real-time applications.

Integrating wireless networks into the Internet is relatively simple, due
to the physical layer independence of the *Internet Protocol* (IP), which
offers a standard interface to higher layers. However, while providing
IP services over wireless links is easy, the resulting performance is often
disappointing due to the relatively high error rates of wireless links. In
addition, while satellite and terrestrial microwave links have long been
part of the Internet, higher layer protocols commonly make assumptions
about link performance that cannot be met by wireless links, leading to
further performance degradation.

This chapter describes an architecture that improves the performance
of diverse Internet applications while also extending Internet QoS support over wireless networks. In Sec. 10.2 we outline the problem and
review previous work on wireless link enhancements and service differentiation. In Sec. 10.3 we describe the simulation setup that we use
throughout the chapter. In Sec. 10.4 we present single-service simulations showing that different applications favor different link enhancement
schemes. In Sec. 10.5 we present a *multi-service link layer* architecture
that simultaneously enhances the performance of diverse applications by
supporting multiple link mechanisms in parallel. In Sec. 10.6 we present
simulations showing that with this architecture each application achieves
similar gains as when it operates by itself over its preferred link layer
mechanism. We then discuss the integration of this approach with various Internet QoS schemes: Sec. 10.7 covers the traditional best-effort
service, Sec. 10.8 the Differentiated Services architecture, and Sec. 10.9
an advanced dynamic service discovery architecture.

2.   Background and Related Work

IP offers an unreliable packet delivery service between any two Internet hosts. IP packets may be lost, reordered or duplicated. Applications can use the *User Datagram protocol* (UDP) for direct access to
this service. Some applications employ UDP assuming that the network
is reliable enough for their needs, for example, file sharing via NFS over

wired LANs. Delay sensitive applications may also use UDP in order to employ their own customized loss recovery mechanisms. For example, delay sensitive applications may add redundancy to their data so as to tolerate some losses without (slow) end-to-end retransmissions.

Most applications however prefer complete reliability, hence they employ the *Transmission Control Protocol* (TCP), which offers a reliable byte stream service. TCP breaks the application data stream into segments which are reassembled at the receiver. The receiver returns cumulative *acknowledgments* (ACKs) for segments received in sequence, with duplicate ACKs for out of sequence ones. Since IP may reorder packets, the sender retransmits the first unacknowledged segment only after receiving multiple (usually 3) duplicate ACKs. The sender tracks the round trip delay of the connection, so that if an ACK for a segment does not arrive in time, the segment is retransmitted [3]. Due to the high reliability of wired links, TCP assumes that all losses signify congestion. Therefore, after a loss is detected, TCP reduces its transmission rate, and then increases it slowly so as to gently probe the network [3].

With wireless links, some common assumptions about network reliability are no longer valid. The high error rate of wireless links causes UDP application performance to degrade or even become unacceptable, since it is up to the application to recover from any losses. On the other hand, TCP continuously reduces its transmission rate due to wireless errors to avoid what it (falsely) assumes to be congestion. With multiple wireless links on the path, throughput is reduced in a multiplicative manner [4]. For example, WLANs depict losses of up to 1.5% for Ethernet size frames [5], while cellular systems suffer from losses of 1–2% for their much shorter frames [6]. The effects of these losses are dramatic: a 2% frame loss rate over a WLAN halves TCP throughput [7].

The performance of UDP applications over wireless links has not been studied extensively, not only due to their diversity, but also because they were perceived as oriented towards wired LANs, an assumption challenged by media streaming over the Internet. Considerable work has been devoted to TCP however. Most TCP enhancements try to avoid triggering congestion recovery due to wireless errors. Generic TCP enhancements such as *Eifel* [8] and *Selective Acknowledgments* [9] improve TCP performance by reducing the *number* of redundant TCP retransmissions, without however reducing their (end-to-end) *delay*.

One wireless specific approach is to *split* TCP connections into one connection over the wireless link and another one over the wired part of the path, bridged by an agent [10]. Recovery is only performed locally over the wireless link. This scheme violates the end-to-end semantics of TCP and it is incompatible with IP security which encrypts TCP headers [11]. Other approaches maintain TCP semantics by *freezing* TCP state whenever persistent errors are detected [12, 13]. As long as errors persist, TCP does not invoke congestion control. Wireless errors and congestion losses are differentiated using either explicit loss notifications [12] or explicit congestion notifications [13]. Thus, performance is not unnecessarily degraded, but error recovery remains end-to-end and the network must provide explicit loss or congestion notifications.

The main alternative to TCP modifications is local error recovery at the link layer. One option is to use the *Radio Link Protocols* (RLPs) of cellular systems which provide error control customized for the underlying link [6, 14]. Since these RLPs only apply a single error control scheme though, they may be inappropriate for some traffic. For example, retransmissions are beneficial to TCP but problematic for real-time traffic. Another problem is that RLPs may interfere with TCP recovery [15], leading to conflicting retransmissions between the link and transport layers. This is avoided by only performing retransmissions when the link layer recovers from losses much faster than the transport layer [16].

One method of jointly optimizing recovery between layers is to exploit transport layer information at the link layer. By *snooping* inside *each* TCP stream at the wireless base station, we can transparently retransmit lost segments when duplicate ACKs arrive, hiding the duplicates from the sender to avoid end-to-end recovery. This approach avoids both control overhead and adverse cross layer interactions [7]. However, it is also incompatible with IP security as it employs TCP header information and it only works in the direction from the wired Internet towards the wireless host due to its reliance on TCP ACKs. In the reverse direction ACKs are returned late and they may signify congestion losses [4].

In order to prevent RLPS from adversely affecting non-TCP traffic, it has been suggested that link layer traffic should be split in two classes, TCP and UDP-based, so as to provide reliable transmission for TCP traffic only. Such a link layer may be either aware or unaware [17, 18] of TCP semantics, as long as it can distinguish TCP from UDP packets. This method does not improve the performance of UDP applications however, which may then fail over the error prone wireless links.

In general, link layer approaches have the advantage over higher layer ones of providing a local solution that does not require any changes to higher layers. This makes them transparent to the rest of the Internet, enables them to recover faster than transport layer solutions and allows them to exploit lower layer information to optimize recovery [16, 18]. Therefore, for the remainder of this chapter we will focus on link layer mechanisms that are able to provide customized error control to each traffic class and the introduction of such mechanisms into the Internet.

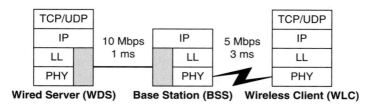

Figure 10.1. Simulation topology.

3. Simulation Setup

To study the interactions between link layer schemes, transport layer protocols and applications, we performed extensive simulations using the ns-2 simulator [19], enhanced with additional error modules, link layer schemes and applications [20]. To compensate for statistical fluctuations, we repeated each experiment 30 times, using the first 30 random seeds embedded in ns-2. The results shown below represent averaged values from all runs, with error bars at plus/minus one standard deviation. We provide elsewhere additional simulation details, as well as results for other network topologies and wireless links [21, 22]. Here we only provide a sample of our results so as to motivate our design.

The simulated network topology is shown in Fig. 10.1. A *Wired Server* (WDS) communicates with a *Wireless Client* (WLC) via a *Base Station* (BSS). The wired link between the WDS and the BSS is a LAN with 10 Mbps of bandwidth and a 1 ms delay. The wireless link between the BSS and the WLC simulates an IEEE 802.11b WLAN with 5 Mbps of bandwidth and a 3 ms delay, using 1000 byte frames. All links are treated as full-duplex pipes for simplicity. To allow comparisons with other studies, the wireless link corrupts bits at exponentially distributed intervals with average durations of 2^{14}, 2^{15}, 2^{16} and 2^{17} bits [7], leading to frame loss rates of 0.8% to 5.9%. We also ran experiments under error-free conditions for reference purposes. The error processes in each wireless link direction are identical but independent. We ignored TCP/UDP/IP headers as they uniformly influence all link layer schemes, but accounted for the *exact* framing overhead required by each link layer scheme.

We tested both TCP and UDP applications, so as to evaluate the suitability of various link layer schemes for diverse applications. For TCP, we simulated file transfers and World Wide Web browsing, using the TCP Reno module of ns-2 with 500 ms granularity timers. For UDP, we simulated continuous media distribution, a delay sensitive but error tolerant application. In all cases, most data flows from the WDS to the WLC, simulating a client-server interaction with the client on a wireless network. Some data also flow in the reverse direction, for example, TCP ACKs. As a baseline, we simulated a *Raw Link* scheme, which does not perform any error recovery, under both TCP and UDP.

For TCP applications, we tested reliable link layer schemes delivering frames in sequence to higher layers so as to avoid TCP retransmissions. In *Go Back N* the sender buffers outgoing frames and retransmits unacknowledged frames after a timeout. The receiver positively acknowledges frames received in sequence, dropping out of sequence ones. After a timeout the sender retransmits *all* outstanding frames. *Selective Repeat* improves upon this by buffering out of sequence frames at the receiver and returning *negative* ACKs (NACKs) when it detects gaps, thus allowing the sender to retransmit lost frames only. Our Selective Repeat variant allows multiple NACKs per loss [23]. In both schemes, under persistent losses the sender may exhaust its window and stall. Each frame includes sequence and acknowledgment numbers (2 bytes).

To prevent conflicts with TCP retransmissions due to multiple link layer retransmissions [15], the sender in *Karn's RLP* abandons frames not received after 3 retransmissions (for TCP) [6]. Thus, the sender never stalls. This scheme uses only NACK and *keepalive* frames during idle periods, thus frames only include a sequence number (1 byte). Finally, *Berkeley Snoop* is a TCP aware scheme [7], included in the ns-2 distribution. A module at the BSS *snoops* inside TCP segments, buffering data sent to the WLC. If duplicate TCP ACKs indicate a lost packet, this is retransmitted by the BSS and the ACKs are suppressed.

For the UDP-based application simulated, low delay is preferable to full reliability. *Forward Error Correction* (FEC) schemes offer limited recovery by adding redundancy to the transmitted stream, allowing the receiver to recover from losses. The *XOR based FEC* scheme sends data frames unmodified, but every 12 frames (a tradeoff between overhead and delay) a *parity* frame is also transmitted, generated by XOR'ing the preceding data frames, collectively called a *block*. If a *single* data frame is lost from a block, we can recover it by XOR'ing the remaining data frames with the parity frame. A timeout is used to prematurely emit a parity frame when the link becomes idle before the current block has been completed. All frames include a sequence number (1 byte).

The UDP-based application was tested with Selective Repeat, in order to examine the interactions between a fully reliable scheme and a delay sensitive application. We also tested Karn's RLP, but with 1 retransmission per loss to keep delay low. In this scheme, frame losses cause

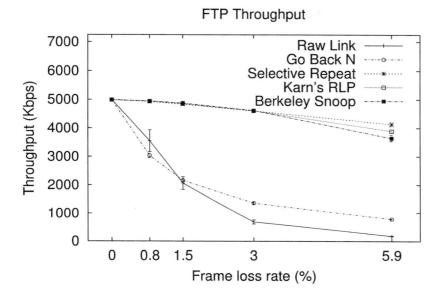

Figure 10.2. Stand-alone file transfer: throughput.

subsequently received frames to wait until the missing one is received or abandoned. This is detrimental to applications using their own rese-quencing buffers, a common case in continuous media distribution. Our *Out of Sequence* (OOS) variant of Karn's RLP immediately releases all received frames to higher layers so as to avoid such delays. This variant was also tested with 1 retransmission per loss.

4. Single-Service Link Layer Performance

The first TCP-based application tested was file transfer over FTP. We simulated a 100 MByte file transfer from the WDS to the WLC. File transfers are unidirectional, with TCP ACKs travelling in the reverse direction. The ns-2 FTP module sends data as fast as possible, with TCP handling flow and congestion control. As TCP completely controls FTP behavior, performance studies usually rely on FTP transfers in the wired to wireless direction, assumed to be the most common case, so as to characterize TCP performance [7]. We measured application throughput, i.e. the amount of *application* data transferred divided by total time. TCP and link layer retransmissions are *not* included as they signify overhead from the application's viewpoint.

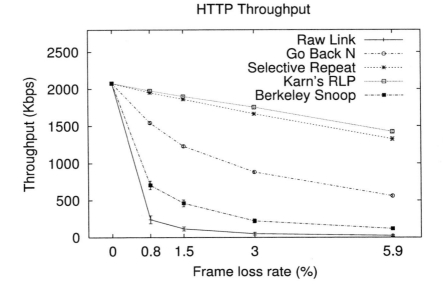

Figure 10.3.   Stand-alone WWW browsing: throughput.

Fig. 10.2 shows the FTP throughput achieved by each link layer scheme tested for a range of error rates, averaged over 30 experiments. Most schemes offer large, but similar, performance gains over Raw Link. The exception is Go Back N, due to its naive strategy of retransmitting all outstanding frames after a loss. Due to the high speed of the wireless link, at any given time there are multiple frames in flight over the link. Therefore, retransmitting all outstanding frames after each loss wastes a lot of bandwidth. The other three enhancement schemes are only differentiated at high loss rates, where the most persistent scheme, Selective Repeat, is ahead of Karn's RLP and Berkeley Snoop.

While most studies of wireless TCP performance employ large file transfers, most real applications make many *short* data exchanges, thus TCP rarely reaches the peak throughput suggested by FTP tests. In addition, most applications are either interactive or employ request/reply protocols, thus data flows in *both* directions, and *each* exchange must complete for the application to proceed. Therefore, we also simulated *World Wide Web* (WWW) browsing over HTTP [24], the most popular Internet application. A WWW client accesses *pages* containing text, links and embedded objects, stored on a WWW server. The client-server interaction consists of *transactions*: the client requests a page

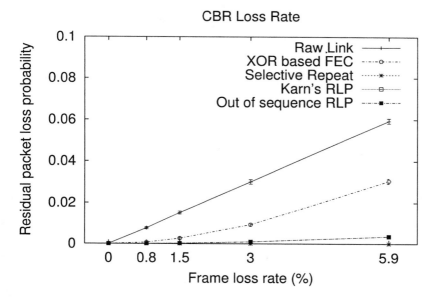

Figure 10.4. Stand-alone continuous media distribution: loss.

from a server, the server returns the page, the client requests all embedded objects, and the server returns them. All transfers are performed over TCP. The ns-2 HTTP module provides empirical distributions for the request, page and embedded object sizes, as well as for the number of objects per page [24]. Only one transaction is in progress at any time and there are no pauses between transactions. WWW browsing was simulated between the WDS and the WLC for 500 s. We measured WWW browsing throughput, i.e. the amount of *application* data transferred from the WDS to the WLC, including both pages and embedded objects, divided by total time.

Fig. 10.3 shows the WWW browsing throughput achieved by each link layer scheme tested. Since the short data transfers of WWW browsing rarely reach high speeds, the naive retransmission strategy of Go Back N does not cause a very dramatic performance deterioration. Karn's RLP performs slightly better than Selective Repeat, again due to the small transfers, since its keepalive feature allows it to recover fast from losses when the link becomes idle, unlike Selective Repeat which has to wait for a timeout to detect losses. The major difference is in the performance of Berkeley Snoop, which provides only minor improvements over Raw Link, as it does not retransmit in the WLC to BSS direction. Thus, even

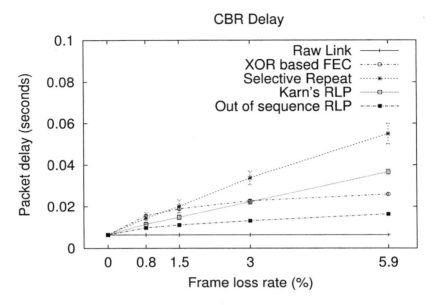

Figure 10.5.   Stand-alone continuous media distribution: delay.

if most traffic flows in the server to client direction, in an interactive application the traffic in the client to server direction is equally critical for overall application performance.

Turning now to UDP, while applications using UDP due to its simplicity would work well with all of the above schemes, delay sensitive applications would face serious problems with a fully reliable retransmission scheme. We thus tested UDP performance using real-time continuous media distribution, i.e. a lecture where a speaker at the WDS sends audio and video to an audience including the WLC. We used a speech model where the speaker alternates between *talking* and *silent* states with exponential durations, averaging 1 s and 1.35 s, respectively [25]. Media are only transmitted in the talking state. When talking, the speaker transmits packets isochronously at a *Constant Bit Rate* (CBR) of 1 Mbps. We simulated continuous media distribution for 500 s. We assumed that the application uses FEC to tolerate some loss without retransmissions and that received packets are buffered until a *playback point* determined by human perception. To characterize performance, we measured the *residual* loss rate at the receiver after link layer recovery. Since packets missing the playback point are dropped, loss rate must be coupled with a delay metric covering *most* packets. We used as a metric mean packet delay plus twice its standard deviation, so as to incorporate the effects of variable delays.

Fig. 10.4 shows residual loss with each scheme. Both RLP schemes dramatically reduce losses, even with a single retransmission per loss. In contrast, XOR based FEC depicts gains that do not justify its overhead. For example, by introducing 8.3% of overhead, a native loss rate of 3% is reduced to 1%. Selective Repeat offers full reliability, but at the cost of very high delays. Fig. 10.5 shows the delay metrics of each scheme with continuous media distribution. The Raw Link curve is flat since no recovery takes place. Selective Repeat exhibits the highest delay due to its full recovery. While both RLP variants perform similarly at low error rates, as link conditions deteriorate only OOS RLP manages to keep delay low, since in sequence delivery causes many packets to be delayed after each loss. Interestingly, XOR based FEC is slower than OOS RLP. The reason is that while OOS RLP requires a NAK and a retransmission to recover form a loss, the FEC scheme requires half a block of frames to be received, on average, in order to recover a lost frame. Since this wireless link has low delay, high bandwidth and uses large frames, retransmission turns out to be faster than this FEC scheme.

5. Multi-Service Link Layer Architecture

The results presented above show that error recovery at the link layer can considerably enhance Internet application performance over wireless links. They also show however that different schemes work best for the TCP and UDP applications tested. This means that a single link layer scheme *cannot* optimize the performance of all TCP and UDP applications. We have therefore developed a *multi-service link layer* architecture, supporting multiple link enhancement services in parallel over a single physical link [26]. Each service fits the needs of an application class, such as TCP-based or real-time UDP-based. Since the adjacent protocol layers expect the link layer to have single entry and exit points, our architecture provides sublayers to assign incoming packets to services and to multiplex outgoing packets from all services into a single stream. These sublayers keep services unaware of the fact that they are operating within a multi-service context.

Fig. 10.6 outlines our architecture, showing data flow in one direction. Incoming packets are classified and passed to the most appropriate service, based on their application class. A simple classifier may use the protocol field of the IP header to distinguish between TCP and UDP, and the port field of the TCP/UDP header to determine the application in use. When *Differentiated Services* (DS) are used, the classifier may

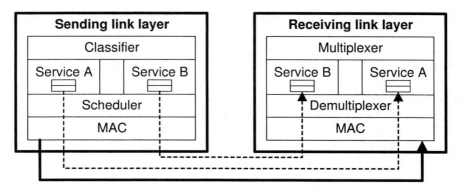

Figure 10.6. Multi-service link layer architecture.

use the DS field of the IP header [1], which remains visible even with IP security [11]. When *Integrated Services* [2] are used, packets can be classified based on flow state maintained by higher layers. Packets from unknown applications are mapped to the default, best-effort, service.

Each service may employ retransmissions, FEC, or any other mechanism desired, keeping its private buffers, counters and timers. Outgoing frames are passed to a scheduler which tags each frame with a service number and eventually delivers it to the MAC sublayer for transmission. At the receiver, frames are passed by the MAC sublayer to the demultiplexer which checks their tags and delivers them to the proper service. Services may eventually release data to the multiplexer, which passes them to the network layer. Therefore, the peer link layer services communicate over *virtual links*.

As each service may arbitrarily inflate its data stream with error recovery overhead, if we simply transmit all frames in a FIFO manner we will penalize the services introducing less overhead. Therefore, we have introduced a frame scheduler to ensure that the link is *impartially* shared between services. Impartiality means that each service should receive the same amount of bandwidth as in a single-service link layer. That is, a service performing no error recovery will get the same bandwidth for data in both cases. In contrast, an error recovery service will get the same *total* bandwidth, but it will have to divide it between its data and its error recovery overhead.

In order to achieve impartiality, we chose a *Self Clocked Fair Queueing* (SCFQ) scheduler [27] which efficiently, but strictly, enforces the desired bandwidth allocations for each service. When some services are

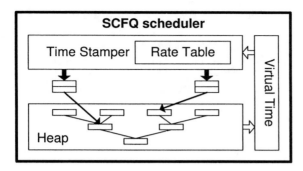

Figure 10.7. Self Clocked Fair Queueing frame scheduler.

idle, their bandwidth is shared among the rest in proportion to their allocations. Fig. 10.7 gives an outline of the scheduler. The *rate table* holds the fraction of link bandwidth allocated to each service. Frames awaiting transmission are buffered in a separate FIFO queue per service. A virtual time variable is maintained, equal to the *time stamp* of the last packet transmitted. To determine the time stamp of an incoming packet, we divide its size by its service rate, and add the result to the time stamp of the preceding frame in its queue. If this queue is empty, we add the result to the current virtual time. When the link is idle, the frame with the *lowest* virtual time is dequeued, its time stamp becomes the system's virtual time, and the frame is transmitted.

Since the frames entering each of the FIFO queues have increasing time stamps, we only need to check the head of each queue to determine which frame has the lowest virtual time. The scheduler keeps all non-empty queues in a heap, sorted by the time stamp of their first frame. The heap is sorted when a frame is dequeued for transmission, based on the new head of the corresponding queue. Empty queues are removed from the heap and are re-inserted when they receive a new frame, which also causes the heap to be sorted. Thus, each frame requires $O(\log_2 n)$ operations (for n services) to sort the heap when the frame leaves (and, possibly, when it enters) the scheduler.

This multi-service link layer architecture can be locally deployed over isolated wireless links, transparently to the rest of the Internet. Additional services can be provided by inserting new modules and extending the mappings of the classifier. Services may be optimized for the underlying link, freely selecting the most appropriate mechanisms. The scheduler ensures that these services fairly share the link, despite their variable overheads, thus keeping services unaware of each other.

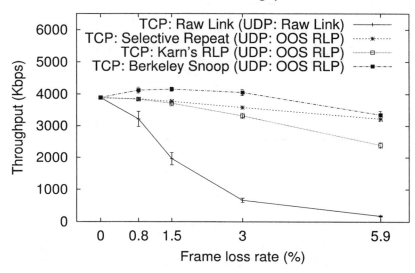

Figure 10.8. All applications: file transfer throughput.

6.   Multi-Service Link Layer Performance

To verify whether our architecture can indeed simultaneously enhance the performance of diverse applications, we repeated the simulations of Sec. 10.4 with all applications, i.e. file transfer, WWW browsing and continuous media distribution, executing in parallel but over different link layer services. The TCP servers and UDP sender were at the WDS and the TCP clients and UDP receiver were at the WLC. All applications started together and the simulation ended when the 100 Mbyte file transfer completed. Our multi-service link layer module for ns-2 provided one TCP and one UDP service, using the link layer schemes that performed best in single application experiments. Both TCP applications used the *same* link layer service. The classifier assigned packets to services based on their DS fields [1] which were set by the applications.

The multi-service link layer module used the SCFQ scheduler described. The rate table was set statically so that continuous media distribution was guaranteed its *peak* bandwidth. Although the scheduler allocated 1 Mbps to the UDP service, the average bandwidth available to the TCP service was 4.575 Mbps, since the UDP application was not constantly active. File transfer is generally more aggressive in terms of

HTTP Throughput

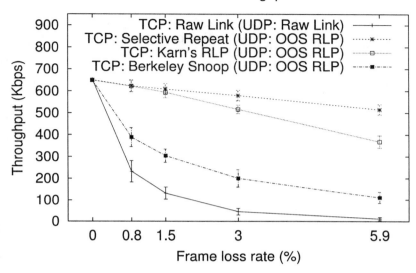

Figure 10.9. All applications: WWW browsing throughput.

competing for bandwidth than WWW browsing, as it performs a bulk transfer. This is handled by existing TCP mechanisms however.

Fig. 10.8 presents file transfer throughput results, with each curve showing the service used for TCP (in parentheses, the service used for UDP). These throughput curves are generally similar to those of single application experiments, but with lower average throughput due to contention with other applications. Selective Repeat performs better than Karn's RLP, especially at higher loss rates, due to its higher persistence. Berkeley Snoop performs best, with its throughput initially increasing with higher loss rates. This is due to a corresponding degradation in WWW browsing throughput with this scheme, as shown below, which leaves more bandwidth available for file transfer.

Results for WWW browsing throughput, shown in Fig 10.9, are also similar to those of single application experiments. The performance of Selective Repeat and Karn's RLP is quite similar to the file transfer throughput results, with very large performance gains compared to Raw Link. Selective Repeat performs better than Karn's RLP as, due to its persistence, it competes for bandwidth more aggressively. Berkeley Snoop fails to offer significant gains due to its topological limitations, exactly as in single application experiments. This drop in WWW browsing

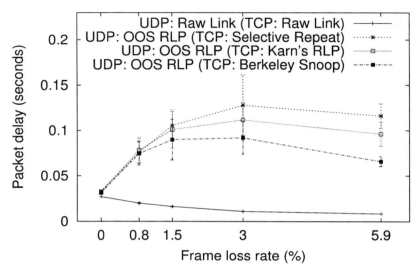

Figure 10.10.   All applications: continuous media distribution delay.

throughput with Berkeley Snoop explains why file transfer throughput initially increases with higher loss rates.

Residual losses for continuous media distribution were the same as in single application experiments, since error recovery is not influenced by contention. The delay metrics, given in Fig. 10.10, show for each curve the service used for UDP (in parentheses, the service used for TCP). Delay was inflated, an unavoidable effect with our non-preemptive work-conserving scheduler. The scheduler managed however to keep delay at reasonable levels, especially considering that the additional delay over Raw Link is partly due to the OOS RLP scheme itself. The reduced delays at higher native loss rates were due to the deteriorating TCP performance which reduced contention. Karn's RLP provided a better balance between TCP throughput and UDP delay than the more persistent Selective Repeat. Berkeley Snoop caused the lowest additional delays, at the cost of reduced aggregate TCP performance.

7.   Best-effort Service Interface

The network layer of the Internet provides a single, best-effort, packet delivery service. Higher layer protocols can extend this service so as to satisfy additional application requirements. Our simulations show however that multiple link layer services are needed in order to enhance Internet application performance over wireless links. Since higher layer protocols are not aware of multi-service links, they cannot map their requirements to available services. To retain compatibility with existing infrastructure, our architecture must perform this mapping transparently. This implies that performance should be at least as good as with a single service, that is, applications should not be mapped to inappropriate services and enhancing the performance of one application should not hurt the performance of other applications. These requirements allow services to be introduced gradually to selectively enhance the performance of some applications.

Since our link layer architecture emulates a single service, the only data that can be used for service selection are the contents of incoming IP packets. We can match application requirements to services using a heuristic classifier which recognizes applications based on IP, TCP and UDP packet header fields. Heuristic packet classification starts with the protocol field of the IP header, indicating TCP, UDP, or another protocol. TCP applications can be all mapped to a single service. For UDP applications however, decisions must be made on a per application basis. Known applications can be detected by examining the source and destination port fields of the UDP header. Another field that may be used is the *Type of Service* (TOS) field of the IPv4 header. Originally, this field was intended to indicate application preferences and packet priorities. This field, however, is rarely used and it has been redefined to support Differentiated Services. This also applies to the *Traffic Class* field of the IPv6 header.

Fig. 10.11 shows data flow in such a classifier for IPv4 packets. Headers are masked to isolate the fields used for classification. These fields pass through a hashing function that produces an index to a lookup table, whose entries point at the available services. Unrecognized applications are mapped to the default (native) link service which offers the same performance as a single service link layer. The header mask, hashing function and lookup table are provided by an external entity, i.e. an administrator, which is aware of application requirements, header fields and link services.

Higher layers expect the link layer to allocate bandwidth as if a single service was offered, therefore link services should respect this (implicit) allocation. As packets are assigned to services, the classifier measures their size to deduce the share of the link allocated to each service. Over regular time intervals, the classifier divides the amount of data assigned to each service i, by the total amount of data seen, to get a fraction r_i, where $\Sigma_{i=1}^{n} r_i = 1$, for n services. These fractions, or normalized

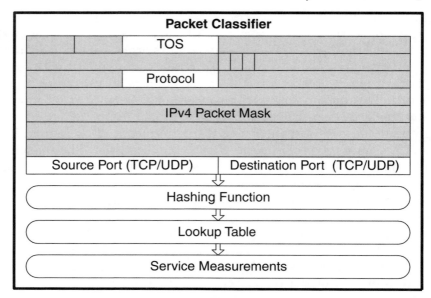

Figure 10.11. Heuristic packet classifier.

service rates, are entered into the rate table. Applications mapped to the default service are allocated the same bandwidth as with a single service. The remaining services trade-off throughput for error recovery.

One drawback of heuristic classifiers is the effort required to construct and maintain them. Applications must be manually matched to services over each type of wireless link whenever new applications or services are added. Many applications will be unrecognized, because they do not use well-known ports. Although some higher layer QoS schemes combine the transport protocol and the source/destination port/address fields to classify packets [28], maintaining state for all these combinations requires end-to-end signaling. Another important problem is that IP security mechanisms encrypt the port fields of TCP and UDP headers and put the identifier of the IP security protocol in the protocol field [11], thus hindering heuristic classification.

8.   Differentiated Services Interface

The best-effort service provided by IP is becoming inadequate as real-time multimedia applications are being deployed over the Internet. To support these applications, some type of performance guarantees must be introduced into the Internet. The main issue for Internet QoS is generally considered to be congestion control, since applications may not be able to get their required throughput due to contention for bandwidth and their end-to-end delay may increase due to queueing delays. One scheme for Internet QoS provision is the *Integrated Services* architecture [2], which has been criticized in two ways. First, it must be widely deployed over the Internet to be useful, since its guarantees rely on actions at every router on a path. Second, it mandates resource reservations on a per flow basis, where a flow is a data stream between two user processes. Since a huge number of flows exists, the scalability of this scheme is questionable.

An alternative scheme that attempts to avoid these limitations is the *Differentiated Services* architecture [1], in which flows are aggregated into a few classes, either when entering the network or when crossing network domains. At these points flows may be rate limited, shaped or marked to conform to specific traffic profiles. For neighboring domains, traffic profiles represent large traffic aggregates, while at network entry points they represent user requirements. Within a domain, routers only need to select a *Per-Hop Behavior* (PHB) for each packet, based on its class, denoted by the 8-bit Differentiated Services (DS) field of the IP header, which subsumes the IPv4 TOS and the IPv6 Traffic Class fields.

The services provided by this architecture are meant to provide generic QoS levels, not application specific guarantees. For example, the expedited forwarding PHB provides guaranteed bandwidth at each router, for traffic that was rate limited when entering the network or the domain so as not to exceed this bandwidth. This PHB provides low delay and loss by eliminating congestion for this restricted traffic class. Only network entry points are aware of both application requirements and PHB semantics so as to perform flow aggregation. Similarly, only domain entry points are aware of the semantics of PHBs available in their neighboring domains so as to perform appropriate translations.

Differentiated Services and multi-service link layers solve orthogonal but complementary problems. Differentiated Services are concerned with congestion and its impact on throughput, delay and loss. The services offered are link independent and are supported by IP level mechanisms. Multi-service link layers are concerned with recovery from link errors, customized to each type of application. The services offered are link dependent and local, as the frame scheduler only protects services from each other. By only providing Differentiated Services over wireless links we can offer a nominal IP level QoS, but the actual performance will be limited by link losses. Multi-service link layers provide adequate recovery to fully utilize wireless links, but they need higher layer guidance to allocate link bandwidth.

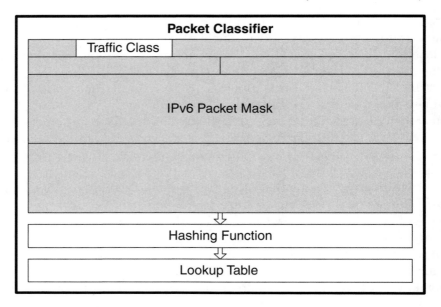

Figure 10.12.   Differentiated Services packet classifier.

The two architectures can be implemented so as to complement each other. Differentiated Services provide congestion control, while multi-service link layers add application dependent error control. They both offer few services (PHBs or link mechanisms) for traffic classes with common requirements. They can be combined by extending the DS field to also specify the error requirements of each traffic class. For example, a DS traffic class could be subdivided into two subclasses with different error recovery requirements by using one more bit of the DS field. Applications could indicate their requirements when injecting their traffic into the network, with boundary routers translating them to local equivalents. The result is a simplified classifier, shown in Fig. 10.12 for IPv6 packets. The DS field is isolated via a header mask, and then a hashing function and a lookup table map it to an appropriate service.

This classifier does not rely on multiple header fields and complex rules to determine application requirements. More importantly, the DS field is not obscured by IP security mechanisms. Since the Differentiated Services module performs scheduling, traffic entering the multi-service link layer already obeys the required bandwidth allocations. Therefore, the subclasses of different DS classes can share the same link service without introducing congestion. The service rates can be set in two

Table 10.1. Service characterization metrics.

Name	Symbol	Definition
Goodput	g_i	$\dfrac{\text{higher layer data transmitted}}{\text{link layer data transmitted}}$
Loss	l_i	$\dfrac{\text{higher layer data lost}}{\text{higher layer data transmitted}}$
Delay	d_i	Average packet delivery delay
Effective Goodput	$e_i = g_i * (1 - l_i)$	$\dfrac{\text{higher layer data received}}{\text{link layer data transmitted}}$

ways. If the subclasses of each DS traffic class have separate bandwidth allocations at the IP level, then the bandwidths for all subclasses mapped to the same link service are added to set its service rate. Otherwise, a measurement module can be inserted after the lookup table, as in Fig. 10.11, to determine service rates as explained in Sec. 10.7.

9. Advanced Quality of Service Interface

The performance of the services available at each wireless link may vary widely, depending both on the underlying link and on transient conditions. If we could provide a characterization of the end-to-end performance of a network path, applications could verify if a given service was suitable for their needs [29]. If these characterizations were updated whenever path characteristics changed significantly, they would also enable adaptive protocols and applications to modify their policies accordingly. To describe the services offered over diverse network paths however, we must dynamically discover what is provided at each multi-service wireless link on the path. This can be achieved if each service dynamically characterizes its performance with a set of standardized metrics. Dynamic characterization means that performance may be evaluated as often as needed, while metric standardization means that higher layers will be able to assess arbitrary services without any knowledge of the mechanisms employed. This would enable end-to-end QoS modules to compose end-to-end path metrics from local link metrics.

To support such services, we have defined three link independent metrics, reflecting some common trade-offs of error recovery schemes:

- *Goodput* (g_i, for service i) is the ratio of higher layer data transmitted during the measurement interval to link layer data transmitted, including all overhead. Note that the amount of higher layer data transmitted may differ from the amount received due to residual losses. Goodput can be calculated at the sender without any receiver feedback.

- *Loss* (l_i) is the ratio of higher layer data lost to higher layer data transmitted (lost plus received). Loss is calculated by the receiver based on the sequence of data released to higher layers. Residual loss can be greater than zero for limited recovery schemes.

- *Delay* (d_i) is the one way average delay for higher layer packets using this service. Delay can be estimated at the receiver based on knowledge of the implemented recovery scheme and wireless link characteristics. For example, retransmission schemes could add one round trip delay for each actual retransmission to their one way delay estimate.

The delay metric only reflects error recovery delay, so it should be added to IP level queueing delay to give the total delay for a node. Delay sensitive traffic subclasses should choose the lowest delay service whose residual loss falls within their tolerance limits. Goodput can be combined with loss to get *Effective Goodput* (e_i), defined as $e_i = g_i * (1 - l_i)$, or, the ratio of higher layer data received to link layer data transmitted. Essentially, e_i shows how much of the bandwidth allocated to a service is used by data actually received, after subtracting error recovery overhead and residual losses. If the link bandwidth is B and service i is allocated a normalized service rate r_i in the scheduler, its throughput is $B * r_i * e_i$. This expression may be used to estimate the throughput for each service given a set of service rates, or to calculate the service rate needed to achieve a target throughput. All metrics are summarized in Table 10.1.

These metrics may be used at multiple layers to serve different needs, as shown in Fig. 10.13. The physical layer provides hardware information, such as the fixed one-way delay, that may be used by link layer services to provide their link independent metrics. At the network layer, scheduling mechanisms such as *Class Based Queueing* (CBQ) may use those metrics to set bandwidth allocations for each service. End-to-end QoS schemes may use the *Resource ReSerVation Protocol* (RSVP) to gather information about node services so as to estimate end-to-end path characteristics. These can in turn be used by transport protocols

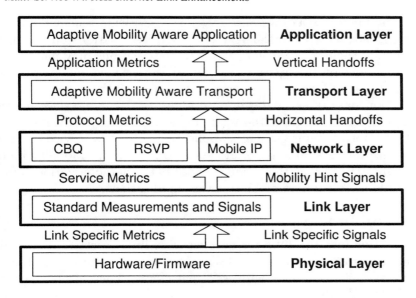

Figure 10.13. Propagation of service measurement and mobility feedback.

and applications to adapt their operation to prevailing conditions. For example, video conferencing applications may select encoding schemes appropriate to the residual loss characteristics of the path.

To assist higher layers in dealing with mobility, we can extend this interface to provide *mobility hints* to interested parties via upcalls, as shown in Fig. 10.13. The link layer can combine hardware signals with its own state to detect events like connections and disconnections. These link independent upcalls can be used by IP mobility extensions to allow fast detection of handoffs, instead of relying on periodic network layer probes [30]. Higher layers may be notified by the network layer via further upcalls of horizontal and vertical handoffs, i.e. handoffs between radio cells employing the same technology and handoffs between radio cells employing different technologies, respectively. TCP may be notified of pending horizontal handoffs to temporarily freeze its timers and avoid timeouts during disconnection. A video conferencing application may be notified of vertical handoffs so as to change the encoding scheme used to a higher or lower resolution one, depending on available bandwidth.

This interface fits the requirements of *One-Pass With Advertising* (OPWA) [29] resource reservation mechanisms. One-pass resource reservation schemes cannot specify a desired service in advance as they do not know what is available on a path [28], thus the resources reserved may prove to be inadequate. Two-pass resource reservation schemes specify a service in advance but make very restrictive reservations on the first pass, relaxing them on a second pass. Such reservations may fail due to tight restrictions on the first pass. In an OPWA scheme, an advertising pass is first made to discover the services available on the path, and then a reservation pass reserves the resources needed. Our interface allows OPWA schemes to discover the restrictions imposed by wireless links. After that information is gathered, applications choose a service and the reservation pass sets up appropriate state to provide it. Mobility hints notify higher layers that they should revise path characterizations after handoffs, so as to support mobility aware applications.

10.   Summary

While extending the Internet over wireless links is straightforward, application performance over wireless links using the Internet protocols is often disappointing due to wireless impairments. These problems considerably complicate the provision of Quality of Service guarantees over wireless Internet links. In order to investigate application performance over wireless links, we have simulated a number of link layer error control schemes. Our results show that different applications favor different approaches, so we have developed a link layer architecture that supports multiple services simultaneously over a single link, allowing the most appropriate link service to be used by each traffic class.

Our approach is easy to optimize for each underlying wireless link and efficient to operate. It can be locally deployed, transparently to the rest of the Internet, and it is easy to extend to address future requirements. It can be incorporated into the Internet QoS architecture in three stages. First, in the current best-effort only Internet, heuristic classifiers can be employed to select the services provided over individual multi-service links, so as to enhance the performance of recognized applications. Second, in the context of the Differentiated Services architecture, which focuses on end-to-end congestion control, multi-service link layers can provide application dependent error control, respecting higher layer scheduling decisions despite the introduction of error recovery overhead. Finally, multi-service link layers can support an advanced QoS application interface, offering dynamic end-to-end service discovery.

References

[1] S. Blake, D. Black, M. Carlson, E. Davies, Z. Wang, and W. Weiss, "An architecture for Differentiated Services," RFC 2475, December 1998.

[2] D. D. Clark, S. Shenker, and L. Zhang, "Supporting real-time applications in an integrated services packet network: architecture and mechanism," in *Proc. of the ACM SIGCOMM '96*, August 1996, pp. 243–254.

[3] W. Stevens, "TCP slow start, congestion avoidance, fast retransmit, and fast recovery algorithms," RFC 2001, January 1997.

[4] G. Xylomenos and G. C. Polyzos, "Internet protocol performance over networks with wireless links," *IEEE Network*, vol. 13, no. 5, pp. 55–63, July/August 1999.

[5] G. T. Nguyen, R. H. Katz, B. Noble, and M. Satyanarayanan, "A trace-based approach for modeling wireless channel behavior," in *Proc. of the Winter Simulation Conference*, 1996.

[6] P. Karn, "The Qualcomm CDMA digital cellular system," in *Proc. of the USENIX Mobile and Location-Independent Computing Symposium*, 1993, pp. 35–39.

[7] H. Balakrishnan, V. N. Padmanabhan, S. Seshan, and R. H. Katz, "A comparison of mechanisms for improving TCP performance over wireless links," in *Proc. of the ACM SIGCOMM '96*, 1996, pp. 256–267.

[8] R. Ludwig and R. H. Katz, "The Eifel algorithm: making TCP robust against spurious retransmissions," *Computer Communications Review*, vol. 30, no. 1, pp. 30–36, January 2000.

[9] M. Mathis, J. Mahdavi, S. Floyd, and A. Romanow, "TCP selective acknowledgment options," RFC 2018, October 1996.

[10] B. R. Badrinath, A. Bakre, T. Imielinski, and R. Marantz, "Handling mobile clients: A case for indirect interaction," in *Proc. of the 4th Workshop on Workstation Operating Systems*, 1993, pp. 91–97.

[11] S. Kent and R. Atkinson, "IP encapsulating security payload (ESP)," RFC 2406, November 1998.

[12] T. Goff, J. Moronski, D.S. Phatak, and V. Gupta, "Freeze-TCP: A true end-to-end TCP enhancement mechanism for mobile environments," in *Proc. of the IEEE INFOCOM '00*, 2000, pp. 1537–1545.

[13] J. Liu and S. Singh, "ATCP: TCP for mobile ad hoc networks," *IEEE Journal on Selected Areas in Communications*, vol. 19, no. 7, pp. 1300–1315, July 2001.

[14] S. Nanda, R. Eljak, and B. T. Doshi, "A retransmission scheme for circuit-mode data on wireless links," *IEEE Journal on Selected Areas in Communications*, vol. 12, no. 8, pp. 1338–1352, October 1994.

[15] A. DeSimone, M. C. Chuah, and O.C. Yue, "Throughput performance of transport-layer protocols over wireless LANs," in *Proc. of the IEEE GLOBECOM '93*, 1993, pp. 542–549.

[16] Q. Pang, A. Bigloo, V.C.M. Leung, and C. Scholefield, "Performance evaluation of retrnasmission mechanisms in GPRS networks," in *Proc. of the IEEE WCNC '00*, 2000, pp. 1182–1186.

[17] R. Ludwig and B. Rathonyi, "Link layer enhancements for TCP/IP over GSM," in *Proc. of the IEEE INFOCOM '99*, 1999, pp. 415–422.

[18] C. Parsa and J.J. Garcia-Luna-Aceves, "Improving TCP performance over wireless networks at the link layer," *Mobile Networks and Applications*, vol. 5, no. 1, pp. 57–71, 2000.

[19] The VINT Project, "UCB/LBNL/VINT Network Simulator - ns (version 2)," Available at http://www.isi.edu/nsnam.

[20] G. Xylomenos, "Multi-service link layer extensions for ns-2," Available at http://www.mm.aueb.gr/~xgeorge/.

[21] G. Xylomenos and G. C. Polyzos, "Multi-service link layer enhancements for the wireless Internet," in *Proc. of the IEEE Symposium on Computers and Communications '03*, 2003, pp. 1147–1152.

[22] G. Xylomenos and G. C. Polyzos, "Wireless link layer enhancements for TCP and UDP applications," in *Proc. of the International Parallel and Distributed Processing Symposium '03*, 2003, pp. 225–232.

[23] P. T. Brady, "Evaluation of multireject, selective reject, and other protocol enhancements," *IEEE Transactions on Communications*, vol. 35, no. 6, pp. 659–666, June 1987.

[24] B. A. Mah, "An empirical model of HTTP network traffic," in *Proc. of the IEEE INFOCOM '97*, 1997, pp. 592–600.

[25] S. Nanda, D. J. Goodman, and U. Timor, "Performance of PRMA: a packet voice protocol for cellular systems," *IEEE Transactions on Vehicular Technology*, vol. 40, no. 3, pp. 584–598, August 1991.

[26] G. Xylomenos and G. C. Polyzos, "Quality of service support over multi-service wireless Internet links," *Computer Networks*, vol. 37, no. 5, pp. 601–615, 2001.

[27] S. Golestani, "A self-clocked fair queueing scheme for broadband applications," in *Proc. of the IEEE INFOCOM '94*, 1994, pp. 636–646.

[28] L. Zhang, S. Deering, D. Estrin, S. Shenker, and D. Zappala, "RSVP: A new resource reservation protocol," *IEEE Network*, vol. 7, no. 5, pp. 8–18, September 1993.

[29] S. Shenker and L. Breslau, "Two issues in reservation establishment," in *Proc. of the ACM SIGCOMM '95*, October 1995, pp. 14–26.

[30] C. Perkins, "IP mobility support," Internet Request For Comments, October 1996, RFC 2002.

Chapter 11

PORTABILITY ARCHITECTURE FOR NOMADIC WIRELESS INTERNET USERS AND SECURITY PERFORMANCE EVALUATION

Mustafa M. Matalgah[1], Jihad Qaddour[2], Omar S. Elkeelany[3] and Khurram P. Sheikh[4]

[1] *University of Mississippi :* [2] *Illinois State University :* [3] *University of Missouri, Kansas City :* [4] *Sprint*

Abstract   A value added service to broadband wireless network is the remote access virtual private network (VPN). The corporate legitimate portable users can connect to their offices through a wireless network from different locations and get secure services as if they were connected to the corporate local area network. One of the main challenges is to block illegitimate wireless users' requests. Registration and authentication functions should be implemented with highly secured wireless connection. These functions are accomplished by tunnelling the user information in a secured form to the corporate authentication server through the Internet traffic. The Corporate Authentication Server then grants or denies the user access. This chapter addresses various portability scenarios, architectures, implementation, and requirement issues for portable wireless Internet access systems. Moreover, performance evaluation and comparison are presented for the state-of-the-art security and authentication techniques.

Keywords: Wireless Internet, portability architecture, remote wireless authentication and security, VPN tunnelling.

1.   Introduction

One of the most important requirements for high-speed fixed wireless Internet services is to support customer premises equipment (CPE) self install feature. This provides the user the ability to move around his location and re-establish radio and Internet protocol (IP) sessions easily. This is the simplest form of portability that can be supported to nomadic users. In mobility it is required to keep the radio and IP sessions connected continuously. Whereas, for portability when a user moves from one regional base station to another then a hard handoff is experienced between the two regional base stations that belong to two different radio network controllers (RNC) in European networks (or mobile switching centers in American networks (MSC)). Different handoff scenarios and requirements need to be defined for both IP session handoff via a routing gateway, and radio network (RN) handoff via the radio link controller (RLC). If the user moves from one base station to another that belongs to the same RNC or MSC, then cell reselection need to be accomplished to reconnect the portable device at the new location. In mobility, cell reconnection for this scenario is performed smoothly without reestablishing a new session with the new cell. However, a new IP address reassignment might need to be established if subnet has changed.

When a portable device moves from one base station to another that belongs to a different RNC or MSC, then a new radio link protocol (RLP) signaling session with the new base station is needed for cell reselection. User registration and authentication is also needed if each RNC has its own server for Remote Authentication Dial In User Service (RADIUS). This server is used for dial-up security management.

This chapter addresses the following : 1) End-to-end element functions that meet portability requirements. 2) Various portability scenarios, architectures and implementation issues in fixed wireless Internet access systems. One of the major issues to be addressed is how to complete a secured time-bounded authentication, authorization and accounting when a user travels from one wireless market area to another through an IP network. 3) Private end-to-end architecture between the home and visited market network to complete a secured time-bounded portability requirement.

2.   System Authentication

When a portable device moves from one base station to another that belongs to a different RNC, two methods are proposed for user registration and authentication (assuming each RNC belongs to a different RADIUS). In one method, since a different RADIUS serves each RNC, the new RNC will verify the user eligibility through its local (primary) RADIUS database, using tunnelling and routing protocols, for registration and authentication. The local RADIUS will then send interrogation signaling messages to all the RADIUSs in the network to verify the eligibility of that user. The RADIUS with the database that the user belongs to, will then reply with the eligibility status of that user to accept or deny access. Interrogation messages can be established through layer 2 tunnelling protocols such as layer 2 tunnelling protocol (L2TP) if private leased lines are utilized or layer 3 IP tunnelling protocols such as IPSec if public IP network is encountered. Tunnelling protocols will be covered in more details later in this chapter.

In the second method, it is assumed that there is secondary RADIUS in the network of the service provider that contains a copy of the entire primary RADIUSs at all the RNCs networkwide. This RADIUS might be co-located in one of the RNCs and connected to all the other RNCs in the network through secure private lines. Unlike the first method, the local (primary) RADIUS in the visited region will need to send authentication messages only to the central (secondary) RADIUS for user registration and authentication. This process saves the burden of interrogation messaging forwarding to all the RNCs networkwide.

In both methods, after the customer eligibility is confirmed, the new RNC will grant the user the authority to establish RLP and IP sessions through the access base station. The user information is then saved in the visited RADIUS database cache memory for future authentication. All the signaling messages and transmissions between and within the RNCs and RADIUSs may be established through the service provider IP network. A tunnelling protocol with security protocol such as a combination of L2TP and IPSec transport mode or IPSec tunnel mode might then be needed to ensure security. If leased private lines are used for transport, then L2TP alone is enough for delivering the messages through a point-to-point (PPP) [25] session.

3.   Portability Architecture Elements

A high level architecture for a wireless Internet access network with the necessary elements to support portability is depicted in Figure 11.1. The definition and functionality of each of the network elements in this architecture are described in the following sections.

3.1   CPE

Customer premises equipment (CPE) is required to monitor few neighbor lists and measure the strongest power among them. The CPE then can send its measurements messages to the current serving base transceiver station (BTS), which will send a request message to the radio access controller for a hard hand-off. This request message is sent through a BTSs Broadcast Channel (BCCH). The user will then experience a disconnection and will be asked to reconnect. No authentication is required in this case if the two BTSs belong to the same RADIUS

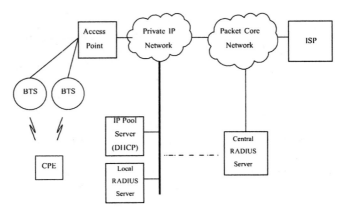

Figure 11.1. Network elements for portable wireless Internet architecture

server. A routing card at the radio controller is required to route the user information to the new sector or BTS through an L2TP session.

3.2   BTS and Radio Access Controller

Base transceiver station (BTS) should have the intelligence to sense user movements in the RF coverage area and between BTSs and inter-operate with the CPE and the radio controller portability requirements. Portable wireless Internet architecture is expected to include a radio controller that is capable of performing bandwidth allocation/management and traffic scheduling. The radio controller might be a part of each BTS or a one common radio controller might serve several BTSs. Its performance is determined by the system capability of supporting QoS and maintaining same data rate per user as power is changing when the user changes his location. Adaptive modulation and coding technique is also very important in bandwidth allocation and traffic scheduling. The radio access controller need to interface with a packet radio services supporting node that will provide the packet session management and handoff between radio access controllers and associated (BTSs) for portable devices.

The following are major functions that should be implemented in radio controller system and the BTS for a portable architecture:

1. Detection and authentication of new subscribers.

2. Dynamic allocation of bandwidth to subscribers and routing.

3. Cell selection and reselection on the wireless access network.

4. User Prioritization.

5. User Bandwidth Distribution.

6. Overload Control.

7. User traffic and quality of service (QoS) scheduling based on users service level agreement (SLA).

8. Advanced medium access controller (MAC) to support real-time QoS-based applications.

9. Ability to provide DiffServ.

10. Minimized Latency.

11. Operation, administration, and management (OAM) configuration capability.

Other functions provided by the resource allocation mechanisms include:

1. Differentiate between users on different tiers of service.

2. Provide fair queuing within these tiers of service so that a particular user cannot dominate resources.

3. Allocate resource so that power control can work efficiently.

4. Accommodate users on the edge of the cell who will probably be on lower transport formats with lower rates and aim to provide them with a reasonable throughput whilst not excessively reducing the overall cell throughput.

Resource allocation parameters that could be operator controlled are:

1. Tier weighting parameter: This provides a means of discriminating between groups of users or tiers and providing different levels of service to each tier.

2. Tier target rate: This allows the operator to define an upper limit on the throughput that a tier can experience. This is so that if a single lower tier user is present he will only experience maximum throughput for his tier of service. This can be disabled during off-peak periods for example.

3. Fairness parameter: This can be set between two extremes, one which divides the resources equally at the expense of the throughput to users on lower transport formats and the other which divides the throughput equally at the expense of overall cell throughput. This parameter provides a means of providing service anywhere between these two extremes.

4. Maximum number of downlink codes: This provides a means of trading coverage with overall system throughput.

5. Allocation window length: This determines how far in advance an allocation can be made.

3.3   IP Pool Server (DHCP)

The main function of the IP Pool or Dynamic Host Configuration Protocol (DHCP) Server is to perform IP address reassignment as needed when the user is moving regionally and nationally or for driving vehicles with portable devices.

3.4   Encryption or Tunnelling Client

An encryption or tunnelling client is required at the RNC or master hub to guarantee and secure the messaging packets transmission through the IP network to the secondary RADIUS for radio access device authorization when the user is moving regionally or nationally. Authorization, control, and registration are required to be established in a secured form. They are also required for the Internet service provider (ISP) authorization. Tunnelling can also be used for IP session hard handoff in ISP network when IP address reassignment is required for driving vehicles portable devices. For example, the maximum speed allowed to achieve portability IP session handoff in Universal Mobile Telecommunication systems (UMTS) is less than 30 km/hr.

Tunnelling options include IPSec, L2TP, ... etc. These protocols will be explained in more details later in this chapter. IPSec is an encryption protocol that encrypts all received packets to be transmitted to the remote location for authentication. No pre setup time is required for this protocol. The only time delay to be experienced is from the encryption process for each packet. IPSec is not a tunnelling protocol, but provides secured encrypted IP packet transmission when used in association with a tunnelling protocol such as L2TP. Further details about IPSec different modes are provided later in this chapter. L2TP is a tunnelling protocol to transmit PPP frames at Layer 2. A pre setup time is required to create the tunnel. TCP control channel (Layer 4) is used in the initial set up for the tunnel creation. After the tunnel creation is complete the PPP frames are transmitted over the IP cloud in which the nodes function as an intermediate hop switching (Layer 2). Tunnelling in L2TP is equivalent to Virtual Path Identifier (VPI) in ATM switching. L2TP can be used as an extension to the PPP connection from the CPE all the way to the ISP router or the RADIUS. More details are provided in the following sections of this chapter.

3.5   Re-encryption or Tunnelling Server

A re-encryption or tunnelling server is required at the remote RADIUS location to complete the tunnelling process or re-encrypt the messaging packets at the remote authentication server (RADIUS) for the radio network authorization or the remote ISP server for ISP authorization.

3.6   IP Network

If the authentication/authorization messaging encrypted packets or tunnelled frames have to be transmitted through an Intranet or Internet, then it is expected to support the encryption or tunnelling requested by the client.

3.7   Local RADIUS Server

A local (or primary) RADIUS [32] server is required in each service market and keeps the database for all customers that belong to a regional service market. Each RNC (or MSC) is expected to belong to a RADIUS server for the registration and authentication.

3.8   Central RADIUS Server

It is expected to have a one central (or secondary) RADIUS server in the operators network that will be used to authenticate portable users (as explained in the second method in Section 2). The central RADIUS server provides backup for the local RADIUS and a master database for authentication of roaming portable users. When a user moves outside his home location, the radio access controller at the visited region will handle the authentication and authorization through the secondary (central) RADIUS server.

4.   Key Factors in Portability Architecture

The following are key factors that should be considered in designing portability network architecture:

4.1   QoS Management

The system should be able to adapt and identify users who are portable in nature. This is necessary on account of the fact that portable users are treated differently than fixed users. The primary reason for this is that data rates vary from one point to another in a given cell coverage area. This is driven by the ratio of the carrier to interference (C/I) distribution. Hence, it is critical that a given users' SLA changes with the available data rates in that region of the cell. Also, optimal market penetration should be maintained in the region of lowest data rates. Hence, the MAC layer should be capable of identifying such users and updating their traffic profile accordingly.

4.2   Service Level Agreement

Since a portable user always changes his/her location, it is not possible to guarantee high data rates for that user. This is primarily because of the fact that any given system which supports adaptive modulation supports data rate based on the C/I ratio at a given location in the cell. Hence, the SLA for a portable user should be less than that of a fixed wireless case, such that a large user population can be supported at the lowest modulation scheme.

4.3   Indoor Penetration

The air interface technology used by the portable network should be optimized for high downloading speeds to the user for carrying Internet Protocol (IP) traffic for Internet access and other portable/ fixed data applications operating in a non line-of sight environment with building penetration. Indoor wall penetration loss in the range 10 to 18 dB should be considered, depending on the technology, for cell design. In a typical network designed with sufficient link margin for building-penetration, the CPE Modem should operate indoors without the requirement for an external antenna.

4.4 Carrier to Interference (C/I) Ratio

The technology transport formats with less redundancy (weaker FEC, higher user data rate) require higher C/I than those with more redundancy (stronger FEC, lower user data rate) in order to meet a specified optimal value for the BLER. The basic rule is that if the block error rate (BLER) meets the specified optimal value, then maximum throughput is obtained. On the other hand, throughput is reduced if either BLER is more than the optimal value and then there are too many retransmissions to occur or if BLER is less than the optimal value and then FEC coding is more than necessary. Simulation model for the cell area coverage as related to C/I in the downlink should be provided in order to achieve an optimal cell design in non-line-of-site (NLOS) environments. Transport formats are defined by the technology in terms of different parameters such as number of codes, modulation type, turbo coding rate, puncturing limit, number of slots, ... etc. For each transport format using the ratio of the bit energy to noise density (Eb/No) requirement and the Processing Gain (Gp), the required C/I to achieve certain data rate can be calculated and then the coverage percentage is determined from the simulation model relating the C/I to the percentage cell coverage area. Figure 11.2 shows an example of a C/I cumulative distribution function (CDF) in a code division multiple access (CDMA) air interface system using a standard 19-cell cluster model with user population density of 3395 user/km^2, 0.75 km radius, 100% frequency reuse, and a chip rate of 3.84 Mcps (in a 2.6 GHz spectrum with 5 MHz channels) [17].

The simulation model and path loss parameters used for calculating the C/I CDF in Figure 11.1 are summarized in Table 11.1

Table 11-1 Simulation parameters

Carrier frequency	2600 MHZ
Chip Rate	3.84Mcps
Frequency Reuse factor	N=1
Pathloss Model	3GPP TS25.942
In-Building Loss	15 dB
Cell Radius	750 m
Subscriber Density	3395 per sqkm
Log-Normal Standard Deviation	10 dB
Antenna Front-to Back Ratio	30 dB
Sectorizaion	Tri-Sectored
Antenna Pattern	cos (θ)
Node-B Max Transmit Power	34 dBm
Node-B Antenna Gain	18 dBi
Antenna Height	25 m
Node-B Tx Feeder Loss	2 dB
Modem Antenna Gain	2 dBi
Modem Receiver Noise Figure	5 dB

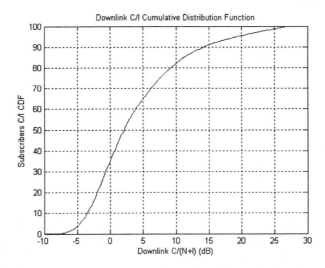

Figure 11.2. Downlink C/I CDF for a population density of 3395 users per square kilometer

5. Portability Scenarios

5.1 In-House Portability

For the in-house portability, the user can move around the house without shutting down the IP session. The house and the surrounding area are most likely assumed to be under same base station coverage. In this scenario, the following factors are considered:

1. CPE is initially provisioned with the best in-house QoS.

2. While the user is moving around the house a soft QoS will be required to keep certain data rate per the customer SLA. This will require maintaining a high SLA for the customer and providing the adaptive coverage per the SLA as the user moves. Adaptive coding and adaptive modulation techniques such as QPSK, 8PSK, 16QAM, or 64QAM are used for providing coverage adaptation.

Figure 11.3 shows a scenario for in-house portability sector architecture where the user is moving from point P1 to point P2 within sector 2.

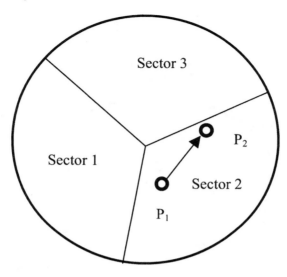

Figure 11.3. Intra-Sector probability architecture

5.2   Inter-Cell Regional Portability (Intra-RNC)

A higher level of regional portability is defined as when a user travels from one BTS to another BTS that belong to the same RNC as shown in Figure 11.4. This is referred to as Inter-Cell or Intra-RNC portability. If the user travels from one BTS to another BTS that belongs to the same RNC, then cell reselection should be established through the BTS and the RNC to establish a new RLP session with the new cell. In this case a new IP address needs to be reassigned (if subnet has changed). In this form of portability the following factors should be considered carefully.

In addition to the QoS, SLA, C/I and indoor penetration requirements mentioned above, virtual IP tunnelling and maybe VPN virtual path network)application will play a major role in this scenario. More details on tunnelling and security applications for portability are provided later in this chapter. Figure 11.4 shows a block diagram for inter-cell portability scenario architecture. The portable user is moving from point (P1) to point (P2) as shown in the figure.

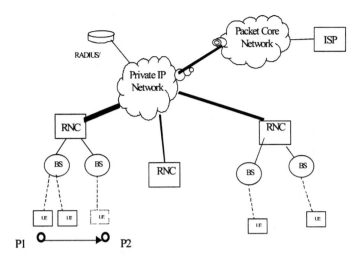

Figure 11.4. Inter-cell architecture for cell-to-cell regional portability within an RNC.

5.3 Inter-RNC Regional Portability (Same RADIUS)

Inter-RNC regional portability occurs when a user travels between two base stations that belong to different RNCs with same RADIUS. In this scenario, cell reselection is accomplished through the new BTS and the two RNCs to complete the RLP signaling process and establish a radio network (RN) session to initiate an IP session. A new IP address is assigned through the DHCP in the new RNC.

This is an on-network value-added hard handoff feature that needs authentication across the network between different RNCs. Figure11.5 shows an architecture scenario for Inter-RNC regional portability. The portable user is moving from point (P1) to point (P2) as shown in the figure.

5.4 National Portability (Inter-RADIUS, On-Network, same technology)

National portability is the most complicated scenario. It can also be referred to as Inter-RADIUS portability. When portable device moves

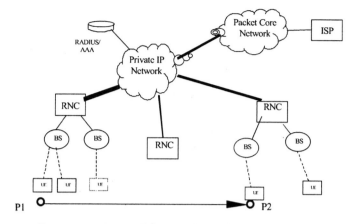

Figure 11.5. Inter-RNC regional portability architecture.

from one cell to another cell in a different RNC that belongs to a different RADIUS then cell reselection signaling and authentication/authorization process will need to occur across the network between the two RADIUSs. In this case the portable device needs to establish a new RLP session with the new cell via the BTS and the RNC. In this scenario the new RNC will check the user eligibility through its local RADIUS database, through tunnelling and routing protocols, for registration and authentication. Then using method I (refer to section 2), the RADIUS sends interrogation signaling messages to all RADIUSs across the network to check eligibility authentication of that user. The home RADIUS that the user belongs to, replies with the eligibility status of that user to confirm authentication and registration. Another authentication scenario can also be used (refer to method II in section 2).

Interrogation messages can be established through layer 2 tunnelling protocols such as L2TP, PPTP, and L2F or layer 3 IP tunnelling protocols such as IP in IP, GRE and IPSec. Other tunnelling mechanisms can be established using MPLS VPNs.

After the customer eligibility is confirmed, the new RNC provides the user the authority to establish an IP session through the visited access base station.

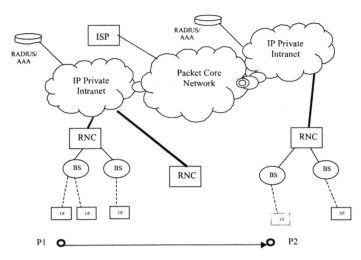

Figure 11.6. Inter-RNC architecture for national portability.

All the signaling messages and transmissions between the RNCs and the RADIUSs may be established through a private IP network if available between the two RADIUSs. A security protocol such as IPSec can be used to ensure security if public IP network is used. National portability architecture is shown in Figure 11.6.

5.5 Ubiquitous National Portability (OFF-Network, alternative carrier operator)

This occurs between two different service provider operators or within a service provider operator but with dual band technologies such as PCS (Personal Communications Service in the 1.9 GHz spectrum) and MMDS (Multichannel Multipoint Distribution Service in the 2.6 GHz spectrum). There are three types of roaming for ubiquitous portability:

1. Roaming from one MMDS service provider network to another MMDS service provider network. This is referred to as single-band dual-mode devices. Network architecture is illustrated in Figure 11.7 (Roaming from P1 to P2).

2. Roaming within the same service provider network but between two different services such as MMDS and PCS. This is referred

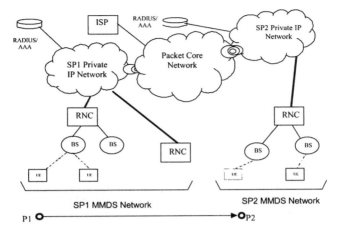

Figure 11.7. Architecture for single-band dual-mode ubiquitous national portability (SP1: Service Provider 1, SP2: Service Provider 2)

to as dual-band dual-mode devices. Network architecture is illustrated in Figure 11.8 (Roaming from P1 to P2).

3. Other roaming scenarios might also be practical such as dual-band dual-mode roaming between two different service providers such as roaming from MMDS services from one operator to PCS services from another operator or vice versa.

User interaction and/or pre-subscription are required for this scenario. By FCC (Federal Communications Commission) regulations the user must be informed of this form of roaming and the new charges rate.

6. IP Tunnelling

Tunnelling is a way of establishing a logical (virtual) path in an internetworking infrastructure to transfer data from source to a destination on the network per an SLA. In a tunnelling protocol, data packets or frames are encapsulated in a new header that contain routing information such as source, destination, QoS level, data rate, ... etc. The virtual (logical) path through which the encapsulated packets travel through the network is called a "tunnel"

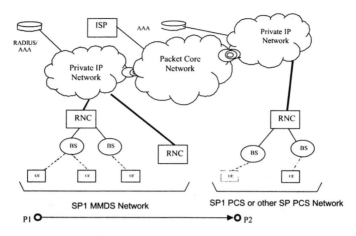

Figure 11.8. Architecture for dual-band dual-mode ubiquitous national portability (SP1: Service Provider 1, SP: Service Provider)

6.1   Tunnelling vs. Source Routing

Internet protocol (IP) source routing is a best effort and non-QoS transport protocol. IP source routing involves defining the path that the IP packet must follow through the network. Unresolved issues related to security, packet delay, packet drop, jitter and other performance metrics are still exist in IP source routing. These issues include examples such as non-guaranteed performance delivery by the IP router in forwarding datagrams for the IP source route options, incorrect processing of IP source route options, exclusion of such packets by firewalls, ... etc.

In IP tunnelling a logical path is set up to guarantee ... etc. In IP tunnelling a logical path is set up to guarantee packets delivery. QoS mechanism is to be utilized along with tunnelling to guarantee certain delivery performance requirements such as data rate (or throughput), time delay, jitter, ... etc.

6.2   IP Tunnelling for Portability Architecture

If a portable node moves from one RNC to another that belongs to different RADIUS, then authentication and registration need to be processed at the new RNC with a specified time interval. Tunnelling along other QoS and Security mechanisms provide a secure and time bounded mechanism for this application. If the tunnel between the two RADIUSs has to go through a public IP/ATM network, then a SLA has to exist between the wireless Internet service provider and the IP/ATM service provider.

The main reason for utilizing tunnelling techniques in portability applications is to achieve the customer authentication and registration in a time bounded delay.

6.3   Layer 2 Tunnelling

Layer 2 tunnelling protocols are based on the well-defined point-to-point protocol (PPP) over L2TP and hence they inherit a suite of useful features like user authentication, dynamic assignment of addresses, data compression and encryption, and management. Layer 2 protocols correspond to the data link layer of the OSI seven layers model and they use frames as their unit of exchange. PPTP, L2TP, Layer 2 Forwarding (L2F) are examples of layer 2 protocols and they encapsulate the payload in a PPP frame to be sent across an internetwork.

6.3.1   L2TP (RFC 2661) .   Remote system or L2TP Access Concentrator (LAC) client initiates a PPP connection across the PSTN cloud to the LAC. The LAC then tunnels the PPP connection across the Internet, Frame Relay, or ATM cloud to LNS whereby access to a Home LAN is obtained. Figure 11.9 shows a general schematic diagram for the L2TP topology [28]. L2TP utilizes two types of messages: control messages and data messages. Control messages are used in the establishment, maintaining and clearance of tunnels and calls. Data messages are used to encapsulate PPP frames being carried over the tunnel. L2TP is therefore considered a signaling protocol as well.

In L2TP protocol, control messages utilize reliable control channel (acknowledgement and retransmitting lost packets is supported) whereas the data channel is unreliable (acknowledgement and retransmitting lost packets is not supported). The following are the main features of L2TP:

1. Layer Two Tunnelling Protocol (L2TP) is a combination of PPTP and Layer 2 Forwarding (L2F), a technology proposed by Cisco Systems, Inc.

2. L2TP combines the best features of both PPTP and L2F.

3. L2TP encapsulates PPP frames to be sent over IP, X.25, Frame Relay, or ATM networks.

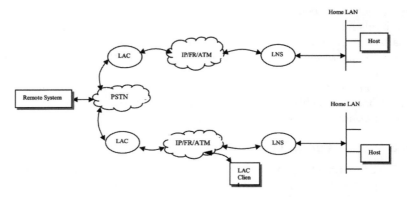

Figure 11.9. L2TP Generic Topology

4. When sent over an IP internetwork, L2TP frames are encapsulated as User Datagram Protocol (UDP) messages.

5. L2TP is used for tunnelling over the Internet as well as over private Intranets.

6. L2TP can be used for the establishment of compulsory as well as voluntary tunnels. Compulsory tunnels are tunnels setup by the LAC without the client knowing about it while voluntary tunnel setup is done by the client without the aid of the LAC as a separate entity. The client has the L2TP software running on it natively.

There are two tunnelling types associated with L2TP tunnelling protocols, Voluntary and Compulsory Tunnelling.

In voluntary tunnelling, a client directly connected to the Internet and running L2TP natively can also participate in tunnelling to the corporate LAN without the use of a separate LAC. A virtual PPP connection is created and the local L2TP client software creates a tunnel to the LNS. The client should have the tunnelling software installed and it functions as the end point of the tunnel. A block diagram that illustrates voluntary tunnelling is shown in Figure 11.10.

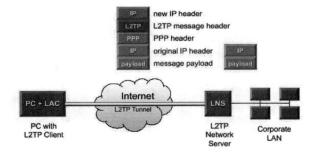

Figure 11.10. Voluntary tunnelling architecture.

In compulsory tunnelling, the remote user (PC) initiates a PPP connection across the PSTN cloud to the LAC. The LAC then tunnels the PPP connection across the Internet, Frame relay or ATM cloud to the L2TP Network Server (LNS) whereby access to the corporate LAN is obtained. The PC gets addresses from the corporate LAN via PPP-NCP negotiation. AAA services are provided by LANs management domain as if the user were connected to the LNS directly. This is illustrated in Figure 11.11.

6.3.2   PPTP (RFC 2637) .   Point-to-Point Tunnelling Protocol (PPTP) defines a protocol, which allows the PPP to be tunnelled through an IP network [14]. There are two parallel components of PPTP. One component is a control connection between each PAC-PNS pair operating over TCP. The second component is an IP tunnel operating between the same PAC-PNS pair, which is used to transport generic routing encapsulation (GRE), encapsulated PPP packets for user sessions between a pair. The most commonly used protocol for remote access to the Internet is point-to-point protocol (PPP). PPTP builds on the functionality of PPP to provide remote access that can be tunnelled through the Internet to destination site. As currently implemented, PPTP en-

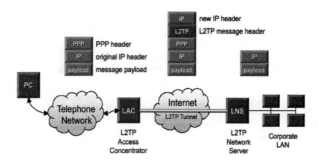

Figure 11.11. Compulsory tunnelling architecture.

capsulates PPP packets using a modified version of GRE protocol, which gives PPTP the flexibility of handling protocols other than IP, such as Internet packet exchange (IPX) and network basic input/output system extended user interface (NetBEUI).

PPTP is one of the first protocols deployed for VPN. It has been a widely deployed as a solution for dial-in VPNs since Microsoft included support for it in RRAS for Windows NT Server 4.0 and offered a PPTP client in a service pack for Windows 95. Microsoft's inclusion of a PPTP client in Windows 98 practically ensures its continued use for the next few years,although it is not likely that PPTP will become a formal standard endorsed by any of the standard bodies (like the IETF). Because of its dependence on PPP, PPTP relies on the authentication mechanisms within PPP namely password authentication protocol (PAP) and CHAP. Because there is a strong tie between PPTP and Windows NT, an enhanced version of CHAP, MS-CHAP, is also used, which utilizes information within NT domains for security. Similarly, PPTP can use PPP to encrypt data, but Microsoft has also incorporated a stronger encryption called Microsoft point-to-point encryption (MPPE) for use with PPTP.

Aside from the relative simplicity of client support for PPTP, one of the protocol's main advantages is that PPTP is designed to run at open systems interconnection (OSI) Layer 2 as opposed to IPSec, which runs at Layer 3. By supporting data communications at Layer 2, PPTP can transmit protocols other than IP over its tunnels. PPTP does have some limitations. For example, it does not provide strong encryption for protecting data nor does it support any token-based methods for authenticating users.

6.3.3   L2F (RFC 2341).   Layer Two Forwarding (L2F) Protocol, a technology proposed by Cisco [31], is a transmission protocol that allows dial-up access servers to frame dial-up traffic in PPP and transmit it over WAN links to an L2F server (a router). The L2F server then unwraps the packets and injects them into the network. Unlike PPTP and L2TP, L2F has no defined client. Note that L2F functions in compulsory tunnels only.

6.3.4   GRE (RFC1701).   A number of different proposals currently exist for the encapsulation of one protocol over another protocol. Other types of encapsulations have been proposed for transporting IP over IP for policy purposes. Generic Routing Encapsulation (GRE) is a general purpose encapsulation protocol and it is the attempt of this protocol to provide a simple, general purpose mechanism which reduces the problem of encapsulation from its current $O(n^2)$ size to a more manageable size [5]. In the most general case, a system has a packet that needs to be encapsulated and delivered to some destination. We will call this the payload packet. The payload is first encapsulated in a GRE packet. The resulting GRE packet can then be encapsulated in some other protocol and then forwarded. We will call this outer protocol the delivery protocol. Overall packet format is explained in Figure 11.12.

In forwarding GRE packet, normally a system, which is forwarding delivery layer packets, will not differentiate GRE packets from other packets in any way. However, a GRE packet may be received by a system. In this case, the system should use some delivery-specific means to determine that this is a GRE packet. Once this is determined, the Key, Sequence Number and Checksum fields if they contain valid information as indicated by the corresponding flags may be checked. If the Routing Present bit is set to 1, then the Address Family field should be checked to determine the semantics and use of the SRE Length, SRE Offset and Routing Information fields. The exact semantics for processing a SRE for each Address Family is defined in other documents. Once all SREs have been processed, then the source route is complete, the GRE header

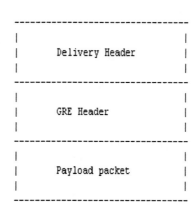

Figure 11.12. GRE packet format.

should be removed, the payload's TTL must be decremented (if one exists) and the payload packet should be forwarded as a normal packet. The exact forwarding method depends on the Protocol Type field.

6.4   Layer 3 Tunnelling

6.4.1   IPSec Tunnel (RFC 2401) .

IPSec is a layer 3 solution for secured tunnelling, which means authenticated and encrypted tunnel [11]. It provides security (Authentication and encryption) at Layer 3 (IP Network Layer) by enabling a system to select required security protocols, and determine the algorithm(s) for encryption. IPSec has two security protocols, IP Authentication Header (IPSec AH) [9], and IP Encapsulating Security Payload (IPSec ESP) [10]. IPSec provides two types of security algorithms: symmetric encryption algorithms (e.g. Data Encryption Standard DES) [29], [30], [15], [27] and, one-way hash functions (e.g., Message Digest MD5, Secured Hash Algorithm SHA-1) [21], [14], [13], [24].

IPSec is used to secure tunnels against false data origins, and encrypt traffic against unwanted network passive or active intruders from listening or modifying actions. IPSec can be used between pairs of Security Gateways, pairs of peer hosts or, combined (e.g. between a host and corporate equipment through security Gateways.) It has two modes of operation: Transport mode, and Tunnel mode. The transport mode provides upper layer protection (transport layer). Transport mode applies to pairs of peer hosts. Also in this mode, traffic goes to the ultimate destination directly. In the Tunnel mode, a tunnelled IP protection is provided (via Gateways). Traffic have to pass through a gateway (i.e. firewall) to access ultimate destination.

In the following sections, we present terminology used in the IPSec context then present the two types of IPSec AH and ESP.

A. Terminology.

- Security Association (SA): a Simplex (unidirectional) secured connection identified by the following:

1. Security Parameter Index (SPI), two bytes saved in IPSec header, which with the following uniquely identifies the SA.

2. IP Destination Address from the IP packet header.

3. Security protocol identifier (50 for AH and 51 for ESP.) One byte saved in the IP packet header.

- Integrity: there are three types of integrity in the context of IPSec:

1. Connectionless integrity, that detects modifications of a single IP packet by the use of an algorithm specific Integrity Check Value (ICV).

2. Anti-replay integrity, that detects duplicate IP packets at the receiver end by the use of sequence numbers in IPSec header.

3. Connection-oriented integrity, that detects lost or reordered IP packets at the receiver end by the use of sequence numbers in IPSec header as well.

B. IPSec Authentication Header (AH).   IPSec Authentication Header (AH) provides Connectionless Integrity, Data Origin Authentication, and Anti-Replay Integrity (optional - not enforced at receivers end). Figure 11.13 depicts the IP AH header format. The Next Header field is of 8 bits size and specifies the type of the transport protocol used in the upper layer. The Payload Length field is also an 8 bits size, and contains the IPSec header length in words (32 bit) minus 2 words (e.g. 3+3-2=4, if Authentication Data is 3 words (96 bits)). The sender always transmits the Sequence Number field (32 bits), but receiver might

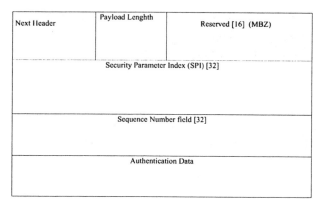

Figure 11.13. IPSec AH Header format.

optionally act on it. Finally, the Authentication Data field (variable size, multiple of 32 bits), contains Integrity Check Value (ICV) for the attached packet (Including the AH header itself). Reserved bits (Must Be Zero) are used for future extensions.

Figure 11.14 depicts the coverage of authentication protection for IP AH in Transport mode and Tunnel mode. Note how is the IP AH header is inserted between the IP Header (IPH1) and the upper header (e.g. TCP Header) in Transport mode. Note also, in tunnel mode the IPSec AH constructs another IP Header (IPH2) for the IP address of destination gateway.

The ICV is computed first at transmitter by the use of a common authentication algorithm that is also known to the receiver. Then ICV is recomputed at receiver and compared to match the received value for authentication integrity. ICV computation excludes non-predictable IP Header (IPH) fields like TTL, Flags, TOS, Fragment offset, Checksum, ... etc. If IP fragmentation occurs at sender, it should be performed after AH processing. The IP reassembly should then be performed before AH processing at the receiver.

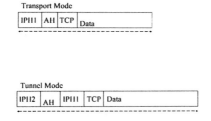

Figure 11.14. IP AH Authentication protection (dotted arrows)

C. IPSec Encapsulating Security Payload (ESP). IPSec encapsulation security payload (ESP) provides Confidentiality (Encryption), Connectionless Integrity (optional, not enforced at receiver end), Data Origin Authentication (optional, not enforced at receiver end), Limited Traffic Flow Confidentiality, and Anti-Replay Integrity. Figure 11.15 depicts the IP ESP header format. The Next Header field is of 8 bits size and specifies the type of the transport protocol used in the upper layer. The Pad length contains the number of pad bytes inserted by the encryption algorithm. The Payload Length field is also an 8 bits size, and contains the IPSec header length in words (32bit) minus 2 words (e.g. 3+3-2=4, if Authentication Data is 3 words (96bits)). The sender always transmits the Sequence Number field (32 bits), but the receiver might optionally act on it. Finally, the Authentication Data field (variable size, multiple of 32 bits) contains Integrity Check Value (ICV) for the encapsulated packet and the ESP header/ trailer (not including the authentication data itself).

Figure 11.16 depicts the coverage of authentication and encryption protections for IP ESP in Transport mode and Tunnel mode. The ESP header is inserted exactly same way as in AH header. The ESP trailer is

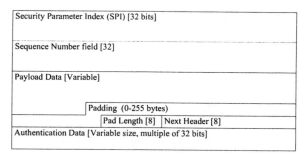

Figure 11.15.   IPSec ESP Header format.

inserted after the payload data and before the ESP authentication data. ICV computation steps are the same as in IPSec AH.

6.5   MPLS Tunnelling

Multi protocol label switching (MPLS) is another technology that can be used in creating Virtual Private Networks (VPN) tunnels. Here, VPN tunnels are created using MPLS and IPSec in a public IP network. Virtually all the interfaces in an IP based network could be implemented via a MPLS VPN tunnels leased from a public carrier. The current efforts in IETF are focused on defining MPLS architecture and associated label distribution protocols. Label switching router (LSR) and label switching path (LSP) or tunnel are the main features of MPLS.

7.   Performance Analysis of IPSec Protocol

IPSec provides two types of security algorithms, symmetric encryption algorithms (e.g. Data Encryption Standard DES) for encryption, and one-way hash functions (e.g., Message Digest MD5 and Secured Hash Algorithm SHA1) for authentication. This section presents performance analysis and comparisons between these algorithms in terms of time

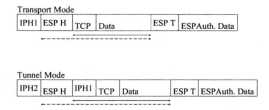

Figure 11.16. IP ESP Authentication protection (dotted arrow) and Encryption protection (solid arrow).

complexity and space complexity. Performance Parameters considered are processing power and input size. The analysis results revealed that HMAC-MD5 can be sufficient for the authentication purposes rather than using the more complicated HMAC-SHA1 algorithm. In encryption applications, authentication should be combined with DES. This section focuses on IPSec authentication and encryption space complexity, computation time complexity, and protocol limitations.

7.1 Space Complexity

IPSec AH/ESP header size is 12/10 bytes fixed header fields respectively, plus the variable size authentication data. Authentication data is algorithm and packet specific field. IPSec authentication uses Keyed-Hashing for Message Authentication (HMAC) [14], [13], [12] combined with Message Digest algorithm (MD5) [24], or Security Hash Algorithm (SHA-1) [13]. All algorithms use secret key distribution. In combined HMAC-MD5, the key size is 128 bit while in the combined HMAC-SHA-1 the key size is 160 bits. In both cases the algorithm works with 64-byte message blocks. They generate a truncated ICV of 96 bits (12 bytes) to conform with the IPSec AH and ESP authentication data size. This means a total header size of 24/22 bytes for AH/ESP respectively, per packet. This overhead is needed for each authentication request transmitted through a security association in transport mode. Additional 20 bytes of IP header are also needed per packet for tunnel destination in tunnel mode. A total overhead of 44 bytes per packet assumes no header compression mechanisms used. Table 11.2 summarizes these results for both Transport and Tunnel modes.

IPSec ESP encryption may use extra padding bytes (0 to 255 bytes). They are algorithm and payload specific.

Table 11-2. IPSEC HEADER FIELD SIZES (BYTES)

Protocol	IPSec AH			IPSec ESP		
	Fixed	Variable	Total	Fixed	Variable	Total
Transport Mode	12	12	24	10	12	22
Tunnel Mode	12+20	12	44	10+20	12	42

7.2   Time Complexity

7.2.1   Encryption.   IPSec ESP Encryption algorithm is triple Data Encryption Standard (3DES) in Cipher Block Chaining (CBC) mode [15]. Encryption using DES algorithm is the most time consuming process. DES uses a 56-bit key, and block sizes of 64 bits. The algorithm has 19 distinct steps. The first step is a key independent transposition on the 64bit input block. The last step is the exact inverse of this transposition. In step 18, a 32 bit SWAP operation is performed. The remaining steps (2 to 17) are functionally identical and are dependent on different portions of the input key. Each of these 16 steps takes two 32 bit inputs, and produce two 32 bit outputs. The left output is a COPY of right input. The right output is an XOR of left input and a function f of right input and the step key. The function f consists of 4 operations. First, a 32:48 bit transposition/expansion of the right input is applied; second a 48 bit XOR of the output with the step key. Then a group mapping is performed to reduce the output size from 48 to 32 bits using two dimensional look-up table of 4 rows by 16 columns for all possible 64 inputs. Finally, 32-bit transposition is performed.

Each of the 16 steps has a special key, which is derived from the 56-bit key using the KS function. For the KS function, an initial 56-bit transposition is performed to the input key. Before each step, the key is divided into two 28- bit sub-keys. Each of which is rotated left using LEFT SHIFT operation. Finally, 56:48 bit transposition/reduction is performed.

Table 11.3 lists DES basic operations and their equivalent simple ones. It also gives space requirement for each operation. Table 11.4 lists the DES and 3DES computations of total number of simple operations needed. 3DES is the chained form of DES with a chain size of 3. For simplicity, 3DES is assumed to have only 3 times number of operations as DES has. 3DES requires an extra random Initialization Vector (IV) of 8 bytes. For complete details of DES and 3DES algorithms see [29] [30].

As shown in table 11.4, the total number of operations per 64-bit block is 8091 operations. Given a packet size of N bits then the number of blocks is given by

$$n = \lceil \frac{N}{64} \rceil \tag{11.1}$$

Where [.] means the smallest integer bigger than or equal to the operand. Therefore, 3DES time complexity is of $O(n)$.

Table 11-3. DES basic operations

Basic Operation	Equivalent simple operation Type	#Times needed	Space needed
b bit transposition	One dimensional table look-up	b	b
Two dimensional table map (for 6:4 bit map.)	Multiply	1	
	Add	1	
	One dimensional table look-up	1	4 rows×16 cols

Table 11-4 DES Operation in one block encryption

Step#	operation	#Times	Equiv.*Total**	Notes
1,19	64 bit transposition	2	64×2	
2-17	32 bit COPY	16	16	16steps
2-17	32 bit XOR	16	16	
2-17	48 bit transposition	16	48×16	f function
2-17	48 bit XOR	16	16	
2-17	6:4 bit Two Dimensional Table Mapping	8×16	3×128	
2-17	32 bit transposition	16	32×16	
1	56 bit transposition	1	56	KS function
2-17	28 bit LEFT SHIFT	2×16	32	
2-17	48 bit transposition	16	48×16	
18	32 bit SWAP	1	1	Pre-out
			2697	DES Total
			8091	3DES Total

*using equivalent simple substitution from table 11.3

Figure 11.17 shows the computed encryption time in microseconds as function of input packet size in blocks of 64 bits and processing power in Millions of Instructions Per Second (MIPS). For example, time complexity of 3DES means approximately 4 Mbps encryption rate with a 500 MIPS.

7.3   Authentication

As presented in the previous section, IPSec Authentication algorithms are HMAC-MD5, or HMAC-SHA-1. These two algorithms are explained in this section.

7.3.1   MD5.   The first step in MD5 algorithm is padding the original message by appending 1 to 512 bits to it. The original message size is also appended in 64 bits. The total padded message with the message size field should have a size, which is multiple of 512 bits (64 bytes).

Table 11-5 MD5 Operations per block

Round #	Base Function♠	#Steps	Operations /step	Total
1	F(X,Y,Z)= X^Y∨¬X^Z	16	12	192
2	G(X,Y,Z)= X^Z∨Y^¬Z	16	12	192
3	H(X,Y,Z)= X⊕Y⊕Z	16	10	160
4	I(X,Y,Z)= Y⊕(X∨¬Z)	16	11	176
			Total(T)	720*

♠ Symbols ∧ for AND, ∨ for OR, ¬ for NOT and ⊕ for EXCLUSIVE OR
* Plus 24 operations per block for initiation and termination

The algorithm uses four 4-byte registers (A, B, C, and D) which keep intermediate and, ultimately, the final value of the message digest of 128 bits. For each block of 64 byte of the message the algorithm performs 4 rounds of 16 steps each. In each step, one of the functions in table 11.5 is computed in terms the registers (A, B, C, or D), the current input block, and a lookup value. By investigating MD5, one can find out that 10 to 12 operations per step are needed. Thus, in table 11.5, the total number of operations (T) is shown as 744 (720 + 24) operations per block. The time complexity is $O(n)$, where n is number of input blocks given by

$$n = \frac{N}{512} \qquad (11.2)$$

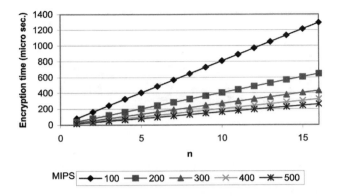

Figure 11.17. 3DES Encryption vs. no. of blocks and processing power.

$$N = X + pad + Size \qquad (11.3)$$

where X is the input text, *pad* is the padding field, *size* is the size field, and N is the total message size.

Figure 11.18 shows the computed algorithm computation time (digest time) in microseconds as a function of input size in blocks of 512 bits (64 bytes) and processing power in MIPS. The algorithm has been tested on a Pentium II machine with 300 MHz rate for one million 1024 byte messages. The digest time computed was 159 seconds. The digest rate then is 51.5 Mbps. This means that every operation counted in table 11.5 executed on the average in 6 cycles.

7.3.2   HMAC-MD5.   The combined HMAC-MD5 algorithm is formulated as follows:

$$MD5(K_o, MD5(K_i, Text)$$

$$K_i = Key \oplus ipad \qquad (11.4)$$
$$K_o = Key \oplus opad \qquad (11.5)$$

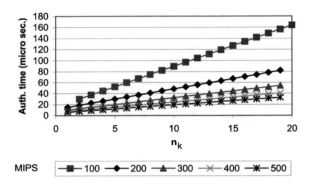

Figure 11.18.   MD5 Digest time vs. no. of input blocks and processing power.

where Ki and Ko are two extended forms (512-bit) of the input Key and are generated by exclusive or the Key with $ipad$ the inner padding (512 bits), and $opad$ the outer padding (512 bits). Key is an arbitrary size secret key shared by sender and receiver. $Text$ is the given input message subject to authentication. For an input text of size X bits, the number of input blocks for the inner MD5, n_k, is

$$n_k \;=\; \frac{N}{512} \tag{11.6}$$

$$N \;=\; X + K\_pad + size \tag{11.7}$$

where K is the size of the extra appended inner form of the key (512 bits).

In the outer MD5, the output of the inner MD5 (128-bit digest) is appended to Ko (512 bit). According to MD5, this is padded to two 512-bit blocks. Thus, the HMAC-MD5 total number of operations (T) is

$$T(n_k) \;=\; 32 + (2 + n_k) \times 744 \tag{11.8}$$

where 32 comes since in (11.4) and (11.5) the XOR operands are of size 512 bits. In a machine of word size 32 bits, this has to be partitioned in

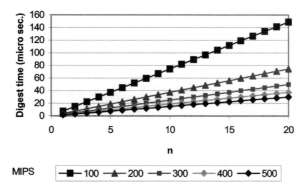

Figure 11.19. HMAC-MD5 Authentication time vs. no. of input blocks and processing power.

16 consecutive XOR operations. A total of 32 extra operations is added. The time complexity is then $O(n_k)$.

Figure 11.19 shows the computed authentication time in microseconds vs. no. of input blocks and processing power in MIPS.

7.3.3   SHA-1.   SHA1 follows exactly the same way of message padding as in MD5 algorithm. However, the algorithm uses five 4-byte intermediate registers instead of four. Thus, the final value of the message digest is 160 bits. For each block of 64 bytes of the message the algorithm performs 4 rounds of 20 steps each. In each step, a functional computation based on the temporary registers, the current input block, and a constant value. By investigating SHA1, one can find out that 10 to 13 operations per step are needed. In table 11.6, the total number of operations (T) is computed as 1110 (=900+210) operations per block. The algorithms time complexity is $O(n_k)$, where n_k is the number of input blocks as in (11.6).

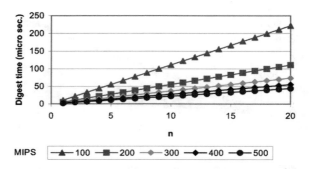

Figure 11.20. SHA1 digest time vs. no. of input blocks and processing power.

Table 11-6. SHA1 Operations per block

Round#	Base Function	Steps	Operation/Step	Total
1	F(X,Y,Z)=X^Y∨¬X^Z	20	12	240
2	G(X,Y,Z)=X⊕Y⊕Z	20	10	200
3	H(X,Y,Z)= X^Y∨X^Z∨Y^Z	20	13	260
4	I(X,Y,Z)=X⊕Y⊕Z	20	10	200
			Total	900*

* Plus 210 operation per block for initiation and termination

Figure 20 shows the computed digest time in microseconds as a function of number of input blocks and MIPS.

7.3.4   HMAC-SHA-1.   The combined HMAC-SHA1 algorithm is formulated as follows:

$$SHA1(K_o, \; SHA1(K_i, Test))$$

The total number of operations (T) needed for HMAC-SHA1 is of $O(n_k)$ where,

$$T(n_k) \quad = \quad 32 + (2 + n_k) \times 1110 \tag{11.9}$$

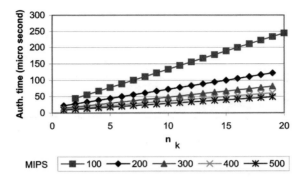

Figure 11.21. HMAC-SHA1 Authentication vs. no. of input blocks and processing power

Figure 11.21 shows the computed authentication time in microseconds as a function of number of input blocks and processing power in MIPS.

In summary Figure 11. 22 shows a comparison between 3DES, HMAC-MD5, and HMAC-SHA1 throughput vs. processing power. It shows how HMAC-MD5 supercedes HMAC-SHA1. It also shows how far the authentication throughput is as compared to encryption throughput.

7.4   Protocol Limitations

1. IPSec is IP based only. It does not provide security for any other network layer protocols.

2. Encryption using DES is based on a secret key of 56-bit size. In todays equipment this is very small key to break even if using brute force approach given the knowledge of plain text- cipher text pair. The addition of extra chaining of 3 levels (3DES) improves this limitation by the addition of an extra Initial Vector (IV) of random 64 bits. Practically, the IV can be set as the last 8 bytes of an encrypted packet for the next encryption process. In this case 3DES can be logically extended over consecutive packets [15]. Encryption should not be offered without the use of data origin

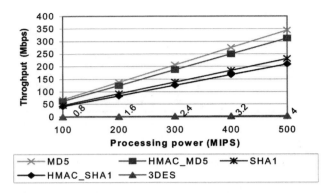

Figure 11.22. Throughput comparison vs. processing power.

authentication. After all, 3DES is much better than sending plain text.

8.   Conclusions

In this chapter, we presented various portability scenarios, architectures, implementation, and requirement issues for nomadic wireless Internet users. Secured and fast connection for authentication and authorization is crucial in today's IP network. We focused on the state-of-the-art VPN tunnelling techniques that are required for remote nomadic wireless users authentication and authorization. IPSec provides two types of security mechanisms symmetric encryption algorithm (e.g., DES), and one way hash functions for authentication (e.g., MD5 and SHA 1). Performance evaluation and comparison between these mechanisms in terms of computation time complexity and space complexity have been presented in this chapter (parameters considered are power and input size). Our results revealed that HMAC-MD5 is sufficient for authentication purpose rather than using the more complicated HMAC-SHA1 algorithm. Moreover, MD5 is having higher throughput as compare to SHA1. Since the IPSec requires only 96 bits of the of the message digest, MD can be sufficient for the authentication purpose. 3DES encryption throughput is very low compared to authentication throughput. It should only be used in critical user information not for regular traffic flow. Encryption if needed should be combined with authentication. In this case if the message fails authentication, decryption process is waived (not performed).

References

[1] R. Droms, "Dynamic Host Configuration Protocol," IETF RFC 1541, October 1993.

[2] O. S. Elkeelany, M. M. Matalgah, K. P. Sheikh, M. Thaker, G. Chaudhry, D. Medhi, and J. Qaddour, "Performance Analysis of IPSec Protocol: Encryption and Authentication," IEEE Communications Conference (ICC 2002), pp. 1164-1168, New York, NY, April 28 - May 2, 2002.

[3] B. Fox and B. Gleeson, "Virtual Private Networks,"IETF work, RFC 2685, September 1999.

[4] K. Hamzeh, G. Pall, W. Verthein, J. Taarud, W. Little, G. Zorn, "Point-to-Point Tunneling Protocol (PPTP)," IETF RFC 2637, July 1999.

[5] S. Hanks, T. Li, D. Farinacci, P. Traina, "Generic Routing Encapsulation (GRE)," IETF RFC 1701, October 1994.

[6] Lin J. Heinanen, G. Armitage and A. Malis, "A framework for IP based Virtual Private Networks", IETF RFC 2764, February 2000.

[7] Walter Honcharenko, Jan P. Kruys, David Y. Lee, and Nitin J Shah," Broadband Wireless Access," IEEE communication magazine, Jan 1997.

[8] International Business Machines Corporation, Using IPSEC to construct secure VPN, 1998. http://www.firstvpn.com/papers/ibm/ipsecvpn.htm

[9] S. Kent and R. Atkinson, " IP Authentication Header," IETF RFC 2402, 1998.

[10] S. Kent and R. Atkinson, " IP Encapsulating security Payload (ESP)," IETF RFC 2406, 1998.

[11] S. Kent and R. Atkinson, " Security Architecture for the Internet Protocol," IETF RFC 2401, 1998.

[12] H. Krawczyk, M. Bellare, and R. Canetti, "HMAC: Keyed-Hashing for Message Authentication," RFC 2104, February 1997.

[13] C. Madson and R. Glenn, "The Use of HMAC-MD5-96 within ESP and AH," RFC 403, November 1998.

[14] C. Madson and R. Glenn, "The Use of HMAC-SHA-1-96 within ESP and AH," RFC 404, November 1998.

[15] C. Madson and N. Doraswamy, "The ESP DES-CBC Cipher Algorithm with Explicit IV," RFC 2405, November 1998.

[16] G. S. Malkin, "Dial-in Virtual Private Networks Using Layer 3 Tunneling," IEEE Computer Society, 22nd Annual Conference on Local Computer Networks, 1997.

[17] M. M. Matalgah, J. Qaddour, A. Sharma, and K. P. Sheikh, "Capacity Planning in UMTS WCDMA TDD: Throughput and Spectral Efficiency Analysis," Proceedings of the 3GWireless'2003 Conference, San Francisco, CA, May 27 - 30, 2003.

[18] G. Mcgregor, "PPP Internet Protocol Control Protocol," IETF RFC 1332, Merit, May 1992.

[19] G. Meyer, "The PPP Encryption Control Protocol (ECP)," IETF RFC 1968, June 1996.

[20] P. Mockapetris, "Domain Names -Concepts and Facilities," IETF RFC 1034, 1987 Conference, San Francisco, CA, May 27 30, 2003.

[21] NIST, FIPS PUB 180-1: "Secure Hash Standard," 1995.

[22] B. Patel, B. Aboba, W. Dixon, G. Zorn, "Securing L2TP using IPSec," Internet-Draft, July 2001.

[23] D. Rand, "The PPP Compression Control Protocol," IETF RFC 1962, June 1996.

[24] R. Rivest, "The MD5 Message-Digest Algorithm," RFC 1321, April 1992.

[25] W. Simpson, "The Point-to-Point Protocol (PPP)," IETF RFC 1661, July 1994.

[26] W. Simpson, "The Point-to-Point Protocol (PPP)," IETF RFC 1661, July 1994.

[27] A. S. Tanenbaum, "Computer Networks," Prentice Hall PTR, 3rd Ed., pp. 588-595, 1996.

[28] W. Townsley, A. Valenciam A. Rubens, G. Pall, G. Zorn, and B. Palter, "Layer Two Tunneling Protocol (L2TP)," IETF RFC 2661, 1999.

[29] US National Bureau of Standards, "DES modes of operation," Federal Information Processing Standard (FIPS) publication 81, December 1980. http://www.itl.nist.gov/fipspubs/fip81.htm

[30] US National Bureau of Standards, "Data Encryption Standard," Federal Information Processing Standard (FIPS) publication 46-2, December 1993. http://www.itl.nist.gov/fipspubs/fip46-2.htm

[31] A. Valencia, M. Littlewood, T. Kolar, "Cisco Layer Two Forwarding (Protocol) L2F," IETF RFC 2341, May 1998.

[32] Willens, S. A. Rubens, W. Simpson, C. Rigency, "Remote Authentication Dial In User Service," IETF RFC 2509, January 1997.

[33] Willens, S. A. Rubens, W. Simpson, C. Rigency, "Remote Authentication Dial In User Service," IETF RFC 2138, April 1997.

Chapter 12

DESIGN AND IMPLEMENTATION OF A SOFTSWITCH FOR THIRD GENERATION MOBILE ALL-IP NETWORK

Vincent W.-S. Feng and Yi-Bing Lin
Department of Computer Science & Information Engineering
National Chiao Tung University
Taiwan, R.O.C.

S.-L. Chou
Computer and Communication Research Laboratory
Industrial Technology Research Institute
Taiwan, R.O.C.

Abstract   This chapter describes the design and implementation of a softswitch for UMTS developed by Computer and Communication Research Laboratory (CCL) of Industrial Technology Research Institute (ITRI). This softswitch can be utilized as call agent (media gateway controller) for the third generation mobile all-IP network such as UMTS. The CCL Softswitch follows the reference architecture proposed by International Softswitch Consortium. In our approach, the Intelligent Network (IN) call model is implemented to interwork with existing IN devices. We design protocol adapter modules and service provider interface to ensure that multiple VoIP protocols can be supported without modifying the core of the softswitch. Furthermore, the message flows of call setup, call transfer and inter-softswitch call are described to show the feasibility of the softswitch.

Key words   H.323, Megaco, MGCP, SIP, Softswitch, UMTS, VoIP

1.   Introduction

Most of today's telecommunications networks are based on the circuit-switched technologies. Next generation networks such as the third generation (3G) mobile telecommunications systems [1, 2] utilize the packet-switched technologies. Many new applications and services have been developed based on such technologies and networks. Among them, Voice over IP (VoIP) is one of the most important applications. In the early deployment of carrier-grade IP telephony, services were provided by gatekeepers [3], SIP proxy servers [4], or call agents that control media gateways [5, 6]. These mechanisms have been evolved into a more intelligent call control architecture called softswitch, which supports multi-protocols and provides interworking among heterogeneous networks. A softswitch is a cost-saving solution (in terms of capital and operation) that can flexibly enable new value-added telecommunications services and support the integration of web and multimedia services [7]. In 3G all-IP networks such as Universal Mobile Telecommunications System (UMTS) [8] the softswitch can be utilized as a 3G call agent or media gateway controller (MGC; see Figure 12.1 (6)). Unlike a traditional switch that includes hardwired-switching fabrics, the softswitch (Figure 12.1 (6)) is a pure software-control mechanism without switching circuit hardware. The circuit hardware is implemented in separate media gateways (see Figure 12.1 (7)). Figure 12.1 illustrates the UMTS all-IP network architecture [1, 2], which consists of following components:

1. The radio access network consists of *User Equipments* (UEs; Figure 12.1 (1)) and *UMTS Terrestrial Radio Access Network* (UTRAN; Figure 12.1 (2)).
2. The GPRS network (Figure 12.1 (3)) consists of *serving GPRS support nodes* (SGSNs) and *gateway GPRS support nodes* (GGSNs) that provide mobility management and packet data services to mobile users.
3. The *home subscriber server* (HSS; Figure 12.1 (4)) is the master database containing all 3G user-related subscription information.
4. The *call state control function* (CSCF; Figure 12.1 (5)) is a SIP server, which is responsible for call control.
5. The softswitch architecture consists of the *softswitch* or *media gateway controller* (Figure 12.1 (6)), *media gateway* (Figure 12.1 (7)) and *signaling gateway* (Figure 12.1 (8)).

In this section, we elaborate on the softswitch architecture (in Figure 12.1 (6)~(8)). Then based on this architecture, we describe the design and implementation of a softswitch we developed in Computer and Communication Research Laboratory (CCL), Industrial Technology Research Institute (ITRI).

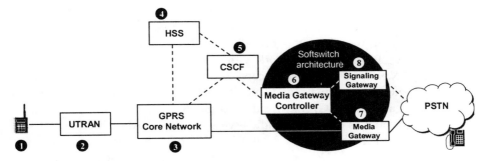

Figure 12.1. UMTS all-IP Network (dashed lines: signaling; solid lines: data and signaling)

1.1 ISC Reference Architecture

The International Softswitch Consortium (ISC) [9] is the premiere forum for the worldwide advancement of softswitch interoperability. This forum promotes Internet-based multimedia communications and applications. Figure 12.2 illustrates the reference softswitch architecture proposed by ISC. This architecture consists of four planes:

1. *Transport Plane* (see (1) in Figure 12.2) is used to transport packets across the IP network. These packets include call control signals and media data. The Transport Plane includes three domains:

   (i)   *IP Transport Domain* consists of devices such as routers and switches that form the IP backbone. The QoS assured mechanisms (such as RSVP, Diffserv, MPLS, etc.) are also included in this domain.

(ii)   *Interworking Domain* consists of devices that interwork with other networks. For example, the trunking gateways (TG) and signaling gateways (SG) interwork with the PSTN and SS7 networks. The VoIP gateways interwork with other VoIP networks.

(iii) *Non-IP Access Domain* consists of devices that interwork with non-IP terminals and mobile networks. For example, access gateways (AG) and residential gateway interwork with non-IP terminals or phones; radio access network (RAN) AGs interwork with mobile network (GPRS/3G) devices; integrated access devices (IADs) and multimedia terminal adaptors (MTAs) interwork with broadband access (e.g. DSL/HFC) devices.

2. *Control & Signaling Plane* (see (2) in Figure 12.2) is the kernel of the softswitch architecture, which receives the signals from the Transport Plane and controls the flow of calls. The signaling protocols of this plane include traditional circuit-switched network protocols such as ISDN User Part (ISUP) [10] and Mobile Application Part (MAP) [11], and VoIP protocols such as H.323 [3], Media Gateway Control Protocol (MGCP) [5], Megaco [6] and Session Initiation Protocol (SIP) [4]. Softswitches reside in this plane.

3. *Service Plane* (see (3) in Figure 12.2) consists of application servers, which provide the service creation environment (SCE) and the service logic execution environments (SLEE). The SCE is a toolkit with graphic user interface (GUI), which allows users to develop services easily and quickly. The SLEE compiles service logics and communicates with the Control & Signaling Plane via standard Application Programming Interface (API; such as JAIN [12] and Parlay [13]) or protocols such as Intelligent Network Application Part (INAP) [14], Customized Application for Mobile Network Enhanced Logic (CAMEL) [15] and SIP. This plane also consists of media servers that provide conferencing and voice mail services.

4. *Management Plane* (see (4) in Figure 12.2) handles the management issues of the above three planes. The issues include service provisioning, network management, billing, etc.

The ISC reference architecture provides an interoperability view of softswitch, which is followed by the softswitch systems developed by Lucent [16], Siemens [17], Telcordia [18], and many other softswitch products and prototypes. In this chapter, we concentrate on the Control & Signaling Plane of the ISC Reference Architecture. Details of other planes are out of the scope of this chapter, and the reader is referred to [9].

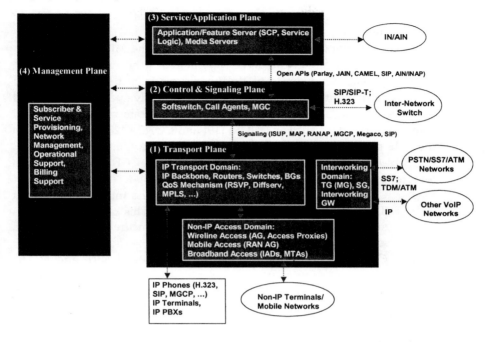

Figure 12.2. ISC Reference Architecture

1.2   CCL Softswitch Architecture

This section describes the softswitch modules and the design philosophy of our approach. Figure 12.3 illustrates the CCL Softswitch, which focuses on the Control & Signaling Plane. This architecture consists of the following modules:

1. *Protocol Adapter* (see (1) in Figure 12.3) provides mapping between the Transport Plane and the Call Server. The adapter converts the signals of various Transport-Plane protocols (such as ISUP, MGCP, Megaco, SIP and H.323) into a uniform format and delivers them to the Call Server through the Softswitch Service Provider Interface (SPI). With the adapter module, new protocols can be easily plugged into the system. Current implementation of the CCL Softswitch adapter module supports the following Transport-Plane protocols:

   (i)   Off-net calls: Call setup between the PSTN and a VoIP system (including H.323, MGCP, Megaco and SIP)

(ii)   On-net calls: Call setup between system of various VoIP protocols

2. *Softswitch SPI* (see (2) in Figure 12.3) is based on the Microsoft Windows Open Service Architecture (WOSA) model. The SPI isolates the knowledge and dependency of VoIP protocols from the Call Server, so that the call control mechanism can be effectively maintained and managed. The CCL Softswitch SPI technology follows the distributed computing standard such as DCOM [19] and CORBA [20]. Mapping from the VoIP protocols to the SPI will be described in Section 3.

3. *Call Server* (see (3) in Figure 12.3) is the core of the CCL Softswitch. Main features of the Call Server include:

   (i)   Call handing: The Call Server exercises the Intelligent Network (IN) Basic Call State Model (BCSM) [21]. This call model ensures that the Softswitch can interwork with incumbent IN/CAMEL service platforms.

   (ii)   Communication with Protocol Adapters and Service Agents: The Call Server receives call control signals from the Protocol Adapters via Softswitch SPI. It also receives service requests from the Service Agents via Softswitch API. With the SPI and API models, the Call Server implementation is flexible, reliable and effective.

   (iii) Database maintaining: The Call Server communicates with databases via standard Open Database Connectivity (ODBC) interface [22]. The Call Server maintains routing tables, call detail records, user profiles, event logs and OA&M information (see (4) in Figure 12.3).

4. *Softswitch API* (see (5) in Figure 12.3) is a proprietary interface, which provides the abstract view of the CCL Softswitch operations to the Service Agents. With this API, new services can be effectively deployed. The Softswitch API features include call control, administration, and network management. Just like Softswitch SPI, Softswitch API follows the DCOM/CORBA standards. Examples of built-in services are unified messaging service, follow me, one number, etc.

5. *Service Agent* (see (6) in Figure 12.3) wraps Softswitch API into standard APIs such as Parlay or JAIN and interworks with the application servers in the Service Plane. With the Service Agent model, new APIs can be easily plugged into the system. Currently, Service Agents for TAPI [23] is provided in the CCL Softswitch. Agents for OSA [24] and INAP [14] are under construction.

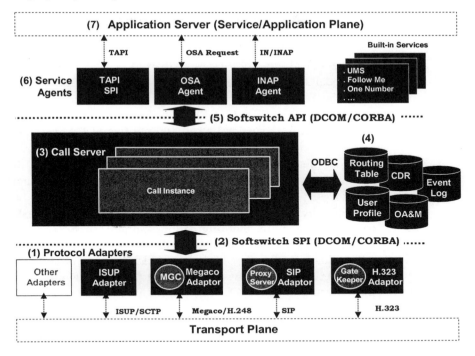

Figure 12.3. CCL Softswitch Architecture

2. The Call Model and Softswitch SPI

In this section, we describe the call model of the CCL Softswitch and the interaction between the Call Server and the Softswitch SPI.

2.1 The Call Model

A call model is a set of call control states, which represent call progress activities in a switch. Call control functionalities of a switch is usually determined by the complexity of the call model. The call models widely used today include IN [21], Q.931 [25] and Computer Supported Telecommunications Application (CSTA) [26]. The CCL Softswitch utilizes the IN call model due to the following reasons:

1. The IN call model supports the most general call model. The call states of Q.931 and CSTA are subsets of that for the IN call model. The number of call states of Q.931 and CSTA are 17 and 7, respectively. On the other hand, the number of call states of IN is more than 19.
2. The IN call model has been effectively used to enhance the service capabilities of the PSTN. The IN separates the service logic from the switch into a *Service Control Point* (SCP). By utilizing the call model, an IN switch has the ability to suspend a call and notify a SCP to executes the service logic. Other call models do not provide this mechanism.
3. Many profitable IN services have been deployed and commercialized. By using the IN call model, our softswitch implementation can leverage existing IN services.

The IN call model divides a call into two portions: the originating BCSM (O-BCSM) that associates with the originating (calling) party and the terminating BCSM (T-BCSM) that associates with the terminating (called) party. Figure 12.4 illustrates the IN call model, which consists of *Points in Calls* (PICs) and *Detection Points* (DPs). A PIC (see (1)-(17) in Figure 12.4) represents a call state in the switch. A DP (see (a)-(o) in Figure 12.4) indicates the point of call processing at which transfer of control from the switch to the IN service logics can occur. A DP must be armed in order to notify an IN service logic that the DP has been encountered, and potentially allow the IN service logic to influence subsequent call processing. If a DP is not armed, the switch continues call processing without the SCP involvement. When an IN service is invoked, the switch will suspend the call process until the execution is returned from the service logic of the SCP. The reader is referred to [21] for more details about PICs and DPs.

We use the originating-screening service to illustrate the usage of DPs. This service checks if the caller identification is on the call-allowed list. The unauthorized people not in the list are not allowed to make special calls such as 1-900 or international calls. The Analysed_Information DP (see (d) in Figure 12.4) is armed if a user subscribes to this service. In this scenario, the user's call will be suspended when the call process enters the Analysed_Information DP. The switch will send an IN request to the SCP to check if the caller identification is valid. If the caller is allowed to make the call, the call process will be resumed.

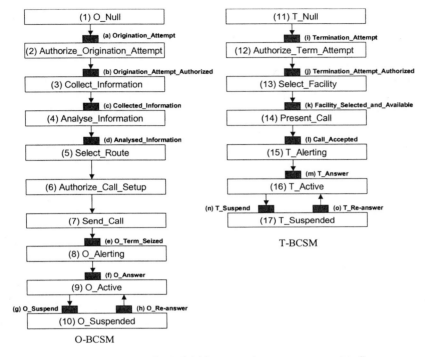

Figure 12.4. IN Call Model (the exception states are omitted)

2.2 Softswitch SPI

Interworking mechanisms between VoIP and IN has been intensively studied [27, 28]. Most of the previous approaches directly map the VoIP protocols into the IN call model. These approaches do not consider interworking among various VoIP protocols. In our approach, the VoIP protocols are transformed into a uniform format and then delivered to the Call Server through the Softswitch SPI. The design of the Softswitch SPI ensures that new protocols can be added to the system without modifying the Call Server.

As shown in Figure 12.5, the Softswitch SPI consists of two interfaces. The ICallControl (see (2) in Figure 12.5) is provided by the Call Server. All actions through the ICallControl interface are initiated by the Protocol Adapters. The ICallControl interface provides the following services:

1. Information retrieving services allow the Protocol Adapter to retrieve device information (GetDeviceInfo), device state (GetDeviceState), digit map (GetDigitMap) and authentication information (GetAuthenticationInfo).
2. Registration services include device registration and de-registration (RegisterDevice and UnregisterDevice).
3. Call control services include AnswerCall, ConsultCall, DropCall, HoldCall, MakeCall, PrepareCall, RejectCall, TransferCall, etc.

We will elaborate on these services later. Upon receipt of VoIP/ISUP signals from the end terminals (see (5) in Figure 12.5), the Protocol Adapter (Figure 12.5 (4)) will pack these signals into one message. This message is sent to the Call Server via the ICallControl interface (Figure 12.5 (2)). After the message is executed, the call states of BCSMs will be changed (see (1) in Figure 12.5).

The IProtocolControl interface (see (3) in Figure 12.5) is provided by the Protocol Adapter. All actions through the IProtocolControl interface are initiated by the Call Server. If the Call Server attempts to notify an end terminal of a specific event, it sends a message to the Protocol Adapter via the IProtocolControl interface. Then the Protocol Adapter translates this message to the corresponding VoIP/ISUP signals and sends them to the end terminal.

Through the IProtocolControl interface, two kinds of actions can be performed. An originating action (the messages prefixed with O) is issued by the O-BCSM and a terminating action (the messages prefixed with T) is issued by the T-BCSM. The IProtocolControl interface provides the following services (details of the services will be described in the next section):

1. Originating services include ORingback, OAnswerCall, ORejectCall, ODropCall, OHoldCall, OPrepareCall, OMakeCall, etc.
2. Terminating services include TRinging, TInitAnswerCall, TanswerCall, TRejectCall, TDropCall, THoldCall, etc.

The call control scenario described above is called the *first-party call control*. The first-party call control provides a direct interface between the user and the telephone switch, which intercepts line signaling between the telephone and the switch, and is thus provided on a single-line basis [29]. Another call control scenario is called the *third-party call control*, which controls call processing through an external agent. The third-party call control mechanism can also be efficiently implemented in the CCL Softswitch. Specifically, the Application Server (Figure 12.3 (7)) can play the role as an external agent. For example, the Application Server may make a call control request to the Call Server through the Service Agent (Figure 12.3 (6)) and the Softswitch API (Figure 12.3 (5)). Based on this request, the Call Server controls the end terminals through the Softswitch SPI.

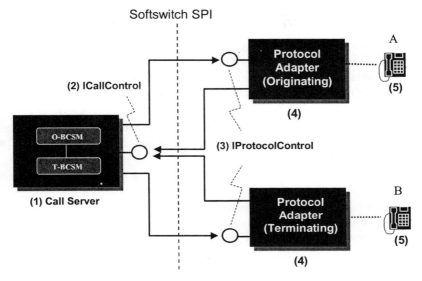

Figure 12.5. Softswitch SPI

3. Message Flows of Call Control

This section uses call setup and call transfer as two examples to describe the interaction between the Call Server and the Softswitch SPI.

3.1 Call Setup

Figure 12.6 illustrates the message flow of call setup from a Megaco end point (Party A) to a SIP user agent (Party B). The state transition of the O-BCSM and the T-BCSM are also shown in this figure (see the ellipses in Figure 12.6). The call setup procedure consists of three phases: preparing call phase (Steps 1.1 ~ 1.3), making call phase (Steps 1.4 ~ 1.7) and answering call phase (Steps 1.8 ~ 1.12). The initial states of the O-BCSM and the T-BCSM are O_Null and T_Null, respectively. The message flow in Figure 12.6 follows the notation of Unified Modeling Language (UML) [30].

Step 1.1: Party A picks up the phone. An off-hook event is sent to the Megaco Adapter through the Megaco message Notify. This message triggers the preparing call procedure, and the Megaco Adapter forwards the call setup request to the Call Server using the PrepareCall message in the ICallControl interface. The Call Server changes the O-BCSM state from O_Null to Authorize_Origination_Attempt and checks the authority and ability of Party A. Then the O-BCSM enters the Collect_Information state.

Step 1.2: The Call Server sends the OPrepareCall message to the Megaco Adapter through the IProtocolControl interface, and prepares to collect the dialed digits from Party A. The Megaco Adapter forwards this request to Party A using the Megaco message Modify that includes the dial tone signal and the digit map information. After this message is executed at the end terminal, Party A will hear a dial tone. The Megaco Adapter replies OPrepareCall_Ack to the Call Server to indicate that the dial tone has been sent to Party A.

Step 1.3: The PrepareCall_ack message is sent from the Call Server to the Megaco Adapter and the preparing call phase is complete.

Step 1.4: Party A sends the Megaco message Notify with the dialed digit string (the address of Party B) to the Megaco Adapter. The Megaco Adapter forwards the MakeCall request to the Call Server. The Call Server processes the call at the originating portion based on the IN call model (see Figure 12.4), which changes the O-BCSM state with the following steps: The Call Server analyzes the collected information (Analyse_Information), searches a route list and selects a route (Select_Route), verifies the authority of the originating party to place this call (Authorize_Call_Setup), and sends an indication to set up the call to the terminating party (Send_Call). Then the Call Server processes the call at the terminating call portion, which includes checking the authority and ability of the terminating party (Authorize_Termination_Attempt), checking if the terminating party is busy (Select_Facility), and presenting a call to the terminating party (Present_Call). Note that if the originating party is an H.323 terminal or a SIP user agent, then there is no need to send extra VoIP messages to the calling party for collecting dialed digits. The SIP INVITE message and the H.323 Setup message already contain the digit information of the terminating party. The related message flows are illustrated in Figures 12.7 and 12.8. In Figures 12.7 and 12.8, the actions taken by the Call Server are exactly the same as that in Figure 12.6, which are not affected by the VoIP protocols exercised at the end terminals.

Step 1.5: The Call Server rings the terminating party. That is, it sends the TRinging message to the SIP Adapter. The SIP Adapter forwards this request to Party B using the SIP message INVITE. Party B replies the SIP Adapter with SIP messages 100 Trying and 180 Ringing. The T-BCSM enters the T_Alerting state.

Step 1.6: The Call Server rings back the originating party. It sends the ORingback message to the Megaco Adapter. The Megaco Adapter forwards this request to Party A using the Megaco message Modify with the ring back signal. The O-BCSM enters the O_Alerting state.

Step 1.7: The MakeCall_ack message is sent from the Call Server to the Megaco Adapter and the making call phase is complete.

Step 1.8: Party B answers the call and sends a SIP message 200 OK with Session Description Protocol (SDP) [31] information to the SIP Adapter. This message triggers the answering call procedure. The SIP Adapter translates this message to the AnswerCall message, and forwards it to the Call Server. At this point, the RTP connection between the two parties is not ready, and the conversation is not available. To establish the RTP connection, the two parties exchange the SDP information in the next three steps (Steps 1.9 ~ 1.11).

Step 1.9: The Call Server sends the TInitAnswerCall message to the SIP Adapter. The SIP Adapter returns the SDP information of Party B. In this example, the SIP Adapter does not send any VoIP message to the user agent. In the Megaco case (see Figure 12.9), the Protocol Adapter will send the Megaco messages Add ($) (where the dollar sign '$' is used to signal media gateway to select an idle end point; since Party B is a media gateway with only one end point, this message signals Party B to select itself) and Modify (to stop ringing) to the terminating party. In the H.323 case (see Figure 12.10), the Protocol Adapter will set terminal capability and open logical channel to the terminating party.

Step 1.10: The Call Server sends the OAnswerCall message to the Megaco Adapter. This message contains the SDP information of Party B. The Megaco Adapter translates this message into the Megaco messages Modify and ADD, which stop the ring back tone and inform Party A to determine the SDP information. Then the Megaco Adapter returns the SDP information of Party A to the Call Server.

Step 1.11: The Call Server sends the TAnswerCall message to the SIP Adapter. The message contains the SDP information of Party A. The SIP Adapter translates this message into the SIP message ACK with the

SDP information of Party A. This message is sent to Party B. At this
point the two parties have exchanged the SDP information.

Step 1.12: The AnswerCall_ack message is sent from the Call Server to the
SIP Adapter and the call setup procedure is complete. The final states
of the O-BCSM and the T-BCSM are O_Active and T_Active,
respectively.

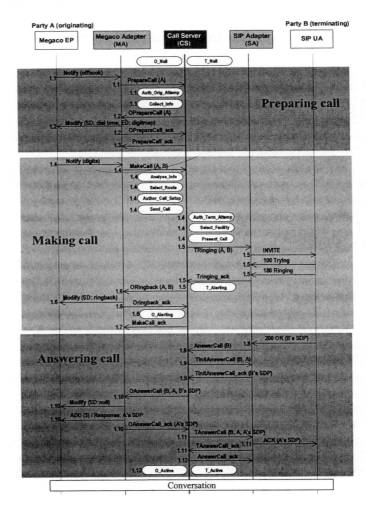

Figure 12.6. The Message Flow and BCSM Transition of Call Setup

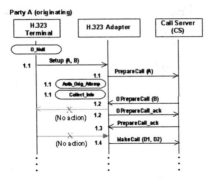

Figure 12.7. H.323 terminal as an originating party

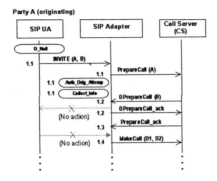

Figure 12.8. SIP UA as an originating party

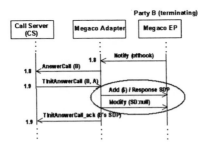

Figure 12.9. Megaco end point as an terminating party

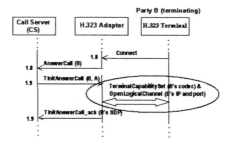

Figure 12.10. H.323 terminal as a terminating party

3.2   Call Transfer

Figures 12.11 and 12.12 illustrate the message flow for call transfer involving a Megaco end point (Party A), a SIP user agent (Party B) and an H.323 terminal (Party C). Initially, a call is set up between Party A and Party B. This call session is referred to as Call 1 (the BCSM states are illustrated by white ellipses in Figure 12.11). Then Party B holds Call 1 and makes a call to Party C (Call 2; where the BCSM states are illustrated by black ellipses in Figure 12.11). Finally, Party B makes a transfer call request (by handing up the phone or pushing the transfer button). Then the Call Server releases Call 1 and Call 2, and connects Party A and Party C (this call session is referred to as Call 3; where the BCSM states are illustrated by dashed ellipses in Figure 12.12). The call transfer procedure consists of four phases: holding call phase (Steps 2.1 ~ 2.4), consulting call phase (Steps 2.5 ~ 2.8), answering call phase (Steps 2.9 ~ 2.13) and transferring call phase (Steps 2.14 ~ 2.23). When call transfer is performed, the initial BCSM states of Call 1 are O_Active and T_Active, respectively.

Step 2.1: Party B sends a call hold event through the SIP INVITE message to the SIP Adapter. By setting the connection address to 0.0.0.0 [32], Party A will be placed on hold. This message triggers the preparing call procedure and the SIP Adapter forwards this request to the Call Server using the HoldCall message in the ICallControl interface.

Step 2.2: The Call Server informs Party B to hold the call. That is, it sends the THoldCall message in the IProtocolControl interface to the SIP Adapter. The SIP Adapter exchanges the SIP messages 200 OK and ACK with Party B. Then the THoldCall_ack is sent back to the Call Server, which triggers the T-BCSM of Call 1 to enter the T_Supsended state.

Step 2.3: The Call Server informs Party A to hold the call. It sends the OHoldCall message to the Megaco Adapter. The Megaco Adapter forwards this message to Party A using the Megaco Subtract message, which disconnects the RTP connection to Party A. The O-BCSM of Call 1 enters O_Supsended.

Step 2.4: The HoldCall_ack message is sent from the Call Server to the SIP Adapter and the holding call phase is complete.

Step 2.5: Party B makes a new call to Party C. That is, Party B sends a SIP INVITE message with the address of Party C to the SIP Adapter. This message triggers the consulting call procedure. The SIP Adapter forwards this request to the Call Server using the ConsultCall message. The Call Server creates the BCSMs of Call 2 and changes the call states step by step (see Step 1.4 in Section 3.1). The final O-BCSM and the T-BCSM of Call 2 are Send_Call and Present_Call, respectively.

Step 2.6: The Call Server rings Party C. It sends the TRinging message to the H.323 Adapter. This message is translated into the H.323 Setup message and is sent to Party C. Party C replies the H.323 Alerting message. The T-BCSM of Call 2 enters T_Alerting.

Step 2.7: The Call Server rings back Party B. It sends the ORingback message to the SIP Adapter and the SIP Adapter forwards this request to Party B using the SIP 180 Ringing message. The O-BCSM of Call 2 enters O_Alerting.

Step 2.8: The ConsultCall_ack message is sent from the Call Server to the SIP Adapter and the consulting call phase is complete.

Steps 2.9-2.13: Party B and Party C answer the call. These steps are similar to Steps 1.8-1.12 in Section 3.1 for answering the call. The final states of the O-BCSM and the T-BCSM of Call 2 are O_Active and T_Active, respectively.

Step 2.14: Party B sends a SIP extension message REFER [33] to transfer the call. The SIP Adapter forwards this message to the Call Server using TransferCall and triggers the transferring call procedure.

Steps 2.15-2.18: The Call Server releases Call 2 and Call 1 with following steps. The Call Server informs Party B to release the originating portion of Call 2 and resets the O-BCSM of Call 2 to O_Null (Step 2.15), then the Call Server informs Party C to release the terminating portion of Call 2 and resets the T-BCSM of Call 2 to T_Null (Step 2.16). Next, the Call Server informs Party A to release the originating portion of Call 1 and resets the O-BCSM of Call 1 to O_Null (Step 2.17), then the Call Server informs Party B to release the terminating portion of Call 1 and resets the T-BCSM of Call 1 to T_Null (Step 2.18). After Call 1 and Call 2 are released, the Call Server creates the BCSMs of Call 3 and changes the call states just like Step 1.4 in Section 3.1. The final O-BCSM and the T-BCSM of Call 3 are Send_Call and Present_Call, respectively.

Step 2.19: Since Party A and Party C already picked up the phones, the Call Server needs not to ring both parties. The O-BCSM and the T-BCSM of Call 3 directly enter O_Alerting and T_Alerting, respectively.

Steps 2.20-2.22: These steps are similar to Steps 2.10-2.12 for answering the call.

Step 2.23: The TransferCall_ack message is sent from the Call Sever to the SIP Adapter and the transferring call phase is complete. The final states of the O-BCSM and the T-BCSM of Call 3 are O_Active and T_Active, respectively.

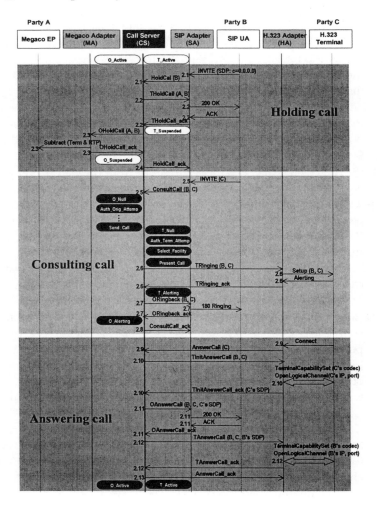

Figure 12.11. The Message Flow and BCSM Transition of Call Transfer (part 1)

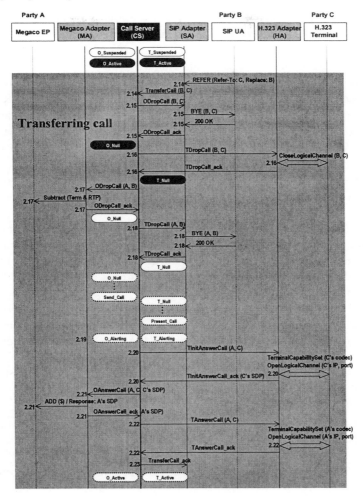

Figure 12.12. The Message Flow and BCSM Transition of Call Transfer (part 2)

3.3 Inter-Softswitch Call

In a call setup, the call parties may be located at different softswitches. The current CCL Softswitch implementation utilizes SIP as the inter-softswitch protocol. That is, the protocol exercised between two SIP Adapters is the same as the protocol exercised between a SIP Adapter and a SIP user agent. The routing information of CCL Softswitch resides at a routing table (see (4) in Figure 12.3). This table is referenced by the Call Server when the O-BCSM of the Call Server enters Select_Route. Based on the routing information, the Call Server will decide whether to route the call to a local extension or to a remote softswitch.

Figure 12.13 illustrates an example of inter-softswitch call, where Party 2001 of Softswitch A makes a call to Party 5001 of Softswitch B. Routing tables A and B (see (a) and (b) in Figure 12.13) list the numbering plans and the routing information of Softswitch A and B, respectively. The extension numbers of Softswitch A are prefixed with 2 and 3, and the extension numbers of Softswitch B are prefixed with 5 and 6.

Steps 3.1: Party 2001 picks up the phone and sends a make call request to the Call Server through the Megaco Adapter. The call setup follows Steps 1.1-1.4 in Figure 12.6.

Step 3.2: The Call Server looks up Routing Table A and decides to route the call to Softswitch B (see Figure 12.13 (a)). The Call Server changes the originating party number "2001" and the terminating party number "5001" to SIP Request-URI (Uniform Resource Identity) "sip:2001@A_IP" and "sip:5001@B_IP", respectively (where A_IP and B_IP indicate the IP address of Softswitch A and Softswitch B). The Call Server of Softswitch A treats this call as being initiated by a local Megaco end point (Extension 2001) to a remote SIP user agent (sip:5001@B_IP).

Step 3.3: The Call Server sends the call setup request to the SIP Adapter. The SIP Adapter translates this request to SIP message INVITE and sends it to Softswitch B.

Step 3.4: The SIP Adapter of Softswitch B receives the INVITE message, and this message triggers the prepare call procedure following Steps 1.1-1.4 in Figure 12.8.

Step 3.5: The Call Sever looks up Routing Table B and decides to route the call to Party 5001 (see Figure 12.13 (b)). The Call Server of Softswitch B treats this call as being initiated by a remote SIP user agent (sip:2001@A_IP) to a local Megaco end point (Extension 5001).

Step 3.6: The Call Server sends a make call request to the Party 5001 through the Megaco Adapter. The call setup follows Step 1.5 in Figure 12.6 except that the terminating party is a Megaco end point.

Step 3.7: Party 5001 answers the call. The answer call procedure follows Steps 1.8 and 1.9 in Figure 12.9.

Step 3.8: The SIP Adapter of Softswitch B sends a SIP message 200 OK to the SIP Adapter of Softswitch A.

Step 3.9: The SIP Adapter of Softswitch A receives SIP message 200 OK, and this message triggers the answering call procedure following Steps 1.8-1.12 in Figure 12.6.

Step 3.10: The SIP Adapter of Softswitch A sends a SIP message ACK to the SIP Adapter of Softswitch B. Then Softswitch B completes the answering call phase following Steps 2.11-2.13 in Figure 12.11 except that the terminating party is a Megaco end point.

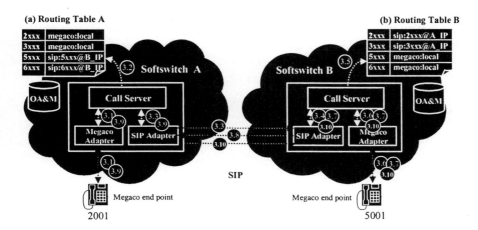

Figure 12.13. Inter-Softswitch Call

4. Conclusion

Softswitch has been proposed as an intelligent call control and multi-protocol supported mechanism for VoIP services, which is utilized in the third generation mobile all-IP networks. In this chapter, we described design and implementation of a softswitch (3G call agent) based on the ISC reference architecture. The ISC architecture provides an interoperability view of softswitch, which has been followed by many softswitch developers. Then we proposed a softswitch design implemented in CCL/ITRI

The CCL Softswitch focuses on the Control & Signaling Plane of the ISC architecture. It consists of Protocol Adapter, Softswitch SPI, Call Server, Softswitch API and Service Agent. The CCL Softswitch uses the IN call model in order to access existing IN services and to interwork with the IN SCP. The design of the Softswitch SPI ensures that new VoIP protocols can be added to the system without modifying the Call Server. We also illustrated the message flows of the interaction between the Call Server and the Softswitch SPI. Through the call setup and call transfer procedures between different VoIP end terminals. We also showed that the inter-softswitch calls can be easily implemented in the CCL Softswitch.

Acknowledgement

S.-Y. Chang, J.-Y. Huang and M.-H. Chen provided useful information regarding CCL Softswitch architecture. H.-C. Hsu and Y.-C. Liao provided information about VoIP protocols and call setup procedures..

References

1 3GPP, "Service and Systems Aspects; Architecture for an All IP Network," Tech. rep. 3G TR 23.922 v. 1.0.0 (1999-10), 1999.
2 3GPP, "Service and Systems Aspects; Network Architecture," Tech. spec. 3G TS 23.002 v. 5.3.0 (2001-06), 2001.
3 ITU, "Packet-Based Multimedia Communications Systems," ITU-T H.323, 2000.
4 J. Rosenberg, et al., "SIP: Session Initiation Protocol," IETF RFC 3261, Jun. 2002.
5 M. Arango, et al., "Media Gateway Control Protocol (MGCP) Version 1.0," IETF RFC 2705, Oct. 1999.
6 F. Cuervo, et al., "Megaco Protocol Version 1.0," IETF RFC 3015, Nov. 2000.
7 I. Stevenson, et al., *Softswitches: The Keys to the Next-Generation IP Network Opportunity*, Ovum Ltd., 2001
8 Y.-B. Lin and I. Chlamtac, *Wireless and Mobile Network Architectures*, Wiley, 2001.
9 http://www.softswitch.org
10 ITU, "Signaling System No. 7 – ISDN User Part general functions of messages and signals," ITU-T Q.762, 1999.
11 ETSI/TC, "Mobile Application Part (MAP) Specification," v. 7.3.0. Tech. Rep. GSM 09.02, 2000.
12 http://www.java.org/jain/
13 http://www.parlay.org
14 ITU, "Interface Recommendation for Intelligent Network Capability Set 2," ITU-T Q.1228, 1999
15 3GPP, "Core Network; Customised Applications for Mobile network Enhanced logic (CAMEL) Phase 3 – Stage 3," Tech. spec. 3G TS 23.078, v. 5.0.0 (2002-06), 2002.
16 http://www.lucent.com

17 F. Erfurt, "How to make a Softswitch Part of the Distributed World of Converged Networks," Intelligent Network Workshop, 2001.

18 http://www.telcordia.com

19 D. Box, *Essential COM*, Addison Wesley, 1997.

20 http://www.corba.com

21 ITU, "Distributed functional plane for intelligent network Capability Set 2," ITU-T Q.1224, 1999.

22 K. Geiger, *Inside ODBC*, Microsoft Press, 1995.

23 C. Sells, Win32 Telephony Programming: A Developer's Guide to TAPI, Addison Wesley, 1998.

24 3GPP, "Core Network; Open Service Access (OSA); Application Programming Interface (API)," Tech. spec. 3G TS 29.198, v. 4.1.0 (2001-06), 2001.

25 ITU, "ISDN user-network interface layer 3 specification for basic call control," ITU-T Q.931, 1999.

26 ECMA, "Services for Computer Supported Telecommunications Applications (CSTA) Phase III," ECMA-269, 1998.

27 K. V. Vemuri, "SPHINX: A Study in Convergent Telephony," IP Telecom Services Workshop 2000, 2000.

28 S. Kapur, "Approach For Services in Converged Network," IP Telecom Services Workshop 2000, 2000.

29 S. L. Chou and Y. B. Lin, "Computer Telephony Integration and Its Applications," *IEEE Communication Surveys & Tutorials*, Vol. 3, 2000.

30 http://www.omg.org/uml

31 M. Handley and V. Jacobson, "SDP: Session Description Protocol," IETF RFC 2327, Apr. 1998.

32 J. Rosenberg, et al., "An Offer/Answer Model with the Session Description Protocol (SDP)," IETF RFC 3264, Jun. 2002.

33 R. Sparks, "The SIP Refer Method," draft-ietf-sip-refer-06, IETF, Jul. 2002.

Chapter 13

CLUSTERING IN MOBILE WIRELESS AD HOC NETWORKS

Issues and Approaches

Ekram Hossain[1], Rajesh Palit[1] and Parimala Thulasiraman[2]

[1] *Department of Electrical and Computer Engineering*
University of Manitoba
Winnipeg, MB, Canada R3T 5V6
{ekram, rpalit}@ee.umanitoba.ca

[2] *Department of Computer Science*
University of Manitoba
Winnipeg, MB, Canada R3T 2N2
thulasir@cs.umanitoba.ca

Abstract

Wireless mobile ad hoc networks consist of mobile nodes which can communicate with each other in a peer-to-peer fashion (over single hop or multiple hops) without any fixed infrastructure such as access point or base station. In a multi-hop ad hoc wireless network, which changes its topology dynamically, efficient resource allocation, energy management, routing and end-to-end throughput performance can be achieved through adaptive clustering of the mobile nodes. Impacts of clustering on radio resource management and protocol performance in a multi-hop ad hoc network are described, and a survey of the different clustering mechanisms is presented. A comparative performance analysis among the different clustering mechanisms based on the metrics such as cluster stability, load distribution, control signaling overhead, energy-awareness is performed.

Keywords:

Mobile ad hoc networks, adaptive clustering, radio resource management, routing.

1.   Introduction

Rapid developments in the portable electronic device technology have made the communication devices more compact, powerful and low-cost. Furthermore, the recent advances in wireless communications technology have spawned an increasing demand for various services to the nomadic users over mobile networks. The aim of the future-generation wireless mobile systems is to achieve seamless services across both wired and wireless networks under global user mobility. This is paving the way towards rapid development of infrastructureless "self-organizing" mobile networks which are expected to complement the infrastructure-based networks in scenarios where the nature of the communication requires the mobile devices to be adaptive and self-organizing [1].

A mobile ad hoc wireless network (also known as a packet radio network) consists of a set of self-organizing mobile nodes which communicate with each other over wireless links without requiring any fixed infrastructure (Fig. 13.1). A wireless node can communicate with another node that is within its radio range or outside its radio range. If two nodes cannot communicate directly, an intermediate node(s) is used to relay or forward data from the source node to the destination node. The wireless nodes (or devices) vary in their size, communication capabilities, computational power, memory, storage, mobility, and battery capacity (Table 13.1). This heterogeneity affects communication performance and the design of communication protocols. The mobile devices in an ad hoc network should not only detect the presence of the connectivity with neighboring devices or nodes, but also identify the devices' types and their corresponding attributes.

The potential applications of wireless ad hoc networks include instant network infrastructure to support disaster recovery communication requirements, mobile patient monitoring, collaborative computing, vehicle-to-vehicle communication, distributed control, and microsensor networking [2]. A microsensor network is a distributed network of thousands of collaborating tiny devices, which gather multidimensional observations of the environment [3]. Many of the necessary components and technologies for microsensor networks are already available. Microscopic micro-electrical mechanical system (MEMS) motion sensors are routinely fabricated on silicon. Entire radio transceivers, including the associated digital electronics, have been fabricated on a single chip. Microsensors promise to revolutionize spatial data gathering. Driven by data aggregation - the fusion of multiple observations from different perspectives - a spatially distributed network of microsensor nodes returns a rich, high resolution, multidimensional picture of the environment. In

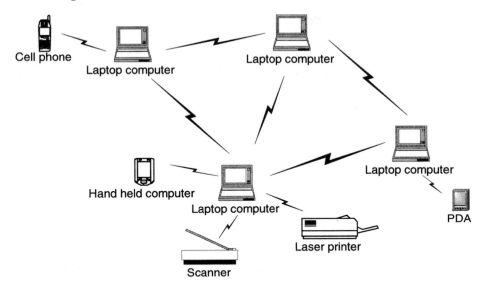

Figure 13.1. A mobile wireless ad hoc network.

Table 13.1. Wireless devices.

Devices	Example	Communication power	Computational power	Battery capacity	Mobility
Sensors	Temperature, humidity sensors	Low	Low	Low	Medium
Audio/video (remotely administerable)	Remote controller	Low	Low	High	Low
Computers	Pocket PC, laptop	Medium	High	Medium	Medium
Communicators	Mobile phone	Medium	Medium	Medium	High
Vehicular devices		High	High	High	High

microsensor networks, the nodes are self-organized into ad hoc networks, and work in a distributed manner.

Unlike a cellular wireless network, there is no base station (BS) to provide coordination among the mobile nodes in a multi-hop ad hoc wireless network. However, this idea can be extended to multi-hop networks by creating clusters of nodes so that access can be controlled and bandwidth can be allocated in each cluster to provide quality of service (QoS) support [4]. Clustering helps to reduce the overhead due to generation and propagation of routing information, supports multicasting, fault-tolerant routing [5] and efficient network management [6]. Due to the topology changes resulting from node insertion, removal, and mobility, the clusters may need to be constructed periodically. Since all the mobile nodes participate in forming the clusters by broadcasting their identity so that the nodes become aware of the changes in their neighborhood, the cluster formation proceeds in a dynamic and distributed fashion.

'K-hop' clustering occurs when every node in a cluster is reachable with at most K hops from any other node in the cluster. The properties of K-hop clustering are as follows [7]:

- Each cluster is designated a leader or *clusterhead* (CH) which can communicate with any node in that cluster within a single hop. Clusterheads are responsible for routing messages to all the nodes within their cluster. A hierarchical routing or network management protocol can be more efficiently implemented with clusterheads.

  In fact, the concept of clusterhead is an application oriented issue. For example, in a wireless sensor network, a clusterhead may be required to control the sensor nodes which act as slaves. In clusters without any clusterhead, a proactive strategy is used for intra-cluster routing while a reactive strategy is used for inter-cluster routing. However, as the network size grows this strategy may incur significant traffic overhead [5].

  The election of cluster heads for two-hop clusters has $O(n)$ time complexity, where n is the number of nodes in the network [8].

- The nodes at the fringe of a cluster are called *gateway* nodes which typically communicate with gateway nodes of other clusters.

- No two CHs can directly reach each other.

Since the available bandwidth is limited in a mobile ad hoc network, one way to increase the network capacity is to reuse the bandwidth (i.e., "spatial reuse" of channel spectrum). The available bandwidth can be efficiently distributed among the nodes in the different clusters to increase the system bandwidth efficiency. Again, based on the clustering information, the transmission power (and hence the transmission range) of the mobile nodes can be carefully limited, and thereby the battery lives of the mobile nodes can be extended. Also, reduction in transmission range results in reduced interference in the system and hence results in increased system capacity. The main obstacle to seamless communications in ad hoc networks is node mobility which may result in frequent link failures. The impact of link failure on network performance can be also minimized by suitable clustering mechanism.

2. Impact of Clustering on Radio Resource Management and Protocol Performance

- *Clustering and Medium Access Control*: In a mobile ad hoc network, node mobility, vulnerability of the radio channel(s), and the lack of any central coordination give rise to the well known *hidden node* and *exposed node* problems. The medium access control protocols must handle these problems.

  A hidden node is a node which is out of range of a transmitter node (node **A** in Fig. 13.2), but in the range of a receiver node (node **B** in Fig. 13.2) [10]. A hidden node does not hear the data sent from a transmitter to a receiver (node **C** is hidden from node **A**). When node **C** transmits to node **D**, the transmission collides with that from node **A** to node **B**. Obviously, the hidden nodes lead to higher collision probability.

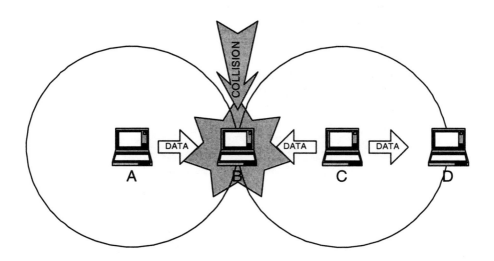

Figure 13.2. Hidden node problem.

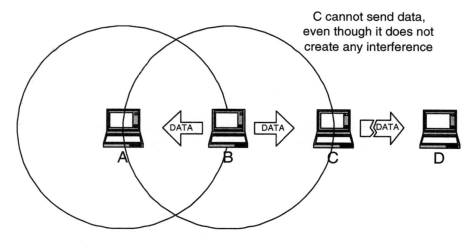

Figure 13.3. Exposed node problem.

An exposed node (node **C** is exposed to **B** in Fig. 13.3) is a node which is out of range of a receiver (node **A**), but in the range of the corresponding transmitter (node **B**). Node **C** defers transmission (to node **D**) upon detecting data from node **B**, even though a transmission from node **C** does not interfere with the reception at node **A**. The link utilization may be significantly impaired due to the exposed node problem.

The hidden node and the exposed node problems have a different look in the multi-channel environments [11]. Let us consider a scenario where two mobile nodes two-hops away from each other are trying to transmit data to a node which is one-hop away from both of the transmitting nodes. If both transmitting nodes use the same channel, the transmissions will be garbled. Even if the targets of these two transmitting nodes are different, there is still a chance of collision. This situation could be avoided if each mobile node would have information on the channels used by the other nodes which are two-hops away so that each transmitting node could use a channel which is different from another transmitting node at least two-hops away. However, realizing this solution in a way so that the least amount of channel bandwidth is used for control signaling is an NP-complete problem [11].

One solution to the above mentioned problem that does not incur huge control overhead is through clustering. By using clustering the network is divided into smaller groups so that the wireless channels can be reused across spatially distributed regions. Most hierarchical clustering architectures for mobile radio networks are based on the concept of CH. The CH acts as a local coordinator for transmissions within the cluster, and is responsible for collecting information on channel use from neighboring clusters and selecting gateways to route packets to them. Therefore, the control message overhead for routing can be significantly reduced. However, sometimes the clusterhead may become a bottleneck of the cluster since it's a central point of administration and a failure in CH degrades the performance of the entire network.

Controlling transmission power is one of the most effective ways to reduce the interference among co-channel transmissions in a multichannel ad hoc wireless network [12]. Based on the clustering of the mobile nodes, the transmission power (and hence transmission range) of the mobile nodes can be controlled. Transmission power control not only reduces interference, but also saves valuable battery power resulting in more energy-efficient MAC protocols. Also, by adapting the transmission range the impact of mobility on the network performance can be reduced to some extent.

- *Clustering and Routing:* The tasks of a routing protocol are to find path/route between a source node and a destination node and maintain the route until the transmission session ends. The most desirable properties of a routing protocol are robustness and stability. There are two extreme routing mechanisms for mobile ad hoc networks – shortest-path routing that is suitable for low rate of topology change, and flooding which is suitable for high rate of topology change. Flooding increases communications overhead and shortest path techniques require each node to maintain up-to-date routing tables. Both of these techniques result in increased co-channel interference and degrade network throughput and response-time performances [13].

Improved system performance can be achieved by using routing protocols designed based on clustering ([13]-[16]). The goal of clustering is to partition the network logically in such a way that slow changes in the actual topology do not affect the logical structure of the network. It helps the routing process by allocating the tasks of establishing and maintaining routes among the mobile nodes. Some of the nodes in each cluster can act as gateways to communicate with the neighboring clusters. The clusterhead, gateways and cluster members collectively participate in the routing process

[2]. A proactive routing algorithm such as Link Vector Algorithm (LVA) [17] can be used for intra-cluster routing where each node in a cluster maintains topology information and routes to every node in its cluster. Routes to destinations outside of a node's cluster are established on a demand basis (i.e., by using reactive routing strategy such as AODV, TORA etc). This type of two level routing strategy reduces the time and control signaling overhead significantly. Clustering also reduces the effect of link failures due to the physical movements of the mobile nodes. If a link along a route fails, the nodes in the particular cluster may establish the link again.

Routing based on clustering may also result in efficient battery power usage at the mobile nodes. Let us consider three mobile nodes **A**, **B**, and **C** which use identical transmitters and receivers (Fig. 13.4) and let T_h be the minimum signal power required at the receiver for successful data reception. If node **A** tries to send data to node **C**, the transmission power must be at least $T_h d_{AC}^n$ (considering signal attenuation only due to path loss with path-loss exponent n, where n is generally ≥ 2). But if the transmission takes place via node **B**, then total power required for the transmission is $T_h d_{AB}^n + T_h d_{BC}^n$. Since $d_{AB}^n + d_{BC}^n < d_{AC}^n$, rather than transmitting directly to the destination node, relaying the transmission through an intermediate node (which serves as a 'gateway' within a cluster) may result in lower transmit power. This trade-off between the transmission power and the number of hops along the route from the source node to the destination node (and hence transmission delay) should be considered while designing a clustering algorithm. Cluster-based routing which involves routing along the gateway nodes (for inter-cluster routing) may therefore result in better usage of battery power.

Again, if a clustering mechanism can exploit the mobility information in a proactive manner, the number of link failures can be reduced and hence better routing performance can be achieved.

- *Clustering and Transport Protocol Performance*

Since the performance of a transport layer protocol is largely impacted by the underlying network and radio link control/medium access control protocol, transport protocol performance improves as the routing and/or wireless channel access performances improve (Table 13.2).

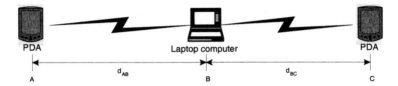

Figure 13.4. Tradeoff between transmision power and number of hops en-route.

Table 13.2. Impact of clustering on protocol performance

	Transport protocol (e.g., TCP)	Since routing and MAC protocol performances improve, transport protocol performance improves.
Clustering	Routing	Cluster-based routing incurs lower control signaling overhead and results in improved system performance.
	MAC	Clustering allows efficient channel resource management and battery power usage.

3. System Model and Performance Metrics for a Clustering Algorithm

In this section, we describe the system model that we use to study the clustering algorithms and the general performance measures for the different clustering algorithms.

3.1 System Model

We consider a multi-cluster multi-hop packet radio network where each node has a unique ID and is capable of adjusting the transmission power. Between a pair of nodes, a direct wireless link can be established only if they are within the transmission range of each other. If a wireless link can be established between two nodes, they are said to be one-hop away from each other. If the nodes have different transmission ranges, there will be both bi-directional and unidirectional links present among the nodes. A mobile node determines the presence of neighboring nodes by using *beacons* (i.e., periodically transmitted broadcast signals). When a node does not receive a certain number of successive beacons from a node, it concludes that the latter node is no longer a neighbor.

Two mobility models, namely, the *random walk mobility model* and the *natural random mobility model* are considered for evaluating the performances of the clustering algorithms. In the random walk mobility model, a mobile node moves from its current location to a new location by randomly choosing a direction and speed to travel [18]. The speed and direction are chosen from the ranges $[minspeed, maxspeed]$ and $[0, 2\pi]$, respectively. Each movement occurs in either a constant time interval or a constant distance traveled, at the end of which new direction and speed are calculated. Since this is a memory-less mobility model, it may generate unrealistic movements.

A more realistic mobility model can be achieved by storing the current mobility information of a node and by allowing only partial change of current mobility pattern. This gives rise to the natural random mobility model for which it is easier to predict the mobility patterns.

The position of a node at the next epoch is determined as follows:
$$x(t + \Delta t) = x(t) + v\Delta t \cos \theta.$$
$$y(t + \Delta t) = y(t) + v\Delta t \sin \theta.$$
For the random mobility case
$$v = \text{random } (maxspeed, minspeed), \text{ and}$$
$$\theta = \text{random}(maxdir, mindir).$$
For the natural mobility model
$$v = \max\{minspeed, min\,(p, maxspeed)\} \text{ and}$$
$$\theta = \max\{mindir, min\,(q, maxdir)\}, \text{ where}$$
$$p = v + v \times \text{random}(acc\_factor, -acc\_factor) \text{ and}$$
$$q = \theta + \theta \times \text{random}(str\_factor, -str\_factor).$$
Here, acc_factor and str_factor denote the maximum amount by which a node change its speed and direction, respectively, in an epoch. In this article we assume that $acc\_factor = str\_factor = 0.2$.

3.2   Performance Metrics

It is desirable that the cluster configuration does not change rapidly when only few nodes are moving and the topology is slowly changing. For more stable clustering the mobility information of the mobile nodes should be exploited. For example, a node with high mobility should not be chosen as a clusterhead. Rather it can be treated as a single node cluster.

Also, since clusterheads involve more computations they deplete the battery power more rapidly compared to the other mobile nodes, therefore, during selection of clusterheads, a clustering algorithm should give preference to nodes which have spent lesser amount of time being cluster-heads. This will result in a more uniform clusterhead load distribution among the nodes.

Since most portable devices are powered by batteries with very limited life time, energy efficiency is one of the biggest factors that need to be considered for designing ad hoc network protocols.

As mentioned earlier, a mobile node needs to broadcast its new cluster information to its neighbors periodically. Therefore, the clustering algorithm should try to minimize the overhead due to control signaling.

The stability, the control signaling overhead involved in maintaining the clusters, and the load distribution among the nodes primarily determine the performance of a clustering algorithm. Generally the following metrics have been used in the literature to evaluate the performance of a clustering algorithm:

- *Number of clusterhead changes*: If a clusterhead moves out of a cluster, a new clusterhead needs to be selected. Frequent changes of clusterhead may incur significant control overhead.

  Let $y_i(t) = \{1, 0\}$ denote whether node i is a clusterhead or not at time t. Then the instability of node i as a clusterhead during a last period of time T can be measured as the number of times the node changes its role between a clusterhead and a cluster member as follows [19]:

  $$z_i(t) = \frac{1}{T} \sum_{k=1}^{n_T} |y_i(t - (k-1)\Delta t) - y_i(t - k\Delta t)| \qquad (13.1)$$

  where $n_T = \frac{T}{\Delta t}$. Then the stability, $s_i(t)$ of node i at time t can be determined as follows:

  $$s_i(t) = \exp\{-z_i(t)\}. \qquad (13.2)$$

  The higher the value of $s_i(t)$ $(0 \leq s_i(t) \leq 1)$, the better is the stability.

- *Number of cluster membership changes*: Each time a node changes its cluster, it has to broadcast this change. Therefore, overhead due to control signaling increases with increasing number of cluster membership changes.

- *Number of link failures among the mobile nodes within a cluster*: Link failures within a cluster generally trigger reclustering which adversely affects system performance. The number of link failures

within a cluster can be reduced by exploiting the mobility information of the nodes in a proactive manner during clustering.

- *Load distribution*: If $q_i(t)$ $(0 \leq q_i(t) \leq 1)$ is the fraction of time a
  node remains a clusterhead and $g(t)$ $(0 \leq g(t) \leq 1)$ is the granularity of clusterheads of the system defined as

$$g(t) = \frac{1}{N} \sum_{i=1}^{N} q_i(t) \qquad (13.3)$$

the clusterhead load distribution $d(t)$ $(0 \leq d(t) \leq 1)$ of the system
can be defined as follows [19]

$$d(t) = 1 - \sqrt{\frac{1}{N} \sum_{i=1}^{N} (q_i(t) - g(t))^2}. \qquad (13.4)$$

A value of $d(t)$ close to 1 indicates that the load distribution among
the nodes is more fair over a certain period of time.

- *Impact of transmission power and node density*: It is also important to observe the impacts of mobile nodes' transmission power
  (and hence transmission range) and node density in the network
  on the performance behavior of a clustering algorithm.

4.   Clustering Algorithms

An ad hoc network can be modeled as a graph $G = (V, E)$, where
two nodes are connected by an edge if they can communicate with each
other. The objective of a clustering algorithm is to find a feasible interconnected set of groups covering the entire node population. A set of
nodes S in $G = (V, E)$ is called a D-hop dominating set if every node
in V is at most D $(D > 1)$ hops away from a vertex in S. For a graph
G and an integer $K \geq 0$, the problem of determining whether G has
dominating set of size $\leq K$ was proven to be NP-complete ([8]-[9]). For
the special family of graphs known as unit disk graphs that represent ad
hoc wireless networks, polynomial time and message complexity approximation solution to the K-clustering problem (where every two wireless
hosts are at most K hops away from each other) can be found [9].

Several of the popular heuristic-based clustering algorithms, namely,
the linked cluster algorithm (LCA) [20], the lowest ID (LID) algorithm
[4], highest connectivity algorithm [7], least cluster change (LCC) algorithm [7], max-min D-clustering algorithm [8], mobility-based adaptive
clustering algorithm (MBAC) [2], and access-based clustering protocol
(ABCP) [5] will be discussed in the following sections.

4.1 Linked Cluster Algorithm (LCA)

The linked cluster algorithm in [20] was proposed as a survivability solution for HF Intra-Task Force (ITF) networks to organize radio-equipped mobile nodes into a reliable network structure and to maintain this structure in the face of arbitrary topological changes. The nodes in a HF ITF network communicate via radio links in the HF band (2-30 MHz) and one important characteristic of such a network is changing topology due to variations in the radio communication range of the nodes.

In the proposed architecture the network is organized into a set of node clusters and each node belongs to at least one cluster. Every cluster has its own clusterhead which acts as a local controller for the nodes in that cluster. The clusterheads control the access of the radio channels. The clusterheads are linked via gateway nodes to connect the neighboring clusters and to provide global network connectivity. The LCA algorithm establishes (from any initial node configuration) and maintains (for mobility of nodes) the logical topology as long as the nodes have not moved so far apart that the network has become disconnected. The LCA algorithm is distributed and does not depend on the existence of any particular node. The algorithm has two logical stages - formation of clusters and linking of the clusters. At the completion of the LCA, each node becomes either an ordinary node, a gateway node or a clusterhead.

The HF band is divided into M subbands and separate runs of the algorithm are performed consecutively for M epochs to obtain the corresponding sets of clusters. The algorithm is run for the ith subband of the HF channel at the ith epoch. During any epoch only one set of linked cluster is organized, the remaining $M-1$ sets are unaffected. Each epoch is divided into two frames (*frame 1* and *frame 2*) and each frame is subdivided into N timeslots, where N is the total number of nodes (Fig. 13.5). The epochs repeat in a cyclic fashion providing a continual updating process. The necessary control messages are transmitted in a separate control channel.

During execution of the LCA algorithm each node maintains the following data structures:

- *heads_one_hop_away* is a list recording those clusterheads that are connected to a node.

- *heads_two_hops_away* is a list of the clusterheads, which are not directly connected, but connected to the neighbors of a node.

- *nodes_heard* is a list that includes all neighboring nodes to which a bi-directional link exists.

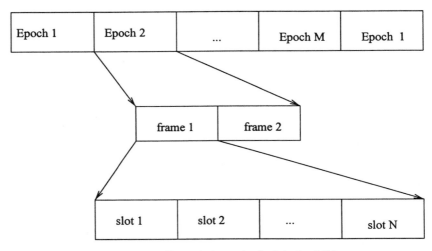

Figure 13.5. Control channel schedule for LCA.

- *connectivity* is a matrix having binary entries. A value of 1/0 in the (i, j) position indicates the existence/absence of a link between nodes i and j.

- *own_head* is the identity of the clusterhead for a given node.

- *node_status* indicates the status (i.e, *ordinary*, *gateway*, or *cluster-head*) of a node.

In each epoch, the algorithm proceeds on a frame-by-frame basis. During the ith slot of *frame 1*, node i broadcasts its *nodes_heard* list (it has heard during the earlier slots of this frame). Therefore, at the end of *frame 1* node i can fill in elements (i, j) in the connectivity matrix where $j > i$. During *frame 2* each node broadcasts its full connectivity information in its assigned slot and node i determines the two-way connectivity of links (i, j) for $j < i$. At the end of *frame 2* all bidirectional links are determined. However, the global connectivity information is not available to every individual node.

Also, at the ith slot of *frame 2* node i also transmits its *node_status*. The clusterhead selection rule is that the node with the highest ID number among a group of nodes is the first candidate to become a clusterhead. At the end of *frame 2* each node is able to fill in its *heads_one_hop_away* and *heads_two_hops_away* lists and each node has at least one clusterhead in its vicinity. Note that, there is only one clusterhead if all nodes are within a distance of one hop from each other.

After the clusters are formed at the end of *frame 2*, a procedure *delete_heads* is used to eliminate redundant clusters. For example, if one cluster covers (i.e., overlaps) another, *delete_heads* eliminates the covered clusterhead.

The clusters are linked by assigning *gateway* status to some nodes that connect adjacent clusters. If two clusterheads are linked directly there is no need for gateways. In case of overlapping clusters where the clusterheads are not directly linked, one gateway node is needed which can be chosen among the nodes in the common intersection region. The gateway selection is performed by procedure *linkup1* and the highest numbered node in the intersection region is chosen to become the gateway. In case of non-overlapping clusters, a gateway node pair must be formed to link the clusterheads and this is performed by procedure *linkup2*. The nodes in a pair with the largest sum of the ID numbers is chosen to be the gateway nodes.

The entire linked cluster algorithm can be described as follows:

```
Process at node i:

Begin
  own_head = self,
  node_status = clusterhead,
  heads_one_hop_away = empty,
  heads_two_hops_away = empty,
  nodes_heard = empty,
  connectivity = identity matrix.

Repeat

  // The following tasks are performed during frame 1 and 2
  // in each epoch
  begin

  // Frame 1 events

  Node i broadcasts its nodes_heard list in slot i.
  In other slots node i receives nodes_heard list
    from its neighboring nodes.
  If node i receives nodes_heard list from node j
    put node j into node i's nodes_heard list.
```

```
If node i was heard by node j
    set connectivity[i, j] = 1.

// Frame 2 events

Node i determines its node status with the information
    collected in frame 1.
In slot i node i broadcasts row i of its connectivity
    matrix and node_status.

If node i receives connectivity/status message from node j
    it fills in row j of its connectivity matrix.
If (j < i)
    connectivity[i, j] = connectivity[j,i].
If status of node j == clusterhead
    include j in heads_one_hop_away list.

End
Until (epoch ends)

// Fill in heads_two_hops_away list

Let node k is the clusterhead for node j.
If node i, which is bidirectionally linked to node j,
is not connected to node k, then add k in heads_two_hops_away
list of node i.
```

The above procedure may produce a few unnecessary clusterheads under some circumstances. A procedure named delete_heads is invoked to remove these redundant clusterheads.

Call procedure linkup1 to select the gateways for overlapping clusters.

Call procedure linkup2 to select the gateways for non-overlapping clusters.

End

The fundamental assumption of the proposed algorithm is that each node maintains a common clock and knows the precise length of each time frame. Again, the number of nodes in the network is assumed to be known *a priori*. If the number of nodes cannot be bounded with certainty, a modification of the algorithm would be necessary to allow the occasional adjustment of the frame length. Another limitation of this algorithm is that it chooses the highest-ID node as clusterhead which may result in an unbalanced load distribution. Since in each frame the nodes have to broadcast their *nodes_heard* list, the control message overhead is relatively high. LCA does not consider the node mobility, adaptive transmission range and power efficiency issues.

4.2   Max-Min D-Clustering Algorithm (MMD)

The max-min D-clustering algorithm proposed in [8] uses a load-balancing heuristic (max-min heuristic) to form D-hop clusters in a wireless ad hoc network so that a fair distribution of load among clusterheads can be ensured. In a D-hop cluster each node is at most D-hops away from the clusterhead. The algorithm aims to avoid the clock synchronization overhead, limit the number of messages sent between nodes to $O(D)$, improve cluster stability and control the density of clusterheads as a function of D. Similar to the LCA, clusterheads are determined based on the node ID.

Execution of the max-min heuristic involves $2D$ rounds of information exchange and each node needs to maintain two arrays *WINNER* and *SENDER* each of size $2D$ node IDs. The *WINNER* and *SENDER* are the winning node ID and the node that sent the winning node ID, respectively, of a particular round. Initially each node sets its *WINNER* to be equal to its ID.

The heuristic has four logical stages - *floodmax*, *floodmin*, determination of clusterheads and linking of clusters. The *floodmax* phase consists of D rounds of information exchange and during each round each node broadcasts its present *WINNER* value to all of its one-hop neighbors and chooses the largest ID as the new *WINNER*. The nodes record the *WINNER* for each round. Therefore, *floodmax* propagates the largest node ID in each nodes D-neighborhood and the node IDs that it leaves at the end are elected as clusterheads. However, it may result in an unbalanced loading for the clusterheads. After *floodmax*, D rounds for the *floodmin* phase start to propagate smaller node IDs. In contrast to *floodmax* each node chooses the smallest rather than the largest value as its new *WINNER*. Any node ID that occurs at least once as a *WINNER* in both phases at an individual node is called a *node pair*. At the end of *floodmin*, each node determines its clusterhead based on the entries in *WINNER* for the $2D$ rounds of flooding using the following rules:

- *Rule 1*: If a node has received its own ID in the second round of flooding, it declares itself a clusterhead. Otherwise Rule 2 is applied.

- *Rule 2*: Among all node pairs, a node selects the minimum node pair to be the clusterhead. If a node pair does not exist for a node then Rule 3 is applied.

- *Rule 3*: The maximum-ID node in the first round of flooding is elected as the clusterhead for this node.

After clusterhead selection each node broadcasts its elected clusterhead to all of its neighbors. After hearing from all neighbors a node can determine whether it is a gateway node or not. If all neighbors of a node have same clusterhead as its own clusterhead, then the node is not a gateway node. If there exist some neighbors with different clusterheads, then the node is a gateway node.

To establish the backbone of the network the gateway nodes begin a converge-cast to link all nodes in the cluster to the clusterhead and link the clusterhead to other clusters. There are certain scenarios, where the max-min heuristic will generate a clusterhead that is on the path between a node and its elected clusterhead. During converge-cast, the clusterhead receiving the message first adopts the node as one of its children and immediately sends a message to the node identifying itself as the new clusterhead.

The time complexity of the heuristic is $O(D)$ rounds and the storage complexity is $O(D)$. Compared to the LCA, the max-min D-clustering algorithm was observed to produce fewer clusterheads, larger sized clusters and longer clusterhead duration on the average. Also, the average clusterhead duration and cluster member duration were observed to be higher for max-min D-clustering compared to those for a connectivity-based clustering. The clusterhead duration was observed to increase with increasing network density. Max-min heuristic results in a backbone with multiple paths between neighboring clusterheads which provides fault tolerance in the network backbone.

Max-min heuristic does not consider node mobility, transmission power and power efficiency explicitly into consideration. More specifically, since it does not take the mobility or node failure into consideration topology changes may cause some nodes to be stranded without clusterheads when the execution of the heuristic is triggered late.

4.3   Lowest ID (LID) Clustering Algorithm

This is a two-hop clustering algorithm. While executing this algorithm, a mobile node periodically broadcasts the list of nodes that it can hear (including itself) (Fig. 13.6). A node which only hears nodes with ID higher than itself from the one-hop neighborhood, declares itself a clusterhead. It then broadcasts its ID and cluster ID (i.e., ID of the clusterhead). A node that can hear two or more clusterhead is a gateway node, otherwise, it is an ordinary node or cluster member.

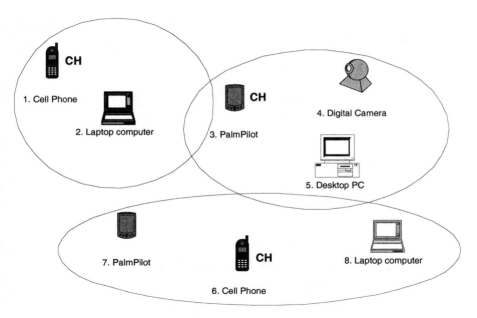

Figure 13.6.   Clustering using lowest-ID algorithm.

The clustering algorithm can be described as follows [4]:

```
Process at node i:
Begin
 // Let S be the set of IDs of one-hop neighbors of node i
```

```
// including itself.
// A clustering message contains node ID and  cluster ID.

Step 1: // This step is executed when node i has the minimum ID in S

        1.a Set i as node i's own cluster ID.
        1.b Broadcast a cluster message with (i, i).
        1.c Remove i from S.

Step 2: // This step is executed when a clustering message is receive
        // at node i.

        2.a Update the clustering information table (at node i) with
            the received data and remove sender's node ID from S.

        2.b If the sender node is a CH and i's cluster ID is UNKNOWN o
            sender's cluster ID is less than i's cluster ID

            set the i's cluster ID to sender's ID.

        2.c If i is the minimum node ID in S and i's cluster ID
            is still UNKNOWN

            set the i's cluster ID to i,
            remove i from S,
            broadcast a cluster message (i, i).

        2.d Repeat step 2 until S is empty.

End
```

4.4 Highest Connectivity (HCN) Clustering Algorithm

Connectivity or degree of a mobile node refers to the number of nodes in its one hop neighborhood. Each node broadcasts the list of nodes that it can hear (including itself). A node is elected as a clusterhead if it is the most highly connected node of all its 'uncovered' neighbor nodes (in case of a tie, the lowest ID node is chosen as the clusterhead) (Fig. 13.7). A node which has not elected its clusterhead yet is an 'uncovered' node, otherwise it is a 'covered' node. A node, which has already elected another node as its clusterhead, gives up its role as a clusterhead.

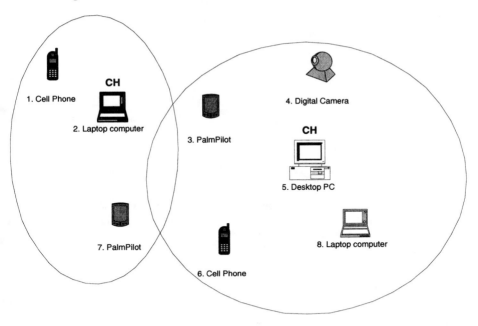

Figure 13.7.   Clustering using highest connectivity algorithm.

A node gathers connectivity information about its neighbors and uses it for cluster formation. Compared to the LID-based clustering the message overhead is higher, because in LID-based clustering only the node ID information are used. However, for a connectivity-based clustering algorithm the clusterhead change is more frequent, and therefore, the load distribution is more fair.

The algorithm can be described as follows:

```
Process at each node i:
Begin
 // Let S be the set of IDs of one-hop neighbors of node i
 // including itself.
 // A clustering message contains node ID and  cluster ID.

 Step 1: // This step is executed when node i has the highest
         // degree in the neighborhood. If there are more than
```

```
// one node with the highest degree, then node i has the
// lowest ID among these nodes.

1.a Set i as node i's cluster ID.
1.b Broadcast a clustering message with (i, i).
1.c Remove i from S.
```

Step 2: `// This step is triggered when a cluster message is received`
`        // at node i.`

> 2.a Update the clustering information table (at node i) with the received data and remove sender's node ID from S.
>
> 2.b If the sender node is a clusterhead and node i's cluster I is UNKNOWN or sender's degree is higher than that of node i's present clusterhead
>
>   set the i's cluster ID to sender's ID.
>
> 2.c If i is the lowest ID which has highest degree in S and node i's cluster ID is still UNKNOWN
>
>   set the i's cluster ID to i.

End

4.5   Least Cluster Change (LCC) Algorithm

Since frequent cluster changes adversely affect the performance of radio resource allocation and scheduling protocols, cluster stability is a major consideration for designing a clustering algorithm. With a view to increasing cluster stability, the LCC algorithm assumes that clusterheads may change only under either of the following two conditions: when two clusterheads come within the transmission range of each other, or when a node loses its membership in any other cluster.

The LCC mechanism can be described as follows:

- At the beginning, the lowest-id or highest connectivity clustering algorithm is used to form initial clusters.

- When a non-clusterhead node in cluster i moves into cluster j, there are no changes for cluster i and cluster j in terms of clusterhead (only the cluster membership changes).

- When a non-clusterhead node moves out of its cluster and does not enter into any existing cluster, it becomes a new clusterhead, forming a new cluster.

- When clusterhead $c(i)$ from cluster i moves into cluster j, it challenges the corresponding clusterhead $c(j)$. Either $c(i)$ or $c(j)$ will give up its clusterhead position according to lowest-id or highest connectivity scheme.

- Nodes which become separated from a cluster, will recompute the clustering according to lowest-id or highest connectivity scheme.

The main objective of this algorithm is to minimize the number of clusterhead changes. The stability of the clusters increases significantly compared to that due to only LID-algorithm or HCN-algorithm. The complexity of the algorithm depends on the technique that is used to form the clusters (i.e., LID or HCN). Since the number of clusterhead changes is minimum here, the load distribution would be more unfair.

4.6 Mobility-Based Adaptive Clustering (MBAC)

The mobility-based adaptive clustering algorithm in [2] uses an estimate of path availability (which changes due to node mobility) for organizing clusters dynamically. This scheme does not use the concept of clusterhead. The proposed adaptive clustering framework supports an adaptive hybrid routing mechanism which can be more responsive and effective when mobility rates are low and more efficient when mobility rates are high. The distributed asynchronous clustering algorithm maintains clusters which satisfy the (α, t) criterion, that is, there is a probabilistic bound α on the mutual availability of paths among all nodes in the cluster over a specified interval of time t.

Let $P_{m,n}^k(t)$ be the status of the path k from node n to node m at time t. $P_{m,n}^k(t) = 1$ if all the links in the path are active at time t, and $P_{m,n}^k(t) = 0$ if one or more links in the path are inactive at time t. The path availability $\pi_{m,n}^k(t)$ between the two nodes n and m at time $t \geq t_0$ is given by: $\pi_{m,n}^k(t) = Pr\left(P_{m,n}^k(t_0 + t) = 1 | P_{m,n}^k(t_0) = 1\right)$.

Path k is defined as an (α, t) path if and only if $\pi_{m,n}^k(t) \geq \alpha$. If nodes n and m are mutually reachable over (α, t) paths, they are said to be available. An (α, t) cluster is a set of (α, t) available nodes.

The cluster parameters α and t are tightly coupled. The parameter α controls the cluster's inherent stability. Larger values of t imply better cluster stability and reduce the computational requirements of cluster maintenance. However, since large values of t will reduce the path availability between nodes of a cluster for the same mobility patterns, they will tend to result in smaller clusters. The parameter α should be chosen considering the traffic intensity and QoS requirements of the connections routed through the clusters. Under the assumption that the path availability is an ergodic process, α represents the average portion of time an (α, t) path is available to carry data, and hence α determines the lower bound on the effective capacity of the path over an interval of length t. Modeling each node as an independent M/M/1 queue, and assuming that t is identical at each node in a cluster, the lower bound on path availability can be found as follows:

$$
t \;\geq\; \frac{1}{\alpha C\mu - \lambda}
$$
$$
\Longrightarrow \alpha \;\geq\; \left(\frac{1 + \lambda t}{\mu t C}\right) \tag{13.5}
$$

where C is the link capacity (in bits/s), $1/\mu$ is the mean packet length (in bits), λ is the aggregate packet arrival rate, $\alpha C\mu$ is the effective service rate, and $\frac{1}{\alpha C\mu - \lambda}$ is the mean packet delay.

The (α, t) cluster algorithm is event-driven and requires the clustered nodes to determine whether or not the (α, t) criteria continues to be satisfied following a topological change. Nodes can asynchronously join, leave, or create clusters. A timer (referred to as the α timer) is maintained at each node which determines the maximum time t for which the node can guarantee path availability to each destination node in the cluster with probability $\geq \alpha$. If any one of the paths is found to be no longer an (α, t) path, the node leaves the cluster. The events that drive the (α, t) cluster algorithm are *node activation*, *link activation*, *link failure*, *expiration of the α timer* and *node deactivation*.

- *Node activation*: To join a cluster, an activating node (or source node) has to obtain the topology information for the cluster from its neighbors and determine the (α, t) availability of all the destination nodes in that cluster. If it is unable to join a cluster, it creates its own cluster (*orphan cluster*).

- *Link activation*: Link activation is triggered when an orphan node attempts to join a cluster. The node receives the cluster topology information and evaluates cluster feasibility and depending upon the outcome of the evaluation it either joins the cluster or returns to its orphan cluster status.

- *Link failure*: When triggered, this event causes a node to determine if the link failure has caused the loss of any (α, t) paths to the destinations in the cluster. A link failure can be detected by a node through the network-interface layer protocol or a topology update which reflects a link failure. The link failure information is forwarded by each node to the remaining cluster destinations and each node receiving this topology update reevaluates the (α, t) path availability. If the node detects that a destination has become unreachable, then it removes that destination from its routing table. If it finds that none of the nodes within the cluster is (α, t) reachable, it leaves the cluster.

- *Expiration of α timer*: Each node in a cluster periodically estimates the path availability to each destination node in the cluster. This is controlled by the α timer and based on the topology information available at each node. The actions taken by a node upon the expiration of its α timer are similar to those due to link or node failure except that the α timer triggers an orphan node to reattempt to join a cluster in a way identical to link activation.

- *Node deactivation*: A node can gracefully deactivate or depart voluntarily from the cluster by announcing its departure through a topology update message to all nodes in the cluster. If a node fails suddenly or becomes disconnected from the cluster due to mobility, it becomes an orphan node and proceeds according to the rules of node activation.

To evaluate path availability, which is the basis for (α, t) cluster management, a random walk-based mobility model is assumed where each node's movement consists of a sequence of random length intervals called mobility epochs during which a node moves in a constant direction at a constant speed. To characterize the availability of a link between two nodes during a period of time $(t_0, t_0 + t)$, mobility distribution of a single node is first determined, and then it is extended to derive the joint mobility distribution which can be used to determine the link availability distribution. Assuming that the links along a path between two nodes fail independently, the path availability can be determined as the product of the individual link availability metrics.

The mobility profile of a node n can be expressed by three parameters: λ_n, μ_n, and σ_n^2, where μ_n and σ_n^2 are the mean and variances of the speed during each epoch, and $1/\lambda_n$ is the mean epoch length (where the epoch lengths and the speeds during each epoch length are assumed to be identically and independently (i.i.d.) distributed). If $\overrightarrow{R}_n(t)$ is the random mobility vector for node n, where the magnitude $R_n(t)$ and the phase angle θ_n represent the aggregate distance and direction of the mobile node, respectively, the distributions of $R_n(t)$ and θ_n are given by

$$Pr(\theta_n \leq \phi) = \frac{1}{2\pi}\phi, 0 \leq \phi \leq 2\pi$$

$$Pr(R_n(t) \leq r) \approx 1 - exp\left(\frac{-r^2}{\alpha_n}\right), 0 \leq r \leq \infty \qquad (13.6)$$

with $\alpha_n = (2t/\lambda_n)(\sigma_n^2 + \mu_n^2)$. Therefore, the random mobility vector has Rayleigh distributed magnitude and uniformly distributed direction.

Now, let $(\lambda_m, \mu_m, \sigma_m^2)$ and $(\lambda_n, \mu_n, \sigma_n^2)$ describe the mobility profiles for two mobile nodes m and n. If the random mobility vectors for these nodes are $\vec{R}_m(t)$ and $\vec{R}_n(t)$, respectively, the random mobility vector of node m with respect to node n is $\vec{R}_{m,n}(t) = \vec{R}_m(t) - \vec{R}_n(t)$ which is approximately Rayleigh distributed (with parameter $\alpha_{m,n} = \alpha_m + \alpha_n = \frac{2t}{\lambda_m}(\sigma_m^2 + \mu_m^2) + \frac{2t}{\lambda_n}(\sigma_n^2 + \mu_n^2)$) and has a uniformly distributed direction.

Based on the above joint node mobility model and the initial status and location the link availability (i.e., the probability that there is an active link between two nodes at time $t_0 + t$, given that there is an active link between them at time t_0) between nodes m and n can be determined. If node m becomes active at time t_0 within a uniform random distance from node n, the distribution of the link availability $A_{m,n}$ between node m and node n can be approximated as follows [2]:

$$A_{m,n} \approx 1 - \Phi\left(\frac{1}{2}, 2, \frac{-4R^2}{\alpha_{m,n}}\right) \qquad (13.7)$$

where $\Phi(a, b, z)$ is the Kummer-confluent hypergeometric function and R is the transmission radius of a mobile node.

If a link activates between n and m at time t_0 (due to node mobility) such that m is located at a uniform random point at distance R from n, then the distribution of link availability is given by

$$A_{m,n}(t) = \frac{1}{2}\left(1 - I_0\left(\frac{-2R^2}{\alpha_{m,n}}\right) exp\left(\frac{-2R^2}{\alpha_{m,n}}\right)\right) \qquad (13.8)$$

where I_0 is a modified Bessel function of the first kind.

Based on the link availability $A_{i,j}$ for link $(i, j) \in$ path k, the path availability between node m and n at time $t_0 + t$ can be found as

$$\pi_{m,n}^k(t) = \prod_{(i,j)\in k} A_{i,j}(t_0 + t). \qquad (13.9)$$

With (α, t) clustering strategy, mean cluster size increases/decreases under low/high node mobility. Lower values of α increase the probability of a node being clustered. However, at high node mobility this probability may drop significantly. The mean node residence time within a given cluster and the mean cluster survival time generally decrease with increased node mobility. The control message processing rate per node does not increase monotonically with increase in node mobility. These control messages include routing updates and those required to join and leave clusters. With increase in node mobility, the control message processing rate increases initially due to increase in topology changes and node clustering activity. However, as mobility increases further, it decreases along with the mean cluster size.

4.7 Access-Based Clustering Protocol (ABCP)

In an attempt to minimize the clustering overhead resulting from the control signaling overhead in a hierarchical ad hoc network the access-based clustering proposed in [5] uses MAC layer process for cluster formation. The system model assumes one time-slotted control channel which is used for the exchange of control messages and multiple data channels which are used for data transport. With an intention to form a cluster each node accesses the control channel to send a clusterhead declaration frame and upon successful transmission of this frame it becomes a clusterhead. A node which receives the clusterhead declaration from its neighbor before it declares itself as a clusterhead becomes a member node. When a node becomes a clusterhead with at least one cluster member, it remains to be a clusterhead until it becomes inactive.

The scheme used by the nodes to access the control channel is a three-phase multiple access (TPMA) scheme consisting of *Request to Send* phase, *Collision Report (CR)* phase, and *Receiver Available* phase. The TPMA scheme provides a distributed method for local broadcast of control messages. The simple broadcast request-response coupled with first-come-first-serve selection constitute the access-based clustering protocol (ABCP). Each node has a unique ID and can act either as an ordinary node or as a clusterhead and the ABCP for these two cases are as follows [5]:

```
Begin
// This algorithm is divided into two cases:
// ordinary node case and clusterhead case.
// When a node is turned on, it becomes active
// with the role of an ordinary node.
```

```
//Ordinary node case:
```

1. At the beginning, the cluster ID of the ordinary node
   is UNKNOWN. It tries to join in an existing cluster by
   sending REQ_TO_JOIN message and waits for response.

   If it receives a HELLO message from a clusterhead, it sets
   its cluster ID to the sender's ID and sends a JOIN message
   confirming its membership.

   If it does not receive any response within a certain time
   period, it tries to become a clusterhead by sending HELLO
   message.

2. If an ordinary node gets a DISCONNECT message from its
   clusterhead (or the link between the node and its
   clusterhead weakens), it sets its cluster ID to UNKNOWN
   and tries to become a member of another cluster.

3. An ordinary node becomes inactive simply by sending a
   DISCONNECT message.

```
//Clusterhead case:
```

1. When a clusterhead receives a REQ_TO_JOIN message from
   an ordinary node, it welcomes the sender node with a
   HELLO message. It waits for an interval TIME_OUT_2 to
   receive a JOIN message. If it does not receive a JOIN
   message, it sends the HELLO message again.

2. A JOIN message comes from a node that may belong to the same
   cluster or another cluster. If it is from the same cluster,
   the clusterhead adds the sender to the clusterhead's member
   list. Otherwise it removes the node from the member list if
   there is any entry for that node. A sender of a HELLO or
   DISCONNECT message is also removed from the member list.

3. A clusterhead becomes inactive by sending a DISCONNECT
   message like an ordinary node.

```
End
```

ABCP incurs less control message overhead compared to the ID-based
clustering protocol and has shorter execution time. This is because, in

ABCP, the node which first transmits the HELLO message successfully becomes a clusterhead and the other nodes that receive the HELLO messages become cluster member by sending JOIN messages. Therefore, each node has to send only one control message to complete the cluster initialization. For cluster maintenance, control message is exchanged between the clusterhead and the cluster member. Also, since the number of control messages is independent of the number of nodes, ABCP scales well with respect to clustering overheads.

4.8 Power-Control-Based Clustering (PCBC)

In [21] a power-control-based two-hop clustering algorithm was proposed in which a clusterhead can adjust the cluster size by exercising power control. To become a member of one or more clusters a mobile node tries to detect the pilot signal transmitted by the clusterheads. Pilot signal carry information such as node ID and transmission power level. A clustered node adjusts its transmission power based on the received pilot signal strength. If a node cannot detect pilot transmission from any clusterhead, it can claim to be a new clusterhead by broadcasting initializing pilot signals which are different from the normal pilot signals transmitted from a functional clusterhead.

The main features of this clustering algorithm can be described as follows:

- *Initial clustering*: All the nodes send out their initialization pilot signals with maximum pilot power to acquire their neighborhood information. It uses LID algorithm to form initial cluster.

- *Cluster maintenance*: When a mobile node goes out of the clustered area, it transmits initialization pilots and attempts to become a clusterhead. When a clusterhead goes into a different cluster area, only one clusterhead (the one with higher degree) survives.

- *Clusterhead power control*: A clusterhead adjusts the pilot signal level when it needs to change the size of the cluster. When necessary it also adjusts the power level so that the furthest node can hear it. If a node in the cluster reports high error rates, the clusterhead increases the data transmission power level. It can adjust the power level of a node by closed loop power control.

- *Cluster member power control*: A cluster member can use both an open loop power control and a closed loop power control. If a node experiences high error rate, it reports to the clusterhead so that a closed loop power control can be initiated.

Adaptive transmission power control can result in substantial energy saving. Also it can provide better channel utilization. By controlling the transmission power at the clusterheads a more adaptive network infrastructure can be achieved.

5. Performance Comparison

To observe the impacts of transmission range, node density and node mobility on the general clustering performance we evaluate the performances of the LID, HCN and LCC algorithms in a unified simulation framework. Since the quantitative performances of MBAC, ABCP and PCBC are more dependent on the specific signaling procedure and/or mobility patterns of the nodes, we have not included these algorithms in our simulation framework.

Total area of simulation is assumed to be $1000\ m \times 1000\ m$ and the total number of nodes is varied from 100 to 800. During the course of a simulation the transmission range of each node is kept fixed; however, it is varied from $25\ m$ to $500\ m$ to examine the effects of variable transmission range on the general clustering performance. The mean speed of all nodes is varied from $1\ m/sec$ to $10\ m/sec$ (i.e., $3.6\ kmph$ to $36\ kmph$). The values of *minspeed, mindir, maxdir* are assumed to be 0, 0, 2π, respectively. The value of *maxspeed* for each node is generated as an exponential random variable around the mean speed. The interval of each epoch is $0.25\ sec$. If a node crossed the simulation boundary due to mobility, it is bounced back to the simulation area.

Stability and load distribution are calculated by using the following algorithm.

```
Begin
  // s: Stability, d: Load distribution \\
  // g: Granularity of CHs
  //    a,b,u,v - temporary variables

  For each epoch do
    For each node do
      If node is a CH then
        NumOfCH[node] = NumOfCH[node] + 1.
      End If
      If CH role changes then
        NumOfCHChange[node] = NumOfCHChange[node] + 1.
      End If
    End For
  End For
```

```
// Total Simulation Time = T

T = TotalEpoch * EpochIntval.
a = b = 0.0.

// q is the fraction of time a node remains as CH

For each node do
  q[node] = NumOfCH[node]/TotalEpoch,
  a = a + q[node],
  b = b + Exp(-NumOfCHChange[node]/T).
End For

g = a/NumOfNodes.
s = b/NumOfNodes.

AvgNumOfCluster = a.    // [= g * NumOfNodes]
AvgClusterSize = NumOfNodes/a.    // [= 1/g]

u = 0.0.
For each node do
  v = sqr(q[node] - g),
  u = u + v.
End For
u = u/NumOfNodes.
d = 1 - sqrt(u).
End
```

5.1   Effect of Transmission Range

For all the four techniques, the average cluster size increases as the transmission ranges of the nodes increase. The average cluster sizes for the least cluster change (LCC) algorithms are higher than those for normal LID and HCN algorithms (Fig. 13.8). Compared to the LID-based algorithms, the average cluster size is observed to be higher for HCN-based algorithms. For inter-cluster routing longer transmission range (and hence larger cluster size) helps in minimizing the delay involved in end-to-end transmission. However, as the number of nodes in a cluster increases the cluster members have to maintain comparatively longer look up tables for intra-cluster routing. Moreover, if the cluster size increases the workload on the clusterhead also increases which may not be desirable. In such a case, a significant overhead will be involved in changing clusterheads.

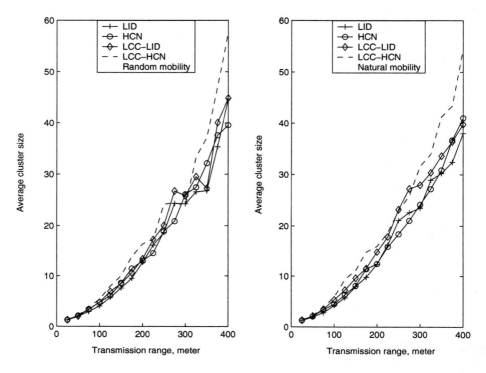

Figure 13.8. Impact of transmission range on average cluster size (for $N = 250$, average node speed $= 2.5$ m/s).

Compared to the pure LID and HCN-based algorithms, both of the LCC algorithms provide better stability across different transmission ranges (Fig. 13.9). Since the LCC algorithm is designed to reduce the number of the clusterhead changes, the average duration for which a node remains clusterhead is higher for the LCC algorithms. The normal HCN algorithm is observed to provide the worst stability. The simulation results reveal that the stability decreases up to a certain transmission range then it again increases.

As the transmission range increases, the probability that a mobile node experiences change in its neighborhood increases. As a result, more clusterhead changes take place for normal LID and HCN algorithms. But the LCC is immune to neighborhood changes unless a node becomes disconnected from its clusterhead. Therefore, in case of LCC the stability increases as the transmission range increases. Note that, at very low

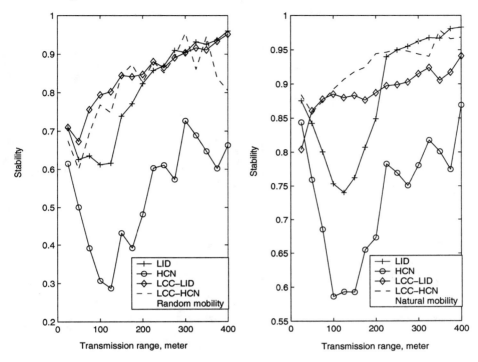

Figure 13.9. Impact of transmission range on cluster stability (for $N = 250$, average node speed $= 2.5\ m/s$).

transmission range the cluster size is very low especially if the node density is not very high. Small clusters experience higher stability and for this reason we observe that for all the algorithms stability is high at very low transmission ranges. Better stability is observed for the natural mobility model compared to that for the random mobility model.

For all the algorithms, as the transmission range increases, the load distribution metric increases except for very small transmission range. Actually at very low transmission range, the cluster size is very small (about one). Therefore, in general, as the range increases, the load distribution becomes more fair (Fig. 13.10). The HCN based algorithms perform better since they are not biased by node ID. In LID or LCA, node ID rather than the neighborhood change biases the clusterhead selection, and therefore, the load distribution may become more uneven.

It is observed that the HCN based LCC algorithm performs better in terms of both stability and load distribution. Better stability and load distribution are achieved with longer transmission range. However,

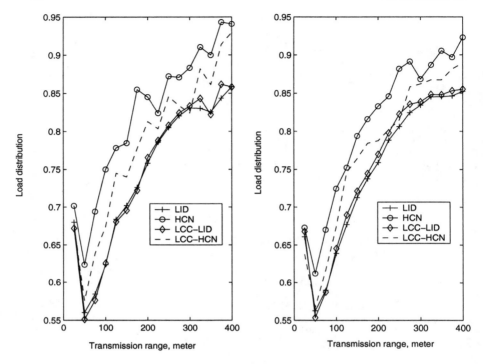

Figure 13.10. Impact of transmission range on load distribution (for $N = 250$, average node speed $= 2.5\ m/s$).

longer transmission range has adverse effect on power consumption and may cause more interference.

5.2 Effect of Node Density

As the node density increases, the average cluster size increases. It is observed that the average cluster size is the largest for LCC-HCN algorithm while it is the smallest for LID algorithm (Fig. 13.11). Also, with normal LCC or HCN algorithms, the stability of the clusters decreases quite rapidly as node density increases. With higher node density, changes in neighborhood of the mobile nodes take place more frequently and since the connectivity of a node is very sensitive to the neighborhood changes, normal HCN algorithms perform the worst in stability measure as the node density increases (Fig. 13.12). The stability measures are observed to be higher for the natural mobility case compared to those in the case of random mobility (Fig. 13.12).

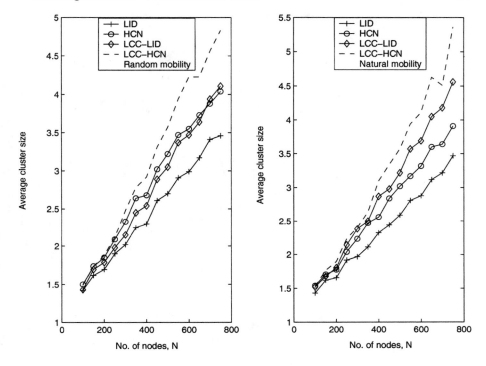

Figure 13.11. Impact of node density on average cluster size (for transmission range = 50 m, average node speed = 2.5 m/s).

Generally, the load distribution becomes more fair as node density increases. Since with the HCN-based algorithms more clusterhead changes take place with increasing node density, the load distribution is observed to be the best for these algorithms (Fig. 13.13). The LID based LCC algorithm performs the worst due to fewer number of clusterhead changes and their biasness towards lowest-ID nodes in selecting clusterheads. The load distribution measures are observed to be lower for the natural mobility case compared to those in the case of random mobility.

5.3   Effect of Node Mobility

In general, under different node speeds the HCN-based algorithms are observed to result in larger clusters compared to those due to the LID-based algorithms (Fig. 13.14). Again, LCC-based algorithms result in larger cluster sizes compared to the normal LID or HCN-based algorithms.

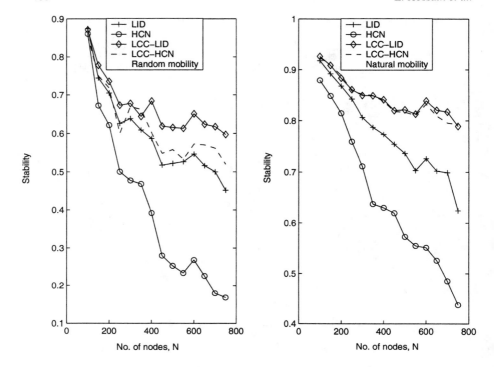

Figure 13.12. Impact of node density on cluster stability (for transmission range = 50 *m*, average node speed = 2.5 *m/s*).

As the mobility/speed of the nodes increases, stability of the clusters decreases in general. The LID-based algorithms provide better stability compared to the HCN-based algorithms as user mobility increases (Fig. 13.15). The performance trend for load distribution is just opposite to that for stability. The HCN-based algorithms provide a more fair load distribution (Fig. 13.16). The HCN-LCC algorithm seems to be a good choice if we consider both the stability and load distribution metrics.

5.4   Summary

The comparative performance among the different clustering algorithms are described qualitatively in Table 13.3.

The LCC algorithm is developed to minimize the clusterhead changes and is known to be the most stable two-hop clustering algorithm. With the LID-based clustering, the clusterhead load is not uniformly distributed among all the nodes. The lower the node ID the more likely

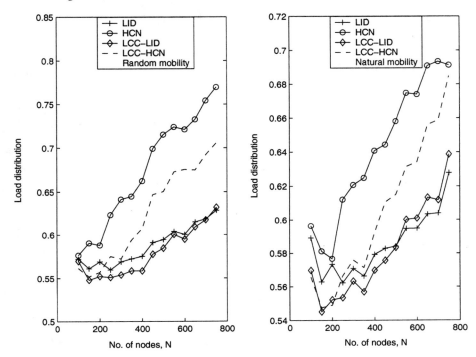

Figure 13.13. Impact of node density on load distribution (for transmission range = 50 *m*, average node speed = 2.5 *m/s*).

it is for the node to become a clusterhead. Since ordering of the node ID plays an important role in this approach, the number of clusterheads in a network varies with the distribution of the node ID. The highest connectivity mechanism aims at reducing the number of clusters at a given time by favoring nodes with largest number of neighbors when it comes to electing clusterheads. As the connectivity of a node may change rapidly as it moves, the connectivity-based algorithm also tends to be less stable.

One of the major limitations of all the above algorithms is that, none of these attempts to use the node mobility information for clustering. In case of link failure due to mobility, the clusters are recomputed for the entire (or some part) of the network.

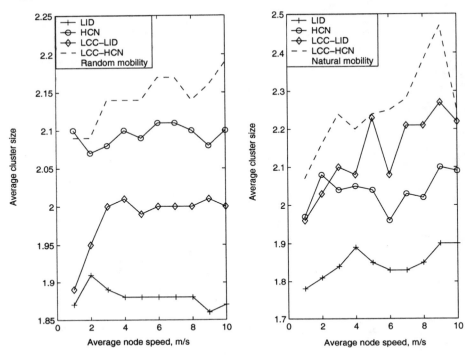

Figure 13.14. Impact of average node speed on average cluster size (for $N = 50$, transmission range $= 50$ m).

Table 13.3. Performance comparison among the clustering algorithms.

Algorithm	Cluster size	Stability	Load distribution	Message complexity	Power awareness	Asynchronous/ synchronous
LCA	No control	Medium	Less fair	High	No	Syn
LID	No control	Medium	Less fair	Low	No	Asyn
HCN	No control	Low	Less fair	High	No	Asyn
MMD	Controlled	High	More fair	High	No	Asyn
LCC	No control	High	More fair	Low	No	Asyn
MBAC	Controlled	Controlled	More fair	High	No	Asyn
ABCP	No control	High	More fair	High	No	Syn
PCBC	Controlled	High	More fair	Low	Yes	Asyn

6.   Chapter Summary

The benefits of network stability, better resource sharing and power efficiency resulting from clustering can be exploited at the MAC, routing

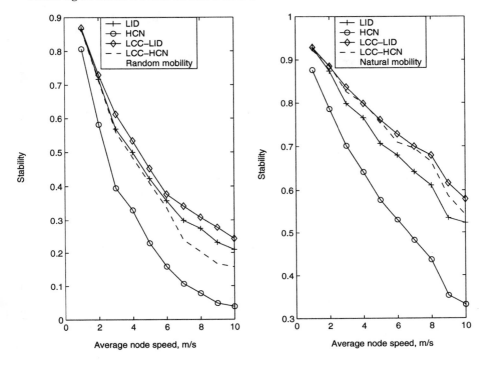

Figure 13.15. Impact of average node speed on cluster stability (for $N = 50$, transmission range $= 50\ m$).

and transport levels to improve the transmission protocol stack performance in a mobile ad hoc network. In this article we have described the performance measures of a clustering algorithm, impact of clustering on protocol performance and provided a summary of the major clustering algorithms proposed in the literature. Quantitative performance results for the three major clustering algorithms, namely, the lowest-ID, highest connectivity and the least cluster change algorithms have been obtained through a unified simulation framework. A qualitative performance comparison among all the clustering algorithms has been presented.

Except MBAC all the clustering algorithms discussed here are reactive (rather than proactive) to the node mobility. Mobility-based proactive clustering mechanisms can provide better fault-tolerant network infrastructure by establishing standby routes and hence better routing performance. Except PCBC all the clustering algorithms use static transmission power (hence fixed transmission range) at the mobile nodes. However, in a clustered architecture the nodes should be able to dy-

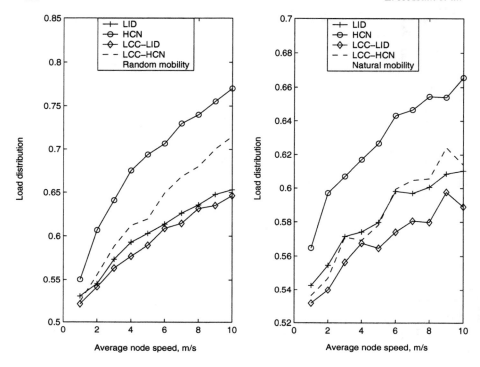

Figure 13.16.   Impact of average node speed on load distribution (for $N = 50$, transmission range $= 50$ m).

namically adjust their transmission power so that battery power can be conserved at the nodes and also unintended interference can be reduced. Clustering mechanisms which are proactive to node mobility and use adaptive transmission power control are desirable. Again, with adaptive transmission range in a clusterhead-based network architecture the load distribution should be made uniform so that the battery power in the nodes deplete in a more fair manner. Therefore, future research in clustering should be directed towards

- designing mobility-aware proactive clustering algorithms under dynamic transmission range control at the mobile nodes

- designing resource management and QoS framework under proactive clustering in wireless mobile ad hoc networks

- optimizing clustering from the transmission protocol stack performance point of view

- evaluating TCP/IP performance under adaptive clustering in mobile ad hoc and integrated cellular-ad hoc networks.

References

[1] Hubaux, J.-P, T. Gross, J.-Y. L. Boudec, and M. Vetterli, "Towards self-organized mobile ad hoc networks: The terminodes project," *IEEE Communications Magazine*, pp. 118-124, Jan. 2001.

[2] McDonald, A. B., and T. F. Znati, "A mobility-based framework for adaptive clustering in wireless ad hoc networks," *IEEE Journal on Selected Areas in Communications*, vol. 17, no. 8, pp. 1466-1487, Aug. 1999.

[3] Min, R., M. Bhardwaj, S.-H. Cho, N. Ickes, E. Shih, A. Sinha, A. Wang, and A. Chandraksan, "Energy-centric enabling technologies for wireless sensor networks," *IEEE Wireless Communications*, Aug. 2002.

[4] Richard, C. L., and M. Gerla, "Adaptive clustering for mobile wireless networks," *IEEE Journal on Selected Areas in Communications*, vol. 15, no. 7, pp.1265-1275, Sept. 1997.

[5] Hou, T.-C., and T.-J. Tsai, "An access-based clustering protocol for multihop wireless ad hoc networks," *IEEE Journal on Selected Areas in Communications*, vol. 19, no. 7, July 2001.

[6] Chen, W., N. Jain, and S. Singh, "ANMP: Ad hoc network management protocol," *IEEE Journal on Selected Areas in Communications*, vol. 17, no. 8, pp. 1506-1531, Aug. 1999.

[7] Gerla, M., and J. T.-C. Tsai, "Multicluster, mobile, multimedia radio network," *ACM/Baltzer Wireless Networks*, vol. 1, no. 3, pp. 255-265, 1995.

[8] Amis, A. D., R. Prakash, T. H. P. Vuong, and D. T. Huynh, "Max-min D-cluster formation in wireless ad hoc networks," *Proc. IEEE INFOCOM'2000*, pp. 32-41.

[9] Fernandess, Y., and D. Malkhi, "K-clustering in wireless ad hoc networks," *Proc. ACM POMC '2002*, Oct. 30-31, 2002, Toulose, France, pp. 31-37.

[10] Chhaya, H. S., and S. Gupta, "Performance of asynchronous data transfer methods for IEEE 802.11 MAC protocol," *IEEE Personal Communications Magazine*, pp. 8-15, Mar. 1996.

[11] Garcia-Luna-Aceves, J. J., and J. Raju, "Distributed assignment of codes for multihop packet-radio networks," *Proc. IEEE MILCOM '1997.*

[12] Rodoplu, V., and T. H. Meng, "Minimum energy mobile wireless networks," *IEEE Journal on Selected Areas in Communications*, pp. 1333-1344, Aug. 1999.

[13] Chiang, C.-C., H.-K. Wu, and M. Gerla, "Routing in clustered multihop, mobile wireless networks with fading channel," Computer Science Department, UCLA.

[14] Krishna, P., N. H. Vaidya, M. Chatterjee, and D. K. Pradhan, "A cluster-based approach for routing in dynamic networks," *Computer Communications Review*, vol. 17, no. 2, Apr. 1997. http://www.acm.org/sigs/sigcomm/ccr/archive/1997/apr97/ccr-9704-krishna.pdf

[15] Sinha, P., R. Sivakumar, and V. Bharghavan, "CEDAR: A core-extraction distributed routing algorithm," *Proc. IEEE INFOCOM '1999*, vol. 1, pp. 202-209.

[16] Krishnan, R., R. Ramanathan, and M. Steenstrup, "Optimization algorithms for large self-structuring networks," *Proc. IEEE INFOCOM '1999*, vol. 1, pp. 71-78.

[17] Toh, C.-K., *Ad Hoc Mobile Wireless Networks: Protocols and Systems*, Prentice Hall, 2001.

[18] Camp, T., J. Boleng, and V. Davies, "A survey of mobility models for ad hoc network research," *Wireless Communications and Mobile Computing*, vol. 2, pp. 483-502, 2002.

[19] Siddiqui, A., and R. Prakash, "Modeling, performance measurement, and control of clustering mechanisms for multi-cluster mobile ad hoc networks," Technical Report (UTDCS-16-01), Department of Computer Science, University of Texas at Dallas.

[20] Baker, D. J., and A. Ephremides, "The architectural organization of a mobile radio network via a distributed algorithm," *IEEE Transactions on Communications*, COM-29(11), pp. 1694-1701, Nov. 1981.

[21] Kwon, T. J., and M. Gerla, "Clustering with power control," *Proc. IEEE MILCOM '1999.*

Chapter 14

CHARACTERIZING UPLINK LOAD
Concepts and Algorithms

Erik Geijer Lundin
Dept. of Electrical Engineering
Linköpings universitet
SE-581 83 Linköping, Sweden
geijer@isy.liu.se

Fredrik Gunnarsson
Dept. of Electrical Engineering
Linköpings universitet
SE-581 83 Linköping, Sweden
fred@isy.liu.se

Abstract   The maximum capacity of a CDMA cellular system's radio interface depends on the time varying radio environment. This makes it hard to establish the amount of currently available capacity. The received interference power is the primarily resource in the uplink. Ability to predict how different resource management decisions affect this spatial quantity is therefore of utmost importance. The uplink interference power is related to the uplink load through the pole equation. In this chapter, we discuss both theoretical and practical aspects of uplink load estimation.

Keywords: Uplink load, Pole capacity, Soft capacity, Hard Capacity

Introduction

The aim of this chapter is to address uplink capacity related theory within the area of CDMA cellular radio systems. A term central to this theory, the *uplink relative load*, will be introduced and explained in Section 1. Since the uplink is in fact interference power limited, and not limited by the amount of hardware in the system, one separates between load measures based on *hard* resources and

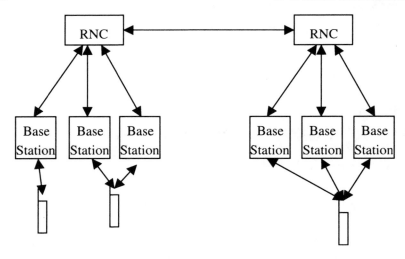

Figure 14.1.   The architecture of a WCDMA system. A radio network controller (RNC) controls
a number of base stations which in turn serve several mobiles.

those based on *soft* resources. What distinguishes a hard resource from a soft is
also explained in the next section. As the total capacity is changing over time,
used capacity should no longer be put in terms of used Erlang capacity, but in
amount of received interference power or a transformation thereof such as the
uplink relative load. Both of these quantities are hard to accurately measure
or estimate. It is therefore much more complicated to determine how loaded a
system is, compared to systems where the capacity is limited by the amount of
hardware.

Sections 2 and 3 study uplink relative load from different angles through theoretical or practical load estimates. Even though the results apply to a general
CDMA cellular system, we will use notation from a WCDMA system throughout the entire chapter. The architecture of a WCDMA system consists of a
number of levels (Holma and Toskala, 2000). Starting from cell level, the services in each cell is provided by a base station, see Figure 14.1. Some resource
management algorithms reside in a base station. An advantage is of course
that a minimum of signaling is required if the decisions can be made on this
intermediate level. Relative load estimates that are designed to reside in base
stations are referred to as *decentralized estimates*. Theory related to such estimates is presented in Section 2.

A base station has no access to information regarding the situation in cells
supported by other base stations. Thus, a decentralized estimate can not predict
how resource management decisions will affect cells supported by other base
stations. Figure 14.1 also shows how signals from different base stations are

combined further up in the hierarchy in a *radio network controller, RNC*. A radio network controller receives information gathered in several cells, even from cells controlled by other RNCs. This more complete information about the system as a whole can be used by relative load estimates in the RNC. Such estimates are herein referred to as *centralized estimates* and are handled in Section 3.

The theoretical total capacity, the *pole capacity*, is interesting to study for comparative purposes, even though it is not achievable in practice. As different estimates of the uplink relative load are presented, corresponding estimates of the pole capacity are discussed. Finally, Section 4 discusses different aspects of the presented approaches to estimating the uplink relative load.

1. System Load and Capacity

A basic requirement for providing services to users is that there is sufficient power available to maintain an acceptable Quality of Service (QoS). In the uplink this, among other things, means that the received interference power in the base station must not be too high. Stability of the system is related to the *uplink noise rise* which will be denoted Λ. Uplink noise rise is defined as the total uplink interference power, I^{tot}, over background noise power, N, i.e.

$$\Lambda \triangleq \frac{I^{tot}}{N}.$$

Since the uplink noise rise is the primary resource, and increasing the number of active users or the active users' quality means a noise rise increase, there is a natural trade off between the number of users and quality. Furthermore, as the users have limited transmission power, a higher noise rise means reduced coverage. An always present trade off is therefore one between the number of users, quality and coverage. Perhaps not useful in practice, but still an educational model is

quality + number of users + coverage = utilized resources

The amount of available soft resources is unknown and time varying. An alternative is to estimate the *uplink relative load* or *uplink fractional load*, L, i.e. the amount of currently utilized resources relative to the total amount of resources

quality + number of users + coverage = $L \cdot$ total amount of resources

Since performance of the system is related to the currently received interference power, a more formal definition of L should incorporate this property of the system. One way is to relate the useful received interference power to the total received interference power. The total received uplink interference power is

$$I^{tot} = N + \sum_{i=1}^{M} C_i, \tag{14.1}$$

where C_i is user i's received carrier power and M is the number of users in the entire network. A definition of L is then

$$L = \frac{\sum_{i=1}^{M} C_i}{N + \sum_{i=1}^{M} C_i} = \frac{I^{tot} - N}{I^{tot}} = 1 - \frac{1}{\Lambda},$$

i.e.

$$L \triangleq 1 - \frac{1}{\Lambda}. \tag{14.2}$$

A rearrangement of the above expression yields the *pole equation* (Holma and Laakso, 1999)

$$\Lambda = \frac{I^{tot}}{N} = \frac{1}{1 - L} \tag{14.3}$$

The equation clearly shows that $L = 0$ implies $I^{tot} = N$, i.e. an empty system with only noise. As L approaches one the system is operated close to the system's theoretical capacity, the *pole capacity*, and the interference power goes to infinity, see Figure 14.2. Since the relative load is a function of the received interference power, it is not purely a function of how many users there are in the system, but also for example where the users are in the radio environment. This means that the pole capacity in terms of e.g. number of users is in general both unknown and time varying. As can be seen in Equation (14.3), a feasible total resource allocation is associated with a relative load between zero and one. As an educational example, consider a simplified scenario with a number of users, all connected to the only base station in the network. The total received interference power can be expressed as in Equation (14.1). The QoS of each user is related to the *carrier-to-total-interference ratio*, β_i,

$$\beta_i \triangleq \frac{C_i}{I^{tot}}. \tag{14.4}$$

Solving for C_i in Equation (14.4) and inserting it into Equation (14.1) yield

$$I^{tot} = N + \sum_{i=1}^{M} \beta_i I^{tot}. \Leftrightarrow \Lambda = \frac{I^{tot}}{N} = \frac{1}{1 - \sum_{i=1}^{M} \beta_i}. \tag{14.5}$$

The equation has the same form as Equation (14.3) and hence

$$L = \sum_{i=1}^{M} \beta_i. \tag{14.6}$$

This shows that the capacity of a CDMA system with conventional receivers (i.e. not utilizing multiuser detection) is in fact *interference power limited* because the interference from other connections limits the capacity. The opposite is a *noise limited* system. Consider for example a system with one user in an isolated cell. The carrier-to-total-interference ratio is

$$\beta = \frac{C}{I^{tot}} = \frac{C}{N + C} < 1,$$

eliminating C gives

$$I^{tot} = \frac{1}{1 - \beta} N.$$

The total interference power, I^{tot} is thus finite for all possible carrier-to-total-interference ratios. In a noise limited system, quality (data rate) is therefore limited by the amount of available transmission power.

Equation (14.5) shows that the number of users in an interference power limited system is, even with unlimited transmission power, limited by the mutual interference power between the connections. Furthermore, the amount of excess interference power caused by admitting a new user depends on the current interference power in the system in this case. Figure 14.2 illustrates the higher interference power contribution of an admitted user at high load, compared to that at low load. In Figure 14.2, ΔL is associated with the amount of relative load a user contributes with. For example in the isolated cell case, $\Delta L = \beta^{tgt}$ for the admitted user according to Equation (14.6).

A traditional definition of relative load is the number of currently used channels over total number of channels,

$$L^{trad} \triangleq \frac{M}{M^{max}}.$$

A capacity defined as a fixed maximum number of channels is an example of a *hard capacity*. The hard capacity of a system is fixed and known, unlike the *soft capacity*. The soft capacity can only be achieved when a soft resource is studied in the resource management algorithms. Uplink noise rise is an example of a *soft resource* since it depends on time varying variables such as the path gains the users experience. As the uplink of a WCDMA system is limited by this spatial resource, the uplink's capacity depends on the situation in several cells. If the load is low in surrounding cells, little interference power is received from these cells. This results in an increased capacity in the own cell compared to when the surrounding cells are more loaded. A centralized resource management algorithm based on a soft resource can, unlike a decentralized algorithm which is studying a hard resource, utilize this additional capacity. For an informative example, see Section 3.

In the cases where the soft capacity equals the hard capacity, the traditional

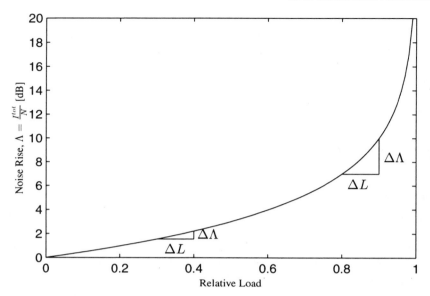

Figure 14.2. The nonlinear relation between relative load, L, and noise rise, Λ. ΔL could be the increase caused by admitting another user. This gives different noise rise contribution, $\Delta\Lambda$, depending on the system's initial load.

definition of relative load also equals the definition in Equation (14.2). If we study a single service single cell situation the pole capacity can be calculated using Equation (14.6) as ($L = 1$ and assuming perfect power control, $\beta_i = \beta^{tgt}$, give the maximum number of users)

$$1 = \sum_{i=1}^{M^{pole}} \beta^{tgt} = M^{pole}\beta^{tgt} \Leftrightarrow M^{max} = M^{pole} = \frac{1}{\beta^{tgt}}.$$

According to Equation (14.6), the relative load in this scenario is merely the number of users times the target carrier-to-total-interference ratio. The relative load can thus be expressed as

$$L = M\beta^{tgt} = \frac{M}{1/\beta^{tgt}} = \frac{M}{M^{max}} = L^{trad}. \tag{14.7}$$

(Huang and Yates, 1996) studies a single service scenario with several cells. The total interference is therein divided into three parts, background noise, *intracell interference* which is interference from users within the cell, I^{own}, and *intercell interference* which is interference from users served by another cell, I^{other}

$$I^{tot} = N + I^{own} + I^{other} \tag{14.8}$$

Since a single service scenario is studied, I^{other} can be converted into a corresponding number of users, $M^{other} = I^{other}/C$, where C is the received power required to maintain β^{tgt}. This once again enables an expression corresponding to Equation (14.7), only with M substituted by $M^{own} + M^{other}$.

2. Decentralized Load Estimates

Some resource management algorithms act in a decentralized node. Therefore, an attractive property of an estimate is that it uses only information which is locally available. For example, the interference caused by users connected to another cell, I^{other}, depends on variables not known in the own cell. Therefore a decentralized estimate has to either measure it or somehow estimate it using local variables only.

2.1 Interference Expansion Factor

One way of eliminating the inter-cell interference from Equation (14.8) is to simply state that it is a nominal fraction, f, of the intra-cell interference, i.e.

$$I^{other} = f \, I^{own}.$$

This is a natural assumption since an increase in interference power in one cell leaks to surrounding cells. Higher interference power, in turn, causes users in these cells to use higher transmission power. This effect would not be captured if I^{other} would have been assumed constant.

Combining the above expression with Equation (14.8) results in an expression for I^{tot} that contains only local variables,

$$I^{tot} = N + (1 + f)I^{own}.$$

According to Equation (14.4), the received power of user i is $C_i = \beta_i I^{tot}$. The interference power from the own cell, I^{own}, is simply the sum of these user individual carrier powers. An alternative expression for the total interference power is therefore

$$I^{tot} = N + (1 + f) \sum_{i=1}^{M^{own}} C_i = N + (1 + f) \sum_{i=1}^{M^{own}} \beta_i I^{tot}.$$

Solving for I^{tot} yields

$$I^{tot} = \frac{N}{1 - (1 + f) \sum_{i=1}^{M^{own}} \beta_i}.$$

Once again comparing with Equation (14.3) we see that an estimate of the uplink relative load in a multi cell scenario is

$$L = (1 + f) \sum_{i=1}^{M^{own}} \beta_i. \qquad (14.9)$$

A comparison with Equation (14.6) shows that one should not consider cells as isolated, since this gives an underestimation of the relative load. This is perhaps obvious since considering cells as isolated corresponds to completely ignoring the intercell interference, I^{other} in Equation (14.8). Using this technique, the requirement for pole capacity that the relative load should equal one corresponds to

$$\sum_{i=1}^{M^{own}} \beta_i = \frac{1}{1 + f}$$

I.e., an estimate of the pole capacity is once again put in terms of combined carrier-to-total-interference ratio. In case of a single service scenario, an estimate of the maximum number of users would be

$$M^{max} = \left\lfloor \frac{1}{(1 + f)\beta^{tgt}} \right\rfloor.$$

The intercell-to-intracell factor is widely used throughout the literature. Examples of where it is used in relative load expressions are (Ying et al., 2002; Zhang and Yue, 2001; Boyer et al., 2001; Hiltunen and Binucci, 2002; Holma and Laakso, 1999; Sanchez et al., 2002). The design parameter f is usually chosen somewhere between 0.5 and 0.6 if uniform traffic is expected.

2.2   Interference Power Measurement Based Estimates

Another way of considering the entire intercell interference power using only local information is to simply assume that it is measurable. As concluded in Section 1, the increase in interference power due to an admitted user depends on the interference level. (Holma and Laakso, 1999) uses measurements of I^{tot} and estimates the additional interference power a new user would cause through derivatives of Equation (14.2)

$$\frac{\partial I^{tot}}{\partial L} = \frac{N}{(1 - L)^2} = I^{tot} \frac{1}{1 - L}.$$

An estimate of the interference power increase due to an additional load of ΔL is then

$$\Delta I^{tot} = \frac{\partial I^{tot}}{\partial L} \Delta L = I^{tot} \frac{\Delta L}{1 - L_0},$$

where L_0 is the relative load before admitting the new user. An alternative estimate of the interference power increase is derived as the integrated difference in I^{tot}

$$\Delta I^{tot} = \int_{L_0}^{L_0+\Delta L} \frac{\partial I^{tot}}{\partial L} dL = I^{tot} \frac{\Delta L}{1 - L_0 - \Delta L}.$$

Motivated by the calculations leading to Equation (14.6), ΔL can be estimated as the new user's target carrier-to-total-interference ratio. Using measurements of the total interference power inherently catches the variations in intercell interference. However, it relies heavily on somewhat accurate measurements of the current uplink interference power.

3. Centralized Load Estimates

The estimates in Section 2 can not predict the effects of a decision made in one cell will have on other cells. Especially important are the users located close to the cell border. These users can occasionally introduce considerable interference power in other cells. As an example, study a prospective user in cell k. Power control will force him to use a transmission power p_i which satisfies

$$p_i g_{ik} = \beta_i^{tgt} I_k^{tot} \Leftrightarrow p_i = \beta_i^{tgt} I_k^{tot} \frac{1}{g_{ik}},$$

where $g_{ik}(< 1)$ is the path gain between user i and cell k and I_k^{tot} is the total received interference power in cell k. User i's signal will be received in cell j with a power of

$$C_{ij} = p_i g_{ij} = \beta_i^{tgt} I_k^{tot} \frac{g_{ij}}{g_{ik}}.$$

C_{ij} will in cell j be a part of the interference from other cells, I^{other}. How large C_{ij} will be depends on where the user is located. If user i is close to the cell border, g_{ij} and g_{ik} will be of the same order. The received carrier power, C_{ij}, will in this case contribute to the uplink load in cell j to a greater extent, compared to the case where user i is close to base station k. From a resource management point of view it is quite interesting to have an idea of how much load user i will actually induce in cell j. This, however, requires information which is gathered in several cells and an estimate of the uplink load using this information would therefore have to reside in a more centralized site than a base station serving just a few cells.

3.1 Power Control Feasibility

The uplink relative load may also be put in terms of feasibility and stability of the associated power control problem. This is more thoroughly discussed in Chapter 7 from a control perspective. Some aspects are also brought up here to relate to other uplink load estimates. Inspired by the presentation in

(Hanly and Tse, 1999) we study a multi-cell CDMA system. Assume that user i is power controlled solely from one cell, denoted cell j_i. Consider a power control algorithm that at each time instant t sets user i's transmission power according to

$$p_i(t) = \beta_i^{tgt} \frac{I_{j_i}^{tot}(t-1)}{g_{ij_i}} = \beta_i^{tgt}\left(\frac{N_{j_i}}{g_{ij_i}} + \sum_{\ell=1}^{M} p_\ell(t-1)\frac{g_{\ell j_i}}{g_{ij_i}}\right),$$

where β_i^{tgt} is user i's target carrier-to-total-interference ratio, N_{j_i} is the background noise power user i experiences (which is assumed constant) and M is the total number of users in the entire network. A matrix expression for all users update at time t is therefore

$$P(t) = B(\tilde{N} + Z^T P(t-1)), \tag{14.10}$$

where

$$P(t) \triangleq [p_i(t)], \ B \triangleq \mathrm{diag}(\beta_1^{tgt}, \beta_2^{tgt}, \dots, \beta_M^{tgt}),$$

$$\tilde{N} = [N_i] \triangleq [\frac{N_{j_i}}{g_{ij_i}}], \ Z = [z_{\ell k}] \triangleq [\frac{g_{\ell j_k}}{g_{k j_k}}].$$

The ith position in the vector $j \in \mathbb{R}^M$ contains the number of the cell user i is connected to. This notation makes the B and Z matrices have size M x M. It is well known from the theory of linear time series that the recursion in Equation (14.10) will converge if and only if all the eigenvalues of BZ^T are within the unit circle, $|\lambda_i| < 1$, $i = 1, 2, \dots, M$. Thus, a measure of the convergence rate is

$$\lambda^* = \max |\lambda_i|. \tag{14.11}$$

The fact that $0 < \lambda^* < 1$ for tractable power control problems relates the convergence rate to the definition of the relative load. Furthermore, this is comparable to the feasibility relative load in Chapter 7. (Hanly and Tse, 1999) shows that, for a single cell case, λ^* equals the resulting L in Equation (14.6), i.e.

$$\lambda^* = \sum_{i=1}^{M} \beta_i^{tgt}.$$

To exemplify the theory, study a system consisting of three users and two cells. The G-matrix is

$$G = \begin{pmatrix} 0.8 & 0.1 \\ 0.5 & 0.3 \\ 0.2 & 0.6 \end{pmatrix}.$$

Denote the cell represented in the first column A and the second column B. Users 1 and 2 are connected to cell A, see Figure 14.3, and the third user applies

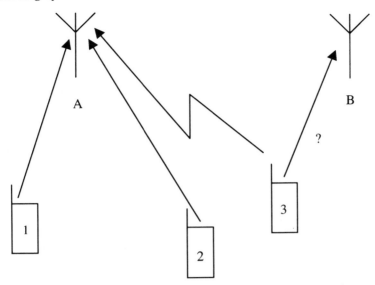

Figure 14.3. Admission of a new user. User 3 applies for a channel in cell B. Users connected to other cells will then experience increased intercell interference power.

for a connection to cell B yielding a j-vector of $j = [1\,1\,2]$. The corresponding Z-matrix is

$$Z = \begin{pmatrix} 1 & 1.6 & 0.17 \\ 0.625 & 1 & 0.5 \\ 0.25 & 0.4 & 1 \end{pmatrix}.$$

At first, consider target carrier-to-total-interference ratios according to

$$B = \operatorname{diag} \begin{pmatrix} 0.5 & 0.3 & 0 \end{pmatrix}.$$

This yields $\lambda^* = 0.8$, which is in correspondence with the result that λ^* equals the sum of the users' β_i^{tgt} values in a single cell scenario. This is also the true load given by Equation (14.2), see Figure 14.4. The top plot of Figure 14.4 shows how the users' transmission powers converge. After 100 iterations, user 3 is connected to cell B with $\beta_3^{tgt} = 0.4$. Naturally this affects the transmission powers for all users in the network, as indicated in Figure 14.4. The load according to Equation (14.11), λ^*, increases up to 0.87. Thus, the fact that admitting a user in one cell requires additional transmission powers for users connected to other cells is reflected in an increased λ^*.

The bottom plot of Figure 14.4 shows what the different estimates provide in this specific example. Note how the estimate defined by Equation (14.6) stationary is completely insensitive to the situation in other cells. We have not used any

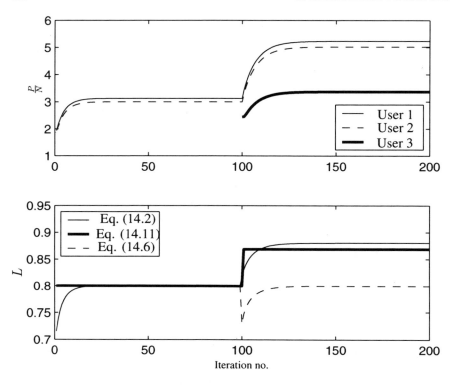

Figure 14.4. Intercell effects. Top: Ratio between users' transmission power and the background noise power. Bottom: True relative load in cell A and two different estimates of it. A new user is admitted in cell B at iteration number 100.

intercell-to-intracell factor here. In this scenario the true intercell-to-intracell factor when the power control has settled is

$$f = \frac{I^{other}}{I^{own}} = \frac{g_{31}p_3}{g_{11}p_1 + g_{21}p_2} \approx 0.1.$$

The example also shows the trade off between coverage and throughput. When a new user is admitted in cell B, the throughput increases but the higher transmission power of the users contributes to a decreased coverage.

3.2   Link Based Estimates

An estimate which considers the number of links each user has can to some extent capture the gain from using soft handover. If we assume that maximum ratio combining is used in the combining of the locally received information (softer handover), a simple way of utilizing the information regarding number of links is to assume that each user's contribution to the relative load is inversely proportional to the number of soft handover links. The relative load estimate would then become

$$L_j = (1 + f) \sum_{i}^{M_j} \frac{\beta_i}{n_i},$$

where M_j is the number of users power controlled in cell j and n_i is user i's number of soft handover links. There has been some studies where link based admission control has been compared with interference power based admission control algorithms. For example (Ishikawa and Umeda, 1997; F. Gunnarsson, E. Geijer Lundin, G. Bark, N. Wiberg, 2002) show that using a link based resource compared to using an interference based resource in the call admission control yields approximately the same performance, but the link based algorithm is far more sensitive to changes in the radio environment. Since these types of estimates indirectly consider the variations in the radio environment the users experience, the estimated pole capacity will also change over time.

3.3   Path Gain Measurement Based Estimates

Clearly, the amount of interference power from other cells depends on where users who are not power controlled in the cell are located. Using information regarding the path gain between these users and the base stations, it is possible to get an estimate of how much interference power they will in fact induce. A set of alternative estimates can be derived from Equation (14.1). In cell j at time t, Equation (14.1) is

$$I_j^{tot}(t) = N_j + \sum_{i=1}^{M} C_{ij}(t) \tag{14.12}$$

Let β_{ik} denote user i's received carrier-to-total-interference ratio in cell k. In case of soft handover, a user's total carrier-to-total-interference ratio, β_i, is a function of the locally obtained ratios, β_{ik}. In the special case of softer handover, maximum ratio combining is utilized. The combined β_i is then approximately the sum of the locally obtained β_{ik},

$$\beta_i(t) \approx \sum_{k \in K_i} \beta_{ik} = \sum_{k \in K_i} \frac{g_{ik}(t) p_i(t)}{I_k^{tot}(t)} = p_i(t) \sum_{k \in K_i} \frac{g_{ik}(t)}{I_k^{tot}(t)},$$

where K_i is the active set, i.e. the set of cells user i is currently connected to. This rather complicated relation between p_i and β_i is due to the fact that a user receives power control commands from all connected cells. Solving for $p_i(t)$, using $C_{ij} = p_i g_{ij}$, assuming perfect power control ($\beta_i = \beta_i^{tgt}$) and inserting it into Equation (14.12) yield

$$I_j^{tot}(t) = N_j + \sum_{i=1}^{M} p_i(t) g_{ij}(t) = N_j + \sum_{i=1}^{M} \beta_i^{tgt}(t) \frac{g_{ij}(t)}{\sum_{k \in K_i} \frac{g_{ik}(t)}{I_k^{tot}(t)}}. \quad (14.13)$$

This defines a set of nonlinear equations. Different techniques have been applied in order to solve these approximately. One way is to temporarily assume that the total interference power in all cells are approximately equal, i.e. $I_k^{tot} = I_j^{tot}$. The nonlinear coupled equations then become a set of decoupled linear equations in I_j^{tot}. Solving for I_j^{tot} yields an approximative solution to Equation (14.13)

$$I_j^{tot} = \frac{N_j}{1 - \sum_{i=1}^{M} \beta_i^{tgt}(t) \frac{g_{ij}(t)}{\sum_{k \in K_i} g_{ik}(t)}}.$$

There is a strong resemblance between the above expression for I_j^{tot} and Equation (14.3). It is thus natural to associate the uplink relative load to

$$L = \sum_{i=1}^{M} \beta_i^{tgt}(t) \frac{g_{ij}(t)}{\sum_{k \in K_i} g_{ik}(t)}.$$

Since this estimate considers the actual contribution from users not only within the own cell it will to a greater extent capture variations in intercell interference power. The fact that a user close to a cell border will provide more load to neighbouring cells is reflected in the estimate through the ratio between g_{ij} and $\sum_i g_{ik}$. This estimate has been used for admission control purposes in (F. Gunnarsson, E. Geijer Lundin, G. Bark, N. Wiberg, 2002) where it is shown that it is quite insensitive to varying traffic situations. Also, if a resource management decision is made based on the estimate in several cells, users with a more demanding service (higher carrier-to-total-interference ratio) will be given a reduced coverage compared to users requesting a less demanding service.
Equation (14.13) may also be solved through fix point iterations. This requires that either all N_j are known or that they can be assumed equal in different cells. Assuming that the background noise power in all cells equals a nominal level, N, enables the following rearrangement of Equation (14.13)

$$\frac{I_j^{tot}(t)}{N} = 1 + \sum_{i=1}^{M} \beta_i^{tgt}(t) \frac{g_{ij}(t)}{\sum_{k \in K_i} \frac{g_{ik}(t)}{\frac{I_k^{tot}(t)}{N}}}$$

Substituting $\frac{I^{tot}}{N}$ with $\hat{\Lambda}_j$ according to Equation (14.3) and using the previous estimate of $\Lambda_k(t)$ in the right hand equation yield the following fix point iteration

$$\hat{\Lambda}_j[n] := 1 + \sum_{i=1}^{M} \beta_i^{tgt}(t) \frac{g_{ij}(t)}{\sum_{k \in K_i} \frac{g_{ik}(t)}{\Lambda_k[n-1]}}.$$

A natural initialization of the iterations is the estimate of the uplink noise rise in the different cells at the previous time instant, so $\hat{\Lambda}_j[0] = \hat{\Lambda}_j(t-1), \forall j$. $\hat{\Lambda}_j(t)$ is set to the convergence value of the fix point iterations.

Path gain between a user and a number of base stations can be made available through mobile-assisted measurements. These measurements are reflecting the downlink path gain. The fast fading is filtered out by the mobile phones, therefore the measurement reports only concerns propagation loss and shadow fading. These two components can perhaps be assumed independent of the direction; uplink or downlink. The reports are readily available for hand over purposes. Furthermore, they can be scheduled either periodically or in an event based matter in conjunction with soft handover requests (Geijer Lundin et al., 2003a). Performance evaluation of these estimates has been carried out through simulations in (Geijer Lundin et al., 2003b).

4. Discussion

Theory regarding uplink load in cellular CDMA systems is presented in this chapter. It has been shown that, despite unlimited transmission power, the capacity of an interference limited system is in fact bounded by a finite time varying capacity, the pole capacity. Since the maximum capacity of the system is generally both unknown and time varying a quantity called uplink relative load has been introduced. The uplink relative load relates the current amount of used capacity to the current maximum capacity, even though both are unknown. Properties of the uplink relative load has been explored through a survey of different approaches to estimating it.

Decentralized Estimates.   One group of estimates, *decentralized estimates*, use information locally available in each cell. An advantage with these is of course that they can directly be used in local resource management algorithms, such as packet scheduling. However, these estimates have no real knowledge of the effects that resource management decisions have on the surrounding cells. Furthermore, a decentralized estimate which makes the assumption that the other cells are equally loaded as the own cell can not fully utilize the soft capacity of a WCDMA system. In fact, it can be argued that these estimates are purely related to the hard capacity of the system. Measuring the current total interference power is one way of locally estimate how loaded the surrounding cells are. This, however, requires accurate interference power estimates – something which should not be taken for granted.

Centralized Estimates.   If a centralized estimate is used, information from several cells can be considered. The type of information may be simply the number of soft handover links each user has or, more advanced, it may be the path gains each user experiences. Some simplifying assumptions still have to be made also in the centralized cases. In the estimate where the number of links is used, the assumption is that the base stations received signals are combined using maximum ratio combining, which is not true in general. Another proposed technique, based on measurements of path gain between users and base stations, captures the soft capacity of the network to a greater extent. These estimates, however, suffers from non-ideal path gain measurements.

A more theoretical approach has also been handled in the chapter. One can show that there is a strong connection between uplink relative load and feasibility of the power control problem. Using techniques from system theory, it can be shown that stability of the power control problem corresponds to a relative load below one. As the power control problem is concerned with the entire network, this relative load applies to the entire network as opposed to just one cell.

References

Boyer, P., Stojanovic, M., and Proakis, J. (2001). A simple generalization of the CDMA reverse link pole capacity formula. *IEEE Transactions on Communications*, pages 1719–1722.

F. Gunnarsson, E. Geijer Lundin, G. Bark, N. Wiberg (2002). Uplink Admission Control in WCDMA Based on Relative Load Estimates. In *Proc. International Conference on Communications, New York*.

Geijer Lundin, E., Gunnarsson, F., and Gustafsson, F. (2003a). Uplink Load Estimates in WCDMA with Different Availability of Measurements. In *Proc. IEEE Vehicular Technology Conference*, Cheju, South Korea.

Geijer Lundin, E., Gunnarsson, F., and Gustafsson, F. (2003b). Uplink Load Estimation in WCDMA. In *Proc. IEEE Wireless Communications and Networking Conference*, New Orleans, Lousiana, USA.

Hanly, S. and Tse, D.-N. (1999). Power control and capacity of spread spectrum wireless networks. *Automatica*, 35:1987–2012.

Hiltunen, K. and Binucci, N. (2002). WCDMA downlink coverage:interference margin for users located at the cell coverage border. In *Proc. IEEE Vehicular Technology Conference*, Birmingham, AL, USA.

Holma, H. and Laakso, J. (1999). Uplink admission control and soft capacity with MUD in CDMA. In *Proc. IEEE Vehicular Technology Conference*, Amsterdam, the Netherlands.

Holma, H. and Toskala, A. (2000). *WCDMA for UMTS, Radio Access For Third Generation Mobile Communications*. John Wiley & Sons, Ltd.

Huang, C. Y. and Yates, R. D. (1996). Call admission in power controlled CDMA systems. In *Proc. IEEE Vehicular Technology Conference*, Atlanta, GA, USA.

Ishikawa, Y. and Umeda, N. (1997). Capacity design and performance of call admission control in cellular CDMA systems. *IEEE Journal on Selected Areas in Communications*, 15(8).

Sanchez, J., Perez-Romero, J., Sallent, O., and Agusti, R. (2002). Mixing conversational and interactive traffic in the UMTS radio access network. In *Proc. IEEE 4th International Workshop on Mobile and Wireless Communications Network, 2002*, pages 597–601.

Ying, W., Jingmei, Z., Weidong, W., and Ping, Z. (2002). Call admission control in hierarchi cell structure. In *Proc. IEEE Vehicular Technology Conference*, Birmingham, AL, USA.

Zhang, Q. and Yue, O. (2001). UMTS air interface voice/data capacity - part 1:reverse link analysis. In *Proc. IEEE Vehicular Technology Conference*, Rhodes, Greece.

Chapter 15

PERFORMANCE ANALYSIS AND OPTIMIZATION OF MULTI-HOP COMMUNICATION SYSTEMS

Mazen O. Hasna

Department of Electrical Engineering
University of Qatar
P.O.Box 2713, Doha, Qatar
Tel: (974) 485-1074, Fax: (974) 392-0044

hasna@qu.edu.qa

Mohamed-Slim Alouini

Department of Electrical and Computer Engineering
University of Minnesota, Twin Cities
Minneapolis, MN 55455, USA
Tel: (612) 625-9055, Fax: (612)625-4583

alouini@ece.umn.edu

Abstract   End-to-end performance analysis of multi-hop transmissions over Rayleigh fading channels is presented. Several types of relays for both regenerative and non-regenerative systems are considered. In addition, optimum power allocation over these hops is investigated as it is considered a scarce resource in the context of relayed transmission. Numerical results show that regenerative systems outperform non-regenerative systems specially at low average signal-to-noise ratios, or when the number of hops is large. They also show that power optimization enhances the system performance, specially if the links are highly unbalanced in terms of their average fading power or if the number of hops is large. Interestingly, they also show that non-regenerative systems with optimum power allocation can outperform regenerative systems with no power optimization.

Keywords: Multi-hop systems, co-operative/collaborative diversity, Rayleigh Fading, outage probability, average bit error rates.

1.   Introduction

Multi-hop transmission consists of relaying the information from a source to a destination via many intermediate relaying terminals in between. This relaying of information on several hops extends the coverage and reduces the need to use large power at the transmitter, which results in extended battery life and lower level of interference [16]. The dual-hop transmission special case was encountered originally in bent-pipe satellites where the primary function of the spacecraft is to relay the uplink carrier into a downlink [8]. It is also common in various fixed microwave links by enabling greater coverage without the need of large power at the transmitter. More recently, this concept has gained new actuality in collaborative/cooperative wireless communication systems and more generally in multi-hop-augmented networks in which packets propagate through many hops before reaching their destination (see for example [25] and references therein).

The two most common performance assessment criteria in literature are outage probability and average bit error rate (BER). Outage probability is defined as the probability that the link quality falls below a predetermined threshold. This threshold is chosen to guarantee a certain quality of service which essentially depends on the type of modulation employed and the type of application supported. Average BER, on the other hand, is the ratio of erroneous bits at the receiver, on average. Both criteria are functions of the fading model of the channel and the type of receivers employed. Moreover, average BER is also a function of the type of modulation used at the transmitter. However, as we will see later on, this problem can be bypassed using the moment generating function (MGF) approach [22] for the performance evaluation over fading channels. Those two performance criteria will be used to asses the performance of both regenerative and non-regenerative multi-hop communication systems (to be defined shortly).

As power is considered a critical resource in the context of multi-hop transmission, optimizing the usage of this resource is essential. As such, Chiang et al. [5] considered recently the problem of optimizing resource allocation in wireless ad hoc networks for regenerative systems. In [5] the general problem was formulated as a geometric programming problem and the interior point algorithm was required to reach the optimal solution. In this chapter, the problem of both regenerative and non-regenerative systems is studied, and closed-form solutions are introduced that are easy to implement in practice.

2.   Classification of Relayed Transmission

Depending on the nature and complexity of the relays, relayed transmission systems can be classified into two main categories, namely, regenerative or non-regenerative systems. In regenerative systems, the relay fully decodes the signal that went through the preceding hop and retransmits the decoded version into the next hop. This is also referred to as decode-and-forward [16] or digital [25] relaying. In these systems, noise propagation from hop to hop is prevented while risking the probability of making an error in detecting the signal at each hop. On the other hand, non-regenerative systems use less complex relays that just amplify and forward the incoming signal without performing any sort of decoding. That is why it is sometimes referred to as amplify-and-forward [16] or analog [25] relaying. This kind of relaying is more useful when the carried information is time sensitive, such as voice and live video. As a further categorization, relays in non-regenerative systems can in their turn be classified into two subcategories, namely, (i) channel state information (CSI)-assisted relays and (ii) "blind" relays. Non-regenerative systems with CSI-assisted relays use instantaneous CSI of the preceding hop to control the gain introduced by the relay and as a result fix the power of the retransmitted signal. In contrast, systems with "blind" relays do not need instantaneous CSI of the preceding hop at the relay but rather employ at these relays amplifiers with fixed gains and consequently result in a signal with variable power at the relay output. Although systems with such kind of blind relays are not expected to perform as well as systems equipped with CSI-assisted relays, their low complexity and ease of deployment (together with their comparable performance to CSI-assisted relays as shown later on in the chapter), make them attractive from a practical standpoint.

3.   Is Relaying Useful?

In order to find out whether we gain from relayed transmission or not (beside the advantages that we talked about in the introduction), we have to compare relayed transmission with an *equivalent* direct transmission. Before doing that, it maybe helpful to see the effect of relaying on the *amount of fading* (AF) of the channel. The AF is defined as the ratio of the variance power of the fading to the square of its mean, and hence it is a measure of the severity of the fading independent of the fading power. Fig. 15.1 shows the AF as a function of the number of hops. It is clear that even with one relay (i.e. dual-hop systems), the AF is reduced by 20 percent (0.8 compared to an AF of 1 for the direct transmission under Rayleigh fading). It is clear also that increasing the

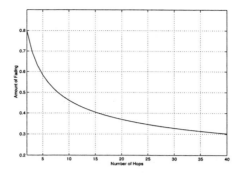

Figure 15.1. Effect of number of hops on the AF of the channel.

number of hops has a diminishing effect on reducing the AF, which is an expected diversity trend. Next we compare relayed transmission with equivalent direct transmission.

3.1 Method 1: Equivalent Coverage/Power Consumption

Consider the two systems shown in Fig. 15.2. Here, terminals A and

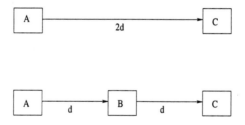

Figure 15.2. Direct vs relayed transmission systems.

C are communicating either directly, or via terminal B. In the case of direct transmission, we assume that the output power of terminal A is $2P$, while in case of relayed transmission, we split the power into P at the output of terminal A, and P at the output of terminal B. Assume that the transmitted signals encounter the basic path loss model where the received signal power, P_R, is related to the transmitted signal power, P_T, by $P_R = P_T/D^\alpha$, where D is the travelled distance, and α is the path loss exponent which is equal to 2 in free space. Now, for direct transmission with a distance of $2d$, the average received power, assuming free space propagation, is $P/2d^2$. For the relayed transmission, the average received power at B and C, assuming free space propagation and that B has a regenerative relay, and is placed in the middle distance between A and C, is P/d^2. Assume that all the terminal are subject to an additive white Gaussian noise (AWGN) that has a power of σ^2. Hence, the average SNRs are given as:

$$\overline{\gamma}_D = \frac{2P}{\sigma^2(2d)^\alpha} \tag{15.1}$$

$$\overline{\gamma}_R = \frac{P}{\sigma^2 d^\alpha} \tag{15.2}$$

where $\overline{\gamma}_D$ refers to the direct link, and $\overline{\gamma}_R$ refers to each hop of the relayed system. For simplicity, denote $\overline{\gamma} = P/\sigma^2 d^\alpha$. Note that if $\alpha = 2$, then both systems will have the same performance outage probability wise (which is a function of the average SNRs of the links as will be shown later on in this chapter), and relaying seems to be useless. Using more realistic values for α, e.g. 3, then the advantage of relaying is more clear as shown in the Fig. 15.3. Here, outage probability is plotted for

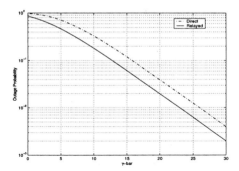

Figure 15.3. Comparison between relayed and equivalent direct transmission.

both scenarios. Note the advantage of relayed transmission over that of direct transmission. This advantage increases as the path loss exponent, α, increases.

3.2   Method 2: Equivalent Average SNR

The second method of comparing direct and relayed transmission is more appropriate to systems with non-regenerative relays where an equivalent link to the relayed link can be found. Hence, the average

SNR of this equivalent link can be used as the average SNR of the direct link. For non-regenerative systems employing CSI-assisted relays, it is shown in [11] that the average SNR of the link is $\overline{\gamma}/3$, given that the two links have averages of $\overline{\gamma}$ each. These values for averages can be interpreted as follows. Using a relay, the resulting link is able to combat fading that if a direct link was to be used, it would result in an average SNR of $\overline{\gamma}/3$ using the same amount of power at the transmitter. Now, we compare a direct link with an average power of $\overline{\gamma}/3$ with a relayed link with an average power of $\overline{\gamma}$ on each hop. Fig. 15.4 compares a relayed non-regenerative link with an equivalent direct link. It is clear again that

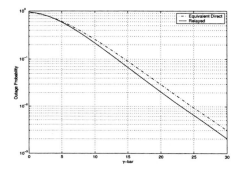

Figure 15.4.   Comparison between relayed and equivalent direct transmission.

relaying enhances the performance of the system in comparison with an equivalent direct transmission.

4.   Dual-Hop Systems

An interesting special case of multi-hop transmission is the dual-hop case. The dual-hop transmission special case was encountered originally in bent-pipe satellites where the primary function of the spacecraft is to relay the uplink carrier into a downlink [8]. It is also common in various fixed microwave links by enabling greater coverage without the need of large power at the transmitter. More recently, this concept has gained new actuality in collaborative/cooperative wireless communication systems [21, 16, 17, 10, 18, 7, 6, 14].

4.1   System and Channel Models

Consider the wireless communication system shown in Fig. 15.5. Here, terminal A is communicating with terminal C through terminal B which

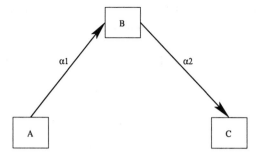

Figure 15.5. A wireless communication system where terminal B is relaying the signal from terminal A to terminal C.

acts as a relay. Assume that terminal A is transmitting a signal $s(t)$ which has an average power of $\mathcal{E}_1$. The received signal at terminal B can be written as

$$r_b(t) = \alpha_1 s(t) + n_1(t), \tag{15.3}$$

where α_1 is the fading amplitude of the channel between terminals A and B, and $n_1(t)$ is an AWGN with a power of N_{0_1} at the input of B. In non-regenerative systems, the received signal and noise are multiplied by the gain of the relay at terminal B, G, and then retransmitted to terminal C. The received signal at terminal C can be written as

$$r_c(t) = \alpha_2 G(\alpha_1 s(t) + n_1(t)) + n_2(t), \tag{15.4}$$

where α_2 is the fading amplitude of the channel between terminals B and C, and $n_2(t)$ is the AWGN with power N_{0_2} at the input of C. The overall SNR at the receiving end can then be written as

$$\gamma_{eq} = \frac{\frac{\mathcal{E}_1 \alpha_1^2}{N_{0_1}} \frac{\alpha_2^2}{N_{0_2}}}{\frac{\alpha_2^2}{N_{0_2}} + \frac{1}{G^2 N_{0_1}}}. \tag{15.5}$$

It is clear from (15.5) that the choice of the relay gain defines the equivalent end-to-end SNR of the two hops. In case of available instantaneous CSI at B, a gain of

$$G^2 = \frac{\mathcal{E}_2}{\mathcal{E}_1 \alpha_1^2 + N_{0_1}}, \tag{15.6}$$

where $\mathcal{E}_2$ is the power of the transmitted signal at the output of the relay, was proposed in [16]. The choice of this gain aims to invert the fading effect of the first channel while limiting the output power of the relay if the fading amplitude of the first hop, α_1, is low. However, this CSI-assisted relay requires a continuous estimate of the channel fading amplitude which may make this choice of gain not always feasible from a practical point of view. Substituting (15.6) in (15.5) leads to γ_{eq_1} given by

$$\gamma_{\text{eq}_1} = \frac{\gamma_1 \gamma_2}{\gamma_1 + \gamma_2 + 1}, \tag{15.7}$$

where $\gamma_i = \mathcal{E}_i \alpha_i^2 / N_{0_i}$, $(i = 1, 2)$ is the per hop SNR. The performance of systems employing such relays over Rayleigh fading channels was first studied in [16] and then in [6, 2]. In addition, [14, 11] presented tight lower bounds on the performance of all CSI-assisted relays for Rayleigh and Nakagami fading channel, respectively.

Blind relays, on the other hand, introduce fixed gains to the received signal regardless of the fading amplitude on the first hop. Let $C = \mathcal{E}_2/(G^2 N_{0_1})$, then (15.5) can be rewritten as

$$\gamma_{\text{eq}_2} = \frac{\gamma_1 \gamma_2}{C + \gamma_2}, \tag{15.8}$$

where C is a constant for a fixed G. The performance of these relays can be found in [15].

The two hops are assumed to be subject to independent not necessarily identically distributed Rayleigh fading. One of the characteristics of a signal that is being perturbed by Rayleigh fading is that its power is exponentially distributed. Consequently, γ_1 and γ_2 are exponentially distributed with parameters $\overline{\gamma}_1 = \mathcal{E}_1 \Omega_1 / N_{0_1}$ and $\overline{\gamma}_2 = \mathcal{E}_2 \Omega_2 / N_{0_2}$ respectively, where $\Omega_i = \overline{\alpha_i^2}$ $(i = 1, 2)$ is the average fading power on the ith hop.

4.2   Outage Probability

4.2.1   Non-Regenerative Systems.

In noise-limited systems, outage probability is defined as the probability that the instantaneous SNR, γ, falls below a predetermined protection ratio, γ_{th}, namely

$$P_{\text{out}} = \text{P}\left[\gamma \leq \gamma_{\text{th}}\right] = \int_0^{\gamma_{\text{th}}} p_\Gamma(\gamma) \, d\gamma = P_\Gamma(\gamma_{\text{th}}). \tag{15.9}$$

In (15.9), the predetermined protection ratio γ_{th} is a threshold SNR above which the quality of service is satisfactory and which essentially depends on the type of modulation employed and the type of application supported. For non-regenerative systems equipped with CSI-assisted relays as per (15.6), outage probability is given as [6]

$$P_{\text{out}} = 1 - \frac{2\sqrt{\gamma_{\text{th}}^2 + \gamma_{\text{th}}}}{\sqrt{\overline{\gamma}_1 \overline{\gamma}_2}} K_1 \left(\frac{2\sqrt{\gamma_{\text{th}}^2 + \gamma_{\text{th}}}}{\sqrt{\overline{\gamma}_1 \overline{\gamma}_2}}\right) e^{-\gamma_{\text{th}}\left(\frac{1}{\overline{\gamma}_1} + \frac{1}{\overline{\gamma}_2}\right)}. \tag{15.10}$$

where $K_1(\cdot)$ is the first order modified Bessel function of the second kind defined in [1, Eq.(9.6.22)]. For non-regenerative systems equipped with blind relays, outage probability is given as [15]

$$P_{\text{out}} = 1 - 2\sqrt{\frac{C\gamma_{\text{th}}}{\bar{\gamma}_1\bar{\gamma}_2}} K_1 \left(2\sqrt{\frac{C\gamma_{\text{th}}}{\bar{\gamma}_1\bar{\gamma}_2}}\right) e^{-\left(\frac{\gamma_{\text{th}}}{\bar{\gamma}_1}\right)}. \qquad (15.11)$$

4.2.2 Regenerative Systems. In regenerative systems, an outage occurs if either one of the links is in outage. Equivalently, it is the complement event of having both links operating above the threshold, γ_{th}. Hence, outage probability is given by

$$\begin{aligned}
P_{\text{out}} &= 1 - \left(\int_{\gamma_{\text{th}}}^{\infty} \frac{1}{\bar{\gamma}_1} e^{-\frac{\gamma}{\bar{\gamma}_1}} d\gamma\right) \left(\int_{\gamma_{\text{th}}}^{\infty} \frac{1}{\bar{\gamma}_2} e^{-\frac{\gamma}{\bar{\gamma}_2}} d\gamma\right) \\
&= 1 - e^{-\gamma_{\text{th}}\left(\frac{1}{\bar{\gamma}_1} + \frac{1}{\bar{\gamma}_2}\right)}.
\end{aligned} \qquad (15.12)$$

4.2.3 Comparison. Fig. 15.6 compares the performance of non-regenerative systems with that of regenerative systems. It is clear

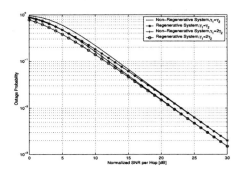

Figure 15.6. Comparison of the outage probability of regenerative and non-regenerative systems.

that regeneration improves the performance at low average SNR in exchange of an increased complexity. At high average SNR, the two systems are equivalent outage probability wise. Note that these results can be proven analytically by comparing the second terms in (15.10) and (15.12) which differ only by the scaling factor $\left[\frac{2\gamma_{\text{th}}}{\sqrt{\bar{\gamma}_1\bar{\gamma}_2}} K_1\left(\frac{2\gamma_{\text{th}}}{\sqrt{\bar{\gamma}_1\bar{\gamma}_2}}\right)\right]$ in front of the exponential function. At high average SNR, this scaling factor converges to 1 since the $K_1(x)$ function converges to $1/x$ when x approaches 0 [1, Eq.(9.6.9)]. Also, note that this behavior holds for both balanced or unbalanced links as shown in the same figure.

4.2.4   Systems with Semi-Blind Relays.   The performance analysis presented in the previous section concerning fixed gain relays applies to blind relays with arbitrary fixed gains. In this section, we still consider fixed gain relays and we still assume that these relays do not have access to instantaneous CSI of the first hop. However, we assume that they have statistical CSI about the first hop and have in particular knowledge of the average fading power Ω_1 which changes slowly (relative to α_1) and as such does not imply continuous monitoring of the channel (as it is the case in CSI-assisted relays). We call these relays "semi-blind" and we study in what follows their performance in comparison with that of CSI-assisted relays as per (15.6).

The relay gain in the semi-blind scenario is chosen such as

$$G^2 = \mathbf{E}\left[\frac{\mathcal{E}_2}{\mathcal{E}_1\alpha_1^2 + N_{0_1}}\right]. \tag{15.13}$$

This way, both relays in (15.6) and (15.13) consume the same power on average. For Rayleigh fading, G^2 in (15.13) can be shown to be given as

$$G^2 = \frac{\mathcal{E}_2}{\mathcal{E}_1\Omega_1}e^{\frac{1}{\overline{\gamma}_1}}E_1\left(\frac{1}{\overline{\gamma}_1}\right). \tag{15.14}$$

where $E_1(\cdot)$ is the exponential integral function defined in [1, Eq. (5.1.1)]. Consequently, C is given by

$$C = \frac{\overline{\gamma}_1}{e^{\frac{1}{\overline{\gamma}_1}}E_1(\frac{1}{\overline{\gamma}_1})}, \tag{15.15}$$

which when substituted back in (15.8) results in an equivalent end-to-end SNR of

$$\gamma_{\text{eq}} = \frac{\gamma_1\gamma_2}{\gamma_2 + \frac{\overline{\gamma}_1}{e^{\frac{1}{\overline{\gamma}_1}}E_1(\frac{1}{\overline{\gamma}_1})}}. \tag{15.16}$$

Fig. 15.7 compares the outage probability performance of a non-regenerative dual-hop system employing a fixed gain relay as per (15.14) with that of an equivalent (in terms of average power consumption) system with a relay as per (15.6), whose exact performance was derived in [6]. The figure also plots, as a benchmark, the performance of the more complex regenerative systems studied in [14]. For the medium to large average SNR region, systems with variable gain relays (15.6) outperform those

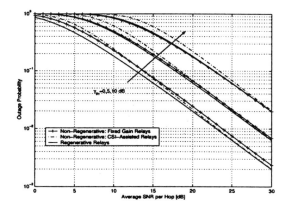

Figure 15.7. Outage probability of a dual-hop system for different relay configurations.

with fixed gain relays (15.14). However, the surprising and interesting result is that the gap in performance is not as much as one would expect in comparison with the difference in implementation and complexity of both relays. Note also that systems with fixed gain relays can even slightly outperform systems with variable gain relays at low average SNR. This is due to the fact that the variable gain relay has a gain floor of $\mathcal{E}_2/N_{0_1}$ when α_1 is too small, which is a relatively frequent event in the low average SNR region. Finally, note that as γ_{th} is increased, the range of average SNR in which fixed gain relays outperform variable gain relays extends to the right. One should bear in mind, however, that the performance of fixed gain relays shown in the figure can be thought of as an upper bound on the performance of practical relays. This due to the fact that fixed gain relays suffer from saturation that, as shown in [15], degrades the performance of these relays at low average SNR ranges.

4.2.5   Systems with Diversity.   With the outage probability expressions in hand, we can also easily get the outage probability performance of a receiver that employs selection combining (SC) when a collaborative diversity system is employed as suggested in [16]. More specifically, consider a collaborative diversity system in which the mobile station (MS) communicates with the base station (BS) through a direct path with SNR γ_0 and L collaborating paths (i.e. via other relaying MS's), each with equivalent SNR γ_l, $l = 1 \cdots L$. Then, outage probability of a SC receiver is given by

$$
\begin{aligned}
P_{\text{out}} &= \left(1 - e^{-\frac{\gamma_{\text{th}}}{\overline{\gamma}_0}} \right) \\
&\times \prod_{l=1}^{L} \left(1 - \frac{2\sqrt{\gamma_{\text{th}}^2 + \gamma_{\text{th}}}}{\overline{\gamma}_l} e^{-\frac{2\sqrt{\gamma_{\text{th}}^2 + \gamma_{\text{th}}}}{\overline{\gamma}_l}} K_1 \left(\frac{2\sqrt{\gamma_{\text{th}}^2 + \gamma_{\text{th}}}}{\overline{\gamma}_l} \right) \right)
\end{aligned}
\tag{15.17}
$$

where we assumed that the collaborating paths use relay gains as per (15.6). Fig. 15.8 shows the effect of the number of collaborative paths on the overall system performance, assuming independent and identically distributed links on each collaborating path for simplicity. Note the

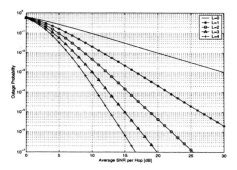

Figure 15.8. The effect of the number of collaborative paths, L, on the end-to-end outage probability.

significant improvement in performance even with a single collaborative path (i.e. $L = 1$) over that with no diversity paths (i.e. $L = 0$). Also, the figure shows a diminishing return in the performance gain as L, the number of collaborative paths, is increased, which is an expected diversity trend.

4.3   Average BER

4.3.1   Non-Regenerative Systems.
In order to use the MGF based approach for the performance evaluation of digital modulations over fading channels [22], the MGF of the equivalent SNR has to be evaluated. The MGF of non-regenerative systems equipped with CSI-assisted relays as per (15.6), (i.e. the MGF of (15.7)), is given by [2]

as

$$\mathcal{M}_{\gamma_{eq}}(s) = \frac{\sigma\delta - 4p}{\rho^2} - \frac{sp(\rho - 2p)}{\rho^3} e^{\left(\frac{\delta - \rho}{2p}\right)} E_1 \left(\frac{\delta - \rho}{2p}\right)$$
$$- \frac{sp(\rho + 2p)}{\rho^3} e^{\left(\frac{\delta + \rho}{2p}\right)} E_1 \left(\frac{\delta + \rho}{2p}\right) \qquad (15.18)$$

where $\rho = \sqrt{\delta^2 - 4p}$, $\delta = \sigma - ps$, $\sigma = \bar{\gamma}_1 + \bar{\gamma}_2$, $p = \bar{\gamma}_1\bar{\gamma}_2$. The MGF for the equivalent SNR in (15.8) for blind relays is given as [15]

$$\mathcal{M}_{\gamma_{eq}}(s) = \frac{1}{(\bar{\gamma}_1 s + 1)} + \frac{C\bar{\gamma}_1 s e^{\left(\frac{C}{\bar{\gamma}_2(\bar{\gamma}_1 s + 1)}\right)}}{\bar{\gamma}_2(\bar{\gamma}_1 s + 1)^2} E_1 \left(\frac{C}{\bar{\gamma}_2(\bar{\gamma}_1 s + 1)}\right), \qquad (15.19)$$

Having the MGF of γ_{eq} in closed form as given in (15.18) and (15.19), and using the MGF approach for the performance evaluation of digital modulations over fading channels [22] allows to obtain the average bit and symbol error rates for a wide variety of M-ary modulations (such as M-ary phase shift keying (M-PSK), M-ary differential phase shift keying (M-DPSK), and M-ary quadrature amplitude modulation (M-QAM)). For example, the average bit error rate (BER) of binary differential phase-shift keying (DPSK) is well known to be given by $P_b(E) = \frac{1}{2}\mathcal{M}_{\gamma_{eq}}(-1)$.

4.3.2 Regenerative Systems.

In regenerative systems, the relaying node decodes the signal and then transmits the detected version to the destination node. This means that the transmitted signal undergoes two states of decoding in cascade, and the overall probability of error is given by [8, Eq. (11.4.12)]

$$P_b(E/\gamma_1, \gamma_2) = P_b(E/\gamma_1) + P_b(E/\gamma_2) - 2P_b(E/\gamma_1)P_b(E/\gamma_2). \qquad (15.20)$$

which when averaged over the two independent RVs γ_1 and γ_2 reduces to

$$P_b(E) = P_b(E_1) + P_b(E_2) - 2P_b(E_1)P_b(E_2), \qquad (15.21)$$

where $P_b(E_i)$ is the average BER of link i, $(i = 1, 2)$. Note that the average BER expression given in (15.21) takes into account the probability of having two consecutive errors in detecting the signal which results in a correct decision at the end of the second hop. For DPSK over independent Rayleigh fading channels, (15.21) can be rewritten as

$$P_b(E) = \frac{1}{2}\left[\left(\frac{1}{1 + \bar{\gamma}_1}\right) + \left(\frac{1}{1 + \bar{\gamma}_2}\right) - \left(\frac{1}{1 + \bar{\gamma}_1}\right)\left(\frac{1}{1 + \bar{\gamma}_2}\right)\right]$$
$$= \frac{1 + \bar{\gamma}_1 + \bar{\gamma}_2}{2(1 + \bar{\gamma}_1)(1 + \bar{\gamma}_2)}. \qquad (15.22)$$

4.3.3   Comparison.   Fig. 15.9 compares the average BER performance of DPSK with regenerative systems (15.22) and with the non-regenerative systems (15.18). It is clear from the figure that (i) regen-

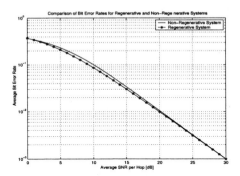

Figure 15.9.   Comparison between the average BER of non-regenerative and regenerative systems (balanced links, $\overline{\gamma}_1 = \overline{\gamma}_2 = \overline{\gamma}$).

erative systems perform better at low average SNR, and (ii) at high average SNR, the two systems are equivalent BER wise. Note that this is in contrast to the AWGN scenario in which there is always (over the whole SNR range) a 3 dB performance gain of regenerative systems over non-regenerative systems when the two hops are balanced (in terms of SNR). Fig. 15.10 compares the the average BER of DPSK performance of a dual-hop non-regenerative system employing several types of relays. It is clear again that the gap in performance is very small for all ranges of average SNR.

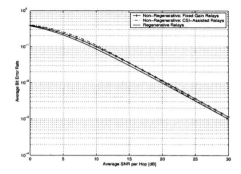

Figure 15.10.   Average BER of a dual-hop system for different relay configurations.

4.3.4   Systems with Diversity.   With the MGF of γ_{eq} in hand, we can also easily get the MGF of the SNR of the output of a maximal ratio combiner (MRC) at the receiver when a collaborative diversity system is employed. More specifically, consider again a collaborative diversity system in which the MS communicates with the BS through a direct path with SNR γ_0 and L collaborating paths (i.e. via other relaying MS's), each with equivalent SNR γ_l, $l = 1 \cdots L$. Using MRC at the receiving BS, the overall SNR can be written as [22]

$$\gamma_t = \gamma_0 + \sum_{l=1}^{L} \gamma_l. \tag{15.23}$$

Consequently, the MGF of the overall SNR is given by

$$\mathcal{M}_{\gamma_t}(s) = \mathcal{M}_{\gamma_0}(s) \prod_{l=1}^{L} \mathcal{M}_{\gamma_l}(s). \tag{15.24}$$

For example, the average BER of binary DPSK is given by $P_b(E) = \frac{1}{2} \mathcal{M}_\gamma(-1)$ for a non-regenerative system with no diversity and $P_b(E) = \frac{1}{2} \mathcal{M}_{\gamma_t}(-1)$ for a system with L collaborative diversity paths and MRC reception. As an example, Fig. 15.11 studies the effect of the number of collaborative paths on the overall system BER. Note again the significant

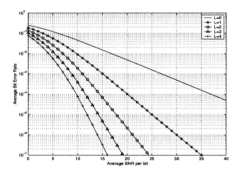

Figure 15.11.   The effect of the number of collaborative paths, L, on the end-to-end average BER.

improvement in performance even with a single collaborative path (i.e. $L = 1$) over that with no diversity paths (i.e. $L = 0$), and the diminishing return in the performance gain as L is increased.

4.4   Optimum Power Allocation

Up to this point, we implicitly assumed that power is distributed uniformly across the various hops. However, in the context of cooperative/collaborative diversity, the process of choosing appropriate relaying nodes encounters links parameters' estimation. Consequently, these information can be taken advantage of to enhance the system performance by optimizing the allocation of power. That is to say, we are interested in designing a power allocation algorithm that minimizes the outage probability of the system subject to the two constraints of a total power budget P_T and a maximum power per hop $P_{\max}$ which is assumed to be related to P_T as $P_{\max} = KP_T$, $0.5 < K < 1$. The total power constraint corresponds to the maximum power that a given packet is allowed to consume throughout its propagation from source to destination, while the maximum power per hop corresponds to the maximum power that each relay node can provide. Consequently, the problem is formulated as

$$
\begin{aligned}
&\text{min} && P_{\text{out}} \\
&\text{subject to} && \left\{ \begin{array}{ll} \sum_{n=1}^{N} p_n = P_T & \\ p_n \leq P_{\max}, & n = 1, \cdots, N. \end{array} \right.
\end{aligned} \tag{15.25}
$$

It should be noted here that the problem formulated above as per (15.25) is a convex problem, which means that it has a single global solution. This is due to the fact that the objective function is convex and that the constraints constitute a convex set. Clearly, as the objective function is an outage probability function, it is convex (when drawn in a linear scale). Also, since all the constraints are linear, they form a convex set [20, Sec. 3.4], which leads to a convex problem, and therefore a unique optimal solution.

4.4.1   Regenerative Systems.

Systems Without Diversity.   The problem in this case is formulated as

$$
\begin{aligned}
&\text{min} && P_{\text{out}} = 1 - e^{-\gamma_{\text{th}}\left(\frac{1}{\bar{\gamma}_1} + \frac{1}{\bar{\gamma}_2}\right)} \\
&\text{subject to} && \left\{ \begin{array}{ll} p_1 + p_2 = P_T & \\ p_n \leq P_{\max}, & n = 1, 2. \end{array} \right.
\end{aligned} \tag{15.26}
$$

Minimizing the objective function in (15.26) is equivalent to maximizing the term $e^{-\gamma_{\text{th}}\left(\frac{1}{\bar{\gamma}_1} + \frac{1}{\bar{\gamma}_2}\right)}$ which in turn is equivalent to maximizing its logarithm. Neglecting the second constraint for the moment and using the

Lagrange multiplier maximization method leads to the following optimal solution for p_1 and p_2

$$p_1^* = P_T \left[1 + \sqrt{\frac{G_1}{G_2}} \right]^{-1}$$

$$p_2^* = P_T \left[1 + \sqrt{\frac{G_2}{G_1}} \right]^{-1}. \qquad (15.27)$$

Now, if the solution in (15.27) leads to a power allocation that exceeds P_{max} on either link, clipping is applied, and the rest available power (i.e., $P_T - P_{max} - p_2^*$, assuming p_1^* was clipped) is added to the other link (link 2 in this case). It should be noted here that P_T in practice satisfies $P_T < 2P_{max}$ (i.e. $0.5 < K < 1$), otherwise, no optimization is required as the optimal solution would be to drive the relays feeding the two links to their maximum allowable power. The use of clipping in the solution of (15.26) preserves optimality as the Karush-Kuhn-Tucker (KKT) condition is applied in this case. Moreover, the optimal solution in (15.27) can be modified to take clipping into account, i.e.

$$\begin{array}{ll} p_1^* = P_{max} & \frac{G_1}{G_2} < \mathcal{K} \\ p_1^* = P_T \left[1 + \sqrt{\frac{G_1}{G_2}} \right]^{-1} & \mathcal{K} < \frac{G_1}{G_2} < \frac{1}{\mathcal{K}} \\ p_1^* = P_T - P_{max} & \frac{G_1}{G_2} > \frac{1}{\mathcal{K}} \end{array} \qquad (15.28)$$

where $p_2^* = P_T - p_1^*$, and $\mathcal{K} = \frac{1 - 2K + K^2}{K^2}$.

Systems with Diversity. Consider that in addition to the relayed signal, the destination is also receiving the original signal transmitted by the source. Assume that the destination keeps monitoring both the relayed and the direct transmitted signal, and at any instant of time only the signal with the highest instantaneous SNR is processed and detected. This type of selection diversity bypasses the need of synchronizing and coherently combining the signals received from the direct and relayed paths. Hence, outage probability (the objective function) is given in this case as

$$P_{out} = \left(1 - e^{-\gamma_{th} \left(\frac{1}{\bar{\gamma}_1} + \frac{1}{\bar{\gamma}_2} \right)} \right) \left(1 - e^{-\frac{\gamma_{th}}{\bar{\gamma}_3}} \right), \qquad (15.29)$$

where $\bar{\gamma}_3 = G_3 p_1$ is the average SNR of the direct link, and we assumed that the transmitter is using an omni directional antenna. Using this expression for outage probability, and repeating the optimization steps results in the following optimal power allocation

$$
p_1 = P_T \left[1 + \left(\frac{G_2}{G_3} \frac{1 - e^{\gamma_{th}\left(\frac{1}{\overline{\gamma}_1} + \frac{1}{\overline{\gamma}_2}\right)}}{e^{\frac{\gamma_{th}}{\overline{\gamma}_3}} - 1} + \frac{G_2}{G_1} \right)^{-\frac{1}{2}} \right]^{-1} . \tag{15.30}
$$

Note that (15.30) reduces to (15.27) when G_3 approaches 0, corresponding to the no diversity case (i.e. no direct link). This can be easily proven by expanding $e^{\frac{\gamma_{th}}{\overline{\gamma}_3}}$ and checking the limits as $(G_3 \to 0)$. The form of the optimal solution in (15.30) makes it suitable to be solved by the successive approximation method [19]. In this method, the iteration $p_1^{k+1} = f(p_1^k)$, where $f(\cdot)$ is the right hand side of (15.30), is used after an initial solution is assumed (e.g. uniform allocation). Using this method, no divergence was encountered, and less than eight iterations were enough to reach an accuracy of 10^{-3} in power assignment. This is good enough for engineering purposes since fine values of power do not affect the outage probability significantly.

4.4.2   Non-regenerative Systems.   The problem is formulated in this case as

$$
\begin{array}{ll}
\min & P_{\text{out}} = 1 - \frac{2\sqrt{\gamma_{th}^2 + \gamma_{th}}}{\sqrt{\overline{\gamma}_1 \overline{\gamma}_2}} K_1\left(\frac{2\sqrt{\gamma_{th}^2 + \gamma_{th}}}{\sqrt{\overline{\gamma}_1 \overline{\gamma}_2}}\right) e^{-\gamma_{th}\left(\frac{1}{\overline{\gamma}_1} + \frac{1}{\overline{\gamma}_2}\right)} \\
\text{subject to} & \left\{ \begin{array}{ll} p_1 + p_2 = P_T \\ p_n \le P_{\max}, & n = 1, 2. \end{array} \right.
\end{array}
$$
$$\tag{15.31}$$

It is clear again that minimizing the objective function of (15.31) is equivalent to maximizing the logarithm of its second term. Using the Lagrange multiplier maximizing method together with the expression for the derivative of the modified Bessel function given in [9, Eq. (8.486.12)], the optimum power allocation in this case can be shown to be given by

$$
\begin{aligned}
p_1^* = & \left[\frac{G_1}{G_2} \frac{1}{(P_T - p_1^*)^2} + \sqrt{\frac{G_1}{G_2}\left(1 + \frac{1}{\gamma_{th}}\right)} \frac{K_0\left(\frac{2\sqrt{\gamma_{th}^2 + \gamma_{th}}}{\sqrt{\overline{\gamma}_1 \overline{\gamma}_2}}\right)}{K_1\left(\frac{2\sqrt{\gamma_{th}^2 + \gamma_{th}}}{\sqrt{\overline{\gamma}_1 \overline{\gamma}_2}}\right)} \right. \\
& \times \left. \left(\frac{1}{\sqrt{p_1^*(P_T - p_1^*)^3}} - \frac{1}{\sqrt{p_1^{*3}(P_T - p_1^*)}} \right) \right]^{-\frac{1}{2}} ,
\end{aligned}
\tag{15.32}
$$

where $K_0(\cdot)$ is the zeroth-order modified Bessel function of the second kind defined in [1, Eq.(9.6.21)]. It is clear that (15.32) is a transcendental function in p_1^* and a similar equation can be written in terms of p_2^* only. Consequently, (15.32) has to be solved numerically. Numerical techniques such as the bisection or the Newton algorithms can be used and the required outage probability accuracy can be set to stop the algorithm. In [3, Section 8.1.4] an algorithm is introduced which uses a combination of the Newton and bisection methods to reach the optimal solution. In each step, it first uses the Newton method, and if the search does not improve the solution, it switches to the bisection method. In this way, the speed advantage of the Newton method is exploited without the fear of its divergence. In practice, (15.32) needs to be solved for p_1 only, and p_2 can be found as $p_2 = P_T - p_1$. As we will see in the numerical results section that follows, it is interesting to note that the closed-form solution corresponding to the regenerative systems case in (15.27) can be used to approximate the solution for the non-regenerative case. This is due to the fact that, as mentioned in [14], both regenerative and non-regenerative systems have similar performance at high average SNR. In addition, the former systems behave as a relatively tight lower bound on the latter ones at low average SNR. The performance of this approximate solution will be studied shortly. Also, for non-regenerative systems with diversity reception, equations are complicated and hence we ought to use again regenerative systems as an approximate solution here.

4.4.3 Numerical Examples. Fig. 15.12 shows the effect of optimizing the allocation of power on a dual-hop scenario for both regenerative and non-regenerative systems in comparison with an equivalent direct transmission. To carry out this comparison, it is assumed that in

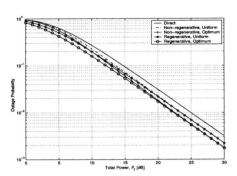

Figure 15.12. Effect of power optimization on the performance of dual-hop regenerative and non-regenerative systems ($G_1 = 1$ and $G_2 = 10$).

the case of relayed transmission, the source, the relay, and the destination nodes are placed on a straight line, and that all links are affected by the same shadowing environment following a simple path loss model with $\alpha = 3$. In this case, it can be shown that $G_3 = G_2/(1 + (G_2/G_1)^{-1/3})^3$. The figure shows the advantage of using relayed over direct transmission, even without power optimization. Note that for both systems, and in comparison with a uniform power allocation policy, optimizing the power allocation reduces outage probability. This implies that the same outage requirements can be met with less power, which in its turn implies less introduced interference to other nodes. On the other hand, it is clear from this same figure that non-regenerative systems with optimum power allocation can outperform regenerative systems with uniform power allocation. Fig. 15.13 shows the result of using the closed-form solution (15.27) to allocate the non-regenerative systems power (to bypass the use of the Newton/bisection iterative method). It is clear from the figure

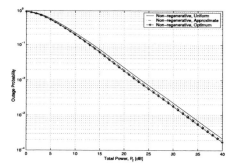

Figure 15.13. Comparison between the uniform, approximate, and optimal power allocation for non-regenerative systems ($G_1 = 1$ and $G_2 = 10$).

that this solution closely approximate the optimal solution, and hence may be attractive to employ from a practical point of view. Fig. 15.14 compares the ratio of power assigned to the weaker link for regenerative systems and non-regenerative systems as a function of the links unbalance ratio. It is clear that both systems split the power in relatively the same fashion between the two links. Consequently, this confirms that the simple solution for regenerative systems can be used to approximate that of non-regenerative systems. Note also that both systems devote larger power to the weaker link to reduce the overall outage probability, in contrast to optimal power allocation within the transmit-diversity context where larger power is allocated to better links [4, 12, 23]. Fig. 15.15

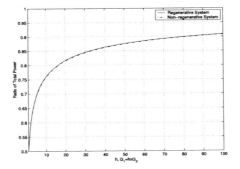

Figure 15.14. Ratio of the total power assigned to the weaker link as function of the links unbalance ($P_T = 30$ dB).

shows that power allocation is more beneficial if diversity reception is used at the destination. It is noted here that in the case of no diversity,

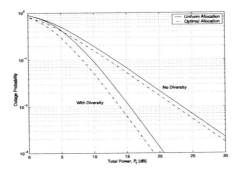

Figure 15.15. Effect of power optimization on the performance of dual-hop regenerative system with and without diversity reception ($G_1 = 1$ and $G_2 = 10$).

the system allocates most of the power to the weakest link, regardless of it being the first or the second, and independent of the total power. For the case of diversity reception, the system will add more power to p_1 as the first relay is driving both the first hop and the direct link. This is clear form Fig. 15.16 which compares the allocation of power in the diversity reception case with that of the simple relayed case. It is shown that more power is devoted to the source, compared with the no diversity case, and that this power increases with the increased total power budget.

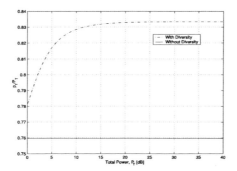

Figure 15.16. Ratio of the total power assigned to the weaker link as function of the total available power ($G_1 = 1$ and $G_2 = 10$).

5.   Multi-Hop Systems

Consider now the wireless communication system shown in Fig. 15.17. Here, signals propagate through N channels/hops before arriving to their

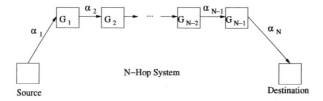

Figure 15.17.   The N-hops System.

destinations. Using the relay gain as per (15.6), the end-to-end SNR can be shown to be given by [13]

$$\gamma_{\text{eq}_1} = \left[\prod_{n=1}^{N} \left(1 + \frac{1}{\gamma_n} \right) - 1 \right]^{-1}, \qquad (15.33)$$

where $\gamma_n = \mathcal{E}_n \alpha_n^2 / N_{0,n}$ is the SNR of the nth hop. The N-hop end-to-end SNR expression in (15.33) is a generalization for the one given in (15.7) for the dual-hop end-to-end SNR. It should be noted that the inverse of the end-to-end SNR can be written in an alternative form which involves the sum of the inverse of the $\gamma(K)$ terms which have the following form

$$\gamma(K) = \prod_{k=1}^{K} \gamma_k, \quad K = 1, \cdots, N. \qquad (15.34)$$

For example, in a triple-hop system, the end-to-end SNR γ_{eq1} can be written as

$$\frac{1}{\gamma_{eq1}} = \frac{1}{\gamma_1} + \frac{1}{\gamma_2} + \frac{1}{\gamma_3} + \frac{1}{\gamma_1\gamma_2} + \frac{1}{\gamma_1\gamma_3} + \frac{1}{\gamma_2\gamma_3}. \tag{15.35}$$

This way, it is easy to see that the end-to-end SNR in (15.33) can be upper bounded by

$$\gamma_{eq2} = \left[\sum_{n=1}^{N} \frac{1}{\gamma_n} \right]^{-1} \tag{15.36}$$

which can be shown to be the end-to-end SNR of an N-hop system in which the relay gain is set to

$$G_n^2 = \frac{\mathcal{E}_n}{\mathcal{E}_{n-1}\alpha_n^2}. \tag{15.37}$$

This corresponds to an ideal relay capable of inverting the channel in the previous hop (regardless of the fading state of that hop). As such the performance of multi-hop systems employing such relays can serve as a benchmark for the performance of all practical non-regenerative systems. In addition, it should be noted here that the form of the equivalent SNR as per (15.36) is related to the harmonic mean of the individual links SNRs. Namely,

$$\gamma_{eq2} = \frac{\mu_H}{N} \tag{15.38}$$

where μ_H is the harmonic mean of the individual link SNRs and N is the number of hops.

5.1   Outage Probability

5.1.1   Non-Regenerative Systems.   For the non-regenerative multi-hop systems of interest, the cumulative distribution function (CFD) of (15.33) has to found to get the outage probability performance. Unfortunately, it seems hard if not impossible to get the CDF of (15.33) or even (15.36) in closed form. Hence, the end-to-end outage probability can be expressed in an alternative form as

$$P_{out} = 1 - \mathcal{L}^{-1}\left(\frac{\mathcal{M}_{1/\gamma_{eq}}(s)}{s} \right)\Big|_{\frac{1}{\gamma_{th}}} \tag{15.39}$$

where $\mathcal{L}^{-1}(\cdot)$ denotes the inverse Laplace transform and $\mathcal{M}_{1/\gamma_{eq}}(\cdot)$ is the moment generating function (MGF) of the inverse of the end-to-end SNR. It is clear from (15.39) that provided this MGF is available in closed-form, using any numerical technique for the Laplace transform inversion (such as, for example, the Euler numerical technique used in [24]), the end-to-end outage probability can be evaluated for many scenarios of interest.

Under the assumption that the hops are subject to independent fading, the MGF of $1/\gamma_{eq_2}$ in (15.36) is simply the product of the MGFs of $1/\gamma_n$, $n = 1, \cdots, N$ which for Rayleigh fading can be evaluated with the help of [9, Eq. (3.471.9)] as

$$\mathcal{M}_{1/\gamma_1}(s) = 2\sqrt{\frac{s}{\bar{\gamma}}}K_1\left(2\sqrt{\frac{s}{\bar{\gamma}}}\right), \qquad (15.40)$$

In contrast, the MGF of $1/\gamma_{eq_1}$ in (15.33) requires evaluating the MGF of the sum of dependent random variables, some of which is the reciprocal of the product of several random variable (i.e. for $K > 1$), which is a more difficult problem. This together with the fact that, from an outage probability standpoint, the form of the equivalent SNR as per (15.36) leads to a tight lower bound on that of (15.33) (as shown by numerical examples that follow) make it attractive to use for studying the performance of multi-hop non-regenerative systems.

5.1.2   Regenerative Systems.   In regenerative systems, outage decisions are taken on a per hop basis, and the overall system outage is dominated by the weakest hop/link. Consequently, outage probability is given by

$$P_{\text{out}} = P[\text{Min}(\gamma_1, \cdots, \gamma_N) < \gamma_{\text{th}}]. \qquad (15.41)$$

For Rayleigh fading, this can be easily found to be given by

$$P_{\text{out}} = 1 - \prod_{n=1}^{N} e^{-\frac{\gamma_{\text{th}}}{\bar{\gamma}_n}} \qquad (15.42)$$

5.1.3   Numerical Examples.   Fig. 15.18 compares the outage probability performance of a non-regenerative triple-hop system with the two relay gains of (15.6) and (15.37). It is clear from the figure that for all practical purposes the two curves are essentially indistinguishable. Consequently, the relay gain of (15.37) will be used for the remaining numerical examples presented in this chapter. Fig. 15.19 studies the system outage probability as a function of the total number of identically distributed hops. Note that a diminishing increase in outage probability results from increasing the number of hops. Finally, Fig. 15.20 compares the performance of non-regenerative systems with that of regenerative systems as a function of the number of hops. It is clear that there is an increasing gap in the performance between the two systems as the number of hops increases which indicates that regeneration is more crucial if the number of hops is large.

5.2   Average BER

5.2.1   Non-Regenerative Systems.   Since the statistics (the MGF in this case) of (15.33) are hard to find, average BER in this case can be bounded using a bound on the equivalent end-to-end SNR. It can

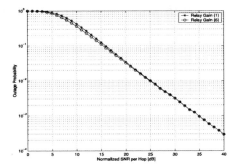

Figure 15.18. Comparison of the end-to-end outage probability of a triple-hop system with the relay gain in (15.6) and (15.37).

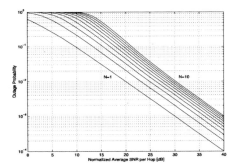

Figure 15.19. Effect of increasing the number of hops on the performance of non-regenerative systems.

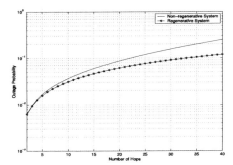

Figure 15.20. Effect of increasing the number of hops on the performance of regenerative and non-regenerative systems (normalized average SNR per hop $\bar{\gamma}/\gamma_{\text{th}} = 25$ dB).

be shown that γ_{eq} in (15.33) can be bounded by

$$\gamma_{\text{eq}} = \left[\prod_{n=1}^{N} \left(1 + \frac{1}{\gamma_n} \right) - 1 \right]^{-1} < \text{Min}(\gamma_1, \cdots, \gamma_N), \tag{15.43}$$

whose MGF can be shown to be given by

$$\mathcal{M}_{\gamma_{\text{eq}}}(s) = \frac{1}{1 + s \left[\sum_{n=1}^{N} \frac{1}{\bar{\gamma}_n} \right]^{-1}} \tag{15.44}$$

It should be noted here that this bound is not useful in studying the outage probability of non-regenerative systems as it corresponds to the exact performance of regenerative systems, and hence is not useful in carrying any comparison between these two systems.

5.2.2   Regenerative Systems.   Let Pe_n denotes the average BER of hop n. Then, the end-to-end average BER is given by [26]

$$P_b(E) = \sum_{n=1}^{N} (-2)^{n-1} \mathcal{S}_N^n, \tag{15.45}$$

where $\mathcal{S}_N^n$ denotes the sum of the product of all n distinctive terms selected from $Pe_1, Pe_2, \cdots, Pe_N$, i.e.,

$$\mathcal{S}_N^n = \sum_{\alpha_1 \cdots \alpha_n = 1}^{N} Pe_{\alpha_1} Pe_{\alpha_1} \cdots Pe_{\alpha_n}. \tag{15.46}$$

When $Pe_1 = Pe_2 = \cdots = Pe_N = Pe$, (15.45) reduces to

$$P_b(E) = \sum_{n=1}^{N} (-2)^{n-1} \binom{N}{n} Pe^n. \tag{15.47}$$

5.3   Optimum Power Allocation

5.3.1   Analysis.   For regenerative systems, the problem in this case is formulated in a similar fashion to (15.26) as

$$\begin{aligned} &\text{min} && P_{\text{out}} = 1 - \prod_{n=1}^{N} e^{-\frac{\gamma_{\text{th}}}{\gamma_n}} \\ &\text{subject to} && \begin{cases} \sum_{n=1}^{N} p_n = P_T \\ p_n \leq P_{\text{max}}, && n = 1, \cdots, N. \end{cases} \end{aligned} \tag{15.48}$$

This problem can be solved using the Lagrange multiplier maximization method to yield the optimum power allocation as

$$p_n^* = P_T \left[1 + \sqrt{G_n} \sum_{k=1, k \neq n}^{N} \frac{1}{\sqrt{G_k}} \right]^{-1} \tag{15.49}$$

The solution given in (15.49) was obtained by neglecting the second constraint in (15.48). If the solution of the optimization problem as per (15.49) leads to a power allocation that violates ($p_n \leq P_{\max}, n = 1, \cdots, N$), then clipping is applied to all violating nodes, and re-optimization takes place again over the rest of the nodes with the first constraint modified as $\left(\sum_{t=1}^{N-T} p_n = P_T - TP_{\max}\right)$, where T is the number of the violating nodes. Clipping means that the KKT condition is fulfilled for all violating nodes, which preserves optimality of the solution.

For non-regenerative systems, it is not easy if not impossible to get a closed-form expression for the end-to-end outage probability [13]. Consequently, the optimization problem is difficult to formulate in a tractable form. Rather, using the fact that regenerative systems performance is a lower bound on that of non-regenerative systems and that this bound is tight for high average SNR, the solution for regenerative systems given in (15.49) can also be used as an approximate solution for non-regenerative systems.

In both systems, optimizing the allocation of power assuming diversity reception seems impractical as the required number of parameters needed to be estimated grows substantially as the number of hops increases. A suboptimal, yet more practical, solution is to optimize the allocation of power assuming no diversity reception. Then, another optimization process can take place locally in clusters of size two to refine the distribution of power as per (15.30).

5.3.2   Numerical Examples.   Fig. 15.21 shows the case of having a six-hop system in which, starting with $G_1 = 2$, hop n, ($n = 2, \cdots, 6$), has a parameter G_n which is twice the preceding one. The

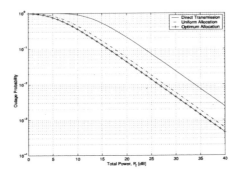

Figure 15.21.   Effect of power optimization on a six-hop system ($G_1 = 2$).

unbalance in terms of G_n refers to the case of having unequally spaced nodes, or links with different shadowing severity. It is clear that power optimization is more pronounced here since we can see a gain of almost

2 dB over uniform power allocation. Also, the advantage of relaying over equivalent direct transmission (in terms of coverage as we assume again that the relays are placed on a straight line, and $\alpha = 3$) is more clear here regardless of power optimization. Fig. 15.22 confirms that optimizing the allocation of power is more essential as the number of hops increases. Here, we start with two hops, and we successively add a new hop with

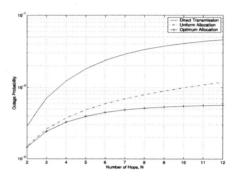

Figure 15.22. Effect of increasing the number of hops on the difference in performance between uniform and optimal power allocation ($G_1 = 2$).

parameter G_n which is twice the preceding one. It is clear from this figure that (i) uniform allocation results in a fast deterioration in the system performance, while optimal power allocation limits the loss in performance resulting from increasing the number of hops and (ii) the advantage of relayed transmission over equivalent direct transmission increases with increasing the number of hops.

6.   Conclusion

We studied in this chapter the end-to-end performance of multi-hop communication systems in terms of average bit error rate and outage probability over Rayleigh fading channels. More specifically, several types of relaying systems were considered and compared. For instance, we studied non-regenerative systems consisting of relays which operate as a simple transparent repeater by amplifying the received signal and retransmitting it to the final destination. We also looked into the performance of systems in which mobile relay units operate with regenerative capabilities. Numerical results show that while dual-hop systems with regenerative relays outperform those with non-regenerative relays at low average SNR, these two systems are essentially equivalent at high average SNR. However the performance gap between the two systems increases as the number of hops increases. We also looked at the optimum power allocation over these multi-hop communication systems and showed that this optimization enhances their performance in particular if the links are highly unbalanced or if the number of hops is large.

References

[1] M. Abramowitz and I. A. Stegun, *Handbook of Mathematical Functions with Formulas, Graphs, and Mathematical Tables.* New York, NY: Dover Publications, ninth ed., 1970.

[2] P. Anghel, and M. Kaveh, "Analysis of two-hop transmission over Rayleigh fading channels," in *Proc. IEEE International Symposium on Advances in Wireless Communications (ISWC'02), Victoria, BC, Canada*, pp. 155–156, September 2002.

[3] C. F. Van Loan, *Introduction to Scientific Computing.* Upper Saddle River, New Jersey: Prentice Hall Publishers, 2000.

[4] J. K. Cavers, "Optimized use of diversity modes in transmitter diversit systems," in *IEEE Veh. Technol. Conf. (VTC'99), Houston, TX*, pp. 1768–1773, May 1999.

[5] M. Chiang, D. ONeill, D. Julian, and S. Boyd, "Resource allocation for QoS provisioning in wireless ad hoc networks," in *Proc. IEEE Global Commun. Conf. (GLOBECOM'01), San Antonio, TX*, pp. 2911–2915, November 2001.

[6] V. Emamian, P. Anghel, and M. Kaveh, "Outage probability of a multi-user spatial diversity system in wireless networks," in *Proc. IEEE Veh. Technol. Conf. (VTC'02), Vancouver, BC, Canada*, pages 573 –576, September 2002.

[7] V. Emamian and M. Kaveh, "Combating shadowing effects for systems with transmitter diversity by using collaboration among mobile users," in *Proceedings of the International Symposium on Communications, (ISC'01), Taiwan*, pp. 105.1–105.4, November 2001.

[8] R. M. Gagliardi, *Introduction to Communications Engineering.* New York, NY: John Wiley & Sons, Inc., 1988.

[9] I. S. Gradshteyn and I. M. Ryzhik, *Table of Integrals, Series, and Products.* San Diego, CA: Academic Press, fifth ed., 1994.

[10] M. Grossglauser and D. Tse, "Mobility increases the capacity of ad-hoc wireless networks," in *Proc. IEEE Conference on Computer Communications (INFOCOM'01), Anchorage, AK*, pp. 1360–1369, April 2001.

[11] M. O. Hasna and M. -S. Alouini, "Application of the harmonic mean statistics to the end-to-end performance oftransmission systems with relays," in *Proc. IEEE Global Commun. Conf. (GLOBECOM'02), Taipei, Taiwan*, November 2002. Journal version to appear in *IEEE Trans. Commun.*

[12] M. O. Hasna and M. -S. Alouini, "Optimum power allocation for selective transmit-diversity systems over Nakagami fading channels," in *Proc. IEEE Int. Conf. on Acoustics, Speech, and Signal Processing (ICASSP'02), Orlando, FL*, pp. 2193 –2196, May 2002.

[13] M. O. Hasna and M. -S. Alouini, "Outage probability of multihop transmissions over Nakagami fading channels," *IEEE Commun. Letters*, vol. 7, pp. 216–218, May 2003.

[14] M. O. Hasna and M. -S. Alouini, "Performance analysis of two-hop relayed transmission over Rayleigh fading channels," in *Proc. IEEE Veh. Technol. Conf. (VTC'02), Vancouver, BC, Canada*, pp. 1992 – 1996, September 2002. Journal version to appear in *IEEE Trans. Wireless Commun.*, vol. 2, November 2003.

[15] M. O. Hasna and M. -S. Alouini, "A performance study of dual-hop transmissions with fixed gain relays," in *Proc. IEEE Int. Conf. on Acoustics, Speech, and Signal Processing (ICASSP'03), Hong Kong*, April 2003.

[16] J. N. Laneman and G. W. Wornell, "Energy efficient antenna sharing and relaying for wireless networks," in *Proc. IEEE Wireless Com. and Net. Conf. (WCNC'00), Chicago, IL*, pp. 7–12, October 2000.

[17] J. N. Laneman and G. W. Wornell, "Exploiting distributed spatial diversity in wireless networks," in *Proc. of the 40th Allerton Conference on Communication, Control, and Computing (Allerton'00), Allerton Park, IL*, pp. 775–785, September 2000.

[18] J. N. Laneman, G. W. Wornell, and D. N. C. Tse, "An efficient protocol for realizing cooperative diversity in wireless networks," in *Proc. IEEE Int. Symposium Inform. Theory (ISIT'01), Washington, DC*, p. 294, June 2001.

[19] T. K. Moon and W. C. Stirling, *Mathematical Methods and Algorithms for Signal Processing*. Upper Saddle River, New Jersey: Prentice Hall Publishers, 2000.

[20] R. L. Rardin, *Optimization In Operations Research*. Upper Saddle River, New Jersey: Prentice Hall Publishers, 1998.

[21] A. Sendonaris, E. Erkip, and B. Aazhang, "Increasing uplink capacity via user cooperation diversity," in *Proc. IEEE Int. Symposium Inform. Theory (ISIT'98), Cambridge, MA*, p. 156, August 1998.

[22] M. K. Simon and M. -S. Alouini, *Digital Communication over Fading Channels*. New York, NY: John Wiley & Sons, Inc., 2000.

[23] Y. -C. Ko and M. -S. Alouini, "Estimation of Nakagami fading channel parameters with application to optimized transmitter diversity systems," *IEEE Trans. on Wireless Commun.*, vol. 2, pp. 250–259, March 2003.

[24] Y. -C. Ko, M. -S. Alouini, and M. K. Simon, "An MGF-based numerical technique for the outage probability evaluation of diversity systems," *IEEE Trans. Commun.*, vol. 48, pp. 1783–1787, November 2000.

[25] H. Yanikomeroglu, "Fixed and mobile relaying technologies for cellular networks," in *Proc. of the Second Workshop on Applications and Services in Wireless Networks (ASWN'02), Paris, France*, pp. 75–81, July 2002.

[26] R. Zhang and M. -S. Alouini, "A channel-aware inter-cluster routing protocol for wireless ad hoc networks," in *Proc. IEEE Int. Symposium on Commun. and Applications (ISCTA'01), Ambelside, UK.*, pp. 46–51, 2001.

Chapter 16

MOBILITY MANAGEMENT FOR WIRELESS NETWORKS: MODELING AND ANALYSIS

Yuguang Fang and Wenchao Ma
Department of Electrical and Computer Engineering
University of Florida
435 Engineering Building, P.O.Box 116130
Gainesville, FL 32611
Tel: (352) 846-3043, Fax: (352) 392-0044
fang@ece.ufl.edu

Abstract   Mobility management plays a significant role in the current and the future wireless mobile networks in effectively delivering services to the mobile users on the move. Many schemes have been proposed and investigated extensively in the past. However, most performance analyses were carried out either under simplistic assumptions on some time variables or via simulations. Recently, we have developed a new analytical approach to investigating the modeling and performance analysis for mobility management schemes under fairly general assumption. In this chapter, we present the techniques we have developed for this approach lately and summarize the major results we have obtained for a few mobility management schemes such as *Movement-based Mobility Management, Pointer Forwarding Scheme (PFS), Two-level location management scheme*, and *Two-Location Algorithm (TLA)*.

Keywords: Mobility management, location management, registration, paging, cellular, wireless networks

1.   Introduction

In wireless mobile networks, in order to effectively deliver a service to a mobile user, the location of a called mobile user must be determined within a certain time limit (before the service is blocked). When the call connection is in progress while the mobile is moving, the network has to follow the mobile users and allocate enough resource to provide seamless continuing service without user awareness that in fact the network facility (such as base station) is changing. *Mobility management* is used to track the mobile users that move from place to place in the coverage of a wireless mobile network or in the coverage of multiple communications networks working together to fulfill the grand vision of ubiquitous communications. Thus, *mobility management* is a key component for the effective operations of wireless networks to deliver wireless Internet services (see [6] and references therein).

Wireless networks provide services to their subscribers in the coverage area. Such area is populated with base stations, each of which is responsible for relaying communication services for the mobiles traveling in its coverage called *cells*. A group of cells form a *registration area (RA)*, which is managed by *mobile switching center* (MSC) connecting directly to the *Public Switched Telephone Networks (PSTN)*. In this chapter, we will use *a mobile, a mobile user, a mobile subscriber* and *a mobile terminal*, interchangeably. The netshell communications architecture is shown in Figure 16.1.

When a mobile communicates with another user (either wired user or wireless user), the mobile will connect to the nearest base station (or the base station which provides the strongest receiving signal strength), which then establishes the communication with the other user over the existing communication infrastructure (such as PSTN). When a mobile user engaging a communication and moving from one cell to another, the base station in the new cell will allocate a channel to continue to provide the service to the mobile user without interruption (if possible). The switch from one channel to another (or from one base to another) is called *handoff*. One of important aspect of the mobility management is the handoff management: how to achieve a smooth handoff without degrading the service currently in progress over the air. When users do not engage any communications and move around, the system has to track them in order to deliver possible services to them. This requires mobiles to inform the network their whereabouts when they move, thus the system can locate them based on the previously reported information. This process is called *location management* (sometimes is also called *mobility management*). In this chapter, we focus on the location management.

Location management is a unique feature for wireless networking, and needs to be addressed carefully. In second generation wireless systems, the wireless networks standards IS-41 ([15]) and GSM MAP ([16]) use two-level strategies for mobility management in that a two-tier system consisting of Home Location Register (HLR) and Visitor Location Register (VLR) databases is deployed (see Figure 16.2). Although there are some modifications on the mobility management for the third gen-

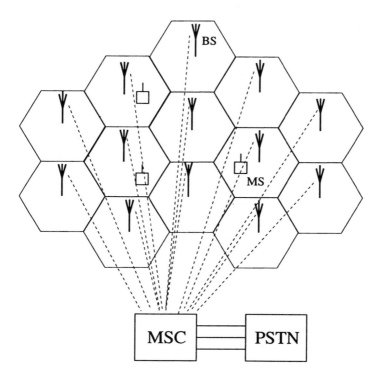

BS: Base Station   MS: Mobile Station

MSC: Mobile Switching Center

PSTN: Public Switched Telephone Network

Figure 16.1.   The communication architecture for wireless cellular systems

eration wireless systems (such as the introduction of *gateway location register* or *GLR*), the fundamental operations in mobility management remains more or less the same. Hence, we will use the IS-41 standard as the baseline study here. In two-tier mobility management architecture shown in Figure 16.2, the HLR stores the user profiles of its registered mobiles, which contain the user profile information such as the mobile's identification number, the type of services subscribed, the quality of service (QoS) requirements and the current location information. The VLR stores the replications of user profiles and the temporary identification number for those mobiles which are currently visiting the associated RA. There are two basic operations in location management: *registration* and *location tracking*. The registration (also called *location update* in many situations) is the process that a mobile informs the system of its current location, the location tracking (also called *call delivery*) is the process that the system locates its mobile in order to deliver a call service to the mobile. When a user subscribes a service to a wireless network, a record of the mobile user's profile will be created in the system's HLR. Whenever and wherever the mobile user travels in the system's coverage, the mobile's location will be reported to the HLR (registration) according to some strategies. Then the location in HLR will be used to locate (*find*) the mobile. When the mobile visits a new area (called *registration area (RA)*), a temporary record will be created in the VLR of the visited system, and the VLR will then sends a registration message to the HLR. All the signaling messages are exchanged over the overlaying signaling network using signaling system 7 (SS7) standard.

To deliver a call service to a mobile from an originating switch (MSC), the calling mobile contacts its MSC, which then initiates a query to the caller's HLR to find where the callee is last reported (i.e., the VLR the callee was sighted). When the HLR receives this query, it will send a query to the VLR. The VLR, upon finding the mobile in its charging area, will return a routable address called the *temporary-location directory number (TLDN)* to the originating switch (MSC) through the HLR. Based on the TLDN, the originating switch (MSC) will set up the trunk (e.g., voice circuit with enough bandwidth for voice service) to the mobile. Thus, the call delivery process consists of two parts: *find*, the method to locate the mobile, and the trunk setup for the mobile using TLDN. The signaling traffic due to *find* and *registration (location update)* can be significant. Various schemes to reduce such traffic have been proposed. A very excellent survey on this topic is presented in [6]. There are mainly two approaches to reduce the signaling traffic. One is to manage the mobility databases physically or logically, the other is to control the location update frequency more intelligently. The funda-

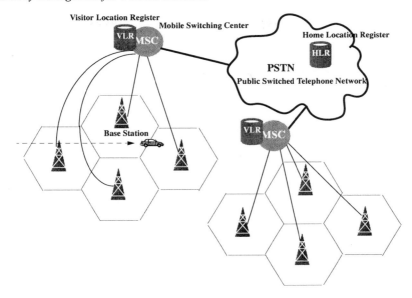

Figure 16.2. Two-tier location management architecture ([44])

mental idea in both approaches is to localize the signaling traffic. Here are a few representative schemes proposed in the past. In [35], the *location cache scheme* was proposed and shown significant improvement over IS-41 scheme when the frequency of the incoming calls is high with respect to the mobility. To reduce the signaling traffic from the mobile to the HLR, the *pointer forwarding scheme* ([34]) was proposed based on the observation that it may be better to setup a forwarding pointer from the previous VLR to avoid more expensive registration from the mobile. By storing location profile, registration traffic can also be reduced, this idea leads to the *alternative location algorithm (ALA)* and *two location algorithm (TLA)* ([42] and references therein). By observing that signaling cost can be significantly affected by the location database distribution, Ho and Akyildiz proposed the *local anchor scheme* to localize the registration traffic ([30]) and the dynamic hierarchical database architecture using *directory registers* ([29]). Recently, Ma and Fang ([47]) proposed a two-level pointer forwarding scheme to combine the advantages of pointer forwarding scheme and local anchor scheme. To maintain the freshness of the location information, active location update (also known as *autonomous registration*) was also suggested in the standard as an option ([16]). How often the active location update should be done is still an open question. Three schemes were investigated in the past ([7]): *timer-based, movement-based* and *distance-based*. It has been shown that the most effective scheme among the three is distance-based: whenever a mobile moves away from the previously reported spot in certain distance, the mobile will carry out the location update. However, the implementation is complicated as most mobiles do not have GPS receivers equipped. The time-based scheme is simple to use, which has been used in most mobiles, however, it is not efficient in terms of signaling traffic, particularly for the mobiles with low mobility. It turns out that the movement-based (MB) scheme strikes the balance between the other two in both efficiency and implementation, which will be the one we investigate in this chapter. The problem is how to choose the movement threshold. The large the the threshold, the larger the uncertainty region of the mobile; while the smaller the threshold, the smaller the uncertainty region, but the more the signaling traffic. All above schemes are for location management. Another major part of signaling traffic comes from the the last segment during the call delivery process: the paging. When the callee's VLR receives a query for the callee, a paging process is initiated. Depending on the information available and resource consideration, sequential paging, parallel paging or selective paging can be chosen ([56]).

To effectively evaluate a mobility management scheme, we have to deal with signaling traffic caused by both location update and call delivery. How to quantify the signaling traffic and the cost analysis become important. Signaling traffic cost relies on many factors such as the terminating call arrivals and the users' mobility. In the past, cost analysis of most mobility management schemes were carried out under the assumption that most time variables are exponentially distributed. For example, the time between two *served* calls, which we called *inter-service time* ([20]), was usually assumed to be exponentially distributed. Although the call arrivals terminating at a mobile, say, $\mathcal{T}$, may be approximately modeled by Poisson process, the inter-service time is not identical to the inter-arrival time due to the *busy-line effect* ([20]), i.e., some call arrivals for the mobile $\mathcal{T}$ may not be connected because $\mathcal{T}$ is serving another call, this is particularly true for data connections that tend to connect to the wireless systems longer than voice connection on average. Hence the served calls are in fact a "sampled" Poisson process, thus will be most likely not Poisson process. The reason we are interested in the inter-service times is that during the call connection of a mobile, no signaling traffic is necessary from this mobile since the network knows where the mobile is, only when the mobile is idle, does it need to make location update to inform the system. Thus, the location management is affected by the inter-service times rather than the inter-arrival times to the mobile. Moreover, due to the new trend of applications and user habits, even the inter-arrival time for the terminating calls to a mobile may not be exponentially distributed anymore. Some adaptive or dynamic schemes for choosing some location management parameters depend on the explicit form of cost ([43]), whether such schemes are still effective or not when the inter-service times are not exponentially distributed is a question, there is no justification in the literature.

Recently, we have developed a new analytical approach to systematically analyze the performance of a few mobility management schemes under more realistic assumptions ([17, 18, 47]). In this chapter, we will present this analytical approach for the modeling and performance analysis for mobility management. We focus on the general analytical results for the signaling cost analysis for a few mobility management schemes we have studied in the past to illustrate our approach. Obviously, our results can be used to design the dynamic location management schemes.

This chapter is organized as follows. In the next section, we present the descriptions of a few known location management schemes we will study in this chapter. We then present a general framework for the evaluation of mobility management. The crucial analytical result on the probability distribution of the number of area boundary crossings are given in the fourth section. In the fifth section, we present the general analytical results for the signaling costs for the mobility management schemes we are interested in. We will conclude this chapter in the last section.

2. Mobility Management Schemes

There are many location management schemes in the literature, [6] presents a very comprehensive survey on all aspects of mobility management. In this chapter, we only concentrate on the schemes for which we have analytical results under fairly general assumptions. For some interesting schemes such as per-user caching scheme ([35]), location anchoring scheme ([30]), and location profile based scheme ([52]), we do not have the analytical results as yet, we will not include them in this section. More details can be found in [6].

2.1 IS-41 Scheme

To set the baseline comparison study, we first briefly go over the IS-41 scheme (or GSM MAP), the de facto standards for second generation wireless systems. We use the terminology used in [34]. An operation *move* means that a mobile user moves from one Registration Area (RA) (also called Location Area (LA)) to another while an operation *find* is the process to determine the RA a mobile user is currently visiting. The *move* and *find* in second generation location management schemes (such as in IS-41 or GSM MAP) are called *basic move* and *basic find*. In the *basic move* operation, a mobile detects if it is in a new RA. If it is, it will send a registration message to the new VLR, the VLR will send a message to the HLR. The HLR will send a de-registration message to the old VLR, which will, upon receiving the de-registration message, send the cancellation confirmation message to the HLR. The HLR will also send a cancellation confirmation message to the new VLR. In the *basic find*, call to a mobile $\mathcal{T}$ is detected at a local switch. If the called party is in the same RA, the connection can be setup directly without querying the HLR. Otherwise, the local switch (VLR) queries the HLR for the callee, then HLR will query the callee's VLR. Upon receiving callee's location, the HLR will forward the location information to the caller's local switch, which will then establish the call connection with the callee. Periodic location update is optional for improving the efficiency of IS-41 and GSM MAP ([15, 16]).

2.2   Movement-based Mobility Management

There are many ways to improve the performance of the mobility management detailed in the IS-41. To minimize the signaling traffic due to the location update while keeping the location information fresh, we need to determine when we would need location update. In the current literature, three location update schemes were proposed and studied ([3, 7]): *distance-based location update, movement-based location update* and *time-based location update*. In distance-based locate update scheme, location update will be performed when a mobile terminal moves d cells away from the cell in which the previous location update was performed, where d is a distance threshold. In the movement-based location update scheme, a mobile terminal will carry out a location update whenever the mobile terminal completes d movements between cells (whenever the mobile moves from one cell to another, we count it as one move), where d is the movement threshold. In the time-based location update scheme, the mobile terminal will update its location every d time units, where d is the time threshold. It has been shown ([7]) that the distance-based location update scheme gives the best result in terms of signaling traffic, however, it may not be practical because a mobile terminal has to know its own position information in the network topology. The time-based location update scheme is the simplest to implement, however, unnecessary signaling traffic may result (imagine a terminal stationary for a long period may not need to do any update before it moves). The movement-based location update scheme seems to be the best choice in terms of signaling traffic and implementation complexity. We will analyze the movement-based mobility management scheme here.

2.3 Pointer Forwarding Scheme (PFS)

The PFS modifies the *move* and *find* used in the IS-41 in the following fashion ([34]). When a mobile $\mathcal{T}$ moves from one RA to another, it will inform its local switch (and VLR) at the new RA, which will then determine whether to invoke the *basic move* or the *forwarding move*. In the *forwarding move*, the new VLR exchanges messages with the old VLR to setup pointer from the old VLR to the new VLR, but does not involve the HLR. A subsequent call to the mobile $\mathcal{T}$ from some other switches will invoke the *forwarding find* procedure to locate the mobile: queries the mobile's HLR as in the *basic find*, and obtains a "potentially outdated" pointer to the old VLR, which will then direct the *find* to the new VLR using the pointer to locate the mobile $\mathcal{T}$. To ensure that the time taken by the *forwarding find* is within the tolerable time limit, the length of the chain of the forwarding pointers must be limited. This can be done by setting up the threshold for chain length to be a number, say, K, i.e., whenever the mobile $\mathcal{T}$ crosses K RA boundaries, it will register itself through the *basic move* (i.e., basic registration with HLR). In this way, the signaling traffic between the mobile and HLR can be curbed potentially.

2.4 Two-Level Pointer Forwarding Scheme

Two-level pointer forwarding scheme (TLPDS) is a generalization of PFS, which attempts to localize the location update signaling traffic and is consistent with the current hierarchical architecture of wireless system design ([47]). Instead of using one-level pointers in PFS, we activate a new layer of management entities, called *Mobility Agent (MA)*, the VLR which is in charge of multiple RAs, reflecting the regional activities of mobile users. The Two-Level Pointer Forwarding Scheme modifies the basic procedures used in IS-41 as follows. When a mobile moves from one RA to another, it informs the switch (and the VLR) at the new RA about the old RA. It also informs the new RA about the previous MA it was registered at. The switch at the new RA determines whether to invoke the *BasicMOVE* (update to the HLR) or the *TwoLevelFwdMOVE* (update to either the previous RA or the previous MA). In *TwoLevelFwd-MOVE*, the new VLR exchanges messages with the old VLR or the old MA to set up a forwarding pointer from the old VLR to the new VLR. If a pointer is set up from the previous MA, the new VLR is selected as the current MA. The *TwoLevelFwdMOVE* procedures do not involve

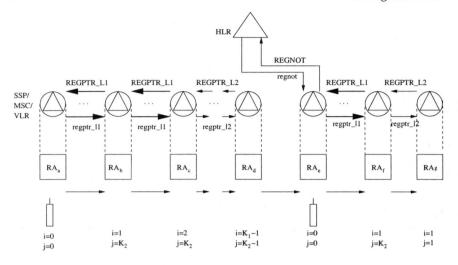

Figure 16.3. TLPDS location management architecture ([47])

the user's HLR. For example, Figure 16.3 shows a *Two-Level Forward MOVE* procedures with level_1 pointers chain threshold limited to 3. Assume that a user moves from RAa to RAg (these RAs are not necessary to be adjacent) and RAa is the user's MA. When the user leaves RAa but before enters RAb, the user informs the new VLRs and the level_2 pointers are built from the old VLR to the new VLR. When the user enters RAb, the chain threshold for level_2 pointer is reached, so RAb is selected as the user's new MA and a level_1 pointer is set up from the old MA to the new MA. At the same time, level_2 pointer chain is reset. The similar procedures are used at RAc. A level_1 pointer is set up from RAb to RAc, and the VLR in RAc is the new user's MA. As the user keeps moving. In RAe, the threshold of level_2 pointer chain is reached again, while this time the threshold of the level_1 pointer chain is reached too. Instead of exchanging information with the previous MA, the *BasicMOVE* is invoked. The HLR is updated with the user current location. The messages REGPTR_L1 and REGPTR_L2 are messages from the new VLR to the old VLR specifying that a level_1 or level_2 forwarding pointer is to be set up; messages regptr_l1 and regptr_l2 are the confirmations from the old VLR (or MA). In this figure, the VLRs in RAa, RAb, RAe and RAf are selected as the user's MAs.

To locate a mobile user, the *Find* procedure, called *TwolevelFwdFIND* procedure, is invoked for the subsequent calls to the mobile user from some other switches. The user's HLR is queried first as in the basic strategy, and a pointer to the user's potentially outdated MA is obtained.

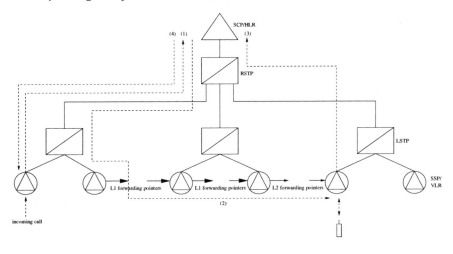

Figure 16.4. TLPDS Find procedure ([47])

The pointer chain is followed to find out the user's current location (see Figure 16.4). As we can see, in the two-level forwarding scheme, the chain length can be longer than that in the basic pointer forwarding scheme without increasing the *Find* penalty significantly. The previous study [34] shows that more saving can be obtained with longer chain. However, the pointer chain length is limited by the delay restriction requirement. By appropriately tuning the two thresholds in our schemes, we can mitigate the signaling cost without too much increase in the call setup delay.

2.5   Two Location Algorithm (TLA)

In the TLA ([42]), a mobile $\mathcal{T}$ has a small built-in memory to store the addresses for the two most recently visited RA's. The record of $\mathcal{T}$ in the HLR also has an extra field to store the two corresponding two locations. The first memory location stores the most recently visited RA. The TLA guarantees that the mobile $\mathcal{T}$ is in one of the two locations. When the mobile $\mathcal{T}$ joins the network, it registers with the network and stores the location in its memory and updates the HLR with its location. When the mobile $\mathcal{T}$ moves to another registration area, it checks whether the RA is in the memory or not. If the new RA is not in the memory, the most recently visited (MRV) RA in the memory is moved to the second memory entry while the new RA is stored in the MRV position in the memory of $\mathcal{T}$. At the same time, a registration operation is performed to make the same modification in the HLR record. If the address for the new RA is already in the memory, swaps the two locations in the memory of $\mathcal{T}$ and no registration is needed and no action is taken in HLR record. Thus, in TLA, no registration is performed when a mobile moves back and forth in two locations. The consequence is that the location entries in the mobile and HLR may not be exactly the same: the MRV RA in HLR may not be the MRV RA in reality!

When a call arrives for the mobile $\mathcal{T}$, the two addresses are used to find the actual location of the mobile $\mathcal{T}$. The order of the addresses used to locate $\mathcal{T}$ will affect the performance of the algorithm. If $\mathcal{T}$ is located in the first try (i.e., a *location hit*), then the *find* cost is the same as the one in IS-41 scheme. Otherwise, the second try (due to the *location miss* in the first try) will find $\mathcal{T}$, which incurs additional cost. This additional cost has to be lower than the saved registration cost in order to make the TLA effective, which will demand a tradeoff analysis.

3.   A General Framework for Performance Evaluation

Consider a wireless mobile network with cells of the same size. A mobile terminal visits a cell for a time interval which is generally distributed, then moves to the neighboring cell with equal probability (we are interested in this paper the homogeneous wireless networks in which all cells are statistically identical and all registration areas (RAs) are statistically identical). When a mobile is engaging a service, the network knows the location of the mobile, hence the basic idea to evaluate the performance of a mobility management scheme is to study the overall signaling traffic caused by location update and call delivery (*Find* or *Paging*) when the mobile is NOT engaging communication through the network, i.e., we only need to study the overall signaling traffic during the inter-service time. To carry tradeoff analysis, the signaling traffic has to be quantified, hence certain cost mapping may be necessary. Based on the cost structure for signaling traffic, performance comparison can be made possible. Moreover, some mobility management schemes have some tunable parameters, which need to be chosen to optimize the performance. By quantifying the signaling traffic, optimization problems can be formulated and solved analytically.

We take the movement-based location update scheme as one example to illustrate the framework for the cost analysis. Let d denote the movement threshold, i.e., a mobile terminal will perform a location update whenever the mobile terminal makes d movements (equal to the number of serving cell switching) after the last location update. When an incoming call to a mobile terminal arrives, the network initiates the paging process to locate the called mobile terminal. Thus, both location update and terminal paging will incur signaling traffic, the location updates consume the uplink bandwidth and mobile terminal power, while the terminal paging mainly utilizes the downlink resource, hence the cost factors for both processes are different. When uplink signaling traffic is high or power consumption is a serious consideration, it may be better to use terminal paging instead. Different users or terminals may have different quality of service (QoS) requirement, location update and paging may be designed to treat them differently. All these factors should be considered and a more general cost function, which reflects these considerations, is desirable. A reasonable cost should consider two factors: location update signaling and paging signaling, which can be captured by the following general cost function:

$$\mathcal{C}(d) = \mathcal{C}_u(N_u(d), \lambda_u, q_s) + \mathcal{C}_p(N_p(d), \lambda_p, q_s) \qquad (16.1)$$

where $N_u(d)$ and $N_p(d)$ are the average number of location updates and the average number of paging messages under the movement-based location update and a paging scheme with movement threshold d, respectively, during a typical period of inter-service time, λ_u is the signaling rate for location updates at a mobile switching center (MSC) and λ_p is the signaling rate for paging at an MSC, q_s indicates the QoS factor, $\mathcal{C}_u$ and $\mathcal{C}_p$ are two functions, reflecting the costs for location updates and paging. Depending on the choice of the two functions, we can obtain different mobility management schemes, particularly, we can obtain the optimal movement threshold d to minimize the total cost $\mathcal{C}(d)$, which gives us the best tradeoff scheme.

In the current literature, the total cost is chosen to be the linear combination of the location update and the paging signaling traffic, i.e., $\mathcal{C}_c(N_u(d), \lambda_u, q_s) = UN_u(d)$, $\mathcal{C}_p(N_p(d), \lambda_p, q_s) = PN_p(d)$, where U is the cost factor for location update and P is the cost factor for terminal paging ([4, 40, 42]). When the signaling traffic for location updates in an area is too high, some mobile terminals may be advised to lower their location updates, this can be done by choosing a function $\mathcal{C}_u$ to contain a factor $1/(\lambda_{\max} - \lambda_u)$ where $\lambda_{\max}$ is the maximum allowable signaling rate for location update. For example, if we choose $\mathcal{C}_u = U/(\lambda_{\max} - \lambda_u)$, when λ_u is approaching $\lambda_{\max}$, the update cost will become huge, the optimization will prefer to use paging more, leading to fewer location updates. How to choose the appropriate cost functions is a challenging problem to be addressed in the future.

For other schemes, the total cost will consist of two parts as well. The first part reflects the cost resulting from the location update, while the second part is contributed by the call delivery cost (*Find* operation). In the movement-based mobility management scheme we discussed above, since the call delivery cost mainly comes from the terminal paging, hence we can use the terminal paging cost. Thus, the remaining task is to find the quantities we need in the cost function for each scheme.

4. Probability of the Number of Area Boundary Crossings

In order to carry out the performance analysis of mobility management schemes, we need to model some of the time variables appropriately and compute the probability distribution of the number of RA boundary crossings or cell boundary crossings. For example, in the mobility-based mobility management, we need to find the average number of location updates during the inter-service time, hence need to determine the distribution of cell boundary crossings. In other schemes (such as the PFS or TLPDS), we need to find the probability distribution of RA crossings to determine the cost for location update and call delivery. The results in this subsection have been presented in [20]. Although the results there are for RA crossings, they are valid for the cell crossings too, the only change to be made is to replace the RA residence time by the cell residence time. As we will show that this distribution plays a significant role in our analytical approach. For the simplicity of presentation, in this section, we use *an area* to denote either a registration area (RA) or a cell so that we can use the results to determine the probability of the number of RA boundary crossings and the probability of the number of cell boundary crossings as long as we use the corresponding residence time.

Assume that the incoming calls to a mobile, say, $\mathcal{T}$, form a Poisson process. The time the mobile stays in an *area* is called the *area residence time (ART)*. The time the mobile stays in a *registration area (RA)* is called the *RA residence time (RRT)*. The time the mobile stays in a cell is called the *cell residence time (CRT)*. Thus, ART can be either RRT or CRT for later applications. We assume that the ART is generally distributed with a non-lattice distribution (i.e., the probability distribution does not have discrete component). The time between the end of a call served and the start of the following call served by the mobile is called *the inter-service time*. It is possible that a call arrives while the previous call served is still in progress ([42]). In this case, the mobile $\mathcal{T}$ cannot accept the new call (the caller senses busy tone in this case and may hang up). Thus, the inter-arrival (inter-call) times for the calls terminated at the mobile $\mathcal{T}$ are different from the inter-service times. This phenomenon is called the *busy line* effect. We emphasize here that

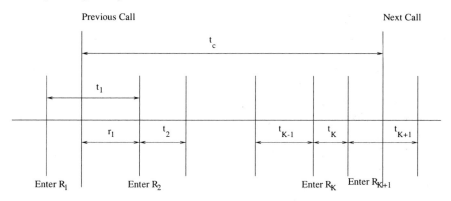

Figure 16.5. The time diagram for K area boundary crossings ([20])

for mobility management, it is the inter-service times, not the inter-arrival times of calls terminating at the mobile, that affects the location update cost, because when a mobile is engaging a call connection, the wireless network knows where the mobile $\mathcal{T}$ is, hence the mobile does not need to carry out any location update during the call connection. Although the incoming calls form a Poisson process (i.e., the interarrival times are exponentially distributed), the inter-service times may not be exponentially distributed. Inter-service time bears the similarity to the inter-departure time of a queueing system with blocking, which has been shown to be non-exponential in general. In this chapter, we assume that the service time is negligibly shorter than the inter-service time (i.e., ignoring the service time) and derive the probability $\alpha(K)$ that a mobile moves across K *areas* between two served calls arriving to the mobile $\mathcal{T}$. In what follows, we will use the *area residence time (ART)* to denote either the RRT or the CRT, depending on the mobility management schemes we evaluate. By ignoring the busy line effect and the service time, assuming that the inter-service time is exponentially distributed, Lin ([42]) is able to derive an analytical formula for $\alpha(K)$, which has been subsequently used ([5, 4, 42, 55]) for tradeoff analysis for location update and paging. In this section, we assume that the inter-service times are generally distributed and derive an analytic expression for $\alpha(K)$.

Let $t_1, t_2, \ldots$ denote the area residence times and let r_1 denote the residual area residence time (i.e., the time interval between the time instant the mobile $\mathcal{T}$ registers to the network and the time instant the mobile exits the first *area*). Let t_c denote the inter-service time between two consecutive served calls to a mobile $\mathcal{T}$. Figure 16.5 shows the time diagram for K area boundary crossings. Suppose that the mobile is in an

area R_1 when the previous call arrives and accepted by $\mathcal{T}$, it then moves
K *areas* during the inter-service time, and $\mathcal{T}$ resides in the jth *area* for
a period t_j $(1 \leq j \leq K+1)$. In this chapter, we consider a homogeneous wireless mobile network, i.e., all *areas* (either registration areas or
cells) in the network are statistically identical, hence ARTs $t_1, t_2, \ldots$ are
independent and identically distributed (iid) with a general probability
density function $f(t)$. We want to point out that the independence assumption is crucial for our derivation, it is not known yet in the current
literature how all performance analysis in the wireless networks can be
carried out analytically without this assumption. Let t_c be generally
distributed with probability density function $f_c(t)$, and let $f_r(t)$ be the
probability density function of r_1. Let $f^*(s)$, $f_c^*(s)$ and $f_r^*(s)$ denote
the *Laplace-Stieltjes (L-S) transforms* (or simply *Laplace transforms*) of
$f(t)$, $f_c(t)$ and $f_r(t)$, respectively. Let $E[t_c] = 1/\lambda_c$ and $E[t_i] = 1/\lambda_m$.
From the residual life theorem ([38]), we have

$$
f_r(t) = \lambda_m \int_t^\infty f(\tau)d\tau = \lambda_m \left[1 - F(t)\right],
$$

$$
f_r^*(s) = \frac{\lambda_m}{s}[1 - f^*(s)], \tag{16.2}
$$

where $F(t)$ is the distribution function of $f(t)$. As a remark, the residual
life theorem is valid only when we deal with the steady-state case, we
can independently model the area residence time r_1 in the initiating
area, i.e., we treat r_1 as a random variable which does not have any
relationship to the ART t_i. The results we present here can be applied
to this very general case. It is obvious that the probability $\alpha(K)$ is given
by

$$
\alpha(0) = \Pr[t_c \leq r_1], \quad K = 0, \tag{16.3}
$$

$$
\alpha(K) = \Pr[r_1 + t_2 + \cdots + t_K < t_c \leq r_1 + t_2 + \cdots + t_{K+1}], \quad K \geq 1 \tag{16.4}
$$

Applying the inverse Laplace transform, we can compute $\alpha(0)$ from
(16.3) as follows:

$$
\alpha(0) = \int_0^\infty \Pr(r_1 \geq t)f_c(t)dt = \int_0^\infty \frac{1}{2\pi j} \int_{\sigma-j\infty}^{\sigma+j\infty} \frac{1 - f_r^*(s)}{s} e^{st} ds\, f_c(t)dt
$$

$$
= \frac{1}{2\pi j} \int_{\sigma-j\infty}^{\sigma+j\infty} \frac{1 - f_r^*(s)}{s} \int_0^\infty f_c(t)e^{st}\, dt\, ds
$$

$$
= \frac{1}{2\pi j} \int_{\sigma-j\infty}^{\sigma+j\infty} \frac{1 - f_r^*(s)}{s} f_c^*(-s)ds
$$

$$
\tag{16.5}
$$

where σ is a sufficiently small positive number which is appropriately chosen for the inverse Laplace transform.

For $K > 0$, $\alpha(K)$ is computed as follows. First, we need to compute $\Pr(r_1 + t_2 + \cdots + t_k \leq t_c)$ for any $k > 0$. Let $\xi = r_1 + t_2 + \cdots + t_k$. Let $f_\xi(t)$ and $f_\xi^*(s)$ be the probability density function and the Laplace transform of ξ. ¿From the independence of $r_1, t_2, t_3, \cdots$, we have

$$f_\xi^*(s) = E[e^{-s\xi}] = E[e^{-sr_1}] \prod_{i=2}^{k} E[e^{-st_i}] = f_r^*(s)(f^*(s))^{k-1}.$$

Thus the probability density function is given by

$$f_\xi(t) = \frac{1}{2\pi j} \int_{\sigma-j\infty}^{\sigma+j\infty} f_r^*(s)(f^*(s))^{k-1} e^{st} ds.$$

Noticing that the Laplace transform of $\Pr(\xi \leq t)$ is $f_\xi^*(s)/s$, applying the inverse Laplace transform, we have

$$\Pr(r_1 + t_2 + \cdots + t_k \leq t_c) = \int_0^\infty \Pr(\xi \leq t) f_c(t) dt$$

$$= \int_0^\infty \frac{1}{2\pi j} \int_{\sigma-j\infty}^{\sigma+j\infty} \frac{f_r^*(s)[f^*(s)]^{k-1}}{s} e^{st} ds \, f_c(t) dt$$

$$= \frac{1}{2\pi j} \int_{\sigma-j\infty}^{\sigma+j\infty} \frac{f_r^*(s)[f^*(s)]^{k-1}}{s} f_c^*(-s) ds.$$

Taking this into (16.4), we obtain

$$\alpha(K) = \Pr(t_c \geq r_1 + t_2 + \cdots + t_K) - \Pr(t_c \geq r_1 + t_2 + \cdots + t_{K+1})$$

$$= \frac{1}{2\pi j} \int_{\sigma-j\infty}^{\sigma+j\infty} \frac{f_r^*(s)[f^*(s)]^{K-1}[1 - f^*(s)]}{s} f_c^*(-s) ds$$

$$(16.6)$$

where σ is a sufficiently small positive number. In summary, we obtain

THEOREM 16.1 *If the probability density function of the inter-service time has only finite possible isolated poles (which is the case when it has a rational Laplace transform), then the probability $\alpha(K)$ that a mobile moves across K areas (either RAs or cells) during the inter-service time is given by*

$$\alpha(0) = \frac{1}{2\pi j} \int_{\sigma-j\infty}^{\sigma+j\infty} \frac{1 - f_r^*(s)}{s} f_c^*(-s) ds$$

$$\alpha(K) = \frac{1}{2\pi j} \int_{\sigma-j\infty}^{\sigma+j\infty} \frac{f_r^*(s)[1 - f^*(s)][f^*(s)]^{K-1}}{s} f_c^*(-s) ds, \quad K > 0$$

$$(16.7)$$

where σ is a sufficiently small positive number and $f_r^*(s) = \lambda_m(1 - f^*(s))/s$.

If the inter-service time is exponentially distributed, then we can apply the Residue Theorem ([39]) to obtain the following simple result:

$$\alpha(0) = 1 - \frac{1 - f^*(\lambda_c)}{\rho},$$

$$\alpha(K) = \frac{1}{\rho}[1 - f^*(\lambda_c)]^2[f^*(\lambda_c)]^{K-1}, \quad K > 0,$$

where $\rho = \lambda_c/\lambda_m$ is the *call-to-mobility ratio*. This result has been obtained in [42] using a different approach.

5.   Performance Evaluation

Signaling traffic in mobility management schemes incur from the operations of *move* (may cause location update) and *find* (may need message exchanges and terminal paging). Most performance evaluation for *move* and *find* were carried out under the assumption that some of the time variables are exponentially distributed. The conclusions drawn from such results may not be extended to the cases when such an assumption is not valid, and the adaptive schemes for choosing certain parameters (such as the threshold for the number of pointers or the threshold in movement-based mobility management scheme) may not be appropriate accordingly. There are not much works handling the non-exponential situations. Recently, we have developed a new approach which handles the non-exponential situations. In this section, we present our analytical results for signaling cost.

5.1   Cost Analysis for IS-41

As a baseline comparative study, we present the cost analysis for IS-41 first. We use the same notation as before, t_c denotes the inter-service time (the inter-arrival time for calls terminated at a mobile $\mathcal{T}$ if we neglect the busy-line effect) with average $1/\lambda_c$ and t_i denotes the RA residence time with average $1/\lambda_m$. Let M and F denote the total cost for *basic moves* during the inter-service time and the total cost for *basic find*, respectively (i.e., the costs incurred in the IS-41 scheme). Since all location management schemes will go through the *move* and *find* whenever a terminating call to a mobile $\mathcal{T}$ arrives, the inter-service time forms the fundamental regenerative period for cost analysis, thus we only need to consider the signaling traffic incurred during this period.

For the IS-41 scheme, whenever the mobile crosses a RA boundary, a registration is triggered. We assume that the unit cost for a basic registration (i.e., *basic move*) is m. From Theorem 16.1, M will be equal to the product of m and the average number of registrations incurred during the inter-service time, given by ($f(t)$ and $f^*(s)$ are for the RA residence time)

$$
\begin{aligned}
M &= m \sum_{K=0}^{\infty} K\alpha(K) \\
&= \frac{m}{2\pi j} \int_{\sigma-j\infty}^{\sigma+j\infty} \frac{f_r^*(s)[1-f^*(s)]}{s} \left(\sum_{K=1}^{\infty} K(f^*(s))^{K-1} \right) f_c^*(-s)ds \\
&= \frac{m}{2\pi j} \int_{\sigma-j\infty}^{\sigma+j\infty} \frac{\lambda_m[1-f^*(-s)]^2}{s^2} \cdot \frac{1}{[1-f^*(-s)]^2} f_c^*(-s)ds \\
&= \frac{m\lambda_m}{2\pi j} \int_{\sigma-j\infty}^{\sigma+j\infty} \frac{1}{s^2} f_c^*(-s)ds \\
&= m\lambda_m \int_0^{\infty} f_c(t) \left(\frac{1}{2\pi j} \int_{\sigma-j\infty}^{\sigma+j\infty} \frac{1}{s^2} e^{st}ds \right) dt \\
&= m\lambda_m \int_0^{\infty} f_c(t)t\,dt = \frac{m\lambda_m}{\lambda_c} = \frac{m}{\rho}
\end{aligned}
$$

where $\rho = \lambda_c/\lambda_m$, which is called the *call-to-mobility ratio*.

The *basic find* operation consists of two parts. The first part includes the interactions between the originating switch and the HLR while the second part includes the interactions between the HLR, the VLR, the MSC (terminating switch) and the mobile. Thus, $F \geq m$. The *basic find* will be the paging cost in one RA. Thus, we obtain the total cost for IS-41 during the inter-service time is

$$
\mathcal{C}_{IS-41} = M + F = \frac{m}{\rho} + F. \tag{16.8}
$$

¿From this result, we observe that the total cost for IS-41 only depends on the CMR (i.e., the first moment of RA residence time and inter-arrival time of terminating calls).

5.2 Cost Analysis for Movement-based Mobility Management (MB)

The periodic location update (autonomous registration) is only an option in the IS-41 or the GSM MAP. The operation for such location update is simple, however, the scheme may not be efficient. The movement-based mobility management (MB) is a better choice in both the efficiency and simplicity. As we mentioned, the location update cost relies on the average number of location updates during the inter-service time. In this section, we present the analytical results for this scheme.

5.2.1   Average Number of Location Updates.   Let d be the threshold for the movement-based location update scheme, i.e., a mobile will make a location update every d crossings of cell boundary when the mobile does not engage in service. Obviously, the average number of location updates during the inter-service time will determine the locate update cost. So, we first want to find this quantity. We assume that $f(t)$ denote the probability density for the cell residence time (CRT) in this section, all other notation will remain the same as before. Then, the average number of location update during an inter-service time interval under movement-based location update scheme can be expressed as ([4])

$$N_u(d) = \sum_{i=1}^{\infty} i \sum_{k=id}^{(i+1)d-1} \alpha(k) \qquad (16.9)$$

where $\alpha(k)$ in this subsection denotes the probability that a mobile crosses k cells during the inter-service time. In what follows in this subsection, we present the computation for $N_u(d)$. Let

$$S(n) = \sum_{k=1}^{n-1} \alpha(k),$$

then, from Theorem 16.1, we obtain

$$\begin{aligned}
S(n) &= \sum_{k=1}^{n-1} \frac{1}{2\pi j} \int_{\sigma-j\infty}^{\sigma+j\infty} \frac{f_r^*(s)[1-f^*(s)][f^*(s)]^{k-1}}{s} f_c^*(-s)ds \\
&= \frac{1}{2\pi j} \int_{\sigma-j\infty}^{\sigma+j\infty} \frac{f_r^*(s)[1-f^*(s)]}{s} \left(\sum_{k=1}^{n-1}[f^*(s)]^{k-1}\right) f_c^*(-s)ds \\
&= \frac{1}{2\pi j} \int_{\sigma-j\infty}^{\sigma+j\infty} \frac{f_r^*(s)[1-(f^*(s))^{n-1}]}{s} f_c^*(-s)ds \qquad (16.10)
\end{aligned}$$

Moreover, we have

$$\begin{aligned}
\sum_{k=1}^{N} S(kd) &= \sum_{k=1}^{N} \frac{1}{2\pi j} \int_{\sigma-j\infty}^{\sigma+j\infty} \frac{f_r^*(s)[1-(f^*(s))^{kd-1}]}{s} f_c^*(-s)ds \\
&= \frac{1}{2\pi j} \int_{\sigma-j\infty}^{\sigma+j\infty} \frac{f_r^*(s)}{s} \left\{\sum_{k=1}^{N} \left[1-(f^*(s))^{kd-1}\right]\right\} f_c^*(-s)ds \\
&= \frac{1}{2\pi j} \int_{\sigma-j\infty}^{\sigma+j\infty} \frac{f_r^*(s)}{s} \left\{N - \frac{(f^*(s))^{d-1}[1-(f^*(s))^{Nd}]}{1-(f^*(s))^d}\right\} f_c^*(-s)d
\end{aligned}$$

$$(16.11$$

Thus, from (16.9), (16.10) and (16.11), after some mathematical manipulation (details can be found in [17]), we obtain

$$
\begin{aligned}
N_u(d) &= \sum_{i=1}^{\infty} i\left[S((i+1)d) - S(id)\right] \\
&= \lim_{N \to \infty} \left\{ \sum_{i=1}^{N} i\left[S((i+1)d) - S(id)\right] \right\} \\
&= \lim_{N \to \infty} \left\{ NS((N+1)d) - \sum_{i=1}^{N} S(id) \right\} \\
&= \frac{1}{2\pi j} \int_{\sigma-j\infty}^{\sigma+j\infty} \frac{f_r^*(s)(f^*(s))^{d-1}}{s[1 - (f^*(s))^d]} f_c^*(-s) ds
\end{aligned}
$$

where $\sigma > 0$ is a sufficiently small positive number.

In summary, we obtain

THEOREM 16.2 *If the inter-service time $f_c^*(s)$ has a finite number of poles (such as a proper rational function) and the cell residence time (CRT) is generally distributed with probability density function $f(t)$ and L-S transform $f^*(s)$, then the average number of location updates $N_u(d)$ is given by*

$$
N_u(d) = \frac{1}{2\pi j} \int_{\sigma-j\infty}^{\sigma+j\infty} \frac{f_r^*(s)(f^*(s))^{d-1}}{s[1 - (f^*(s))^d]} f_c^*(-s) ds \qquad (16.12)
$$

where $f_r^*(s)$ is the L-S transform of the probability density of the residual cell residence time. If $f_c^*(s)$ has a finite number of poles, which is the case when it is a rational function, then we have

$$
N_u(d) = - \sum_{p \in \sigma_c} \operatorname*{Res}_{s=p} \frac{f_r^*(s)(f^*(s))^{d-1}}{s[1 - (f^*(s))^d]} f_c^*(-s), \qquad (16.13)
$$

where σ_c is the set of poles of $f_c^*(-s)$.

5.2.2   Tradeoff Cost Analysis: A Case Study.   In this subsection, we present some results for the tradeoff analysis under movement-based location update scheme and some paging scheme with the linear cost functional for illustration purpose.

Location Update Cost:

Let m denotes the unit cost for location update (i.e., the cost for a *basic move* as in IS-41), then total location update cost is given by

$$C_u(d) = -m \sum_{p \in \sigma_c} \operatorname*{Res}_{s=p} \frac{f_r^*(s)(f^*(s))^{d-1}}{s[1 - (f^*(s))^d]} f_c^*(-s), \qquad (16.14)$$

If the inter-service time t_c is exponentially distributed with parameter λ_c, then $f_c^*(s) = \lambda_c/(s + \lambda_c)$, from (16.14), we can easily obtain

$$C_u(d) = \frac{m f_r^*(\lambda_c)[f^*(\lambda_c)]^{d-1}}{1 - [f^*(\lambda_c)]^d}. \qquad (16.15)$$

This result was obtained in [40] (noticing that $f_r^*(s) = \lambda_m(1 - f^*(s))/s$) via a different approach.

Paging Cost:

We consider the paging strategy used in [40] for our case study. Consider the hexagonal layout for the wireless network, all cells are statistically identical. According to the movement-based location update scheme, a mobile terminal moves at most d cells away from the previous position where it performs the last location update. Thus, a mobile terminal will surely be located in a cell which is less than d cell away from the previously reported position. If we page in the circular area with d cells as radius and with the previously reported position as the center, then we can definitely find the mobile terminal. Thus, this paging scheme is the most conservative among all paging. If we let P denote the unit cost for each paging in a cell, the maximum paging cost for this paging scheme is given by ([40])

$$C_p(d) = P(1 + 3d(d - 1)), \qquad (16.16)$$

Total Cost:

The unit costs (cost factors) m and P can be chosen to reflect the significance of the signaling (which may be significantly different from each other because they use different network resources). Given m, P and the movement threshold d, the total cost for location update and paging will be given by

$$C_{MB}(d) = C_u(d) + C_p(d)$$
$$= -m \sum_{p \in \sigma_c} \operatorname*{Res}_{s=p} \frac{f_r^*(s)(f^*(s))^{d-1}}{s[1 - (f^*(s))^d]} f_c^*(-s) + P(1 + 3d(d - 1)).$$

$$(16.17)$$

We observe that as long as we find the probability distributions of cell residence time and inter-service time, we can find the total cost using 16.17. Many numerical results ([17]) have shown that the cost function $\mathcal{C}_{MB}(d)$ is a convex function of d. If this is true, then we can find the unique optimal threshold d^* to minimize the cost. Unfortunately, we have not proved the convexity of $\mathcal{C}(d)$ yet.

5.3   Cost Analysis for Pointer Forwarding

In this subsection, we evaluate the performance of pointer forwarding scheme (PFS). We need to quantify the signaling traffic incurring in this scheme. Let M' and F' denote the corresponding costs for the pointer forwarding scheme during the inter-service time, in which every K moves will trigger a new registration, where K is the maximum pointer chain length. In this subsection, a move means the crossing from one RA to another. Let S denote the cost of setting up a forwarding pointer between VLRs during a pointer forwarding *move* and let T denote the cost of traversing a forwarding pointer between VLRs during a pointer forwarding *find*. We first derive M' and F'.

Suppose that a mobile $\mathcal{T}$ crosses i RA boundaries during the inter-service time, then there are $i - \lfloor i/K \rfloor$ pointer creations (every K moves require $K - 1$ pointer creations) and the HLR is updated $\lfloor i/K \rfloor$ times (with pointer forwarding, because the mobile $\mathcal{T}$ registers every Kth move). Here, we use $\lfloor x \rfloor$ to denote the floor function, i.e., the largest integer not exceeding x. Let $f(t)$ and $f^*(s)$ denote the probability density function and its corresponding L-S transform of the RA residence time (RRT). Let $\alpha(k)$ denote the probability that the mobile crosses k RAs during the inter-service time. Then, we obtain

$$
\begin{aligned}
M' &= \sum_{i=0}^{\infty} \left[\left(i - \left\lfloor \frac{i}{K} \right\rfloor \right) S + \left\lfloor \frac{i}{K} \right\rfloor m \right] \alpha(i) \\
&= S \sum_{i=0}^{\infty} i\alpha(i) + (m - S) \sum_{i=0}^{\infty} \left\lfloor \frac{i}{K} \right\rfloor \alpha(i) \\
&= S \sum_{i=0}^{\infty} i\alpha(i) + (m - S) \sum_{r=0}^{\infty} r \left(\sum_{i=rK}^{(r+1)K-1} \alpha(i) \right) \quad (16.18)
\end{aligned}
$$

Let

$$
X(K) = \sum_{r=1}^{\infty} r \left(\sum_{i=rK}^{(r+1)K-1} \alpha(i) \right)
$$

$$S(n) \;=\; \sum_{k=1}^{n-1} \alpha(k)$$

then, from Theorem 16.1 and following the same procedure we used in the last subsection, we obtain

$$S(n) = \frac{1}{2\pi j} \int_{\sigma-j\infty}^{\sigma+j\infty} \frac{f_r^*(s)[1-(f^*(s))^{n-1}]}{s} f_c^*(-s)ds, \qquad (16.19)$$

$$\sum_{i=1}^{N} s(iK) = \frac{1}{2\pi j} \int_{\sigma-j\infty}^{\sigma+j\infty} \frac{f_r^*(s)}{s} \left\{ N - \frac{(f^*(s))^{K-1}[1-(f^*(s))^{NK}]}{1-(f^*(s))^K} \right\} f_c^*(-s) \cdot \qquad (16.20)$$

and

$$\begin{aligned} X(K) &= \sum_{r=1}^{\infty} r\left[S((r+1)K) - S(rK)\right] \\ &= \frac{1}{2\pi j} \int_{\sigma-j\infty}^{\sigma+j\infty} \frac{f_r^*(s)(f^*(s))^{K-1}}{s[1-(f^*(s))^K]} f_c^*(-s)ds \qquad (16.21) \end{aligned}$$

Noticing that $X(1) = \sum_{i=0}^{\infty} i\alpha(i)$, from (16.18), we obtain

$$\begin{aligned} M' &= SX(1) + (m-S)X(K) \\ &= \frac{1}{2\pi j} \int_{\sigma-j\infty}^{\sigma+j\infty} \left[\frac{Sf_r^*(s)}{s[1-f^*(s)]} + \frac{(m-S)f_r^*(s)(f^*(s))^{K-1}}{s[1-(f^*(s))^K]} \right] f_c^*(-s)ds \end{aligned}$$

$$\qquad (16.22)$$

Next, we derive F'. After the last *basic move* operation (if any), the mobile $\mathcal{T}$ crosses $n = i - K\lfloor i/K \rfloor$ RA boundaries. Let $\Theta(n)$ denote the number of pointers to be tracked in order to find the mobile $\mathcal{T}$ in the pointer forwarding *find* operation. If the mobile visits a RA more than once (i.e., a "loop" exists among n moves), then $\Theta(n)$ may not need to trace n pointers, thus, $\Theta(n) \le n$. From this argument and applying Theorem 16.1, we obtain

$$\begin{aligned} F' &= \sum_{i=0}^{\infty} T\Theta\left(i - K\lfloor i \rfloor\right)\alpha(i) + F = T\sum_{r=0}^{\infty}\sum_{k=0}^{K-1} \Theta(k)\alpha(rK+k) + F \\ &= T\sum_{k=0}^{K-1} \Theta(k)\left(\sum_{r=0}^{\infty} \alpha(rK+k)\right) + F \\ &= T\sum_{k=0}^{K-1} \Theta(k)\frac{1}{2\pi j}\int_{\sigma-j\infty}^{\sigma+j\infty} \frac{f_r^*(s)[1-f^*(s)](f^*(s))^{k-1}}{s[1-(f^*(s))^K]} f_c^*(-s)ds + F \end{aligned}$$

$$= \frac{T}{2\pi j} \int_{\sigma-j\infty}^{\sigma+j\infty} \frac{f_r^*(s)[1 - f^*(s)]}{s[1 - (f^*(s))^K]} \left(\sum_{k=0}^{K-1} \Theta(k)(f^*(s))^{k-1} \right) f_c^*(-s)ds + F$$

In summary and applying the Residue Theorem ([39]), we finally arrive at

THEOREM 16.3 *If the inter-service time* $f_c^*(s)$ *has a finite number of poles (such as a proper rational function) and the RA residence time is generally distributed with probability density function* $f(t)$ *and L-S transform* $f^*(s)$, *then we have*

$$M' = \frac{1}{2\pi j} \int_{\sigma-j\infty}^{\sigma+j\infty} \left[\frac{Sf_r^*(s)}{s[1 - f^*(s)]} + \frac{(m - S)f_r^*(s)(f^*(s))^{K-1}}{s[1 - (f^*(s))^K]} \right] f_c^*(-s)ds$$

$$= -\sum_{p\in\sigma_c} \mathop{\mathrm{Res}}_{s=p} \left[\frac{Sf_r^*(s)}{s[1 - f^*(s)]} + \frac{(m - S)f_r^*(s)(f^*(s))^{K-1}}{s[1 - (f^*(s))^K]} \right] f_c^*(-s)$$

$$F' = F + \frac{T}{2\pi j} \int_{\sigma-j\infty}^{\sigma+j\infty} \frac{f_r^*(s)[1 - f^*(s)]}{s[1 - (f^*(s))^K]} \left(\sum_{k=0}^{K-1} \Theta(k)(f^*(s))^{k-1} \right) f_c^*(-s)ds$$

$$= F - \sum_{p\in\sigma_c} \mathop{\mathrm{Res}}_{s=p} \frac{f_r^*(s)[1 - f^*(s)]}{s[1 - (f^*(s))^K]} \left(\sum_{k=0}^{K-1} \Theta(k)(f^*(s))^{k-1} \right) f_c^*(-s)$$

where σ_c *denotes the set of poles of* $f_c^*(-s)$ *and* $\mathrm{Res}_{s=p}$ *denotes the residue at the pole* $s = p$.

If the inter-service time is exponentially distributed with parameter λ_c, then we have $f_c^*(-s) = \lambda_c/(-s + \lambda_c)$, from Theorem 16.3, we obtain

$$M' = \frac{Sf_r^*(\lambda_c)}{1 - f^*(\lambda_c)} + \frac{(m - S)f_r^*(\lambda_c)(f^*(\lambda_c))^{K-1}}{1 - (f^*(\lambda_c))^K}$$

$$F' = F + \frac{Tf_r^*(\lambda_c)[1 - f^*(\lambda_c)]}{1 - (f^*(\lambda_c))^K} \left(\sum_{k=0}^{K-1} \Theta(k)(f^*(\lambda_c))^{k-1} \right)$$

which were obtained in [34] with a slight different form.

The worst case for F' would be when all pointers are traced, i.e., when $\Theta(n) = n$. In this case, we have the following result:

$$F' = F + \frac{T}{2\pi j} \int_{\sigma-j\infty}^{\sigma+j\infty} \frac{f_r^*(s)[1 - f^*(s)]}{s[1 - (f^*(s))^K]} \left(\sum_{k=0}^{K-1} k(f^*(s))^{k-1} \right) f_c^*(-s)ds$$

$$= F - T\sum_{p\in\sigma_c} \mathop{\mathrm{Res}}_{s=p} \frac{f_r^*(s)[1 - K(f^*(s))^{K-1} + (K - 1)(f^*(s))^K}{s[1 - (f^*(s))^K][1 - f^*(s)]} f_c^*(-s)$$

Thus, the total cost for PFS during the inter-service time can be computed as follows:

$$\mathcal{C}_{PFS} = M' + F' = F + \frac{1}{2\pi j} \int_{\sigma-j\infty}^{\sigma+j\infty} g(s) f_c^*(-s) ds = F - \sum_{p \in \sigma_c} \operatorname*{Res}_{s=p} g(s) f_c^*(-s)$$

(16.23)

where

$$g(s) = \frac{S f_r^*(s)}{s[1 - f^*(s)]} + \frac{(m - S) f_r^*(s)(f^*(s))^{K-1}}{s[1 - (f^*(s))^K]}$$
$$+ \frac{T f_r^*(s)[1 - f^*(s)]}{s[1 - (f^*(s))^K]} \left(\sum_{k=0}^{K-1} \Theta(k) \right).$$

5.4   Cost Analysis for Two-Level Pointer Forwarding

Two-Level Pointer Forwarding (TLPFS) is a generalization of PFS in the sense that a new level of mobility agents (MA) is added between the home systems and visiting systems to regionalize (or localize) the signaling traffic. In [47], we derive the signaling cost under exponentially distributed inter-service time. Although we could obtain more general analytical results for this scheme under more general assumption in a similar way used for PFS, for simplicity, we only present the results in [47] and leave the derivation of more general results to the readers.

We use the same notation as before, we let t_c denote the inter-service time with average $1/\lambda_c$ and let $f(t)$ denote the probability density function of the RA residence time with average $1/\lambda_m$. The call-to-mobility ratio (CMR) is then given by $\rho = \lambda_c/\lambda_m$. We assume that the inter-service time is exponentially distributed. We define $\mathcal{C}_{IS-41}$ and $\mathcal{C}_{TLPFS}$ as the total costs of updating the location information (location update) and locating the user (location tracking) during the inter-service time in IS-41 and in the TLPFS, respectively. The following parameters used in the TLPFS:

$m =$ the cost of a single invocation of *BasicMOVE*.

$M =$ the total cost of all the *BasicMOVESs* during the inter-service time.

$F =$ the cost of a single *BasicFIND*.

$M' =$ the expected cost of all *TwoLevelFwdMOVEs* during the inter-service time.

$F' =$ the average cost of the *TwoLevelFwdFIND*.

S_1 = the cost of setting up a forwarding pointer (level_1 pointer) between MAs during a *Two-LevelFwdMOVE*.

S_2 = the cost of setting up a forwarding pointer (level_2 pointer) between VLRs during a *Two-LevelFwdMOVE*.

T_1 = the cost of traversing a forwarding pointer (level_1 pointer) between MAs during a *Two-LevelFwdFIND*.

T_2 = the cost of traversing a forwarding pointer (level_2 pointer) between VLRs during a *Two-LevelFwdFIND*.

K_1 = the threshold of level_1 pointer chain.

K_2 = the threshold of level_2 pointer chain.

$\alpha(k)$ = the probability that there are k RA crossings during the inter-service time.

Then, we have

$$
\begin{aligned}
\mathcal{C}_{IS-41} &= M + F = m/\rho + F, \\
\mathcal{C}_{TLPFS} &= M' + F'.
\end{aligned}
$$

Thus, we need to derive formulas for M' and F'.

In the TLPFS, we observe that the HLR is updated only every $K_1 \cdot K_2$ moves (K_1 and K_2 are the level_1 and level_2 pointer chain length thresholds, respectively), while forwarding pointers are set up for all other moves. If a mobile user crosses i RA boundaries during the inter-service time, then the HLR is updated $\lfloor \frac{i}{K_1 K_2} \rfloor$ times, there are also $\lfloor \frac{i}{K_2} \rfloor - \lfloor \frac{i}{K_1 K_2} \rfloor$ level_1 pointer creations (every K_2 moves may require a level_1 pointer creation but sometimes the HLR is updated and level_1 pointer is not set up), and there are the level_2 pointers created for all the rest $i - \lfloor \frac{i}{K_2} \rfloor$ moves. Thus, we obtain

$$
\begin{aligned}
M' &= \sum_{i=0}^{\infty} \left\{ \left\lfloor \frac{i}{K_1 K_2} \right\rfloor m + \left(\left\lfloor \frac{i}{K_2} \right\rfloor - \left\lfloor \frac{i}{K_1 K_2} \right\rfloor \right) S_1 \right. \\
&\quad \left. + \left(i - \left\lfloor \frac{i}{K_2} \right\rfloor \right) S_2 \right\} \alpha(i) \\
&= \sum_{i=0}^{\infty} i S_2 \alpha(i) + \sum_{i=0}^{\infty} \left\lfloor \frac{i}{K_2} \right\rfloor (S_1 - S_2) \alpha(i) \\
&\quad + \sum_{i=0}^{\infty} \left\lfloor \frac{i}{K_1 K_2} \right\rfloor (m - S_1) \alpha(i).
\end{aligned} \tag{16.24}
$$

The cost F' is derived as follows. After the last *BasicMove* operations (if any), the mobile user traverses $\left\lfloor \dfrac{i - \left\lfloor \frac{i}{K_1 K_2} \right\rfloor K_1 K_2}{K_2} \right\rfloor$ level_1 pointers and

$i - \left\lfloor \dfrac{i}{K_1 K_2} \right\rfloor K_1 K_2 - \left\lfloor \dfrac{i - \left\lfloor \frac{i}{K_1 K_2} \right\rfloor K_1 K_2}{K_2} \right\rfloor K_2$ level_2 pointers. Thus, we obtain

$$
\begin{aligned}
F' &= F + \sum_{i=0}^{\infty} \left\{ \left\lfloor \frac{i - \left\lfloor \frac{i}{K_1 K_2} \right\rfloor K_1 K_2}{K_2} \right\rfloor T_1 \right. \\
&\quad \left. + \left(i - \left\lfloor \frac{i}{K_1 K_2} \right\rfloor K_1 K_2 - \left\lfloor \frac{i - \left\lfloor \frac{i}{K_1 K_2} \right\rfloor K_1 K_2}{K_2} \right\rfloor K_2 \right) T_2 \right\} \alpha(i) \\
&= F + (T_1 - K_2 T_2) \sum_{i=0}^{\infty} \left\lfloor \frac{i - \left\lfloor \frac{i}{K_1 K_2} \right\rfloor K_1 K_2}{K_2} \right\rfloor \alpha(i) \\
&\quad + T_2 \sum_{i=0}^{\infty} \left(i - \left\lfloor \frac{i}{K_1 K_2} \right\rfloor K_1 K_2 \right) \alpha(i).
\end{aligned}
\tag{16.25}
$$

For simplicity, we let $g = f_m^*(\lambda_c)$. We observe that all summations in both equations (16.24) and (16.25) bear close similarities to those in subsection 16.5.3, hence applying a similar procedure, we can find the analytical results for M' and F'. The details can be found in [47]. In summary, we obtain

THEOREM 16.4 *If the inter-service time is exponentially distributed and the RA residence time is generally distributed with probability density function $f(t)$ and L-S transform $f^*(s)$, then the total signaling cost for TLPFS is given by*

$$
\mathcal{C}_{TLPFS} = M' + F'
$$

where $g = f_m^*(\lambda_c)$ and

$$
M' = \frac{S_2}{\rho} + \frac{(1-g)g^{K_2-1}(S_1 - S_2)}{\rho(1 - g^{K_2})} + \frac{(1-g)g^{K_1 K_2-1}(m - S_1)}{\rho(1 - g^{K_1 K_2})}
$$

$$
\begin{aligned}
F' &= F + \frac{[1 - K_1 K_2 g^{K_1 K_2-1} + (K_1 K_2 - 1)g^{K_1 K_2}]T_2}{\rho(1 - g^{K_1 K_2})} \\
&\quad + \frac{(T_1 - K_2 T_2)(1-g)[g^{K_2} - K_1 g^{K_1 K_2} + (K_1 - 1)g^{(K_1+1)K_2}]}{\rho g(1 - g^{K_1 K_2})(1 - g^{K_2})}.
\end{aligned}
$$

5.5   Cost Analysis for TLA

In the Two Location Algorithm (TLA), if HLR has a location miss for a call termination, i.e., the two-location for the called mobile are not the same as the one stored in the HLR, additional signaling traffic will be necessary to setup the call connection. Thus, the probability that the HLR has a location miss for a call setup, say, w, is important to capture the signaling traffic. In [42], this probability is derived under the assumption that the inter-service time is exponential. In this subsection, we first derive a more general analytical result to compute this quantity under the assumption that the inter-service time is generally distributed, then we present the cost analysis for the TLA.

¿From the argument in [42], $(1 - w)$ is the probability that the HLR has the correct view of the latest visited RA when a call arrives (i.e., a location hit occurs and the *find* cost for TLA is the same as that for IS-41). A location hit occurs either when the mobile has not moved since last served call arrival, or when the last location update is followed by an even number of moves during the inter-service time, or when there are an even number of moves with no location update during the inter-service time. Here, in this subsection, a move means a boundary crossing from one RA to another. Let w_1 denote the probability that there is no move during the inter-service time, let w_2 denote the probability that the last served call is followed by an even number of moves without location update during the inter-service time, and let w_3 denote the probability that there are an even number of moves without location update during the inter-service time. Then, we have $1 - w = w_1 + w_2 + w_3$, i.e., $w = 1 - w_1 - w_2 - w_3$. Let $f(t)$ denote the probability density function for the RA residence time (RRT) and other related notations remain the same as before. Let $\alpha(k)$ denote again the probability that a mobile user crosses k RA boundaries during the inter-service time. For w_1, we have

$$w_1 = \Pr(r_1 \leq t_c) = \alpha(0) \tag{16.26}$$

Let θ denote the probability of a move without location update. Let $w_2(K)$ denote the probability that the last registration followed by an even number of moves without location update during the inter-service time given that there are K moves (RA crossings) during the inter-service time. We can easily obtain that ([42])

$$w_2(K) = (1 - \theta) \sum_{i=0}^{\lfloor (K-1)/2 \rfloor} \theta^{2i} = \frac{1 - \theta^{2\lfloor (K-1)/2 \rfloor + 2}}{1 + \theta}, \quad K > 0$$

where $\lfloor x \rfloor$ indicates the floor function, i.e., the largest integer not exceeding x. Noticing that $w_2(2i + 2) = w_2(2i + 1)$ for $i \geq 0$, we have

$$w_2 = \sum_{K=1}^{\infty} w_2(K)\alpha(K) = \sum_{i=0}^{\infty} w_2(2i + 1)[\alpha(2i + 1) + \alpha(2i + 2)]$$

$$= \frac{1}{1+\theta} \sum_{i=1}^{\infty} (1 - \theta^{2i})[\alpha(2i-1) + \alpha(2i)] \tag{16.27}$$

The probability w_3 can be computed as follows:

$$w_3 = \sum_{i=1}^{\infty} \theta^{2i} \alpha(2i).$$

Thus, applying Theorem 16.1 and after some mathematical manipulations similar to the techniques we used for PFS and TLPFS (the details can be found in [18]), we obtain

$$
\begin{aligned}
w &= 1 - w_1 - w_2 - w_3 \\
&= 1 - \alpha(0) - \frac{1}{1+\theta}(1 - \alpha(0)) + \frac{1}{1+\theta} \sum_{i=1}^{\infty} \theta^{2i} - \frac{\theta}{1+\theta} \sum_{i=1}^{\infty} \theta^{2i} \alpha(2i) \\
&= \frac{\theta}{1+\theta} - \frac{\theta}{(1+\theta)2\pi j} \int_{\sigma-j\infty}^{\sigma+j\infty} \frac{1 - f_r^*(s)}{s} f_c^*(-s) ds \\
&\quad + \frac{\theta^2}{(1+\theta)2\pi j} \int_{\sigma-j\infty}^{\sigma+j\infty} \frac{f_r^*(s)[1 - f^*(s)]}{s[1 - \theta^2 f^{*2}(s)]} f_c^*(-s) ds \\
&\quad - \frac{\theta^3}{(1+\theta)2\pi j} \int_{\sigma-j\infty}^{\sigma+j\infty} \frac{f_r^*(s)[1 - f^*(s)]f^*(s)}{s[1 - \theta^2 f^{*2}(s)]} f_c^*(-s) ds \\
&= \frac{1}{2\pi j} \int_{\sigma-j\infty}^{\sigma+j\infty} \frac{\theta f_r^*(s)}{s[1 + \theta f^*(s)]} f_c^*(-s) ds \tag{16.28}
\end{aligned}
$$

Applying the Residue Theorem ([39]), we obtain

THEOREM 16.5 *If the inter-service time is distributed with rational Laplace transform $f_c^*(s)$ and the RA residence time (RRT) is generally distributed with probability density function $f(t)$ and with the L-S transform $f^*(s)$, the probability of a location miss in TLA is given by*

$$w = \frac{1}{2\pi j} \int_{\sigma-j\infty}^{\sigma+j\infty} \frac{\theta f_r^*(s)}{s[1 + \theta f^*(s)]} f_c^*(-s) ds \tag{16.29}$$

where σ is a sufficiently small positive number and $f_r^*(s)$ is the L-S transform of the residual RRT. If σ_c denotes the set of poles of $f_c^*(-s)$, then we have

$$w = - \sum_{p \in \sigma_c} \operatorname*{Res}_{s=p} \frac{\theta f_r^*(s)}{s[1 + \theta f^*(s)]} f_c^*(-s). \tag{16.30}$$

where $\operatorname{Res}_{s=p}$ denotes the residue at the pole $s = p$.

If the inter-service time is exponentially distributed, which is the case studied in [42], then we have $f_c^*(s) = \lambda_c/(s + \lambda_c)$, hence from Theorem 16.5, we obtain

$$w = -\operatorname*{Res}_{s=\lambda_c} \frac{\theta f_r^*(s)}{s[1 + \theta f^*(s)]} \cdot \frac{\lambda_c}{-s + \lambda_c} = \frac{\theta f_c^*(\lambda_c)}{1 + \theta f^*(\lambda_c)},$$

which is exactly the same as in [42] after some simplification of the result in [42].

Now, we are ready to carry out the cost analysis for TLA. We still assume that the cost for *basic move* is m and the cost for the *basic find* is F. The *basic find* consists of two parts: the first part is the message exchange from a mobile to the HLR, which is more or less the cost for the *basic move*; the second part is the message forwarding from the HLR to the callee plus the possible terminal paging. Hence, usually we have $F \geq m$ and the the cost for the second part in the *basic find* is $F - m$. Suppose that the mobile moves across K RAs during the inter-service time. The conditional probability $\Pr[I = i|K]$ that i *location update* operations are performed among the K moves has a Bernoulli distribution:

$$\Pr[I = i|K] = \binom{K}{i} \theta^{K-i}(1 - \theta)^i$$

where θ is the probability that when a mobile $\mathcal{T}$ moves, the new RA address is in the mobile's memory. Then the average number of location update during the inter-service time for TLA is given by

$$
\begin{aligned}
n_{TLA} &= \sum_{K=0}^{\infty} \sum_{i=0}^{K} i \Pr[I = i|K]\alpha(K) = \sum_{K=0}^{\infty} \left(\sum_{i=0}^{K} i \binom{K}{i} \theta^{K-i}\theta^i \right) \alpha(K) \\
&= \sum_{K=0}^{\infty} K(1 - \theta)\alpha(K) = (1 - \theta)X(1) \\
&= \frac{1 - \theta}{2\pi j} \int_{\sigma-j\infty}^{\sigma+j\infty} \frac{f_r^*(s)}{s[1 - f^*(s)]} f_c^*(-s) ds \\
&= -(1 - \theta) \sum_{p \in \sigma_c} \operatorname*{Res}_{s=p} \frac{f_r^*(s)}{s[1 - f^*(s)]} f_c^*(-s) \qquad (16.31)
\end{aligned}
$$

where we have used (16.21). Thus, the total cost for registration during inter-service time is $c_1 = mn_{TLA}$.

For the operations of the second part in the *basic find* for TLA, if we have a location hit (i.e., the location entries in the memories of both HLR and the mobile are identical), the *find* cost will be the same as in *basic find*; if there is a location miss, extra cost from HLR to the VLR (second part of the *basic find* operation) will incur, thus the total cost for the *find operation* in TLA will be

$$c_2 = (1 - \omega) \cdot F + \omega[F + (F - m)] = F + (F - m)\omega.$$

In summary, the total signaling cost during the inter-service time for TLA is given by

$$
\begin{aligned}
C_{TLA} &= c_1 + c_2 = \delta n_{TLA} + F + (F - m)\omega \\
&= F + \frac{1}{2\pi j} \int_{\sigma-j\infty}^{\sigma+j\infty} \left[\frac{(1-\theta)m}{s[1 - f^*(s)]} + \frac{(F-m)\theta}{s[1 + \theta f^*(s)]} \right] f_r^*(s) f_c^*(-s) ds \\
&= F - \sum_{p \in \sigma_c} \operatorname*{Res}_{s=p} \left[\frac{(1-\theta)m}{s[1 - f^*(s)]} + \frac{(F-m)\theta}{s[1 + \theta f^*(s)]} \right] f_r^*(s) f_c^*(-s)
\end{aligned}
$$

$$(16.32)$$

where σ_c is the set of poles of $f_c^*(-s)$.

6.   Some Interesting Special Cases

We notice that all general results presented so far can be easily applied when the inter-service time is distributed with rational Laplace transform. Since distributions with rational Laplace transforms are dense in the space of probability distributions, we can always use a distribution model with rational Laplace transform to approximate any given distribution to any desired accuracy. Some well-known distribution models used in queueing theory, such as Coxian model and phase-type distribution model, can be used to model both inter-service time and the area residence time (RRT or CRT), from which we can easily obtain the analytical results for the performance evaluation of mobility management schemes we investigated in this chapter. Recently, a few distribution models have been proposed to model various time variables in the wireless cellular networks, the hyper-Erlang model and the SOHYP (Sum of Hyper-Exponential) model are two of the important models (see [23] and references therein). They all belong to the phase-type distribution model, however, they are simple enough for analytical analysis and general enough to capture the statistics of the random time variables of interest. Due to the simplicity and generality of the hyper-Erlang model, we will concentrate on this model in this subsection.

The *hyper-Erlang* distribution (a more appropriate term may be *mixed-Erlang distribution*) has the following probability density function and Laplace transform:

$$f_{he}(t) = \sum_{i=1}^{M} p_i \frac{(m_i \eta_i)^{m_i} t^{m_i - 1}}{(m_i - 1)!} e^{-m_i \eta_i t} \quad (t \geq 0),$$

$$f_{he}^*(s) = \sum_{i=1}^{M} p_i \left(\frac{m_i \eta_i}{s + m_i \eta_i}\right)^{m_i}, \qquad (16.33)$$

where

$$p_i \geq 0, \quad \sum_{i=1}^{M} p_i = 1,$$

and $M, m_1, m_2, \ldots, m_M$ are nonnegative integers, $\eta_1, \eta_2, \ldots, \eta_M$ are positive numbers.

It has been demonstrated ([23, 37]) that the hyper-Erlang distribution can be used to approximate the distribution of any nonnegative random variable. Many distribution models, such as the exponential model, the Erlang model and the hyper-exponential model, are special cases of the hyper-Erlang distribution. More importantly, the moments of a hyper-Erlang distribution can be easily obtained. If ξ is hyper-Erlang distributed as in equation (16.33), then its kth moment is given by

$$E[\xi^k] = (-1)^k f_{he}^{*(k)}(0) = \sum_{i=1}^{M} p_i \frac{(m_i + k - 1)!}{(m_i - 1)!}(m_i \eta_i)^{-k}.$$

The parameters p_i, m_i and η_i ($i = 1, 2, \ldots, M$) can be found by fitting a number of moments from field data in practice.

Assuming now that the inter-service time is hyper-Erlang distributed with distribution given in equation (16.33) with

$$\lambda_c = \left(\sum_{i=1}^{M} p_i/\eta_i\right)^{-1},$$

then, applying the Residue Theorem ([39]), we can easily obtain

$$\begin{aligned}
\mathcal{C}_{MB}(d) &= m\sum_{i=1}^{M} p_i \frac{(-1)^{m_i-1}(m_i\eta_i)^{m_i}}{(m_i-1)!} d^{(m_i-1)}(m_i\eta_i) \\
&\quad + P(1 + 3d(d-1)) \\
\mathcal{C}_{PFS} &= F + \sum_{i=1}^{M} p_i \frac{(-1)^{m_i-1}(m_i\eta_i)^{m_i}}{(m_i-1)!} g^{(m_i-1)}(m_i\eta_i) \\
\mathcal{C}_{TLA} &= F + \sum_{i=1}^{M} p_i \frac{(-1)^{m_i-1}(m_i\eta_i)^{m_i}}{(m_i-1)!} h^{(m_i-1)}(m_i\eta_i)
\end{aligned}$$

where

$$d(s) = \frac{f_r^*(s)(f^*(s))^{d-1}}{s[1 - (f^*(s))^d]},$$

$$g(s) = \frac{Sf_r^*(s)}{s[1 - f^*(s)]} + \frac{(m - S)f_r^*(s)(f^*(s))^{K-1}}{s[1 - (f^*(s))^K]}$$
$$+ \frac{Tf_r^*(s)[1 - f^*(s)]}{s[1 - (f^*(s))^K]}\left(\sum_{k=0}^{K-1}\Theta(k)\right),$$

$$h(s) = \left[\frac{(1 - \theta)m}{s[1 - f^*(s)]} + \frac{(F - m)\theta}{s[1 + \theta f^*(s)]}\right]f_r^*(s)$$

and $x^{(i)}(s)$ denotes the ith derivative of the function $x(s)$ at point s.

If the inter-service time is hyper-exponentially distributed with Laplace transform

$$f_c^*(s) = \sum_{i=1}^{M} p_i \frac{\eta_i}{s + \eta_i},$$

then we have

$$\mathcal{C}_{MB}(d) = m \sum_{i=1}^{M} p_i \eta_i d(\eta_i) + P(1 + 3d(d - 1))$$

$$\mathcal{C}_{PFS} = F + \sum_{i=1}^{M} p_i \eta_i g(\eta_i)$$

$$\mathcal{C}_{TLA} = F + \sum_{i=1}^{M} p_i \eta_i h(\eta_i)$$

If the inter-service time is exponentially distributed with Laplace transform

$$f_c^*(s) = \frac{\lambda_c}{s + \lambda_c},$$

then we have

$$\mathcal{C}_{MB}(d) = m\lambda_c d(\lambda_c) + P(1 + 3d(d - 1))$$
$$\mathcal{C}_{PFS} = F + \lambda_c g(\lambda_c)$$
$$\mathcal{C}_{TLA} = F + \lambda_c h(\lambda_c)$$

7.   Conclusions

In this chapter, we have presented a new analytical approach to carrying out the performance evaluation for mobility management in wireless mobile networks. We focus on a few mobility management schemes we have investigated in the past few years and present the analytical results for the signaling traffic under very general assumption. These analytical results can be used to choose the design parameters in the mobility management. Moreover, from these results, possible adaptive mobility management schemes can be developed based on the analytical results we present in this chapter. It is our hope that this chapter will inspire more applications of this novel approach to solve other problems in wireless networking.

References

[1] A. Abutaleb and V. O. K. Li, "Paging strategy optimization in personal communication systems," *ACM Wireless Networks*, vol.3, pp.195-204, 1997.

[2] A. Abutaleb and V. O. K. Li, "Location update optimization in personal communication systems," *ACM Wireless Networks*, vol.3, pp.205-216, 1997.

[3] I.F. Akyildiz and J.S.M. Ho, "Dynamic mobile user location update for wireless PCS networks," *ACM Wireless Networks*, vol.1, pp.187-196, 1995.

[4] I.F. Akyildiz, J.S.M. Ho and Y.-B. Lin, "Movement-based location update and selective paging for PCS networks," *IEEE/ACM Trans. Networking*, vol.4, no.4, pp.629-638, 1996.

[5] I.F. Akyildiz and W. Wang, "A dynamic location management scheme for next generation multi-tier PCS systems," To appear in *IEEE Journal on Selected Areas in Communications*.

[6] I.F. Akyildiz, J. McNair, J.S.M. Ho, H. Uzunalioglu and W. Wang, "Mobility management in next-generation wireless systems," *Proc. of the IEEE*, vol.87, no.8, pp.1347-1384, August 1999.

[7] A. Bar-Noy, I. Kessler and M. Sidi, "Mobile users: to update or not update?" *ACM Wireless Networks*, vol.1, no.2, pp.175-186, July 1994.

[8] R. Caceres and V. N. Padmanabhan, "Fast and scalable handoffs for wireless internetworks", *Proc. ACM Mobicom'96*, pp. 56-66, 1996.

[9] A. T. Campbell, J. Gomez, S. Kim, C.-Y. Wan, Z. R. Turanyi and A. G. Valko, "Comparison of IP micromobility protocols," *IEEE Wireless Communications*, February 2002.

[10] A. T. Campbell, J. Gomez, A. G. Valko, "An overview of cellular IP," *IEEE Wireless Communications and Networking Conference (WCNC'99)*, New Orleans, September 1999.

[11] F.M. Chiussi, D. A. Khotimsky and Santosh Krishnan, "Mobility management in third-generation all-IP networks", *IEEE Communications Magazine*, pp. 124-135, September 2002.

[12] D. C. Cox, "Wireless personal communications: what is it?" *IEEE Personal Comm. Mag.*, pp.20-35, April 1995.

[13] D. R. Cox, *Renewal Theory*. John Wiley & Sons, New York, 1962.

[14] S. Das, A. Misra, P. Agrawal and S. K. Das, "TeleMIP: Telecommunications-enhanced mobile IP architecture for fast intradomain mobility," *IEEE Personal Communications*, pp. 50-58, August 2000.

[15] EIA/TIA, "Cellular radio-telecommunications intersystem operations," *EIA/TIA Technical Report IS-41 Revision B*, 1991.

[16] ETSI, *Digital cellular telecommunications system (phase 2+): mobile application part (MAP) specification (GSM 09.02 version 7.51 Release)*, 1998.

[17] Y. Fang, "Movement-based location management and tradeoff analysis for wireless mobile networks," Accepted for publication in *IEEE Transactions on Computers*.

[18] Y. Fang, "General modeling and performance analysis for location management in wireless mobile networks," *IEEE Transactions on Computers*, **51**(10), 1169-1181, October 2002.

[19] Y. Fang, I. Chlamtac and H. Fei, "Analytical results for optimal choice of location update interval for mobility database failure restoration in PCS networks," *IEEE Transactions on Parallel and Distributed Systems*, vol.11, no.6, pp.615-624, June 2000.

[20] Y. Fang, I. Chlamtac and Y. B. Lin, "Portable movement modeling for PCS networks" *IEEE Trans. Veh. Tech.*, **49**(4), 1356-1363, July 2000.

[21] Y. Fang, I. Chlamtac and H. Fei, "Analytical results for optimal choice of location update interval for mobility database failure restoration in PCS networks," *IEEE Transactions on Parallel and Distributed Systems*, Vol.11, No.6, pp.615-624, June 2000.

[22] Y. Fang, I. Chlamtac and H. Fei, "Failure recovery of HLR mobility databases and parameter optimization for PCS networks," *Journal of Parallel and Distributed Computing*, Vol. 60, No. 4, pp. 431-450, April 2000.

[23] Y. Fang and I. Chlamtac, "Teletraffic analysis and mobility modeling for PCS networks," *IEEE Transactions on Communications*, Vol. 47, No. 7, pp. 1062-1072, July 1999.

[24] Y. Fang, I. Chlamtac and Y. B. Lin, "Billing strategies and performance analysis for PCS networks," *IEEE Transactions on Vehicular Technology*, Vol. 48, No. 2, pp. 638-651, 1999.

[25] Y. Fang, I. Chlamtac and Y. B. Lin, "Channel occupancy times and handoff rate for mobile computing and PCS networks," *IEEE Transactions on Computers*, Vol. 47, No. 6, pp. 679-692, 1998.

[26] Y. Fang, I. Chlamtac and Y. B. Lin, "Modeling PCS networks under general call holding times and cell residence time distributions," *IEEE/ACM Transactions on Networking*. Vol. 5, No. 6, pp. 893-906, December, 1997.

[27] Y. Fang, I. Chlamtac and Y. B. Lin, "Call performance for a PCS networks," *IEEE Journal on Selected Areas in Communications*, Vol. 15, No. 8, pp. 1568-1581, October, 1997.

[28] E. Gustafsson, A. Jonsson and C. Perkins , "Mobile IP regional registration," *Internet draft, draft-ietf-mobileip-reg-tunnel-04.txt*, work in progress, March 2001.

[29] J.S.M. Ho and I.F. Akyildiz, "Dynamic hierarchical database architecture for location management in PCS networks," *IEEE/ACM Transactions on Networks*, **5**(5), 646-660, October 1997.

[30] J.S.M. Ho and I.F. Akyildiz, "Local anchor scheme for reducing signaling costs in personal communications networks," *IEEE/ACM Transactions on Networks*, **4**(5), 709-725, October 1996.

[31] J. S. M. Ho and I. F. Akyildiz, "Mobile user location update and paging under delay constraints," *ACM Wireless Networks*, vol. 1, no. 4, pp. 413-425, December 1995.

[32] D. Hong and S. S. Rappaport, "Traffic model and performance analysis for cellular mobile radio telephone systems with prioritized and nonprioritized handoff procedures," *IEEE Trans. Veh. Tech.*, vol.35, no.3, pp.77-92, 198

[33] B. Jabbari, "Teletraffic aspects of evolving and next-generation wireless communication networks," *IEEE Comm. Mag.*, pp.4-9, December 1994.

[34] R. Jain and Y.B. Lin, "An auxiliary user location strategy employing forwarding pointers to reduce network impacts of PCS," *ACM Wireless Networks*, **1**(2), 197-210, 1995.

[35] R. Jain, Y.B. Lin, C.N. Lo and S. Mohan, "A caching strategy to reduce network impacts of PCS," *IEEE Journal on Selected Areas in Communications*, **12**(8), 1434-1445, 1994.

[36] I. Katzela and M. Naghshineh, "Channel assignment schemes for cellular mobile telecommunication systems: a comprehensive survey," *IEEE Personal Communications.* vol.3, no.3, pp.10-31, June 1996.

[37] F. P. Kelly, *Reversibility and Stochastic Networks*, John Wiley & Sons, New York, 1979.

[38] L. Kleinrock, *Queueing Systems: Theory, Volume I*, John Wiley & Sons, New York, 1975.

[39] W. R. LePage, *Complex Variables and the Laplace Transform for Engineers.* Dover Publications, Inc., New York, 1980.

[40] J. Li, H. Kameda and K. Li, "Optimal dynamic mobility management for PCS networks," *IEEE/ACM Transactions on Networking*, vol.8, no.3, pp.319-327, 2000.

[41] W. Li and A. S. Alfa, "A PCS network with correlated arrival process and splitted-rating channels," *IEEE Journal on Selected Areas in Communications*, **17**(7), 1318-1325, 1999.

[42] Y. B. Lin, "Reducing location update cost in a PCS network," *IEEE/ACM Transactions on Networking*, vol.5, no.1, pp.25-33, 1997.

[43] Y.B. Lin, "Determining the user locations for personal communications networks," *IEEE Transactions on Vehicular Technology*, **43**, 466-473, 1994.

[44] Y.B. Lin and I. Chlamtac, *Wireless and Mobile Network Architectures*, John Wiley and Sons, New York, 2001.

[45] Y.B. Lin, Y.R. Haung, Y.K. Chen and I. Chlamtac, "Mobility management: from GPRS to UMTS,", *Wireless Communications and Mobile Computing (WCMC)*, vol.1, no.4, 339-359, 2001.

[46] Y. B. Lin, S. Mohan and A. Noerpel "Queueing priority channel assignment strategies for handoff and initial access for a PCS network," *IEEE Trans. Veh. Technol.*, vol.43, no.3, pp.704-712, 1994.

[47] W. Ma and Y. Fang, "Two-level pointer forwarding strategy for location management in PCS networks," *IEEE Transactions on Mobile Computing*, Vol.1, No.1, 32-45, Jan.-Mar. 2002.

[48] S. Mohan and R. Jain, "Two user location strategies for personal communications services," *IEEE Personal Communications*, pp.42-50, First quarter, 1994.

[49] C. E. Perkins, "Mobile IP," *IEEE Communications Magazine*, pp. 84-99, May 1997.

[50] R. Ramjee, K. Varadhan, L. Salgarelli, S. R. Thuel, S.-Y. Wang and T. La Porta, "HAWAII: a domain-based approach for supporting mobility in wide-area wireless networks," *IEEE/ACM Trans. of Networking*, vol. 10, no. 5, June 2002.

[51] C. Rose and R. Yates, "Minimizing the average cost of paging under delay constraints," *ACM Wireless Networks*, vol.1, no.2, pp.211-219, 1995.

[52] S. Tabbane, "An alternative strategy for location tracking," *IEEE J. on Selected Areas in Comm.* vol.13, no.5, pp.880-892, 1995.

[53] W. Wang and I.F. Akyildiz, "A dynamic location management scheme for next generation multi-tier PCS systems," *IEEE Transactions on Wireless Communications*, Vol.1., No.1 , pp. 178-190, January 2002.

[54] W. Wang and I.F. Akyildiz, "A new signaling protocol for intersystem roaming in next-generation wireless systems," *IEEE Journal on Selected Areas in Communications*, Vol. 19, No. 10, pp. 2040-2052, October 2001.

[55] W. Wang and I.F. Akyildiz, "Intersystem location update and paging schemes for multitier wireless networks," *Proc. of the ACM MobiCom'2000*, Boston, August 2000.

[56] W. Wang, I.F. Akyildiz and G. Stuber, "Effective paging schemes with delay bounds as QoS constraints in wireless systems," *ACM Wireless Networks*, No. 5, pp. 455-466, September 2001.

[57] J. Xie and I. F. Akyildiz, "A distributed dynamic regional location management scheme for mobile IP," *Proc. IEEE INFOCOM 2002,* pp. 1069-1078, 2002.

Chapter #17

EFFICIENT INFORMATION ACQUISITION AND DISSEMINATION IN PERVASIVE COMPUTING SYSTEMS THROUGH CACHING[1]

Mohan Kumar and Sajal K. Das
The University of Texas at Arlington

Abstract: Pervasive Computing applications require continual and autonomous availability of 'what I want' type of information acquisition and dissemination in a proactive yet unobtrusive way. Mobility and heterogeneity of pervasive environments make this problem even more challenging. Effective use of middleware techniques, such as *caching,* can overcome the dynamic nature of communication media and the limitations of resource-poor devices. In pervasive systems data is needed by users, devices, services and applications whereas caching mechanisms developed for mobile and distributed systems cater mainly to devices and in some special cases to users. Pervasive computing environments present entirely new set of challenges because of the fact that data may be acquired and disseminated at various stages within the system. Therefore, novel caching mechanisms are needed that take into account demand-fetched and prefetched (or pulled), as well as broadcast (or pushed) data. In addition, cache maintenance algorithms should consider such features as heterogeneity, mobility, interoperability, proactivity, and transparency that are unique to pervasive environments.

Key words: Pervasive computing, heterogeneity, mobility, data availability, and caching.

[1] This research work was carried out under support from the Texas Advanced Research Program # 003656-0108-2001

1.   INTRODUCTION

The essence of pervasive computing is the creation of smart environments saturated with computing and communication capability, yet gracefully integrated with human users [Wei91]. Pervasive computing encompasses many different technologies including mobile computing, wireless communications, middleware infrastructure, intelligence building, and wearable computers. Pervasive computing faces many challenges - heterogeneity, invisibility, mobility, and interoperability [Sat01]. In pervasive computing environments, hardware and software entities are expected to function autonomously, continually and correctly. Research in pervasive computing has been progressing at a rapid rate over the recent past in several leading universities and industries around the world. The most intriguing aspect of pervasive computing is its applications in almost all areas of human activity. The enabling technologies such as smart sensors (e.g., UC Berkeley Network Sensor Platform) and Radio Frequency ID (RFID) tags and the new generation of intelligent appliances, embedded processors, wearable computers, handheld computers and so on will continue to play a vital role in improving human quality of life through the advancement of pervasive computing applications [Kum03].

Making data available to users, applications, and services proactively is critical to pervasive computing applications. Current caching and content delivery mechanisms developed for mobile and distributed systems are inadequate for such applications. This is because existing caching mechanisms do not take into account the heterogeneity of the devices and the networking environments, and other unique features of pervasive systems.

This chapter presents in Section 2 an overview of the challenges and issues in pervasive computing for acquisition and dissemination of information in pervasive computing environments. The suitability and inadequacies of existing caching mechanisms for information delivery are discussed in Section 3. Then we propose in Section 4 some extensions to the existing caching mechanisms to make them more suitable for pervasive computing. The extensions include: profiling where in the profile of the user is stored and exploited for caching; data staging that involves staging of data on surrogate machines to decrease latency and save bandwidth; content adaptation of documents wherein the format of the documents are adapted to device profiles; and cache relocation to follow the roaming user. Section 5 summarizes the role of cache in pervasive computing and Section 6 presents our proposed architecture. Section 7 concludes the paper.

2. CHALLENGES OF PERVASIVE COMPUTING

This section explores the challenges and issues in pervasive computing environments, in which applications and users access information sources using a variety of mobile and static devices such as sensors, handheld computers, personal computers, servers, etc. The devices possess heterogeneous capabilities with respect to display size, communication mechanisms, memory, battery power, and computing capabilities. Applications and users attached to these devices connect to the external world via cellular radio networks, local area wireless networks, dial-up connections, or broadband connections. These communication systems offer different functionalities and heterogeneous characteristics with respect to bandwidth, delay, and latency. Also dealing with different degrees of smartness in different smart spaces is a unique challenge. According to Weiser [Wei91], pervasive computing is the complete disappearance of computing technology from a user's consciousness. Allowing access to information and services transparently and unobtrusively is one of the most important goals of pervasive computing.

The emerging area of pervasive computing incorporates research findings from many areas of computer science to meet the challenges posed by a myriad of applications. Today, we do have the necessary hardware and software infrastructure to foster the growth of this area. What is necessary is the *glue* to put together existing entities in order to provide meaningful and unobtrusive services to a variety of applications [Kum03a]. There exist excellent visionary papers, highlighting the issues and challenges of pervasive computing [Sat01, Ban00] – that include: invisibility, proactivity, interoperability, heterogeneity, mobility, intelligence, and security.

The development of computing tools hitherto has been in general, 'reactive' or 'interactive'. On the other hand, 'human in the loop' has its limits, as the number of networked computers will surpass the number of users within a few years. Users of pervasive computing applications would want to receive 'what I want' kind of information and services in a transparent fashion. In a thought provoking paper [Ten00], Tenenhouse envisions that a majority of computing devices in future will be proactive. Proactivity can be provided by the effective use of overlay networks. In pervasive computing environments, users interact with various embedded or invisible computers. A proactive environment ensures proactive acquisition and dissemination of context-aware information to users and applications.

Effective caching of information at various stages can enhance proactivity and improve its availability transparently.

Today's computing world is replete with numerous types of devices, operating systems, and networks. Cooperation and collaboration among various devices and software entities is necessary for pervasive computing. At the same time, the overheads introduced by the adaptation software should be minimal and scalable. While it is almost unthinkable (in today's environment) to have homogeneous devices and software, it is however, possible to build software bridges across various entities to ensure interoperability. The limitations of resource-poor hardware can be overcome by exploiting the concepts of agents and services. A major challenge is to develop effective mechanisms, such as caching at various stages in order to mask the uneven conditions and develop portable and lightweight application software.

Mobile computing devices have limited resources (CPU, memory, bandwidth, energy), are likely to be disconnected, and are required to react (transparently) to frequent changes in the environment. Mobile users desire *anytime anywhere* access to information while on the move. Typically, a wireless interface is required for communication among the mobile devices. The interface can be a wireless LAN, a cellular network, a satellite network, or a combination thereof. Techniques developed for routing, multicasting, caching, and data access in mobile environments need to be extended carefully to pervasive environments consisting of numerous invisible devices, anonymous users, and ubiquitous services. Development of effective information delivery mechanisms to mask the heterogeneous wireless networks and mobility effects is a significant challenge. Provisioning uniform services regardless of location is also a vital component of pervasive computing. The challenge here is to provide location-aware information availability in an adaptive fashion, in a form that is most appropriate to the location as well as the situation under consideration [Che02].

Mobility arises in several forms in pervasive environments – such as mobility of devices, mobility of users and mobility of software agents. Because users, devices, and agents change context, tasks like querying, scheduling and caching management become increasingly complex in pervasive computing systems.

To create smart spaces we need an intelligent, situation-aware system augmented by middleware support. The challenge is to incorporate middleware tools into smart environments. In a crisis management situation, for example, the remote sensing system should have abilities to infer the user's intention and current situation to inform of the possible danger prior to the occurrence of the crisis. Smart home projects are being carried out in many universities and research laboratories [Cur99, Pin03]. Smart homes allow the occupants to interact with the home equipment and appliances by exploiting knowledge gained through user behavior. Current smart homes are not yet advanced enough to allow the user to take the (virtual) home with him/her. Dynamic, context-aware information can be pushed into the user's mobile device prior to leaving the smart home. For example, the Sentient computing environment [Cur99] is localized and very limited in terms of functionality. With the middleware services, the same environment can be used for pervasive health care as envisioned in [Had02].

Several methods for knowledge representation in intelligent spaces have been investigated. Provisioning proactive, context-aware or situation-oriented response from the infrastructure to the users' needs is a challenge. As smart spaces grow in sophistication, the intensity of interactions between a user's personal computing space and his/her surroundings increases. This has severe bandwidth, energy, and distraction implications for a wireless mobile user. The presence of multiple users will further complicate this problem. Scalability of applications is a serious issue on many fronts – number of users, physical spread of the client and variety of devices. Additionally, the application data is likely to have real-time component requiring specific delivery deadlines.

3. CURRENT CACHING MECHANISMS

A key requirement of pervasive computing is the ability of the applications to be aware of the characteristics which are part of the context for the application running on devices. For example, if the multimedia streaming application can get information about the lack/availability of wireless bandwidth it may choose to scale down/up the fidelity of the output. Some caching mechanisms do incorporate mobility issues and wireless communication channel characteristics in identifying data items for caching.

Consistency of a mobile client's cached items in disconnection-prone environments has been a subject of investigation [Cao02]. Some current schemes [Kah01] use *invalidation report* (IR) in which an IR is periodically broadcast by the server and clients use IRs to validate their cached items. There exist several problems with regard to IR-based schemes: i) the client or device may be offline or disconnected when the IR arrives; ii) the source should be queried to obtain the updated data; and iii) the frequency of IR reports may be very high with respect to frequency of reference at certain user sites.

Some mechanisms such as the one by Tuah et al.[Tua02] decide on the data to be cached based on next access probability. Data items are cached based on probability of access in the future or popularity. Caching in low bandwidth networks increases data availability and reduces latency. User and device profiles can help determine the utility values of cacheable data. However, current caching mechanisms do not take into account the device profiles while caching data.

In the remainder of this section we discuss cache coherence and cache replacement mechanisms that have been proposed for mobile and distributed systems.

3.1   Cache Coherence

A cache coherence strategy usually executes in two phases - invalidation and update. Most applications in mobile environments will generate more read operations than write operations. Further-more a mobile client usually can accept a slightly out-dated data in return for faster data retrieval.

In the *refresh time method* [Yue00] of invalidation the estimation of the refresh time for an object depends on its update probability. If an object is updated frequently, its refresh time will be shorter. In this method, the refresh time for an object is updated dynamically whenever the object is sent to a client from the server. This approach does not require a client to be always connected in order to invalidate an object. Furthermore, if an object is never accessed again, it will never be refreshed even after its refresh time expires.

In *asynchronous invalidation* [Cao02] reports are broadcast by the server only when some data changes not periodically. Each mobile host maintains its own home location cache (HLC) to deal with the problem of disconnections. The HLC is maintained at a designated home mobile switching station (MSS). It has an entry for each data item cached by the mobile host and needs to maintain only the time-stamp at which the data item was last invalidated. At the cost of this extra memory overhead for maintaining an HLC, the mobile host can continue to use its cache even after prolonged periods of disconnection from the network. A mobile host alternates between active mode and sleep mode. In the sleep mode, a mobile client is unable to receive any invalidation reports sent to it by the HLC. The time stamp is used to decide as to which invalidations to retransmit to the mobile host. When the mobile host wakes up after a sleep, it sends a probe message (along with the time stamp) to its HLC. In response to this probe message, the HLC sends it an invalidation report. Thus a mobile host can determine which data items got changed while it was disconnected. A mobile host defers all queries, which it receives after waking up until it has received the invalidation report from its HLC.

3.2 Cache Replacement Policies

Replacement policies adopt the access probabilities of database items as an indicator for the necessity of replacing a cached item. For each object a *replacement score* indicating the prediction of its access probability is estimated. The higher the score the higher is the estimated access probability and the lower is its chance of being replaced. The most straightforward way to compute the score for each object is by measuring the mean inter-operation arrival duration. The cached object with the highest mean arrival duration is replaced in the mean scheme.

The mean scheme probably does not adapt well to changes in access patterns since every single trace from the beginning of the access history remains in effect. A better approach is to use a window for the statistical measures. Each object is associated with a window of size W storing the access time of W most recent operations. The cached object with the highest mean arrival duration within the window is replaced. This is known as the *window scheme* [Yue00], whose effectiveness depends on the window size. A problem with the window scheme is the amount of storage needed to maintain the window values.

To avoid the need of a moving window and to adapt quickly to changes in access patterns, the exponentially weighted moving average [EWMA] scheme assigns weights to each arrival/duration. In EWMA recent durations have higher weights and the weights tail off as the durations get longer. The replacement score is the EWMA of arrival durations [Su02]. A parameter to the EWMA is the weight, α which ranges from 0 to 1. The current duration receives a weight of 1, the previous duration receives a weight of α the next previous duration receives a weight of α^2 and so on.

In traditional cache replacement policies, access probability is considered as the most important factor that affects cache performance. A probability-based policy is to replace the data with the least access probability. In location-dependent services, besides access probability there are two other factors, namely data distance and valid scope area that should be considered in cache replacement. Data distance refers to the distance between the current location of a mobile client and the valid scope of a data value. In a location-dependent data service, the server responds to a query with the suitable value of the data item according to the client's current location. As such, when the valid scope of a data value is far away from the client's current location, this data will have a lower chance to become usable again since it will take some time before the client enters the valid scope area again and the data is useless before the user reaches the valid scope area. In this respect, we should favor ejecting the *farthest* data when replacement takes place. Valid scope area refers to the geometric area of the valid scope of a data value. For location-dependent data, valid scope areas can somehow reflect the access probabilities for different data values. That is, the larger the valid scope area of the data, the higher the probability that the client requests this data. This is because, generally, the client has a higher chance of being in large regions than small regions. Thus, a good cache replacement policy should also take this factor into consideration. Based on the above analysis, a promising cache replacement policy should choose as its victim, data with low access probability, a small valid scope area and a long data distance.

4. EXTENDING CURRENT CACHING MECHANISMS FOR PERVASIVE COMPUTING

As seen in the previous section the existing caching mechanisms in use today fall short of the requirements for pervasive computing. This section investigates techniques to make cache management strategies more suitable to meet the challenges of pervasive computing. Pervasive environments may comprise devices with varying communication, display and computing capabilities; and software agents representing disconnected users/devices. One of the challenges in pervasive computing is to adapt the environment to the user's needs by predicting the user's location and needs. In the following, we investigate various caching mechanisms and assess their suitability of pervasive environments. In the following, we discuss various caching strategies that can be adapted to pervasive environments.

4.1   Profiling

Data recharging is a service that aims to provide analogous functionality for recharging the device–resident data cache [Che01]. The goal of data recharging is to deliver data in a flexible and geographically independent manner. That is, users transparently recharge their devices with context-aware data by connecting at any point in an available infrastructure such as the Internet using available communication media. The data needed on a particular device is highly dependent on the user of that device and the tasks he/she is performing. Thus, the choice of data that must be sent to a device in order to recharge it is dependent on the semantics of the applications. The data recharging infrastructure maintains user profiles to facilitate transfer of user-specific data.

A user profile contains two types of information – the types of data that are of interest to the user and the user priorities. These profiles can play a very important role in caching as they could be used while caching data and also while choosing victims to eject from the cache.

4.2   Hierarchical Caching – Content Adaptation

In pervasive computing, users access information sources using a wide variety of mobile devices featuring heterogeneous capabilities with respect to size, memory, battery power and computing capability. The key to meeting the demands in such a heterogeneous environment is the adaptation of the representation of documents to the capabilities of the device. This is called content adaptation. This adaptation may include: format transcoding (e.g. XML, to WML, JPEG to WBMP) scaling of images as well as video and audio streams, media conversion and omission or substitution of document parts. Web caching exploits sharing and temporal locality in the access patterns of clients to satisfy requests by cached copies of the required objects. By terminating client requests at local caching proxies, the network traffic on the Internet as well as the server load is reduced whereas the response time is improved. Furthermore, the robustness of web services is enhanced as a client may retrieve a document from the proxy cache if the server is unavailable.

There are three different approaches for performing the adaptations - content server-side adaptation, proxy-based, and adaptation paths. In server-side adaptations, adapted documents are supplied to the clients by the server either by on-demand dynamic adaptation or by having a repository of pre-adapted documents. With proxy-based adaptation the documents are served by the servers in a generic representation (e.g. XML-based). The adaptation is performed by intermediary proxies. Those proxies are usually placed close to the clients. The concept of adaptation paths has been introduced by the UC Berkeley's Ninja project [??]. As opposed to server-side adaptation, proxy based adaptation allows the caches to be placed between the information sources and adaptation proxies.

4.3   Data Staging

Data Staging is a technique that improves the performance of distributed file systems running on small, storage-limited pervasive computing devices by opportunistically caching data on nearby surrogates[Fli02]. The file system tries to reduce access time by caching files on the client machine, but space limitations prevent it from caching all but a portion of the files that the user might potentially read. This surrogate plays the role of a second level cache for a mobile client. By proactively staging data on the surrogate, cache misses from a nearby mobile client can be serviced at low latency.

5.   ROLE OF CACHE IN PERVASIVE COMPUTING

Pervasive computing aims at seamlessly integrating computing and communication with our environment so as to make our day-to-day activities the principal focus rather than the computing or communication devices. As mobile and computing devices become more pervasive, the nature of interaction between users and computers is evolving. Computing applications need to become increasingly autonomous and invisible by placing greater reliance on context-awareness in order to increase user transparency. Moreover due to the current popular emphasis on *anytime anywhere* computing applications must cope with the highly dynamic environments in which resources such as network connectivity, bandwidth, security and software services vary over time and therefore applications behavior and functionality have to adapt to currently available resources and constraints. Thus we see that pervasive computing although taking shape today imposes a number of challenges. We investigate how a cache meets some of these challenges.

The data from the server has to be adapted and formatted before it can be pushed to the device. A combination of effective caching and middleware agents can be used to meet this challenge. A software agent associated with the cache could be employed to adapt data from the server to meet the media handling capabilities of the device. This transformation of the data may include format transcoding, scaling of images as well as audio and video streams, media conversion and also omission or substitution of document parts. By doing this, the server can serve documents requested in a generic representation and these documents could be formatted /adapted by agents in the cache according to the capabilities of the device.

The second challenge is invisibility, i.e., how to make the interaction between the environment and computing devices unobtrusive. The user must be able to obtain just what he wants without being aware of the computing environment. Frequently accessed data should be made available anywhere to the clients through effective caching. Caching frequently accessed data in the device could improve the response time and hide the latency due to scarce bandwidth. The cache could be used to provide the user with exactly the information he requires by constantly replenishing the cache with user profiles. Thus caching makes the interaction between the environment and the user least obtrusive, by pushing only data that is important to the user in the current context.

When the user moves from one location to another, his requirements change, and the data required may not have been already cached. This will result in fetching relevant data items from the server, thus increasing the latency. This problem can be alleviated if user location prediction methods are employed. Based on this location prediction the cache could prefetch items relevant to the user's next location before the user moves there. Thus when the user moves to the next location, the data required by him in that location are already cached, thereby reducing access latency for these items.

One of the very important requirements of an effective pervasive system is proactivity. The environment should be able to sense the needs of the user and deliver information proactively. Caching and profiling algorithms can be integrated to ensure availability of 'what I want' kind of information in a pervasive computing environment.

6. ARCHITECTURE FOR PERVASING CACHING

The architecture of a caching mechanism for pervasive environments is shown in Figure 1. The main components of such a system are the cache on the receiving entity and the cache on the server or network. The receiving entity can be a user's device, user's software agent, a service or an application. A cache can be associated with each of these entities. The information documents are cached based on their utility values. The algorithms for determining utility values, the optimization strategy, and incorporation of user/system inputs comprise the software components of the system. At any point of time, the cache comprises traditionally cached, prefetched and push-cached documents. The utility algorithm updates the utility values of i) cached documents ii) remote documents to be prefetched and iii) documents in the next broadcast cycle.

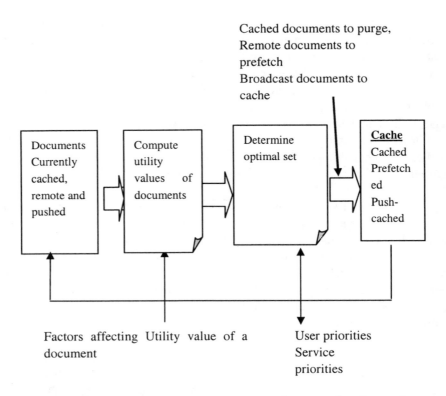

Figure 1 Architecture of a Caching Scheme for Pervasive Computing

6.1   Utility Functions

Each document Di ($0 \le i \le n$) is represented by a tuple $D_i < s_i, u_i>$, where s_i is the document size and ui is the utility of the document. A document's utility ($0 \le u_i \le 1$), varies dynamically based on various parameters and whether the document is cached, broadcast or to be prefetched. The factors pertinent to a document are shown in Table 1 below. The utility of each document is determined by normalizing the effect of several factors by employing a rule similar to those discussed in Greedy-Dual caching strategies [Lau02]. In fact, developing such a rule for content delivery in wireless Internet is one of the main challenges of this project. Furthermore, the computation of utility values can be tuned to user/server requirements by applying weights to various factors in order to provide flexibility.

Table 1. Factors affecting utility value of a document

	Description	Cache	Prefetch	Push
Cost	*The communication cost of transmission*	✔	✔	
Mobility factor	*The rate at which MH moves from node to node*	✔	✔	✔
Hit ratio	*The number of hits/the number of requests*	✔		
Delay saving ratio	*Time to transmit once/time to transmit many times*	✔	✔	✔
Access frequency	*Number of times a document is accessed*	✔		
Dependence	*The dependence relation with documents already in cache*	✔	✔	✔
Bandwidth	*Bandwidth available*		✔	
Client profile	*Utility of the document based on profile*	✔	✔	
Server profile	*Popularity value for a broadcast document*			✔
Device profile	*Capabilities at the receiving device*	✔	✔	✔
Broadcast frequency	*The frequency at which a document is broadcast*			✔
Update frequency	*The frequency at which a document is updated*	✔	✔	✔
Battery power	*Battery power left on device*	✔	✔	

The utility of a remote document D_i to be prefetched can be high if the wireless link between a device and the infrastructure has high bandwidth whereas u_i is low if the wireless link has low bandwidth, unavailable or congested. Likewise, the utility of a cached document decreases every time it is not requested. The mobility of the MH plays a significant role in determining the utility of a document. Thus each cached document's utility is updated continuously and the utility of a remote document is updated whenever it is in contention for a spot in the cache.

6.2   Optimal Set of Documents

The next important step is to determine the optimal set of documents to cache, given document sizes and utilities. The objective is to maximize utility while at the same time make optimal use of the cache space. First, the weight w_i, of the object in the cache is calculated by using $w_i = (u_i \times C/s_i)$, where C is the cache capacity, u_i is the utility and s_i is the size of the document respectively. Then the documents are sorted in order of their weights. In a traditional system, with neither prefetch nor broadcast, documents can be cached based on their weights. However, the situation with pervasive environments is quite complex when one or more of the following events happen: i) wireless bandwidth is available, ii) a broadcast document is received, or iii) an extraordinary event has taken place. On occurrence of such an event, utilities and hence the weights of these documents change. In addition, user-defined and/or server-defined priorities to cache documents including quality of service (QoS) parameters can be incorporated into the architecture. The set of documents to be cached is determined by employing optimization techniques such as greedy algorithms and dynamic programming.

Utility values are updated for all cached documents having high access probability, and documents related to cached documents. Broadcast documents that are in the optimal set are cached and also the procedure for prefetching new documents in the optimal set which are not currently cached is initiated. Once prefetch documents arrive, cached documents that are not in the optimal set are purged to make room. Likewise an optimization algorithm is executed on every server or network node in order to maintain an optimal set of documents. Utility values are computed for all cached documents as well as for those in the prefetch lists. The next broadcast sequence is determined and then requests for transfer of remote documents are sent out on the Internet. Each of these algorithms should ideally execute in $O(n)$ time, where n is the number of documents considered.

6.3   Cache consistency

Assuming successful caching of required data documents, we have to deal with the cache consistency issues. Cached documents may be updated on the infrastructure, when that happens, there will be multiple copies of the same data document. Such copies may be on connected or disconnected mobile devices Therefore we should have special mechanisms to handle such situations that are uncommon in traditional systems. Stale copies of cached documents on mobile devices must be invalidated before they are accessed by the mobile client.

An important criterion for caching is the rate at which a particular data document is updated. If a data document is likely to be updated very frequently, then it's not prudent to cache it. Several researchers have investigated mechanisms to deliver invalidation reports efficiently to mobile clients. Update frequencies of documents should be considered in determining the utility values.

There can be two types of cache consistency maintenance algorithms for mobile computing environments: stateless and stateful. In the stateless approach, the server is unaware of the user's content. Stateless approaches use simple database management and they are less expensive to maintain. Even though, such approaches would suffice for many routine data dissemination applications. On the other hand, stateful approaches require large databases and complex maintenance algorithms. The Scalable Asynchronous Cache Consistency Scheme (SACCS) [Wan03] inherits positive features of both approaches. The SACCS mechanism employs flag bits to indicate data validity at the server and users' ends to minimize cost of consistency maintenance algorithms. An improvisation of SACCS can be used for pervasive environments to incorporate unique features of pervasive environments.

6.4   Multimedia delivery

A continuous or streaming media system allows large multimedia objects to be played out at clients before the entire object is delivered. There are many classes of continuous media (CM) applications relevant to pervasive computing. These classes include multi-party conferencing, peer-to-peer networks, online gaming, multimedia delivery and collaboration. Before CM data is rendered on an end-system, it typically undergoes capturing, encoding, storage, and delivery. There has been much success in using web caches to alleviate network-related bottlenecks. Caches are incrementally deployable solutions that require no support from the network, servers or even clients. As CM usage becomes more prevalent in pervasive systems, it is worthwhile to deploy caching that accommodates special characteristics of CM and related protocols. CM caching research is still in its infancy. Caching can potentially improve system throughput and increase data availability. It is necessary to extend the existing schemes to address wireless multimedia issues such as delay, jitter and bandwidth availability in order to provide a high available, adaptive and consistent system. In [Lau02], a GD-based scheme considered traditional caching for continuous media delivery in wireless networks. For pervasive computing applications it would be necessary to include prefetch and broadcast documents and address heterogeneity and scalability issues.

7.   CONCLUSIONS

Pervasive computing technologies and associated software are being employed to facilitate such applications as telemedicine, education, space endeavors, manufacturing, crisis management, transportation, and defense. In pervasive computing environments, hardware and software entities are expected to function autonomously, continually and correctly. Information acquisition and dissemination by the hardware and software entities in a pervasive computing scenario is a challenging problem. However, existing caching and content delivery mechanisms are inadequate to meet the demands on pervasive environments. This chapter presents an investigation of existing mobile and distributed caching strategies that can be improvised for pervasive computing applications. Further more, an architecture for facilitating information acquisition and dissemination using caching (comprising cached, prefetched and broadcast documents) strategies is presented.

References

[Ban00] G. Banavar, J. Beck and E. Gluzberg, "Challenges: An application model for pervasive computing," in Proceedings of 6th Annual International Conference on Mobile Computing and Networking (MOBICOM 2000), pp.266-274, 2000, Boston MA, USA.

[Cao02] Guohong Cao, On Improving the Performance of Cache Invalidation in Mobile Environments, Mobile Networks and Applications, Volume 7, Issue 4, August 2002.

[Cha01] B.Y. Chan, A. Si, and H. V. Leong, A Framework for Cache Management for Mobile Databases: Design and Evaluation Distributed and Parallel Databases 10 (1):23-57, July 2001.

[Che01] Cherniack, M.,Franklin, M.J., Zdonik, S., Expressing user profiles for Data Recharging , IEEE Personal Communications , Volume: 8 Issue: 4 , Aug 2001 Page(s): 32 -38.

[Che02] Liang Cheng and Ivan Marsic, Piecewise Network Awareness Service for Wireless Mobile Pervasive Computing, Mobile Networks and Applications, Volume 7 , Issue 4, August 2002.

[Cur99] R. Curwen, A. Hopper, P. Steggles and A. Ward, "Sentient Computing," Technical Report 1999.13 (video), AT&T Laboratories Cambridge, 24a Trumpington Street, Cambridge CB2 1QA, England.

[Fli02] Jason Flinn, Shafeeq Sinnamohideen and M. Satyanarayanan , Data Staging on Untrusted Surrogates, Intel Research, Pittsburgh IRP-TR-02-02 May 2002

[Gup01] Gupta, S.K.S. ,Wang-Chien Lee , Purakayastha, A. , Srimani, P.K , An overview of Pervasive Computing IEEE Personal Communications, Volume: 8 Issue: 4, Aug. 2001, Page(s): 8 -9.

[Had02] Stathes Hadjiefthymiades, Vicky Matthaiou, Lazaros Merakos , On Supporting the WWW in Wireless Communications Through Mobile Agents , Mobile Networks and Applications , Volume 7 , Issue 4 , August 2002.

[Kah01] A. Kahol, S. Khurana, S.K.S. Gupta, and P. K. Srimani, A Strategy to Manage Cache Consistency in a Disconnected Distributed Environment, IEEE Trans. on Parallel and Distributed Systems, Vol. 12, No. 7, July 2001, page(s): 686-700.

[Kum03]M Kumar, Pervasive Computing Education, Distributed Systems Online, July 2003, Vol. 4, No.7.

[Kum03a] M Kumar, B Shirazi, S K Das, B Sung, D Levine, and M Singhal, Pervasive Information Communities Organization PICO: A Middleware Framework for Pervasive Computing IEEE Pervasive Computing, July-September 2003, pgs. 10-17.

[Lau02] WHO Lau, M Kumar, and S. Venkatesh, A Cooperative Cache Architecture in Support of Caching Multimedia Objects in MANETs, Atlanta, September 2002.

[Sat01] Satyanarayanan, M., Pervasive Computing: Vision and Challenges, IEEE Personal Communications, August 2001, Pgs. 10- 17.

[Su02] Zhou Su, Jiro Katto, T Nishikawa, M Murakami, T Washizawa and Y Yasuda, An Integrated Scheme for Distributed Segmented Streaming Media over Hierarchical Caches, International Conference on Information Technology and Applications, Bathurst Australia, November 2002.

[Ten00] David Tennenhouse, Proactive Computing, Communications of the ACM, vol. 43, No. 5, May 2000.

[Tua02] N J Tuah M Kumar, S Venkatesh, and S K Das, Performance Evaluation of Speculative Prefetching, IEEE Transactions on Parallel and Distributed Systems, Vol. 13, No. 5, June 2002, pp. 471-484.

[Wan03] Z. Wang, M Kumar, S.K. Das, and H Shen, Investigation of Cache Maintenance Strategies for Multi-cell Environments, 4th International Conference on Mobile Data Management, 21-24 January, 2003, Melbourne, Australia

[Wei91] M. Weiser, "The computer for the 21st Century," Sci. American, Sept.1991.

[Yue00] Joe Chun-Hung Yuen, Edward Chan, Kam-Yiu Lam and HW Leung, Cache Invalidation Scheme for Mobile Computing Systems with Real-time Data, ACM SIGMOD, Dec. 2000.

Chapter 18

SECURITY IN WIRELESS MOBILE AND SENSOR NETWORKS

Sajal K. Das, Afrand Agah and Kalyan Basu
Center for Research in Wireless Mobility and Networking(CReWMaN)
Department of Computer Science and Engineering
The University of Texas at Arlington
Arlington, TX 76019-0015
Tel:(817)272-7409
{das, agah, basu}@cse.uta.edu

Abstract   Security plays a significant role in the current and future wireless mobile networks. The goal of secured communication must ensure that the data we receive is authentic, confidential and have not been altered. Because of the characteristics of wireless medium, limited protection of mobile nodes, nature of connectivity and lack of centralized managing point, wireless mobile networks are too vulnerable and often subject to attacks. In this chapter we discuss security issues and challenges in wireless mobile ad hoc and sensor networks as well as in pervasive computing infrastructures. We also describe security protocols for such environments.

Keywords: Ad hoc networks, authentication, digital signature, intrusion detection, pervasive computing, sensor networks.

1.   Introduction

Rapid growth of wireless mobile and sensor networks has changed the perspective of network security altogether. This is due mainly to the characteristics of wireless medium, severe resource constraints as well as unpredictable mobility, all of which contribute to new vulnerabilities that do not exist in a wired network. Therefore, we can no longer use the same mechanisms that was used for Internet security, to protect wireless mobile and/or sensor networks. First of all, wireless channels are more prone to failures and eavesdropping. Limitations to such resources as CPU, memory, bandwidth and battery power of portable/mobile devices often become a bottleneck in using more sophisticated and computation-intensive key management protocols. Additionally, as mobile nodes are capable of roaming, lack of physical protection results in compromising and capturing these nodes. Clearly, attacks by a compromised node from inside the network is often much more damaging and harder to detect, due to lack of physical protection, than external attacks.

A wireless mobile network is susceptible to a number of compromises both from legitimate users of the system and from intruders. Two general security compromises are host compromise and communication compromise. Host compromises can be categorized as masquerading, unauthorized use of resources, disclosure of information, alteration of information and denial of service. Masquerading is when a user masquerades as another one to gain access to an unauthorized system component. Unauthorized use of resources is when a user accesses system without authorization from real user of the device. Unauthorized reading of stored information is disclosure of information. And unauthorized writing of the information is alteration of information. By denying a resource to an entity, an attacker denies a service. Communication compromises can be categorized as masquerading, interception, fabrication, denial of service, repudiation of actions and unauthorized use of resources. Repudiation of actions is when the sender of a message denies sending it. Interception is when the opponent gains access to the data transmitted over the communication link, so it can obtain the transmitted information or identities and locations of the communication parties.

1.1   Basics of Security Protocols

Any security protocol in a wireless mobile computing environment has the following four goals [5, 11].

Authentication: The portable device (e.g., laptop computer, PDA) used by wireless mobile user and the base station with which the user communicates must be able to mutually authenticate each other. Such authentication enables the base station to prevent unauthorized users from using its services, and enables the user to choose an authorized base station as the access point.

Data privacy: The portable device used by a wireless mobile user and the base station must be able to communicate in a secure way. Data privacy protects the data from being snooped, replayed or any forgery.

Location privacy: Location privacy is a unique goal of wireless mobile communication. It is desirable or critical not to compromise the location and identity of the mobile user.

Accounting: When a mobile user uses a portable device, it establishes communications with a base station, and the service provider needs to know who to charge for the service used.

Physical layer security mainly aims at preventing every unauthorized access to the devices. There are three ways to authenticate devices: (i) What the user knows; it is very useful for small devices. PIN and password entering fall into this category; (ii) What the user owns; it is accessing the control of some devices in the form of key-cards and electronic tokens. Permanent installed devices could use this type of authentication, like SIM cards in GSM; and (iii) What the user is like; it is also known as biometric authentication and is based on checking the user's physical characteristics. Usually a good authentication scheme involves more than one of the above mechanisms. Like ATM machines require both physical card and a PIN.

The following techniques are the basic building blocks of any security protocol [15].

Symmetric Cryptography: It means that the key for the encryption and decryption of messages is the same. Secret-key cryptography (SKC) scheme is an example that is used for confidentiality of stored data.

Asymmetric Cryptography: It uses algorithms with different keys for encryption and decryption. So both parties that want to transfer data, are able to exchange their encryption keys over an even insecure connection link. The general idea behind public key cryptography is that each participant in a communication network has two keys – one public key and one private key. Agent's public key is freely available, or may be obtained from certifying and trusted authority on demand. The agent's private key is meant to be secret: only agent knows this key and letting others know this key will allow them to assume its identity, as well as decrypt all messages, if intercepted, that were addressed to it. Needless to say, public key and private key are different keys.

Certificate Authority: When two parties exchange their public key over an insecure channel, an attacker can substitute the public keys with fake ones. One way to verify that the public key really belongs to the sender, is using a certificate authority (CA), who signs the public key with some information about the owner.

The rest of the chapter is organized as follows. Section 2 discusses intrusion detection and highlights some techniques. Section 3 reviews the security aspects in wireless ad hoc networks and different existing solutions. Section 4 talks about security in pervasive computing. Section 5 describes security in sensor networks. Section 6 concludes the chapter.

2.   Intrusion Detection

By definition, intrusion is any set of actions that attempts to compromise the integrity, confidentiality or availability of resources [4]. As wireless mobile networks do not have a fixed topology (it changes dynamically due to user and device mobility and also due to fluctuations in wireless links) or fixed infrastructure (as in mobile ad hoc networks), standard intrusion detection techniques that rely on real time traffic analysis, will not be sufficient here. For example, in a wired network we can monitor the network at switches or routers, but in mobile ad hoc networks we do not normally have such traffic to get audit data for the whole network. Moreover, a mobile user may often be disconnected which can be due to the environment and nature of the wireless medium, and may not necessarily represent a model of anomaly. Therefore, a good intrusion detection system for mobile wireless networks must be able to recognize anomaly from normalcy. Zhang, et al. [43, 44] examine the vulnerabilities of wireless networks and developed an intrusion detection architecture for such networks. Traffic flow, routing activities and topology patterns are three general type of information that they consider for their approach. In their system, individual intrusion detection agents are placed at each node of the network. Each agent monitors local activities (user and system). If any evidence of intrusion is found, neighboring agents cooperatively help in a more global intrusion detection.

The main assumption in any intrusion detection scheme is that the program activities are under observation. Based on the system audit data, we can broadly categorize intrusion detection systems into three classes:

- *Network-based intrusion detection*: It runs at the gateway of a network and examines all incoming packets. In a wireless network this can be done at the base station. Intrusion prevention measures such as encryption and authentication can be placed at the base station and any new node will be examined accordingly.

- *Host-based intrusion detection*: It receives the necessary audit data from the operating system and analyzes the generated events. In a wireless network, especially in an ad hoc network, the time duration for which a node acts as a host can be short, and also the topology of the network changes very frequently, so the system must act correctly in the new environment.

- *Router-based intrusion detection*: It is installed on routers to prevent intruder from entering into the network. Intrusion detection system gets audit data from operating system to monitor and analyze the events. Wireless mobile networks do not have traffic concentration points that a system can collect audit data. So all the possible audit data are any data available during the communication within the radio range.

Based on the functional characteristics of detection schemes, intrusion detection systems can be categorized into knowledge based (or *misuse detection*)and behavior based systems (or *anomaly detection*). In misuse detection systems, patterns of well-known attacks or weak spots of the system are used to identify known intrusions. For example, if within two minutes interval there are more than four failed login attempts, the system should raise a flag indicating someone is trying to guess the access code or password. The advantage of this approach is accurate detection of knowledge attacks whereas its disadvantage is the lack of the ability to detect newly invented attacks. For example, if someone has stored his credit card number information in his cell phone, and the cell phone gets lost then anyone who uses this phone would be able to misuse the credit card, without being double checked by the system, until the system learns about this attack.

In anomaly detection systems, if an activity differs from the established normal profile, it is flagged as an anomaly. User's normal behavior profile can contain such information as the amount of network traffic, the daily path that a cell phone user follows everyday, resources used, and so on. The advantage of this approach is that it does not require prior knowledge but its disadvantage is high false positive rate.

With the anticipated growth of wireless Internet technologies, it is expected that the mobile commerce transactions will grow significantly. The security of such a highly dynamic environment is a paramount challenge for the information industry. The major vulnerable components in this infrastructure are password, protocol, content, uncontrolled flow and wireless access links. In the Internet domain, it has been found that most of the common attacks have a target, long-term plan and strategy. Indeed the attackers have their loosely connected global organization and pursue over longer time to execute the plan successfully. The strategies used to develop the attack plan include [20]: (i) IP spoofing, (ii) SYN flooding, (iii) sequence number attack, (iv) TCP session hijacking, (v) password guessing, (vi) information theft, and (vii) denial of service. Clearly, the wireless Internet infrastructure will also be prone to these attacks, in addition to other attacks that are specific to wireless networks as detailed in the next section. To protect such environment against such malicious activities, it is necessary to design a robust framework based on systematic learning strategies, than solving the symptoms only. Thus securing the wireless Internet is extremely challenging.

3.   Security in Wireless Ad Hoc Networks

A wireless ad hoc network is characterized by frequent changes in topology. In other words, the trust relationship between the nodes also changes frequently, which can not be achieved with static security configurations. Moreover, the security configuration must be scalable to accommodate hundreds of nodes in the ad hoc network. Routing protocols must not only be be able to cope with dynamically changing topology but also malicious attacks. Since hosts may move in an arbitrary manner in such networks, the routes are subject to frequent disconnections and re-connections.

Traditionally, to secure a wireless network, we consider a wide range of issues: (i) identification of a mobile user, (ii) anonymity of a mobile user, (iii) authentication of a base station, (iv) security of the flow information, (v) prevention of attack, (vi) computational complexity of having authenticity and security at mobile terminal, and (vii) cost of establishing a session key between user and base station.

Getting access to wireless (radio) links is relatively easy, implying the security of wireless communications can be easily compromised, by interception or jamming of radio signals. In particular, if transmissions occur in a large area or when the users are allowed to cross security domains, like permission to use one service in one environment but prohibiting the use of another service in the same area. By using wireless networks we are able to use more flexible communication models than traditional wire-line networks. Mobility and volatility are transparent from the application so each node can communicate with any other node, as it would be in a fixed wired network. In wired networks an attacker must have physical access to the network or passes several fire-walls and gateways, but attacks against a wireless ad hoc network can come from many directions. An attacker may try to steal information, contaminates messages or impersonate a node. Nodes that are roaming in a hostile environment and have poor physical protection have a probability of being compromised, so attacks from inside the network must be considered as important as attacks from outside the network.

If we are to list some of the threats and vulnerabilities of wireless ad hoc networks, the following attacks are the most common ones: (i) all vulnerabilities that exist in a wired network can also be present in wireless networks, (ii) information that is not encrypted and is transmitted between two wireless devices can be intercepted, (iii) malicious user can gain unauthorized access and bypass any firewall protection, (iv) handheld devices with sensitive information can be easily stolen, and (v) data can be corrupted or jammed.

3.1   Digital Signature Based Routing

In most proposed routing protocols for mobile ad hoc networks, routers exchange information to establish routes between different nodes. This information can be very inviting to attackers. Introducing excessive traffic load into the network or advertising incorrect routing information to other nodes are different kinds of attacks to routing protocols.

Dahill, et al. [14] and Sanzgiri, et al. [34] present a secure routing protocol (ARAN) which requires the use of a trusted certificate authority (CA); before a node enters in an ad hoc network, it must get a signed certificate from the CA. The route discovery process is initiated by the source node by flooding a digitally signed Route Discovery packet to its neighbors. When a neighbor receives this message, it sets up a reverse path back to the source node and verifies the signature. ARAN provides end to end and hop by hop authentication of route discovery and reply messages. The certificate includes IP address of node, public key and a time stamp showing when the certificate is created. All nodes are supposed to maintain fresh certificates with the trusted server. This protocol first verifies that the destination is reached, then verifies that a route discovery process has found the shortest path. A source initiates the route discovery process to reach to the destination, by broadcasting a route discovery packet to all its neighbors. Eventually the packet reaches the destination, and the destination node unicast a reply to the source. There is no guarantee that this first received packet at the destination has traversed the shortest path, so the source also broadcasts shortest path confirmation message to its neighbors.

To achieve a highly secure key management service, Zhou and Haas [45] have proposed threshold cryptography, to distribute trust among a set of servers. In this approach a set of arbitrary nodes are chosen as servers. All nodes know the public key but the private key is divided into

n shares between n servers. As shown in Figure 18.1 , each server has a public/private key pair K/k. Private key k is divided into n shares, one for each server. Each server also has a public/private key pair K_i/k_i and knows the public keys of all nodes. A $(n, t+1)$-threshold cryptography scheme allows n parties share the ability to perform a cryptographic operation so that any $t+1$ parties can perform this operation jointly, but it is infeasible to do by at most t parties. The drawback of this mechanism is that in mobile ad hoc networks we rather have the ability to work without any external management or configuration, but the proposed mechanism relies on a subset of nodes playing specific roles at a given point of time, which is undesirable.

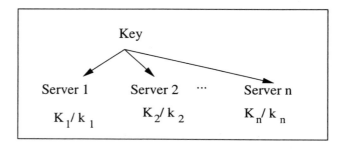

Figure 18.1. Key Distribution

Hubaux, et al. [12, 21] propose a self-organizing public-key management system for mobile ad hoc networks. Certificates are stored and distributed by the users. Each user maintains a local certificate repository, which contains a limited number of certificates. When two users want to verify the public key of each other, they merge their repositories and find a certificate chain, which makes the verification possible. This functionality of this approach depends on the construction of local certificate repositories and certificate graph. A certificate graph is a graph whose edges represent public key certificates issued by the users and whose vertices represent public keys of the users. Each user has several directed and mutually disjoint paths of certificates in its local repository. Each path starts at the node itself and certificates are added to this path such that a new certificate is chosen among those connected to the last node on the path, and the new certificate leads to the node with the highest number of certificates connected to it. When verifying a certificate chain, a node must trust the issuer of the certificate in the chain for checking that the public key in the certificate really belongs to the node named in the certificate. A drawback of this approach is that before performing authentication, each node must build a local repository which is relatively expensive.

Hu, et al. [22] present a secure routing protocol, based on the Destination Sequenced Distance Vector protocol (DSDV). Each node updates its table using the updates it hears, so its route for each destination uses as a next hop the neighbor that advertised the smallest metric for that destination. In this protocol, called secure efficient ad hoc distance vector (SEAD), nodes periodically exchange routing information with each other, so each node always knows a current route to all possible destinations. The protocol uses efficient one way hash chains rather than relaying by using expensive asymmetric cryptography operations. It deals with attacks that modify routing information broadcasted during the update phase of the DSDV protocol. The SEAD protocol assumes a mechanism for a node to distribute an authentic element of the hash chain that can be used to authenticate all the other elements of the chain. This protocol authenticates the sequence number of routing table update messages using hash chains elements. Each node uses an element from its hash chain in each routing update. Based on this element, the one-way hash chain provides authentication in other routing updates for that node.

In [23], Hu, et al. present an on-demand secure ad hoc routing protocol (ARIADNE), based on the Dynamic Source Routing (DSR) protocol, which relies on symmetric cryptography. They present a mechanism that enables the target to verify the authenticity of the ROUTE REQUEST. A per-hop hashing technique is used to verify that no node is missing from the node list in the REQUEST. This protocol guarantees that the target node of a route discovery process authenticates the initiator. To convince the target of the authenticity, the initiator includes a unique data (such as time-stamp) in each packet. The initiator can authenticate each intermediate node to the destination, using PREP message, and no intermediate node is able to remove a node from the node list in the PREQ or PREP messages. When the target wants to authenticate nodes, it will return a REPLY message on a path that only contains legitimate nodes. One drawback of ARIADNE is the high cost of key setup.

3.2   Enforcing Cooperation

Buttyan, et al. [8, 9] present packet purse model. In this model each packet is loaded with nuglets by the source, as depicted in Figure 18.2 , and each forwarding host takes out nuglets for its forwarding service. When a node wants to send a packet as originator, the number of forwarding nodes that are needed to reach the destination is estimated. When a node forwards a packet, its nuglet counter is increased by one. If a node wants to send its own packets, then it must forward packets for benefit of other nodes. So this approach discourages users from flooding the network but the downside is that the source needs to know exactly how many nuglets it has to include in the packet it sends. Another problem with this approach is that intermediate nodes can take out more nuglets than they are supposed to. The solution is based on using nuglet or virtual currency. Another type of nuglet that nodes can use is packet trade. In the packet trade model, each packet is traded for nuglets by the intermediate nodes as follows: each intermediate node buys the packet from the previous node on the path. So the destination has to pay for the packet. Therefore, the source does not need to know how many nuglets are to be loaded in the packet; however malicious flooding can not be prevented either.

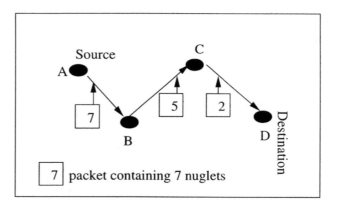

Figure 18.2. Nuglet model

Buchegger and Boudec [6] propose an approach to find selfish and malicious nodes. (A selfish behavior is one that maximizes the utility of a node.) The protocols, called, cooperation of nodes and fairness in dynamic ad hoc networks (CONFIDANT), observes nodes and detects attacks. In this protocol, nodes monitor their neighbors and if there is reason to believe that a node misbehaves, they take action in terms of their own routing and forwarding messages. This protocol encourages cooperation by allowing selfish nodes an incentive to cooperate, which leads to an outcome that is beneficial for the network.

The user cooperation in ad hoc networks has also been studied in [36], which proposes an acceptance algorithm that each node uses, to decide whether to accept or reject a relay request. The system is proved to converge to a Nash equilibrium point in game theoretic sense. Srinivasan, et al. assume that each user has sufficient information about the system, like the number of users in each energy class, and hence users exchange their view of the system. However, they do not consider malicious users. A framework to study the existence of cooperation in packet forwarding in a wireless network is proposed in [18], in which a model is defined and the conditions under which cooperative game strategies can form an equilibrium are identified. This approach does not require each node to keep track of the behavior of other nodes, but it is assumed that all routes are static.

3.3   Denial of Service Attacks

The denial of service (DoS) is an attack that prevents the use of networks by even legitimate users. Recently, DoS attacks have become serious security threats. This is due to the fact that, they can result in massive service disruption. Although it is possible to attack software vulnerabilities, yet DoS attacks usually consume finite resources such as bandwidth, CPU processing power or memory. As nodes use wireless communication, one possible kind of attack on the physical layer is jamming. An intruder may corrupt the network with the radio frequencies that nodes in the network are using. Constant jamming prevents a node to exchange data or even report the attack. If the intruder can jam the entire network, then nodes can go to sleep mode and wake up periodically and check if the jam has ended. Cellular phone networks use code spreading as a defense to DoS attacks [40].

Defense against DoS attacks is hard as the IP address of source is spoofed, which makes it difficult to determine the true origin. There are two kinds of DoS attacks, which use packet floods to consume resources [41].

- *Network resource attack:* when an attacker sends useless packets to a victim server with the intention of wasting bandwidth connecting to the server.

- *Server resource attack:* when an attacker sends useless packets with intention of corrupting the ability of server to process the load, or even taking advantage of ambiguities in protocols, like reservation of resources for half-open connections like SYN flooding.

Yaar, et al. [41] propose a packet marking approach where a path fingerprint is embedded in each packet, thus enabling a victim to identify packets traversing the same paths. They observed that reconstruction of the exact path to the attacker is not necessary. As path information is in each packet, each node can filter packets based on the information carried.

4.   Security in Pervasive Computing

A pervasive computing environment includes numerous secured, often transparent computing devices, which are mobile or embedded in the environment and are connected to an increasingly ubiquitous network infrastructure. A pervasive computing environment is usually more knowledgeable about the users' behavior. And as time passes, the environment becomes more proactive with each individual user [27]. So the user must be able to trust the environment and also the environment must be confident of user's identity. Security is an important concern in the success of a pervasive computing system which must thus be aware of its user's state and surroundings, and must be able to modify its behavior based on this information. A key challenge is to obtain such information. In some cases the information could be part of the user's personal computing space.

Satyanarayanan [35] emphasizes the following research thrusts in the area of pervasive computing.

Effective use of Smart Spaces: The space may be an enclosed area, like a meeting room or a well-defined open area, like a campus. By embedding computing infrastructure, a smart space brings two disjoint worlds together, for examples, adjustment of lighting levels in a room based on the occupant's profile.

Invisibility: This deals with complete disappearance of pervasive computing technology from a user's consciousness. A good tool must be invisible and make us focus on the task and not on the tool. Eyeglass is an example of a good tool through which we look at the world, not the eyeglasses. According to Weiser [39] *"The most profound technologies are those that disappear. They weave themselves into the fabric of everyday life until they are indistinguishable from it."*

Localized scalability: As smart spaces grow, the interactions between a user's personal computing space and his/her surrounding will grow too.

So it has bandwidth and energy distraction for a wireless mobile user.

Masking uneven conditioning: The rate of excerpt of pervasive computing into infrastructure will depend on many non-technical factors like economics and business models. And so when users get used to some level of smartness in one environment, smartness-less in another environment would be very visible and unwanted.

Pervasive computing covers a wide spectrum of applications – from switching on the lights in a conference room to buying and selling stocks, booking airline tickets, providing health-case and telemedicine services, emergency rescues, crisis management, and many more complex ones. This has the potential to create a truly "smart environment".

4.1 Distributed Trust

Pervasive computing involves the interaction, corporation and coordination of several transparent computing devices. One way to satisfy the requirements of pervasive computing is adding distributed trust to the security infrastructure [37]. This approach needs to (i) articulate policies for user authentication and access control; (ii) assign security credentials to individuals; (iii) allow entities to modify access rights of other entities; and (iv) provide access control. Role based control is probably one of the best known methods for access control, where entities are assigned roles, and there are rights associated with each role. This is difficult for systems where it is not possible to assign roles to all users, and foreign users are quite common. Security systems should not only authenticate users, but also allow them to delegate their rights. Agents are authorized to access a certain service if they have the required credentials.

Bharghavan [5] uses additional ontologies that include not just role hierarchies but any properties expressed in a semantic language. For example if an agent in a meeting room is using the projector, it is possible that the agent is a presenter, and must be allowed to use the computer too. So they assign roles dynamically and without making new roles. They extent Agent Management System (AMS) and the Directory Facilitatory (DF) to manage security of the platform. The AMS, DF and agents will follow some security policies to give access rights. They use PKI (public key infrastructure) handshaking protocol between AMS and agents to verify both parties. Once an agent signs a request it will be accountable for it. When an unknown agent attempts to register with a platform, the platform checks the agent's credential and makes a decision. In multi-agent systems, as they are decentralized, it is impossible to have a central database of access rights. Therefore, the policy is checked at two levels: (i) controlling access to the AMS and DF, and (ii) specifying who can access the service. Security can be classified into two levels – platforms and agents. In platform security, the AMS and DF decide to allow an agent to register, search or use other functions. An agent can send security information to the AMS and specifies its category as private, secure or open. When an agent wants to register with the AMS, it signs its request and sends with it the digital identity certificate to the AMS. If it is valid, then the AMS decides about the access rights. If the agent does not have the right to register with the AMS, the latter starts a handshaking protocol.

4.2   Incorporating Vigil System

Based on role based access control with trust management, the Vigil system [25] has five components: (i) service broker, (ii) client, (iii) certificate controller, (iv) role assignment manager, and (v) security agent. The vigil system is divided into *Smart Spaces*, each controlled by a broker. The broker finds matching service for the user. All clients must register with the broker. Each client is only concerned with the trust relationship with its broker. The service brokers establish trust relationships with each other. Trusts between clients are transitive through brokers. When a client registers with a broker, it transmits its digital certificate, a list of roles which can access it, and a flag indicating if the broker should publish that client's presence to other clients. After receiving the registration, the broker verifies the client's certificate in order to send it the digital certificate. The communication manager is a bridge between the broker and the entities in the Smart Space, hence it translates protocols. The certificate controller generates digital certificates.

To get a certificate, an entity sends a request to the certificate controller which signs it and then sends back with its own signed certificate. The role assignment manager maintains access control list for entities. An entity can have more than one role. The security agent manages trust in the Space. All security agents will enforce the global policy and a local policy, which is related to the Space. At the beginning, it reads policies and stores them in a database. All permissions and prohibitions are defined there. Access rights can be delegated, they must be from an authorized entity to another entity. When a user needs to access a service, which does not have the right to do that, it asks another user who has that right. If the asked entity has the permission to delegate the access, it sends a message signed by its own private key with its certificate to the security agent. The security agent verifies that the delegator has the right to delegate, then sets a short period of validity for that permission. After expiring of this period, the delegation must be reprocessed.

Consider a telemedicine scenario as an application for pervasive computing. Suppose Jack has heart attack history, and his doctor wants to have up-to-the minute information about his heart. Jack will use a heart monitor unit, which can be connected to his cell phone. Information will be transfered to his doctor's computer, and accordingly Jack can get diagnostic notification from his doctor. In this case, if someone alternates Jack's information or impersonate the doctor it could have fatal results. Jack and the doctor should not be able to send some information and then deny sending. Jack's information is his privacy; his insurance company has it, but his boss should not be able to access it, so any unauthorized access is forbidden. Having this scenario in mind, let us see how to incorporate security mechanisms in pervasive computing systems.

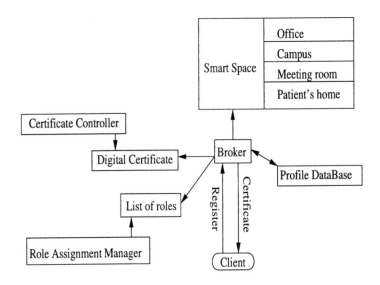

Figure 18.3. Incorporating VIGIL system for a Smart Space scenario

As Figure 18.3 illustrates, we can have different *Smart Spaces*. Each client must register through a broker who gets the digital certificates from the certificate controller and has access to the list of roles provided by the role assignment manager. Then the broker issues certificate to the client. The broker also has access to a profile database. Each different client has a different profile based on his/her previous interaction with the environment. As discussed in [24, 37], we can use the distributed trust concept so entities can modify access rights of other entities by delegating their access rights. It is also possible to have a chain of delegation and then revoke them accordingly. It is worth mentioning that when one entity gives one right to another entity, the entity is able to access services without system creating a new identity for it.

5.   Security in Wireless Sensor Networks

A wireless sensor network consists of different kinds of sensors that monitor the environment. Due to low cost and flexibility, sensor networks are being rapidly deployed for a variety of applications such as ecology and environment monitoring, military, emergency sites, remote sensing, smart homes, health, and commercial areas [3]. However, sensors have very limited resources (CPU, memory, bandwidth, batter power). Thus, designing resource-efficient protocols is a major challenge in sensor network design. Other issues include security, fault-tolerance, survivability, scalability.

As sensor networks occur in the real world, there is a need to protect them from various types of attacks. This is certainly challenging in an environment where the network is designed to be flexible. It is crucial that the security of sensor networks be monitored and diagnosed to ensure their proper behavior. The major challenges in tackling sensor networks security include power conservation for mobile sensors, cooperation among heterogeneous sensors, flexibility in the security level to match the application needs and conserve critical resources, scalability, self organizing and self learning of sensors, trust and security decisions for the application, yet keeping the mobility and volatility transparent and protecting the network from external and internal intrusions. In a nutshell, we must consider three important factors – energy, computation and communication.

Due to resource scarcity of sensors, securing sensor networks is quite different and challenging from traditional schemes that generally involve management and safe keeping of a small number of private and public keys [31]. Disclosing a key with each packet requires too much energy [32]. Storing one-way chain of secret keys along a message route requires considerable memory and computation in the nodes on that route [33]. The key management using a trusted third party again requires an engineered solution that makes it unsuitable for sensor network applications [31]. Although the asymmetric key cryptography does not require a trusted server, key revocation becomes a bottleneck [31] as it involves an authority maintaining a list of revoked keys on a server or requesting the public key directly from the owner.

As mentioned earlier, there are two type of attackers, namely clever or curious outsiders and knowledgeable insiders. An attacker can take over any node within a network. Subverting a single node means that all nodes with the communication radius of that node can be denied receiving any information. Therefore, the goal is to minimize the impact of a subverted node on the rest of the network. The single node should not grant the attacker the ability to subvert the entire network. In a network with a single gateway (i.e., base station), the attacker needs to take over just that one node to render the network inactive. In order to be able to trust sensor data, the source must be authenticated so the malicious sensors can not inject false data. For checking the integrity of data, we must be able to detect data modification. Additionally, data must be confidential so no one else can read it. Thus the threats that a sensor network may encounter are eavesdropping, message injection, message replaying, message modification and denial of service.

The security primitives are message confidentiality, authentication, message integrity, as message freshness are summarized below [31].

Data Confidentiality: A sensor network should not leak sensor readings to other networks. Usually nodes communicate sensitive data. In order to protect sensitive data, they are encrypted with a secret key that only intended receivers have. Moreover, secure channels are set up between the nodes and base stations.

Data Authentication: Since an adversary can easily inject messages, there must be a way that assures receiver about the origin of the data. By authenticating the data we allow the receiver to verify that the data was sent from the claimed sender. In case of two-party communication, the sender and receiver can share a secret key. But this method can not be used in broadcasting as anyone can impersonate the sender and forge messages to other receivers.

Data Integrity: It ensures that the data have not been altered by any adversary while in transit.

Data Freshness: It implies that the data is recent and no adversary replayed old messages.

In most sensor networks, some nodes have special responsibilities. They can act as a coordinator, monitoring access point or cryptographic key manager [29]. So they are more attractive for intruders, as they can provide access to network services. A passive intruder can observe the network traffic and learns the location of these critical resources and act later; this is known as *homing attack*. If all nodes use cryptographic keys, the network encrypt the header of each packet at each hop, and prevents intruder from easily learning about the network. Another kind of attack is *desynchronization*. The intruder forges messages to one or both end points. These messages carry wrong sequence numbers and cause the end points to ask for retransmission of lost packets. Hence a clever intruder can prevent the two end points from exchanging any useful data.

5.1   Sensor Security Schemes

Researchers in UC Berkeley have designed two security building blocks, called SNEP (Secure Network Encryption Protocol) and μTESLA ("micro" version of Timed, Efficient, Streaming, Loss-tolerant Authentication Protocol) [33].

5.1.1   SNEP Protocol.   The SNEP protocol provides data confidentiality, two-party data authentication, data integrity and freshness of data. Data confidentiality is a security primitive which is used in every security protocol. It has low communication overhead since it only adds 8 bytes per message. Also, like many other protocols, it uses a counter. It has semantic security, which ensures an eavesdropper has no information about the plain-text, even if it sees multiple encryptions of the same plain-text [32]. The SNEP protocol uses a cryptographic mechanism that achieved semantic security with no additional transmission overhead. It relies on the value of the counter which is shared between the sender and receiver, and this value is incremented after each block of cipher message. Combination of this and the Message Authentication Code (MAC) forms the SNEP protocol. As the counter is kept at each end point and is not part of the message, it has low communication overhead. The counter value in the MAC also prevents replaying of old messages. In the absence of a counter value in MAC, any adversary may be able to replay old messages [32].

5.1.2 μTESLA Protocol.   The μTESLA provides authentication for data broadcasting. It requires the base station and nodes be loosely time synchronized and each node knows an upper bound on the maximum synchronization error. The base station simply computes a

MAC on the packet with a secret key. When a node receives a packet, it verifies that the corresponding MAC key was not yet disclosed by the base station. The receiving node ensures that the MAC key is known by the base station so no adversary could have altered the packet. The node then stores this packet in a buffer. At the time of key disclosure, the base station broadcasts the verification key to all the receivers. When a node receives a disclosed key, it can verify the correctness of it. If it is correct, the node uses it to authenticate the buffered packet. The μTESLA protocol provides broadcast mechanism, uses asymmetric mechanism to prevent forgery and does not use asymmetric digital signature as it is computationally expensive and needs more storage.

The sender and receiver share the initial TESLA key as follows. The receiver can send a message to base station, asking for the current TESLA key over a secure channel. Since the reply from the base station is encrypted and authenticated, the receiver can bootstrap into the TESLA group. Also by fine-tuning the disclosure interval, limited buffering is needed. In this protocol the sender first generates a sequence of secret keys. The sender chooses the last key randomly, and generates the remaining values by applying a one-way function on it. As time is divided into time intervals, the sender associates each key of the one-way key chain with one time interval. The sender uses the key of current interval to compute MAC. Then it reveals the key after some intervals. The important property of one-way key is that once the receiver has an authenticated key of the chain, subsequent keys of the chain are self-authenticating. Due to loose time synchronization and authenticated key chain, the freshness and point-to-point authentication can be provided. The MAC uses the secret key shared by nodes and the base station, thus authenticating the data. When the receiver receives the packet, it needs to make sure that the sender did not disclose the key of incoming packet. This asymmetric delayed key disclosure requires loosely synchronized clocks between the sender and the receiver.

5.1.3 Key Management.

The limited resources make it undesirable for sensor nodes to use public-key algorithms, such as Diffie-Hellman key agreement [16]. A sensor node may need tens of seconds or even minutes to perform these operations. Usually a large number of sensor nodes are deployed in hostile environments, necessitating them to be low-cost. At the same time it is harder to make them tamper-resistant [17]. An adversary can take control of a sensor node and compromise the cryptographic keys. It is very likely that a sensor node does not know which other nodes will be coming within its communication range. In other words, there is no apriori knowledge of neighboring sensor nodes. Due to limited memory, battery power and other resources, a sensor node can not establish a unique key with every other node in the network. Moreover, a sensor network is vulnerable not only to internal attacks from untrustworthy nodes, but also to external attacks from sources like mobile codes that are used for activating remote software upgrade, configuration management, dynamic resource monitoring, etc. In such an environment any node could claim that it is one hop away from the destination, causing all routes to the destination to pass through. Also a malicious node could corrupt any packet and cause data to be misrouted.

As sensor networks can be deployed in hostile areas, where each node is subject to be captured by adversary, they require protection over their communications. A key-management scheme is proposed in [13], which is designed to satisfy security requirements for the network. This scheme consists of different phases. They generate a large pool of keys, randomly draw a key, load the key in the memory of each sensor node, and save the key identifier on a trusted controller. Each node must discover its neighbors. If a node is compromised, the entire key must be revoked. All such communications need energy, which is very limited to sensor nodes. Meanwhile, the limited computing power is a bottleneck for using more sophisticated protocols.

5.2   Enforcing Cooperation and Reputation

This section describes a game theoretic framework due to Agah, Das and Basu [1], to enforce cooperation and reputation in wireless sensor networks.

Game theory offers a formal tool to analyze interaction among a group of rational players who behave strategically. A game is the interactive situation, specified by the set of players (sensor nodes), the possible actions of each node, and the set of all possible payoffs. What one individual player does directly affects at least one other player in the group. Games in which the actions of the players are directed to maximize the profit without subsequent subdivision of the profit among the players are called *cooperative games*. In such games, the outcome is a result of an agreement among players. These games are compared with respect to the preferability of payoffs [38]. In other words, in a cooperative game, different players form alliance with each other in a way to influence the outcome of the game in his/her favor. In *noncooperative games*, however, no outside authority assures that players stick to the same predetermined rules, and binding agreements are not feasible. It is known that in noncooperative games, there exist sets of optimal strategies, the so called Nash equilibrium [38], used by the players in a game such that no player can benefit by unilaterally changing his or her strategy if the strategies of the other players remain unchanged.

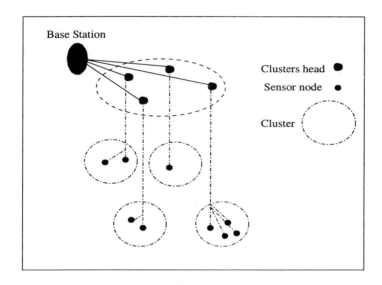

Figure 18.4. Structure of a Sensor Network

5.2.1   Game and Payoff Function.   In [1], a cooperative game is defined between the sensor nodes. A set of strategies for each node is chosen and the payoff function for each node is also defined. Any subset of players (i.e., sensor nodes) will be called a cluster (coalition), as shown in Figure 18.4 The goal of the game is to find the largest payoff that sensor nodes within a cluster can collectively achieve, assuming that all nodes in one cluster use the same key. All sensors play the game in such a way as to have a more populated cluster. But, on the other hand, the total possible number of sensor nodes in each cluster can not be more than what the cluster can handle (this is due to resource constraints). Thus an optimal payoff function is defined based on the security related objectives in the sensor network. This function consists of three factors: (i) cooperation, (ii) reputation and (iii) quality of security.

The payoff between two sensor nodes depends on their distance and each node's transmitter signal strength. The more the transmitter signal strength, the more likely is its cooperation with the close neighbors. In order to determine how much each individual node is useful for the whole network (a measure of reputation), the payoff between should also represent how many packets each node receives and forwards at each time slot. Finally, the payoff between two sensor nodes should also represent the trustworthiness of the traffic. In each cluster, one node with the highest cooperation, best reputation and highest quality of security is chosen as a clusterhead.

5.2.2   The Strategy Set.   In each time slot, a sensor node uses its strategy based on the information obtained in preceding time slots. Consider a packet is coming from a source to a destination node, passing through an intermediate node. This intermediate node decides to forward or not to forward the packet to the destination node, based on three following criteria:

- Reputation: Have these two sensor nodes made enough reputation to trust each other and to cooperate with each other, in regards to forwarding packets on behalf of each other?

- Distance: The closer the two nodes together, the more they trust each other. If they are too far away from each other, it is less likely that they will trust each other.

- Traffic: Have these two sensor nodes a good history of joint operation, and can they trust each other?

If there is enough reputation level, closeness, and good history of joint operation, then the strategy for sensor nodes is to cooperate with each other (where cooperation means receiving or forwarding incoming packets); otherwise the strategy is to defect. For details, refer to [1].

6.   Conclusions

Migration of information from mainframe to laptops introduces more stringent security requirements into our daily lives. Wireless devices can not be treated as gadgets and security must be considered in the design of transport layer, network layer as well as in data link layer. In this chapter we discussed various security related issues, including intrusion detection and denial of service, and surveyed various security protocols in wireless mobile networks such as ad hoc and sensor networks as well as pervasive computing environments. Clearly, due to the nature of each of these networks, a well defined protocol in one infrastructure can not be very useful in another one. We believe security should be taken into account at the early stages of the network design. Furthermore, the security solutions proposed for wireless mobile and sensor networks must consider the limitations on battery life, computing, wireless communication, and storage resources of network components.

Acknowledgment

This work is supported by NSF ITR grant under Award Number IIS-0326505.

References

[1] A. Agah, S. K. Das and K. Basu, "A game theory based approach for security in sensor networks", submitted to IPCCC 2004.

[2] Agah, S. K. Das and K. Basu, "Security in Sensor Networks", submitted to WCNC 2004.

[3] I. F. Akyildiz, W. Su, Y. Sankarasubramanian, and E. Cayrici, "Wireless Sensor Networks: A Survey", *Computer Networks*, vol. 38, pp:393-422, 2002.

[4] B. Balajinath and S. V. Raghavan, "Intrusion detection through learning behavior model", *Computer Communications* vol.24, pp:1202-1212, 2001.

[5] V. Bharghavan, "Secure Wireless LANs". *Proceedings of the ACM Conference on Computers and Communications Security*, pp:10-17, Virginia, 1994.

[6] S. Buchegger and J. Le Boudec, "Performance analysis of the CONFIDANT protocol", In *proceedings of the 3rd ACM International Sympmosium on Mobile Ad Hoc Networking and Computing (MobiHoc)*, pp:226-236, June 2002.

[7] L. Buttyan, J. P. Hubaux and S. Capkun, "A Formal Analysis of Syversons Rational Exchange Protocol" , *CSFW*, pp:181-193, June 2002.

[8] L. Buttyan and J. P. Hubaux, "Stimulating Cooperation in Self-Organized Mobile Ad Hoc Networks", *ACM/Kluwer MONET*, vol.8, no.5, October 2003.

[9] L. Buttyan and J. P. Hubaux, "Nuglets: a virtual currency to stimulate cooperation in self-organized ad hoc networks", Technical Report, 2001.

[10] M. Bykova, S. Ostermann and B. Tjaden, "Detecting Network Intrusions via a Statistical Analysis of Network Packet Characteristics", *IEEE Southeastern Symposium on System Theory*, pp:309-314, 2001.

[11] R. H. Campbell, T. Qian, W. Liao and Z. Liu, "Active Capability: A Unified Security Model for Supporting Mobile, Dynamic and Application Specific Delegation", *Security White Paper: System Software Research Group*, Department of Computer Science, University of Illinois at Urbana-Champaign, 1996.

[12] S. Capkun, L. Buttyan, and J. P. Hubaux, "Self-Organized Public Key Management for Mobile Ad hoc Networks", *IEEE Transactions on Mobile Computing*, vol. 2, no. 1, June 2003.

[13] H. Chan, A. Perrig and D. Song, "Random Key predistribution Schemes for Sensor Networks", In *Proceedings of the 2003 IEEE Symposium on Security and Privacy*, pp:197-213, 2003.

[14] B. Dahill, B. N. Levine, E. Royer and C. Shields, " ARAN: A secure Routing Protocol for Ad Hoc Networks", UMass Tech Report, 2002.

[15] Y. Desmedt, "Threshold Cryptography", *European Transactions on Telecommunications*, vol.5, no.4, pp:449-457, July 1994.

[16] W. Diffie and M. E. Hellman, "New directions in cryptography", *IEEE Trans. Inform. Theory*, pp:644-654, November 1976.

[17] L. Eschenauer and V.D. Gligor, "A Key-management Scheme for distributed Sensor Networks", In *Proceedings of ACM CCS*, pp:41-47, Washington, DC, November 2002.

[18] M. Felegyhazi, L. Buttyan and J.P. Hubaux, "Equilibrium Analysis of Packet Forwarding Strategies in Wireless Ad Hoc Networks-the Static Case", *Proceedings of Personal Wireless Communications (PWC '03)*, Venice, Italy, September 2003.

[19] B. Gelbord, "Graphical Techniques in Intrusion Detection Systems", *IEEE 15th International Conference of Information Networking*, pp:253-258, 2001.

[20] B. Harris and R. Hunt, "TCP/IP security threats and attack methods", *Computer Communications*, vol.22, no.10, pp:885-895, June 1999.

[21] J. P.Hubaux, L. Buttyan and S. Capkun, "The Quest for Security in Mobile Ad Hoc Networks", *Proceedings of the ACM Symposium on Mobile Ad Hoc Networking and Computing (MobiHOC)*, pp:146-155, October 2001.

[22] Y. C. Hu, D. B. Johnson and A. Perrig, "SEAD: Secure Efficient Distance Vector Routing for Mobile Wireless Ad Hoc Networks", in *Proceedings of the 4th IEEE Workshop on Mobile Computing Systems and Applications* (WMCSA), pp:3-13, June 2002.

[23] Y. C. Hu, A. Perrig and D. B. Johnson, "Ariadne: A Secure On-demand Routing Protocol for Ad hoc Networks", *MobiCom*, pp:12-23, 2002.

[24] L. Kagal, J. Undercoffer, F. Perich, A. Joshi and T. Finin, "A Security Architecture Based on Trust Management for Pervasive computing Systems", *Conference - Grace Hopper Celebration of Women in Computing*, 2002.

[25] L. Kagal, J. Undercoffer, A. Joshi and T. Finin, "Vigil: Enforcing Security in Ubiquitous Environments", *http://www.csee.umbc.edu/lkagal1/papers/vigil.pdf.*

[26] M. Kodialam and T.V. Lakshman, "Detecting Network Intrusions via Sampling: A Game Theoretic Approach", *INFOCOM*, pp:1880-1889, 2003.

[27] M. Kumar, B. Shirazi, S. K. Das, B. Sung, D. Levine, and M. Singhal, "PICO: A Middleware Framework for Pervasive Computing", *IEEE Pervasive Computing*, vol. 2, no. 3, pp:10-17, July-Sept 2003.

[28] L. Lankewicz, M. Benard, "Real-time Anomaly Detection Using a Non-Parametric Pattern Recognition Approach", *Proceedings 7th Annual Computer Security Application Conference*, pp:80-89, 1991.

[29] S.J. Lee, W. Su and M. Gerla, "Mobility Prediction in Wireless Networks", In *Proceedings of IEEE ICCCN*, vol. 1, pp:22-25, October 2000.

[30] P. Papadimitratos and Z.J. Haas, "Secure Routing for Mobile Ad hoc Networks", *CNDS*, pp:27-31, January 2002.

[31] A. Perrig, R. Canetti, J. D. Tygar and D. Song, "Efficient Authentication and Signing for Multicast Streams over lossy Channels", In *IEEE Symposium on Security and Privacy*, pp:56-73, May 2000.

[32] A. Perrig, R. Szewczyk, V. Wen, D. Culler and J. D. Tygar, "SPINS: Security Protocols for Sensor Networks", *MobiCom*, pp:189-199, July 2001.

[33] A. Perrig and J. D. Tygar, "Secure Broadcast Communication in Wired and Wireless Networks", *Kluwer Academic Publisher*, ISBN 0792376501, 2003.

[34] B. Sanzgiri, B. Dahill, B. N. Levine, C. Shields and E. M. Belding-Royer, "A secure routing protocol for ad hoc networks", In *Proceedings of the 10th IEEE International Conference on Network Protocols (ICNP)*, pp:12-15, November 2002.

[35] M. Satayanarayanan, "Pervasive Computing: Vision and Challenges", *IEEE Personal Communications*,pp:10-17, August 2001.

[36] V. Srinivasan, P. Nuggehalli, C. F. Chiasserini and R. R. Rao, "Cooperation in Wireless Ad Hoc Networks", *INFOCOM*, vol. 2, pp:808-817, 2003.

[37] J. Undercoffer, F. Peirch, A. Cedilink, L. Kagal and A. Joshi, "A Secure Infrastructure for Service Discovery and Access in Pervasive Computing",*ACM Journal of Mobile Networks and Applications*, Special issue on security in Mobile Computing Environments, vol. 8, no. 2, pp:113-125, April 2003.

[38] N. N. Vorobev, "Game Theory lectures for economists and systems scientists", *Springer-Verlag*, ISBN 0387902384, 1977.

[39] M. Weiser, "The Computer for the 21st Century",*Scientific American*, vol. 265, no. 3, pp:66-75, Sept 1991.

[40] A. D. Wood and J. A. Stankovic, "Denial of Service in Sensor Networks", *IEEE Computer*, vol.35, no.10, pp:54-62, October 2002.

[41] A. Yaar, A. Perrig and D. Song, " Pi: A Path Identification Mechanism to Defend against DoS Attacks", *IEEE Symposium on Security and Privacy*, pp:93-107, May 2003.

[42] N. Ye, S. M. Emran, X. Li and Q. Chen, "Statistical Process Control for Computer Intrusion Detection", *DARPA Information Survivability Conference & Exposition II*, vol. 1, pp:3-14, 2001.

[43] Y. Zhang and W. Lee, "Intrusion Detection in Wireless Ad Hoc Networks", *MOBICOM*, pp:275-283, 2000.

[44] Y. Zhang, W. Lee and Y. Huang, "Intrusion Detection Techniques for Mobile Wireless Networks", *Mobile Networks and Applications*, pp:1-16, 2002.

[45] L. Zhou and Z. Haas, "Securing Ad hoc Networks", *IEEE Network*, vol. 13, no. 6, pp:24-30, Nov 1999.

Chapter 19

WAVEFORM SHAPING TECHNIQUES FOR BANDWIDTH-EFFICIENT DIGITAL MODULATIONS

Hsiao-Hwa Chen

Institute of Communications Engineering

National Sun Yat-Sen University

70 Lien Hai Road, Kaohsiung, Taiwan, ROC

hshwchen@mail.nsysu.edu.tw

Abstract   In this chapter, various issues on pulse shaping waveform design for bandwidth efficient digital modulations will be addressed. Here what we concern about the pulse shaping waveforms is the physical appearances or shapes each information data bit or symbol take before carrier modulations. It is well known that the spectral characteristics of a carrier modulated signal can be uniquely determined by its baseband signal before modulation, and thus the waveform shape applied to pre-modulated signals is of ultimate importance to ensure overall bandwidth-efficiency of a wireless communication system. This chapter starts with an overview on the evolution of pulse shaping technologies from early digital modulation schemes to recently emerging new carrier modulations such as quadrature-overlapped modulations. The impact of the pulse shaping technologies on both bandwidth-efficiency and power-efficiency, which are two essential merit parameters of a digital modem, will also be discussed. The majority of the content in this chapter, however, is dedicated to the discussions on shaping pulse design methodology as well as performance analysis of a pulse-shaped digital modem, in which various shaping pulse design methods (including those commonly referred and those recently proposed in the literature) will be presented. To measure the effectiveness of the pulse-shaping techniques, different pulse shaped digital modulations will be evaluated and then compared in terms of their spectral characteristics, the sensitivity of their performance against synchronization timing jitter as well as their bit error rate performance, where the emphasis will be put on the pulse-shaped quadrature-overlapped (QO) modems under bandpass nonlinear channels, which usually exist in satellite and mobile cellular systems. Several new pulse shaped QO modulation schemes, which adopt various novel pulse shaping waveforms generated by time-domain convolution method, will be presented and their performance analysis will be carried out in comparison with other traditional digital modems, such as QPSK, OQPSK, MSK and so forth.

Keywords: Pulse shaping waveform, digital modulation, bandwidth efficiency, power efficiency

1.   Introduction

In the recent decay, global wireless revolution has significantly changed the life-style of people around the world. The dream for voice communication anywhere and anytime before has already become the reality in many countries, characterized by the great success of the 2nd generation of mobile cellular networks, in particular the Global System for Mobile (GSM) systems, around the world. It is still a bit hard for everyone to believe that over 250 million people in China today are using mobile phones as the part of their daily life. No one can offer an accurate prediction for the number of wireless services subscribers in China after ten years to come. In Taiwan, the penetration rate for the GSM mobile cellular networks has exceeded 120% at the beginning of 2003, meaning that there are virtually more registered GSM hand phones than the total population of Taiwan. The world has entered a new era of wireless communications since NTT DoCoMo of Japan launched the first 3G commercial service in Tokyo in October 2002, soon followed by another 3G operator based on cdma2000 in the same country. The increasing demand on wideband wireless around the world has already made limited radio spectra an extremely precious resource on the earth. Different from the other resources, such as petroleum, minerals, etc., which might be possible to find some replacement as an energy source, the radio spectra are physically non-replaceable, implying that when it is gone no new wireless services could be added. Unfortunately, the suitable spectra for wireless applications (especially mobile cellular services) span only from a few hundreds MHz to less than a few tens GHz, due to the constraints for radio waves to propagate through the air without being sensitively affected by environmental conditions, such as rain, snow, smoke, dust, etc. It is hardly feasible for a future mobile cellular system to run on a RF carrier higher than millimeter wavelength band, in which the radio signal transmissions could never be made weather-proof, due to the fact that its wavelength is almost in the same dimension of rain-drops, snow-flakes, dust, etc., causing a great penetration loss of the signal. In this sense, the bandwidth efficiency of a wireless system has to become a serious concern in the process of today's communication system architecture design.

The seek for bandwidth efficient digital modulations is not a new subject of interest in the area. In fact, it is one of ever-lasting pursuits in communication field research, which can be traced back to as early as the 60th when Baker and several other scholars first proposed $\pi/4$-QPSK modulation scheme [1], followed by lots of discussions on practical QPSK-type modulation schemes in the community [2, 3, 4, 5, 15]. In the 70th the study on other modulation schemes, such as FSK [6, 11], OQPSK [7, 9, 10, 11, 12, 15], MSK [8, 9, 10, 11, 12, 13, 15] was also extremely active. Another important bandwidth-efficient digital modulation scheme, GMSK [14, 16], was introduced in the 80th, which had been successfully applied to the popular GSM networks (a 2G mobile cellular standard) around the world today. A great amount of work had been done to address various related issues [12]-[49] in different schemes, such as error rate analysis, bandwidth efficiency comparison under different operational scenarios and so forth.

In general, as far as a digital modulation scheme is concerned, there are several ways that have been commonly adopted to improve its bandwidth efficiency, which include:

- One of the most straight-forward ways is to prolong baseband bit or symbol time-duration as much as possible under the condition that no excessive inter-symbol interference (ISI) will be introduced in transmission. As the overall bandwidth of modulated signal is basically determined by bit or symbol time-duration, the longer each bit or symbol becomes, the narrower the signal bandwidth will occupy.

- Another way to improve the bandwidth efficiency of a digital modem is to minimize the amount of phase fluctuation or shift between adjacent symbols that take different signs or values. In fact, the phase change between two neighboring symbols does not directly affect bandwidth occupancy of the modulated signal. However, when the signal goes through a bandpass hard-limiting element, such as a band-limited nonlinear RF power amplifier and etc., the modulated signal will suffer so-called *side lobe regeneration* problem [7, 8, 15, 39] if the phase of the modulated signal changes abruptly in the boundary of two symbols. Because most power amplifiers always work in a non-linear range to achieve a higher power utilization efficiency, the *side lobe regeneration* problem can be a serious concern.

- Yet another obvious way to achieve a higher bandwidth efficiency is to employ multi-amplitude or/and multi-phase modulations schemes, however at the price of sacrificing power efficiency.

- Finally, we can also use waveform-shaping or pulse-shaping technologies to improve the spectral characteristics of modulated signal due to the fact that bandwidth occupancy of a modulated signal can be uniquely decided by its pre-modulated baseband signal waveform.

There are pros and cons in all aforementioned methods to improve bandwidth efficiency of a digital modulation scheme. Of course, it will be ideal if all methods can be applied jointly though it is hardly possible due to implementation complexity and other practical concerns. The first method has a technical limitation on how to prolong bit or symbol duration without introducing extra ISI. A typical example in using this method to achieve a high bandwidth efficiency is *partial response waveforms*, which usually extend much longer than the width a symbol really needs. A $sin(\pi t)/\pi t$ pulse has been often taken as one of such partial response waveforms, which span infinitely long in time, and all those pulses will be modulated by different signs to reflect different binary bit information ("+1" or "-1") and overlapped one by one with each offset by a symbol duration. Theoretically speaking, there will be no ISI if the signal is sampled precisely at the tip of each pulse. However, a great ISI will occur if sampling timing jitter exists. Another very successful example in using this method to improve the bandwidth efficiency is *quadrature overlapped* (QO) modulations, which were first proposed in the 70th by [10], followed by [17] in 1981 and our research work [29]-[33] [35]-[42]. The basic idea of the QO modulations is to use a non-square symbol waveform extended twice as long as its original bit or symbol duration and adjacent extended waveforms are overlapped by half with each other in a quadrature modulator. Therefore, overall signal bandwidth is effectively reduced to only the half if compared with that without waveform extension and overlapping. We should discuss more in detail about the QO modulations in this chapter later.

The second way to improve bandwidth efficiency is reflected on the motivation of the research on continuous phase modulation (CPM) schemes in recent 20 years, starting from QPSK and OQPSK, followed by MSK and then GMSK and so on, with their maximal adjacent-symbol phase shift decreased constantly as the new modulations were introduced, from π (for QPSK), $\pi/2$ (for OQPSK) to zero (for MSK and GMSK). To seek for even smoother phase change, many new CPM schemes were proposed but with their implementation complexity increased too. However, GMSK can already offer a very robust performance against the side lobe regeneration problem as shown by its very successful application in GSM system. Therefore, it seems at this moment (maybe not in the future) that there is no much room left for the further improvement of bandwidth efficiency if only through improving the phase continuity of modulated signal.

The use of multi-amplitude/phase modulations will seriously affect power-efficiency or detection-efficiency of a receiver. In other words, we improve bandwidth efficiency in this method at the price of a lower power-efficiency. Therefore, this method will never work well if air-link SNR budget is not sufficiently higher than the threshold, which unfortunately is always the case for most communication systems.

Different from the previous three methods, which are limited by many constraints, the waveform shaping technologies, which is also the focal point of this chapter, leave us much more room or degree-of-freedom to improve the bandwidth efficiency of a digital modulation scheme, without introducing too many negative side-effects to the overall performance of a digital communication system. In addition, the implementation of majority waveform shaping technologies is relatively simple and thus they can be used in different digital modulations effectively.

It is noted that a newly emerging multiple access technology, namely ultra-wideband technology (UWB) [43]-[48], needs also proper pulse shaping waveforms in its data modulation and user multiplexing process. In an UWB system data will be sent via extremely narrow pulses, whose width can be as narrow as a fraction of an nano-second. Obviously, the pulse shapes have a lot to do with the system bandwidth efficiency, which is of ultimate interest to us. In this sense, the discussions about pulse shaping technologies in this chapter can also be found relevant to the UWB system design.

The rest of the chapter are organized as follows. Section 2 will address the issues on the roles the pulse shaping techniques play under the context of bandwidth efficient digital modulations. Section 3 discusses various methodologies used to design pulse shaping waveforms, including intuitive methods, piece-wise fitting methods and time-domain convolution methods. The merit parameters of pulse shaping waveforms (PSW's) and their major performance measures in a digital modulation scheme using designed PSW's are introduced in section 4, followed by the illustrations on how to evaluate spectral efficiency of a pulse-shaped digital modulation in bandpass nonlinear channels in section 5. In section 6, the issues on robustness of a pulse-shaped digital modem against sampling timing jitter will be discussed. Finally, we will give a summary of the chapter.

2. Roles of Waveform Shaping Techniques

The evolution of digital modulations from BPSK to all its later derivatives can be traced in fact from a perspective of waveform-shaping technology. Table (19.1) lists several widely used digital modulation schemes together with their baseband signal waveform shapes and their major spectral characteristics. It is noted from Table (19.1) that, except BPSK, all listed modulation schemes are based on quadrature modulation with I and Q two channels in their modem hardware implementation. Under such a very similar structure, they differ from one another mainly from their different waveform shapes in their baseband signal before carrier modulation process. A generic model for any quadrature modulations is shown in Figure (19.1), where *pulse shaping filter* can be replaced by any particular digital filter to fit for a particular quadrature modulation. For instance, if the output from the filter is a $2T$-long square waveform, then Figure (19.1) shows a OQPSK transmitter. On the other hand, if the output from the filter becomes *half-cycle sinusoidal* waveforms, then the figure gives a MSK modulator. If we replace the filter with a Gaussian waveform generator, Figure (19.1) in fact means the block diagram of a GMSK transmitter. Please also note that the *Post-shaping processor* units, which are reserved for quadrature-overlapped modulations [17, 24, 25, 27], should not function for conventional quadrature modulations, such as OQPSK, MSK and GMSK.

Table 19.1. Several popular digital modulation schemes with their baseband waveform shapes and other spectral characteristics.

	BPSK	QPSK	OQPSK	MSK	GMSK	QORC[a]
Waveform	Square	Square	Square	Half-sinusoidal	Gaussian	Raised cosine
MHC[b]	I	I&Q	I&Q	I&Q	I&Q	I&Q
SWOD[c]	No	No	Yes	Yes	Yes	Yes
SWEO[d]	No	No	No	No	No	Yes
MPS[e]	π	π	$\pi/2$	0	0	0
FNBM[f]	fT	fT	fT	$1.5\,fT$	$1.5\,fT$	fT
SFOR[g]	$(fT)^{-2}$	$(fT)^{-2}$	$(fT)^{-2}$	$(fT)^{-4}$	$(fT)^{-5}$	$(fT)^{-6}$

[a] QORC: Quadrature overlapped raised cosine modulation [17].
[b] MHC: Modem hardware configuration.
[c] SWOD: IQ channel Symbol waveform offset delayed.
[d] SWEO: Symbol waveform extension and overlap.
[e] MPS: Maximum phase shift at the boundary of two adjacent symbols.
[f] FNBM: First null beside the main lobe in unit of fT, where f is the frequency and T is the pre-modulated bit duration.
[g] SFOR: Spectral fall-off rate.

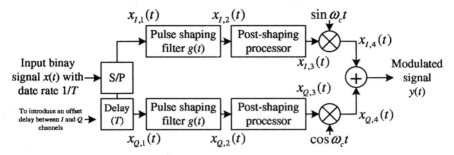

Figure 19.1. A generic quadrature modulator, where *Pulse shaping filter* unit is for particular waveform shaping, *Post-shaping processor* unit is for quadrature-overlapped modulation schemes [17].

It is seen clearly from Table (19.1) that the evolution of bandwidth efficient digital modulations is in fact closely associated with the development of waveform shaping technologies. Then, why the waveform shaping technology can play so much an important role in determining the overall bandwidth efficiency of a digital modulation scheme? This can be explained in the text followed with the same generic model of a digital modulation scheme shown in Figure (19.1) as follows.

From Figure (19.1), we have

$$x_{I,2}(t) = x_{I,1}(t) \otimes g(t) \tag{19.1}$$
$$x_{Q,2}(t) = x_{Q,1}(t) \otimes g(t) \tag{19.2}$$

where the notation $x(t) \otimes y(t)$ stands for convolution operation between two time-domain functions $x(t)$ and $y(t)$, and $g(t)$ is the impulse response of the pulse-shaping filter. If ignoring the *Post-shaping processor* unit in Figure (19.1) for illustration clarity, we have

$$
\begin{aligned}
y(t) &= x_{I,4}(t) + x_{Q,4}(t) & (19.3)\\
&= x_{I,2}(t)sin\omega_c t + x_{Q,2}(t)cos\omega_c t & (19.4)\\
&= (x_{I,1}(t) \otimes g(t))sin\omega_c t + (x_{Q,1}(t) \otimes g(t))cos\omega_c t & (19.5)
\end{aligned}
$$

Therefore, if defining $y(t) \Leftrightarrow Y(f)$, $g(t) \Leftrightarrow G(f)$, $x_{I,1}(t) \Leftrightarrow X_{I,1}(f)$ and $x_{Q,1}(t) \Leftrightarrow X_{Q,1}(f)$ are Fourier transform pairs, we have

$$
\begin{aligned}
&Y(f)\\
&= j\frac{1}{2}(X_{I,1}(f + f_c)G(f + f_c) - X_{I,1}(f - f_c)G(f - f_c))\\
&+ \frac{1}{2}(X_{Q,1}(f - f_c)G(f - f_c) + X_{Q,1}(f + f_c)G(f + f_c)) \quad (19.6)
\end{aligned}
$$

Thus, the power spectral density (PSD) of output modulated signal is

$$
\begin{aligned}
& |\,Y(f)\,|^2 \\
={} & |\,j\frac{1}{2}(X_{I,1}(f+f_c)G(f+f_c) - X_{I,1}(f-f_c)G(f-f_c)) \\
& + \frac{1}{2}(X_{Q,1}(f-f_c)G(f-f_c) + X_{Q,1}(f+f_c)G(f+f_c))\,|^2 (19.7)
\end{aligned}
$$

from which it can be seen that the PSD of the output signal is uniquely determined by the Fourier transform of the pulse shape or $G(f)$ as either $X_{I,1}(f)$ or $X_{Q,1}(f)$ is Fourier transform of a random data sequence which is non-alterable component. Therefore, if we can find a time waveform of $g(t)$ whose Fourier transform $G(f)$ has a very concentrated spectral shape around zero frequency, then we will also have a output modulated signal with a very compact form around the carrier frequency f_c, giving rise to a promising bandwidth efficiency. Please take note that this conclusion is valid for either single channel modulation schemes (such as BPSK) or quadrature modulations with I and Q channels (such as QPSK, etc.). Thus, the key for us to find a bandwidth efficient digital modulation scheme is equivalently to search for a promising time waveform with a very narrow PSD function. This is the most important fact that should be properly underlined throughout this chapter.

The great importance of waveform shaping technologies in the development of digital modulations can also be well explained in a more descriptive way in the sequel. Let us again look at Figure (19.1), which can be viewed as a generic model for many commonly used digital modulators. If only one channel is present, such as BPSK, we could take only I-channel into account without losing generality. Again, let us omit the block of *Post-shaping processor* for illustration clarity. Now we can follow the signal flow from the left hand side to the right hand side in the block diagram. It can be seen from the figure that all signal processing stages, except for the convolution operation taking place in *Pulse-shaping filter* unit, will never alter the spectral characteristics of the input signal. For example, the multiplication of the carrier, either $sin\omega_c t$ or $cos\omega_c t$, with the signal $x_{I,3}(t)$ or $x_{Q,3}(t)$ will not change its original spectral shape at all, only shifting its spectral center from zero to the carrier frequency ω_c. Also, the summation operation at the joint point of I and Q channels will not affect the spectral characteristics of the signals either. In this way, we can easily conclude from here that the single most important part of the modulator that determines the spectral characteristics of the output signal is the impulse response of *Pulse-shaping filter* $g(t)$.

Obviously, the waveform shapes can also sensitively affect many other merit parameters of a digital modulation scheme, as they do with the bandwidth efficiency. For instance, the waveform shapes have a great impact on the opening of the eye-diagrams of a digital modulation scheme. They can also affect the immunity of a digital modem against timing jitter resulted from unstable symbol or bit synchronization circuit. We will address all those issues in the later part of this chapter.

We have to admit that all traditional digital modulations have chosen to use relatively simple bit waveforms and thus little improvement can be achieved in their spectral characteristics as well as other merit parameters, such as power-efficiency and some other time-domain characteristics, such as the robustness against sampling timing jitter, etc. The pulse shapes found in those conventional modulation schemes include square (used by BPSK, QPSK, OQPSK), half-cycle sinusoidal (used by MSK), Gaussian (used in GMSK) and raised cosine (used in QORC [17]) waveforms, etc., which are intuitive or primitive in nature. The correspondence between symbol waveform shapes and overall performance of a modulated signal in both time- and frequence-domains motivates us to search for much more deliberately designed waveform shapes. It is indeed a fascinating research topic worthing our further study.

It is not surprising at all that a carefully designed pulse shaping waveform can be used for any type of digital modulation schemes with immediate improvement in both time- and frequence-domains performance. Let us take the simplest modulation, BPSK, as an example. Replacement of square waveform with raised-cosine pulse in BPSK modulator can directly improve the bandwidth efficiency of BPSK modulated signal, giving rise to a narrower main lobe and faster spectral roll-off rate than those before the replacement.

To have a more objective view on impact of waveform shaping techniques on the overall performance of a digital modem, let us look at some fundamental requirements for a desirable modulation scheme. The following considerations should be taken into account whenever design of a digital modulation scheme is concerned.

- First of all, the bandwidth efficiency of a modem is of primary concern to us. To have an objective measurement of the bandwidth efficiency, there are many different merit parameters which might be used, such as the width of main lobe of power spectral density (PSD) function of a modulated signal, the out-of-band emission defined as the amount of power left outside a specific bandwidth, the spectral roll-off rate defined as the reducing rapidity in PSD side lobe levels with respect to the normalized offset frequency from the carrier fT, etc.

- Another important parameter for a digital modem is its power efficiency, which sometimes can also be called detection efficiency. The power efficiency can be best described by the relation between bit error rate and signal-to-noise-ratio (SNR), which exactly means how efficiently a modem can utilize its initially transmitted power in recovering the information data at a receiver. It is clear that the power efficiency and bandwidth efficiency is a dual, which usually counteract with each other. In other words, it is difficult for a modem to achieve good performance of the both at the same time. A simple example to illustrate this phenomenon is that the use of a multi-amplitude/phase modulation can effectively improve the bandwidth efficiency, but at the price of decreasing the power efficiency at the same time.

- Complexity is another important concern to a digital modulation scheme, which should include both transmitter and receiver hardware and software complexity. It should be stressed that a good modulation scheme does not necessarily mean a high complexity. Our objective is to design a well-performing modulation scheme with a reasonable complexity. However, it should also be pointed out that the *complexity* is only a time-dependent term, which has different measures with the time elapses. It is easy to understand that a circuit that was considered as *very complex* ten years ago is not necessarily the same today due to the fast advancement of microelectronics industry. Therefore, in this sense, the importance of the consideration to the complexity of a digital modem should never be over exaggerated if compared with that for bandwidth efficiency and power efficiency of a modem.

- Yet another concern for a digital modulation is its performance under some particular working environment, such as band-limited non-linear channels, fading channels, multipath channels, noisy channels and other interfering channels, etc.

- When searching for a good digital modulation scheme, we should also pay attention to its performance under sampling time jitter, which is associated with the behaviors of synchronization acquisition and tracking systems at a receiver. This is also a very important issue whenever practical applications will be concerned.

With all aforementioned considerations in mind, we will be ready to discuss the issues on waveform design for well-performing digital modulation schemes. In doing so, we should pay a great attention to the bandwidth efficiency of the schemes in this chapter, although the other performance parameters of the schemes can also be the subject of our interest whenever necessary.

3.   Waveform Design Methodologies

To design promising pulse shaping waveforms to improve the performance of a digital modulation is a traditional research topic, as being evident in the reference list related to the subject and given at the end of this chapter. Some interesting waveforms as well as their generation methods were proposed in the literature [50]-[65], some of which can be summarized as follows.

3.1   Intuitive Methods

The majority of the pulse shaping waveforms used in early digital modulations were introduced in a non-systematic approach or by so called *intuitive methods*, which will be discussed in this section.

The pulse shapes generated using those intuitive methods include half-cycle sinusoidal waveform used in MSK, Gaussian functions for GMSK [14, 16], raised-cosine waveform used in QORC [17], cosine-roll off waveforms and $sinc(t)$ functions, etc, where all of which are finite-time pulses except for Gaussian waveforms and $sinc(t)$ functions defined as $sin(\pi t)/\pi t$. The later has been found in the applications of some partial response transmission systems.

In the early days MSK modulation was often viewed as a special case of continuous phase FSK (CPFSK) modulation due to the fact that its phase change has been made continuous as a very important step forward from OQPSK modulation, whose phase change still abruptly though at a relatively small value $\pi/2$ if compared to π in QPSK. However, from the view point of hardware implementation MSK can also be considered as a special case of OQPSK whose square bit pulse waveforms in both I and Q channels are replaced by half-cycle sinusoidal waveforms. In this sense, the block diagram of a MSK modulator can be shown in Figure (19.2), which in fact is a modified version of Figure (19.1) with only *Post-shaping processor* unit omitted.

The baseband symbol waveform in both I and Q channels in a MSK modulator is shown in Figure (19.3), where it has been assumed that all symbols carry the same sign +1 for representation clarity.

The baseband waveform of a MSK modulator shown in Figure (19.3) will completely determine the spectral characteristic features of MSK modulated signal, which include

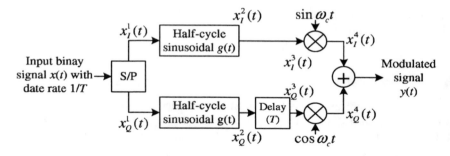

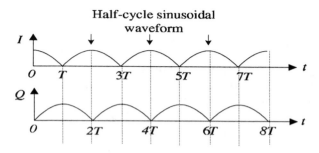

Figure 19.2. Block diagram of MSK modulator presented from waveform shaping perspective.

Figure 19.3. Baseband symbol waveform in both I and Q channels in a MSK modulator.

- continuous phase change with time in the modulated signal, which can effectively reduce side lobe regeneration effect in a band-limited non-linear channel;

- enhanced power concentration in the PSD main labe;

- a fast spectrum fall off at a rate proportional to f^{-4} for large values of f. In contrast, the offset QPSK spectrum falls off at a rate proportional to only f^{-2}, as a direct benefit from using half-cycle sinusoidal waveforms instead of square pulse in OQPSK.

- a truly constant envelope in MSK modulated signal in the time domain, which is important to reduce the distortion in a hard-limiting channel.

Thus, the advantages gained by the transformation from OQPSK to MSK clearly shows true power of waveform shaping technology.

GMSK [14, 16] is another vivid example of bandwidth efficiency enhancement by using waveform shaping techniques. The GMSK modulator structure is very similar to that of MSK, with only difference in the replacement of *half-cycle sinusoidal* pulse shaping filter with *Gaussian* pulse shaping filter in Figure (19.2). However, it has to be mentioned that the Gaussian waveform is not a finite time function, a big difference from the *half-cycle sinusoidal* waveform used in MSK modulator. The impulse response of a Gaussian pulse shaping filter can be written as

$$g_{gaussian}(t) = \frac{\sqrt{\pi}}{\alpha} e^{-\frac{\pi^2 t^2}{\alpha^2}} \tag{19.8}$$

The transfer function or the Fourier transform of (19.8) is

$$G_{gaussian}(f) = e^{-\alpha^2 f^2} \tag{19.9}$$

Therefore, there is an important parameter α in a Gaussian waveform, which controls its waveform extension in the time domain, as shown in Figure (19.4). In (19.8) and (19.9), the parameter α is closely related to 3-dB bandwidth (B_{3dB}) of the Gaussian waveform pulse shaping filter, whose relation can be represented by

$$\alpha = \frac{\sqrt{2ln2}}{B_{3dB}} = \frac{1.1774}{B_{3dB}} \tag{19.10}$$

Obviously, the 3-dB bandwidth of the filter will shrink or the Gaussian waveform shape becomes extended as α increases. GMSK modulated signal has a very good spectral characteristics, which can be even better than that of MSK, especially when it is applied to a system with non-linear power amplifier, which is often used in a mobile handset. Similar to MSK, GMSK modulated signal has also a constant signal envelope and zero phase shift between adjacent symbols. In addition, the side lobes of GMSK PSD function fall off even faster than those of MSK. Therefore, GMSK modulation has been successfully used in GSM system, which is the most popular cellular mobile standard in the world today.

However, it has to be mentioned that GMSK waveform is not a finite time function, as shown in Figure (19.4). Therefore, choosing a wrong (too large) α value will result in serious ISI, greatly impairing detection efficiency at a receiver. On the other hand, as mentioned earlier, a large α helps reduce the overall bandwidth occupancy of modulated signal. Therefore, a careful trade-off should be made when deciding the values of α in GMSK modulation.

Raised-cosine waveform is another widely used pulse shape in digital modulations. The most important example of its application is in QORC modulation, which was first proposed by Mark C. Austin and Ming U. Chang in 1981 [17]. The abbreviation *QORC* means *quadrature-overlapped raised cosine*, which uses raised-cosine waveform as its pulse

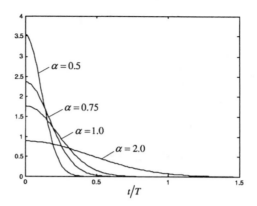

Figure 19.4. Gaussian waveform with different α values used in GMSK modulator, where T is the symbol duration.

shape and quadrature-overlapped scheme as its modulation structure, as shown in Figure (19.1). A $4T$ duration raised-cosine waveform can be represented by

$$g_{rc}(t) = \frac{1}{2}(1 - cos\frac{\pi t}{2T}), \qquad (0 < t \le 4T) \qquad (19.11)$$

where T stands for the symbol duration of the input signal to the modulator, as shown in Figure (19.1). The PSD function of the raised-cosine waveform can be shown as

$$\Phi_{rc}(f) = \mid G_{rc}(f) \mid^2 = 4T^2[2\frac{sin4\pi fT}{4\pi fT} + \frac{sin4\pi(4fT + 1)}{\pi(4fT + 1)}$$

$$+ \frac{sin4\pi(4fT - 1)}{\pi(4fT - 1)}]^2 \qquad (19.12)$$

In fact, the hardware implementation of a QORC modulator is quite similar to a MSK modulator. The main difference lies on its pulse waveform *overlapping*, in addition to their different pulse shapes used: QORC uses raised-cosine and MSK uses half-cycle sinusoidal waveforms. Therefore, the generic model for a digital modulation scheme shown in Figure (19.1) is still applicable to QORC modulation with the *post-shaping processor* unit taking care of waveform overlapping action. To illustrate this point more clearly, we have drawn the baseband symbol waveforms for both MSK and QORC modulations in Figure (19.5), where the time-domain waveforms for QORC appear as their individual symbols (assume to be all +1), instead of the sum of the overlapped waveforms, for illustration clarity.

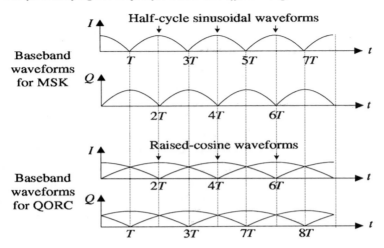

Figure 19.5. Comparison of baseband waveforms of MSK and QORC modulations, where it is assumed that all symbols carry the same information "+1" for illustration clarity.

The QO modulation is of great interest to us in this chapter due to the following reasons:

- The QO modulations can use twice as long symbol waveforms as those in conventional quadrature modulation schemes (such as MSK, GMSK and etc.) to enhance its overall bandwidth efficiency (occupying only about half of the bandwidth required by MSK, GMSK and etc.) without reducing the data transmission rate, as shown in Figure (19.5).

- Although waveform overlapping is used in QO modulations, there should be no ISI if each symbol can be sampled at right the center of the waveform. Therefore, the way used by QO modulation to improve bandwidth efficiency is much more effective than other schemes that also try to use prolonged symbol waveforms to improve the bandwidth efficiency, such as the case in a partial response system.

- The QO modulations retain a relatively simple implementation hardware if compared with many other digital modulation schemes with a similar bandwidth efficiency. The modulator structure is shown in Figure (19.1), which is almost the same as a conventional OQPSK modulator.

As far as bandwidth efficiency is concerned, QORC outperforms MSK with a comfortable margin. Let us take normalized PSD level of -20 dB as a threshold. MSK needs a bandwidth of about $fT = 0.67$ to reach the level; while QORC requires only $fT = 0.41$ to achieve the same level.

However, the QO modulation has also its technical limitations, which can be summarized as follows:

- Usually QO modulated signal is not a constant-envelope signal. The fluctuation in the envelope of QO modulated signal will have some negative impact on its performance in a non-linear channel. Let us define a parameter ρ to measure the severity of such fluctuation as the ratio between the maximum and minimum envelope amplitudes of the modulated signal. QORC gives about $\rho = 0.707$.

- Another concerned issue with a QO modulation is its sensitivity against possible sampling timing jitter. As shown in Figure (19.5), QORC will not produce any ISI only if each symbol is sampled at exactly the center of the symbol. Otherwise, the ISI is inevitable.

Fortunately, the above two major problems with QO modulations can be significantly improved or even eliminated if we can use some other more appropriate pulse shaping waveforms instead of the *raised-cosine* function. Thus, the usefulness of the QO modulations will definitely offer us some more degrees-of-freedom in designing a proper digital modulation scheme for some particular applications. This will be the issue to be discussed in much a great detail in the text followed.

3.2   Piece-Wise Fitting Methods

The piece-wise fitting method was first introduced by D. S-Dias and K. Feher in [20, 24] and used for baseband pulse shaping in $\pi/4$-DQPSK modulation to improve the spectral efficiency of a transceiver with a non-linear amplifier in land mobile environment. It was also demonstrated in [24] that with proposed new pulse shaping method, which resulted in $\pi/4$-QORC (Quadrature Overlapped Raised-Cosine) and $\pi/4$-CTPSK (Controlled Transmission PSK) modulations, the spectral side lobes regeneration after nonlinear amplification can be reduced significantly. Thus, the proposed pulse shaping techniques allow closer channel spacing if compared to a conventional $\pi/4$-DQPSK, the standard modem scheme for the US digital mobile communication system.

The piece-wise fitted pulse shaping technique used in [24] employed the following two-segmented functions:

If $a_n = \pm 1$, then

$$
i_1(t) = \begin{cases} a_{n-1}, & nT < t \le (n + \frac{1}{4})T \\ \frac{a_{n-1}+a_n}{2} - \\ \frac{a_n - a_{n-1}}{2}\cos\frac{4\pi}{3}(\frac{t}{T} - \frac{1}{4}), & (n + \frac{1}{4})T < t \le (n+1)T \end{cases} \tag{19.13}
$$

$$
q_1(t) = \begin{cases} \frac{b_{n-1}+b_n}{2} - \\ \frac{b_n - b_{n-1}}{2}\cos\frac{4\pi}{3}\frac{t}{T}, & nT < t \le (n + \frac{3}{4})T \\ b_n, & (n + \frac{3}{4})T < t \le (n+1)T \end{cases} \tag{19.14}
$$

If $a_n = 0$, then

$$
i_2(t) = \begin{cases} \frac{a_{n-1}+a_n}{2} - \\ \frac{a_n - a_{n-1}}{2}\cos\frac{4\pi}{3}\frac{t}{T}, & nT < t \le (n + \frac{3}{4})T \\ a_n, & (n + \frac{3}{4})T < t \le (n+1)T \end{cases} \tag{19.15}
$$

$$
q_2(t) = \begin{cases} b_{n-1}, & nT < t \le (n + \frac{1}{4})T \\ \frac{b_{n-1}+b_n}{2} - \\ \frac{b_n - b_{n-1}}{2}\cos\frac{4\pi}{3}(\frac{t}{T} - \frac{1}{4}), & (n + \frac{1}{4})T < t \le (n+1)T \end{cases} \tag{19.16}
$$

If $a_n = \pm\frac{1}{\sqrt{2}}$ and $|\, a_n - a_{n-1}\,| > |\, b_n - b_{n-1}\,|$, then

$$
i_3(t) = i_2(t) \tag{19.17}
$$
$$
q_3(t) = q_2(t) \tag{19.18}
$$

If $a_n = \pm\frac{1}{\sqrt{2}}$ and $|\, a_n - a_{n-1}\,| < |\, b_n - b_{n-1}\,|$, then

$$
i_4(t) = i_1(t) \tag{19.19}
$$
$$
q_4(t) = q_1(t) \tag{19.20}
$$

where a and b stand for the $\pi/4$-mapped signal levels at the input to a baseband signal processing unit (which is equivalent to the combination of *Pulse-shaping filter* and *Post-shaping processor* units in Figure (19.1)) in the I and Q channels respectively. The subscripts n and n-1 represent the nth and the $(n-1)$th symbol intervals respectively. The processed output signals in the I and Q channels are given by $i(t)$ and $q(t)$ respectively.

Please note that the above piece-wise fitted pulse shaping technique carries two salient features:

- It is a single-interval pulse shaping technique with no overlapping between adjacent symbols; while the raised-cosine pulse shaping technique used in QORC [17] allows half-symbol overlapping between two adjacent symbols.

- The pulse shaping processes happened to the I and Q channels are not carried out independently; while those in the QORC can be done independently.

Another example of using piece-wise fitting method to implement digital modulation is the two new QO modulations proposed earlier by us in [38], namely *Quadrature-Overlapped Triangular Cosine* (QOTC) modulation and *Modified Quadrature-Overlapped Triangular Cosine* (MQOTC) modulation.

Although these two QO modulations use also the piece-wise fitting technique in their pulse shaping procedure prior to the carrier modulation, as does in the previous case [20, 24], their working principle is quite similar to QORC modulation, which uses raised cosine function as its pulse shaping waveform (PSW). The most important aspect of them is their unique PSWs employed in their modulators. QOTC modulation uses a combination of a triangular and a cosine function as its PSW; and MQOTC modulation uses the second order version of triangular and cosine functions as its PSW.

Therefore, when compared with QORC modulation, which uses simple raised cosine function as its PSW, we can see the delicacy in these new modulations and their unique PSWs that result in much improved bandwidth efficiency and power efficiency. At the same time, the modulator structure of these two new modulations are almost exactly the same as that of QORC modulation, except for their different pulse shaping filters.

The time domain impulse response of the pulse-shaping filter used by QOTC modulation is given by

$$
g_{qotc}(t) = \begin{cases} \frac{t}{2T} - \frac{1}{2\pi} \sin \frac{\pi t}{T}, & 0 < t \leq 2T \\ 2 - \frac{t}{2T} + \frac{1}{2\pi} \sin \frac{\pi t}{T}, & 2T < t \leq 4T \end{cases} \tag{19.21}
$$

where T is the bit duration of input signal to the QOTC modulator, whose hardware structure is similar to that shown in Figure (19.1) that should use equation (19.21) as the impulse response of its *Pulse-shaping filter*.

It can be clearly seen that equation (19.21) in fact can be decomposed into two parts: one part is a triangular waveform as

$$
g_{qotc,1}(t) = \begin{cases} \frac{t}{2T}, & 0 < t \leq 2T \\ 2 - \frac{t}{2T}, & 2T < t \leq 4T \end{cases} \tag{19.22}
$$

and the other part is two back-to-back sinusoidal waveforms whose representation is

$$g_{qotc,2}(t) = \begin{cases} -\frac{1}{2\pi}\sin\frac{\pi t}{T}, & 0 < t \le 2T \\ \frac{1}{2\pi}\sin\frac{\pi t}{T}, & 2T < t \le 4T \end{cases} \tag{19.23}$$

Therefore, we have

$$g_{qotc}(t) = g_{qotc,1}(t) + g_{qotc,2}(t) \tag{19.24}$$

The QOTC time-domain waveform itself is shown in Figure (19.6), where we can see how the triangular pulse and two back-to-back sinusoidal waveforms are combined to form a complete QOTC PSW.

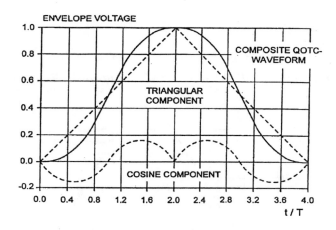

Figure 19.6. QOTC time-domain waveform, which can be decomposed into two components: a triangular pulse and two back-to-back sinusoidal waveforms, where T is the bit duration of input digital signal to the QOTC modem.

Similarly, we have the time-domain impulse response of the MQOTC pulse shaping filter as

$$g_{mqotc}(t) = \begin{cases} \frac{t^2}{2T^2} - \frac{1}{2\pi^2}\sin^2\frac{\pi t}{T}, & 0 < t \le T \\ \frac{2t}{T} - \frac{t^2}{2T^2} - 1 + \frac{1}{2\pi^2}\sin^2\frac{\pi t}{T}, & T < t \le 3T \\ -\frac{4t}{T} + \frac{t^2}{2T^2} + 8 - \frac{1}{2\pi^2}\sin^2\frac{\pi t}{T}, & 3T < t \le 4T \end{cases} \tag{19.25}$$

from which it is observed that, different from the QOTC modulation, the PSW of MQOTC modulation consists of the linear term, the 2nd order non-linear terms and 2nd order sinusoidal terms in its three-segment representation as shown in (19.25), resulting in a more complex equation if compared to that of QOTC. Therefore, in this sense, the MQOTC modulation can be viewed as a *2nd order modified* version of QOTC waveform.

The PSW of MQOTC can also be decomposed into two major parts, one being the rational functions part and the other being the 2nd order trigonometric functions, shown as follows:

$$g_{mqotc,1}(t) = \begin{cases} \frac{t^2}{2T^2}, & 0 < t \leq T \\ \frac{2t}{T} - \frac{t^2}{2T^2} - 1, & T < t \leq 3T \\ -\frac{4t}{T} + \frac{t^2}{2T^2} + 8, & 3T < t \leq 4T \end{cases} \tag{19.26}$$

$$g_{mqotc,2}(t) = \begin{cases} -\frac{1}{2\pi^2} sin^2 \frac{\pi t}{T}, & 0 < t \leq T \\ \frac{1}{2\pi^2} sin^2 \frac{\pi t}{T}, & T < t \leq 3T \\ -\frac{1}{2\pi^2} sin^2 \frac{\pi t}{T}, & 3T < t \leq 4T \end{cases} \tag{19.27}$$

Therefore we also have

$$g_{mqotc}(t) = g_{mqotc,1}(t) + g_{mqotc,2}(t) \tag{19.28}$$

The time-domain PSW of MQOTC is shown in Figure (19.7)

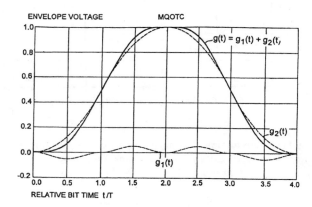

Figure 19.7. MQOTC time-domain waveform, which can be decomposed into two components: 2nd order trigonometric functions and the 1st and 2nd order rational functions, where T is the bit duration of input digital signal to the MQOTC modem.

As the input signal to the pulse-shaping filters in Figure (19.1) is a stream of square bit pulses, it can be shown that the transfer functions for QOTC and MQOTC pulse-shaping filters $g_{qotc}(t)$ and $g_{mqotc}(t)$ should take the forms of

$$G_{qotc}(f) = \frac{sin 2\pi f T}{2\pi f T} \frac{e^{-j2\pi f T}}{(1 - 2fT)} \tag{19.29}$$

and

$$G_{mqotc}(f) = (\frac{sin\pi fT}{\pi fT})^2 \frac{e^{-j2\pi fT}}{1-(fT)^2} \tag{19.30}$$

respectively, in order to obtain the correct PSW's at the output of the filters. Again, we can also see the similarity between (19.29) and (19.30).

It can be shown that the PSD's for the QOTC and MQOTC modulated signals can be written as

$$\Phi_{qotc}(f) = (\frac{sin2\pi fT}{2\pi fT})^4 \frac{1}{(1-(2fT)^2)^2} \tag{19.31}$$

and

$$\Phi_{mqotc}(f) = (\frac{sin2\pi fT}{2\pi fT})^2 \frac{1}{(1-(fT)^2)^2} (\frac{sin\pi fT}{\pi fT})^4 \tag{19.32}$$

respectively, which have quite similar expressions. In Figures (19.8) and (19.9) the PSD's of QOTC and MQOTC modulations are compared with that of MSK.

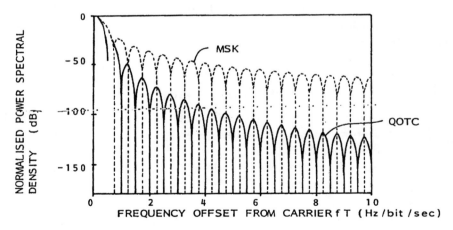

Figure 19.8. PSD function of QOTC in comparison with that of MSK.

It can be seen from Figures (19.8) and (19.9) that the QOTC and MQOTC modulations have the advantages of fast roll-off PSD side lobes and concentrated power nearby the main lobe over MSK. The PSD's of QOTC and MQOTC have a side lobe roll-off rate of $(fT)^{-8}$ and $(fT)^{-10}$, respectively, when compared with that of MSK being $(fT)^{-4}$.

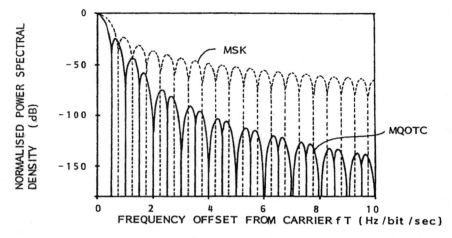

Figure 19.9. PSD function of MQOTC in comparison with that of MSK.

In addition, we can see that the MQOTC has a PSD main lobe of $0.5fT$, which is narrower than that $(0.75fT)$ of MSK. Therefore, it is a quite promising bandwidth efficient digital modulation scheme.

However, we have to admit that the piece-wise fitting method is still not a systematic approach to design suitable PSWs for bandwidth efficient digital modulations. They rely on try-and-error in the whole design process and a lot of time is required to find a suitable PSW for a particular application.

3.3   Time-Domain Convolution Methods

The time-domain convolution (TDC) method is a systematic approach proposed by us, which can be used to generate various promising PSW's to suit for different applications in bandwidth efficient digital modems.

Assume that there are two time-domain functions, $f_1(t)$ and $f_2(t)$, and their Fourier transform pairs, $f_1(t) \Leftrightarrow F_1(f)$ and $f_2(t) \Leftrightarrow F_2(f)$. Letting $f(t) \Leftrightarrow F(f)$, we have

$$f(t) = f_1(t) \otimes f_1(t) \Leftrightarrow F(f) = F_1(f)F_1(f) \tag{19.33}$$

where the operator "$\otimes$" stands for the time-domain convolution between $f_1(t)$ and $f_2(t)$. Thus, $\mid F_1(f) \mid^2$ and $\mid F_2(f) \mid^2$ are normalized PSD for $f_1(t)$ and $f_2(t)$, respectively.

The principle of the TDC method follows the two simple observations, which can be explained in the sequel.

- The normalized PSD of resultant signal $f(t)$ (yielded from time-domain convolution between $f_1(t)$ and $f_2(t)$), $\mid F(f) \mid^2$, always has a faster side lobe roll-off rate than that of either $\mid F_1(f) \mid^2$ or $\mid F_2(f) \mid^2$. Therefore, the $\mid F(f) \mid^2$ usually possesses a better bandwidth efficiency than that of either $\mid F_1(f) \mid^2$ or $\mid F_2(f) \mid^2$.

- If $f_1(t)$ and $f_1(t)$ are two equal length time pulses, the output pulse $f(t)$ from convolution between $f_1(t)$ and $f_2(t)$ has always a duration about twice as long as that of either $f_1(t)$ or $f_1(t)$.

The above two observations in fact are closely related in a sense that a prolonged time waveform can effectively yield a relatively concentrated power distribution in frequency-domain. However, using a prolonged waveform while keeping the data rate unchanged can only be made feasible through application of the QO modulations. This explains the reason why majority of the new PSW's proposed in this chapter are used in QO modulations, not others.

Based on the above two observations, we can generate many interesting PSW's for the applications in QO-based bandwidth efficient digital modulations. Let us first start with conventional MSK modulation, whose PSW is a $2T$-long half-cycle sinusoidal waveform, which can be written as

$$g_{msk}(t) = sin\frac{\pi t}{2T} \quad (0 < t \le 2T) \tag{19.34}$$

where T is the bit duration of input digital signal into the MSK modem.

The normalized output from convolution of two identical $g_{msk}(t)$ becomes

$$g_{qocc}(t) = \begin{cases} \frac{1}{\pi}sin\frac{\pi t}{2T} - \frac{t}{2T}cos\frac{\pi t}{2T}, & 0 < t \le 2T \\ (\frac{t}{2T} - 2)cos\frac{\pi t}{2T} - \frac{1}{\pi}sin\frac{\pi t}{2T}, & 2T < t \le 4T \end{cases} \tag{19.35}$$

whose duration has been extended into $4T$, which is twice as long as that in (19.34). We have employed this PSW in a QO modulation we named as quadrature-overlapped convoluted cosine (QOCC) modulation [38]. The QOCC PSW has been shown in Figure (19.10).

The QOCC PSW can also be viewed as the combination of two waveforms, as shown in Figure (19.10). The normalized PSD of QOCC modulated signal, which is the product of PSD's of two identical half-cycle sinusoidal functions, can be shown as

$$\Phi_{qocc}(f) = \frac{cos^4 2\pi fT}{(1 - (4fT)^2)^4} \tag{19.36}$$

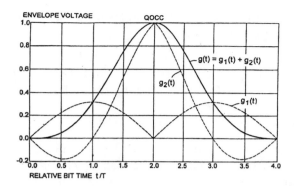

Figure 19.10. Pulse shaping waveform used in QOCC modulation.

from which it can be easily deduced that the PSD side lobe reduction
rate for QOCC modulation is inversely proportional to $(fT)^{-8}$, which
is much faster than that for MSK, $(fT)^{-4}$. The PSD's for both QOCC
and MSK are compared in Figure (19.11).

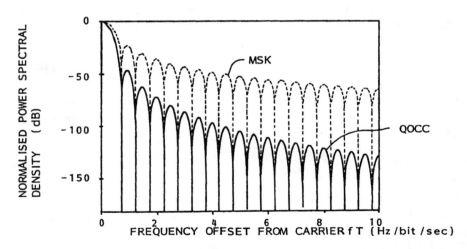

Figure 19.11. PSD function of QOCC in comparison with that of MSK.

Now, let us illustrate another example in using time-domain convolution method to design PSW. This time, we would like to start with two
different $2T$-long waveforms, one being again half-cycle sinusoidal function $g_{msk}(t)$ defined in (19.34) and the other being $2T$-duration QOTC

PSW given by

$$g_{qotc}^{\dagger}(t) = \begin{cases} \frac{t}{T} - \frac{1}{2\pi}sin\frac{2\pi t}{T}, & 0 < t \leq T \\ 2 - \frac{t}{T} + \frac{1}{2\pi}sin\frac{2\pi t}{T}, & T < t \leq T \end{cases} \qquad (19.37)$$

Please note that (19.37) is slightly different from (19.23) because they are defined in different time durations.

The convolution between $g_{msk}(t)$ and $g_{qotc}^{\dagger}(t)$ results in a new normalized PSW, which is used in a new QO modulation scheme named as *Modified Quadrature-Overlapped Convoluted Cosine* (MQOCC) modulation, as

$$g_{mqocc}(t) = \begin{cases} \frac{T sin\frac{2\pi t}{T}}{15\pi^2} - \frac{64T sin\frac{\pi t}{2T}}{15\pi^2} + \frac{2t}{\pi}, & 0 < t \leq T \\ -\frac{T sin\frac{2\pi t}{T}}{15\pi^2} - \frac{128cos\frac{\pi t}{2T}}{15\pi^2} \\ -\frac{64T sin\frac{\pi t}{2T}}{15\pi^2} + \frac{2(2T-t)}{\pi}, & T < t \leq 2T \\ \frac{T sin\frac{2\pi t}{T}}{15\pi^2} - \frac{\sqrt{2}Tcos\pi(\frac{t}{2T}+\frac{1}{4})}{6\pi^2} \\ -\frac{251T cos\frac{\pi t}{2T}}{30\pi^2} + \frac{41T sin\frac{\pi t}{2T}}{10\pi^2} - \frac{4T-2t}{\pi}, & 2T < t \leq 3T \\ -\frac{T sin\frac{2\pi t}{T}}{15\pi^2} + \frac{64T sin\frac{\pi t}{2T}}{15\pi^2} + \frac{8T-2t}{\pi}, & 3T < t \leq 4T \end{cases} \qquad (19.38)$$

whose normalized PSD function can be easily shown to be

$$\Phi_{mqocc}(f) = (\frac{sin\pi fT}{\pi fT})^4 \frac{cos^2 2\pi fT}{(1-(4fT)^2)^2(1-(fT)^2)^2} \qquad (19.39)$$

The PSW and PSD for MQOCC modulation are shown in Figures (19.12) and (19.13) respectively.

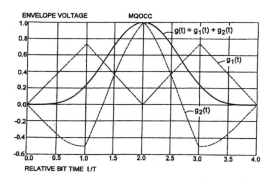

Figure 19.12. Pulse shaping waveform used in MQOCC modulation.

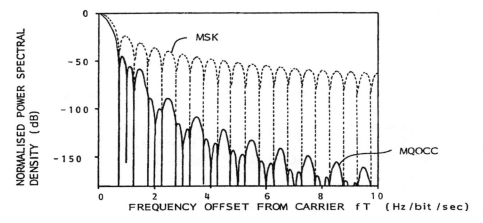

Figure 19.13.   PSD function of MQOCC in comparison with that of MSK.

The TDC method retains several salient features in designing PSW's, which can be summarized as follows:

- It is extremely easy to ensure a specific requirement on PSD side lobes roll-off rate of obtained modulated signal. For instance, if you want to design a PSW with a particular side lobe roll-off rate as $(fT)^{-(m+n)}$, what you need to do is to find two suitable waveforms $f_1(t)$ and $f_2(t)$, each with its PSD side lobe roll-off rate as $(fT)^{-m}$ and $(fT)^{-n}$ respectively. Take the QOCC PSW as an example, which is obtained by convolution of two identical half-cycle sinusoidal waveforms or (19.34), each of which has its PSD side lobe roll-off rate as $(fT)^{-4}$, as seen from its PSD function

$$\Phi_{msk}(f) = \frac{cos^2 2\pi fT}{(1 - (4fT)^2)^2} \tag{19.40}$$

Therefore, we will surely get the QOCC PSW with its PSD side lobe roll-off rate being $(fT)^{-8}$, which is the double of that of MSK. This feature in the TDC method gives us a great degree-of-freedom in controlling a suitable spectral efficiency of resultant PSW, otherwise hardly possible.

- The TDC method guarantees also a controllable PSD main lobe width in the target PSW, which is important when designing a bandwidth efficient digital modulation if some specific requirements on the main lobe exist. For example, if you have known the

PSD functions of two seed time-domain waveforms, $F_1(f)$ with its main lobe width being B_1 and $F_2(f)$ with its main lobe width being B_2 respectively, then you can conclude that the PSD main lobe of obtained PSW B must satisfy $B = min(B_1, B_2)$. As many digital systems often take the main lobe of transmitted signal as the bandwidth of the bandpass filter, this feature in the TDC method can facilitate prediction of bandwidth requirement of a system that uses the obtained PSW.

- The procedure of the TDC method is very straight-forward and computationally simple. There is no complex computations involved, which can help in reducing computational complexity of the PSW design process. What you need to do is to collect as many seed waveforms as possible, which should be relatively simple in their representation and waveform appearance. A data base for those seed waveforms has been created by us, which include the following time functions, just list a few as examples (we assume all seed waveforms to be normalized and equal length of $2T$):

1 Square waveform:

$$g_{sqr}(t) = \begin{cases} 1, & 0 < t \le 2T \\ 0, & \text{e. w.} \end{cases} \tag{19.41}$$

2 Triangular waveform:

$$g_{tri}(t) = \begin{cases} \frac{t}{T}, & 0 < t \le T \\ 2 - \frac{t}{T}, & T < t \le 2T \end{cases} \tag{19.42}$$

3 Half-cycle sinusoidal waveform:

$$g_{sin}(t) = \begin{cases} sin\frac{\pi t}{2T}, & 0 < t \le 2T \\ 0, & \text{e. w.} \end{cases} \tag{19.43}$$

4 Reverse half sinusoidal waveform:

$$g_{rhs}(t) = \begin{cases} 1 - sin\frac{\pi t}{2T}, & 0 < t \le 2T \\ 0, & \text{e. w.} \end{cases} \tag{19.44}$$

5 Raised-cosine waveform:

$$g_{rc}(t) = \begin{cases} \frac{1}{2}(1 - cos\frac{\pi t}{T}), & 0 < t \le 2T \\ 0, & \text{e. w.} \end{cases} \tag{19.45}$$

6 Reverse raised-cosine waveform:

$$g_{rrc}(t) = \begin{cases} \frac{1}{2}(1 + cos\frac{\pi t}{T}), & 0 < t \le 2T \\ 0, & \text{e. w.} \end{cases} \tag{19.46}$$

7 Back-to-back sinusoidal waveform:

$$g_{bbs}(t) = \begin{cases} -sin\frac{2\pi t}{T}, & 0 < t \le T \\ sin\frac{2\pi t}{T}, & T < t \le 2T \end{cases} \qquad (19.47)$$

8 Dual half sinusoidal waveform:

$$g_{dhs}(t) = \begin{cases} sin\frac{\pi t}{T}, & 0 < t \le T \\ -sin\frac{\pi t}{T}, & T < t \le 2T \end{cases} \qquad (19.48)$$

4.   Merit Parameters of Pulse-Shaped Modems

To objectively measure the performance of a pulse-shaped modulation, we need to define some suitable merit parameters.

Since we have already introduced four new QO modulations, QOTC, MQOTC, QOCC and MQOCC, which` were obtained by using various pulse shaping techniques, it is of great interest to us to make a comparison among them with some other conventional digital modulation schemes, such as QPSK and MSK, etc.

Let us first define *out-of-band power* (OOBP) to measure the bandwidth efficiency of different modulations as

$$P_{out}(B) = 10log\left[1 - \frac{\int_{-B}^{B}\Phi(f)df}{\int_{-\infty}^{\infty}\Phi(f)df}\right] \quad (dB) \qquad (19.49)$$

where B is the single-sided bandwidth of the modulated signal of interest and $\Phi(f)$ is its PSD function. Obviously, the value in $P_{out}(B)$ tells us how much power emission outside a particular bandwidth $(-B,B)$, which should be made as small as possible for bandwidth efficiency enhancement. Figure (19.14) shows the OOBP's for all four new QO modulation schemes proposed by us, which are compared to that of MSK.

The bandwidth efficiency of different modulations can also be compared using their normalized single-sided bandwidths, as shown in Table (19.2).

Please note that all data shown in Table (19.2) were obtained with the duration of the PSW of concern defined within $(0,4T)$. The results show a significant improvement in single-sided bandwidth (fT) if the proposed modulations can be used to replace the legacy schemes, such as OQPSK and MSK.

The eye-pattern for QOTC modulation is shown in Figure (19.15), where it is clearly shown that there is a flat opening width in $(-0.1t/T, 0.1t/T)$

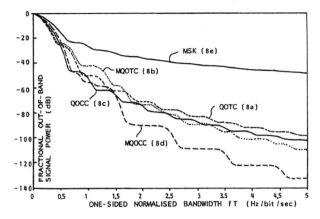

Figure 19.14. OOBP of QOTC, MQOTC, QOCC and MQOCC modulations in comparison with that of MSK.

Table 19.2. Comparison of normalized single-sided bandwidths (in fT) for different digital modulations with varying normalized PSD levels.

PSD (dB)	QPSK	MSK	QORC	QOTC	MQOTC	QOCC	MQOCC
-3	0.2211	0.2968	0.1798	0.1904	0.1933	0.2124	0.2365
-10	0.369	0.5095	0.314	0.3301	0.3341	0.3768	0.4192
-20	1.3406	0.66517	0.4124	0.4274	0.4305	0.5097	0.5648
-30	4.7987	1.1419	0.3989	0.7664	0.7901	0.5948	0.6538
-40	15.7714	2.51	0.6989	0.8623	0.8785	0.6517	0.7064
-50	49.283	4.4887	1.065	1.2426	1.4024	0.8517	0.9245
-60	158.739	7.5661	1.6609	1.3763	1.712	1.1419	1.5055
-70	503.252	14.018	2.4137	1.8653	1.8168	1.6200	1.6308
-80	1591.25	25.001	3.6449	2.7927	2.4246	2.5100	1.693
-90	5032.75	44.031	5.3857	3.7673	2.8084	3.0876	2.4943
-100	15915.8	78.991	7.8884	4.816	3.7687	4.0903	2.6503

of the eye-pattern that is important for QOTC to have a robust performance against sampling timing jitter.

The phaser diagram of QOCC modulation is shown in Figure (19.16), which is to show how much fluctuation in its modulated signal envelope. For a constant-envelope signal, the phaser diagram should be a perfect circle. To characterize the fluctuation in modulated signal's envelope, we have introduced a parameter $0 \le \rho \le 1$, named as *envelope fluctuation factor*, whose value should be made as close to unit as possible to reduce the envelope fluctuation in modulated signal. Clearly, this parameter has

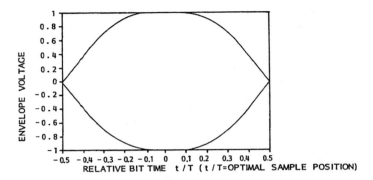

Figure 19.15. Eye-pattern of QOTC modulations.

a great impact on how a modem will perform in a non-linear channels, such as satellite channels and mobile cellular systems, where non-linear C-class power amplifiers are widely used. Such power amplifiers work in a very similar way as a hard-limiter, such that all information carried in the envelope or amplitude will be lost, causing a serious distortion to the modulated signal.

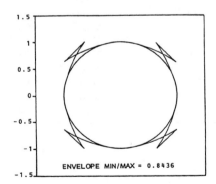

Figure 19.16. Phaser diagram of QOCC modulation.

Another important parameter relevant to pulse-shaping waveforms is defined as

$$\mu = \frac{1}{4T} \int_0^{4T} g^2(t)dt \qquad (19.50)$$

where it has been assumed that the normalized PSW, $g(t)$, of concern has a duration of $4T$. In fact, (19.50) gives the average power of a PSW, which is also important in determining the shapes of a pulse waveform. In some cases we would like also to know the sampling value at a quarter of the pulse duration $(4T)$, or $g(T)$, which decides the opening of the eye-pattern of a PSW. Thus, we have three important parameters in our hand: ρ, μ and $g(T)$, all of which will be used to characterize the exact shape of a particular PSW. Table (19.3) gives their values for four different QO modulation schemes using distinct PSW's.

Table 19.3. Three shape parameters (ρ, μ and $g(T)$) for QORC, QOTC, MQOTC, QOCC and MQOCC modulations

Modulation-type	ρ	μ	$g(T)$
QORC	0.7071	0.375	0.5
QOTC	0.7071	0.3967	0.5
MQOTC	0.7071	0.4024	0.5
QOCC	0.8436	0.2933	0.3183
MQOCC	0.9041	0.2693	0.2363

It is seen from Table (19.3) that MQOCC has its *envelope fluctuation factor* reaching $\rho = 0.9041$, which is very close to unit to become a constant envelope modulation scheme. Therefore, it should be suitable for the applications in non-linear channels, including mobile cellular systems where all handset transmitters always use C-class RF amplifiers with high power-efficiency due to the concern of battery stand-by time.

5. Spectral Efficiency in Band-Limited Nonlinear Channels

The bandwidth-efficiency of a digital modem is always an important issue whenever a practical application is concerned. In this section we will offer a comprehensive discussion on how to evaluate spectral efficiency of a pulse-shaped QO modulation over band-limited nonlinear channels, which can be found in many applications such as satellite communications and mobile cellular systems.

5.1 Four More QO Modulations

It should be pointed out that the evaluation of spectral efficiency for a digital modem over a band-limited nonlinear channels can not be done using only the PSD function of the modulated signal. Instead, the study should take band-limited nonlinear channels into account and it can be a complex procedure. In the text followed there are four QO modulations involved (whose modulator structure retains the same as that shown in Figure (19.1)), which are

1 *overlapped minimum shift keying* (OMSK) modulation, whose PSW can be written as

$$
g_{omsk}(t) = \begin{cases} \frac{1}{\pi} \sin \frac{\pi t}{2T} - \frac{t}{2T} \cos \frac{\pi t}{2T}, & 0 < t \le 2T \\ \left(\frac{t}{2T} - 2\right) \cos \frac{\pi t}{2T} - \frac{1}{\pi} \sin \frac{\pi t}{2T}, & 2T < t \le 4T \end{cases} \quad (19.51)
$$

2 *minimum shift keying triangular cosine* (MSKTC) modulation with its PSW being

$$
g_{msktc}(t) = \begin{cases} \frac{T \sin \frac{2\pi t}{T}}{15\pi^2} - \frac{64T \sin \frac{\pi t}{2T}}{15\pi^2} + \frac{2t}{\pi}, & 0 < t \le T \\ -\frac{T \sin \frac{2\pi t}{T}}{15\pi^2} - \frac{128T \cos \frac{\pi t}{2T}}{15\pi^2} \\ \quad - \frac{64T \sin \frac{\pi t}{2T}}{15\pi^2} + \frac{2(2T-t)}{\pi}, & T < t \le 2T \\ \frac{T \sin \frac{2\pi t}{T}}{15\pi^2} - \frac{\sqrt{2T} \cos[\pi(\frac{t}{2T} + \frac{1}{4})]}{6\pi^2} \\ \quad -\frac{251T \cos \frac{\pi t}{2T}}{30\pi^2} + \frac{41T \sin \frac{\pi t}{2T}}{10\pi^2} - \frac{2(2T-t)}{\pi}, & 2T < t \le 3T \\ -\frac{T \sin \frac{2\pi t}{T}}{15\pi^2} + \frac{64T \sin \frac{\pi t}{2T}}{15\pi^2} + \frac{2(4T-t)}{\pi}, & 3T < t \le 4T \end{cases} \quad (19.5
$$

which should be normalized by a factor of $\frac{128T}{15\pi^2}$;

3 *raised-cosine triangular cosine* (RCTC) modulation, whose PSW is given by

$$
g_{rctc}(t) = \begin{cases} -T\frac{\cos \frac{2\pi t}{T}}{24\pi^2} + 2T\frac{\cos \frac{\pi t}{T}}{3\pi^2} \\ \quad - \frac{5T}{8\pi^2} + \frac{t^2}{4T}, & 0 < t \le T \\ T\frac{\cos \frac{2\pi t}{T}}{24\pi^2} + 2T\frac{\cos \frac{\pi t}{T}}{\pi^2} \\ \quad + \frac{5T}{8\pi^2} - \frac{T}{2} - \frac{t^2}{4T} + t, & T < t \le 3T \\ -t\frac{\cos \frac{2\pi t}{T}}{24\pi^2} + 2T\frac{\cos \frac{\pi t}{T}}{3\pi^2} \\ \quad - \frac{5T}{8\pi^2} + 4T + \frac{t^2}{4T} - 2t, & 3T < t \le 4T \end{cases} \quad (19.53)
$$

which can be normalized by a factor of $\frac{8T}{3\pi^2} + \frac{T}{2}$; and

4 *sinusoidal quadrature overlapped triangular cosine* (SQOTC) modulation, whose PSW can be represented as

$$
g_{sqotc}(t) = \begin{cases}
\frac{t^3}{6T^2} - \frac{t}{2\pi^2} + \frac{5T}{2(2\pi)^3}\sin\frac{2\pi t}{T} \\
\quad - \frac{t}{2(2\pi)^3}\cos\frac{2\pi t}{T}, & 0 < t \le T \\[1ex]
\frac{2T}{3} - 2t + \frac{2t^2}{T} - \frac{t^3}{2T^2} + \frac{6t-8T}{(2\pi)^2} \\
\quad - \frac{15T}{2(2\pi)^3}\sin\frac{2\pi t}{T} + \frac{(\frac{3t}{2}-2T)}{(2\pi)^2}\cos\frac{2\pi t}{T}, & T < t \le 2T \\[1ex]
-\frac{22T}{3} + 10t - \frac{4t^2}{T} + \frac{t^3}{2T^2} + \frac{16T-6t}{(2\pi)^2} \\
\quad + \frac{15T}{2(2\pi)^3}\sin\frac{2\pi t}{T} + \frac{(4T-\frac{3t}{2})}{(2\pi)^2}\cos\frac{2\pi t}{T}, & 2T < t \le 3T \\[1ex]
\frac{32T}{3} - 8t + \frac{2t^2}{T} - \frac{t^3}{6T} + \frac{2t-8T}{(2\pi)^2} \\
\quad - \frac{5T}{2(2\pi^3)}\sin\frac{2\pi t}{T} + \frac{(\frac{t}{2}-2T)}{(2\pi^2)}\cos\frac{2\pi t}{T}, & 3T < t \le 4T
\end{cases}
\tag{19.54}
$$

whose normalizing factor is $\frac{2T}{3} + \frac{5T}{4\pi^2}$.

All above four PSW's can be generated with the help of the TDC method by choosing some proper seed pulse waveforms, as explained in the previous section. The four different PSW's are shown in Figure (19.17), where the raised-cosine waveform is also plotted for easy comparison.

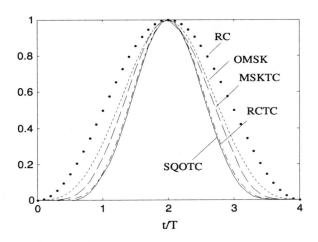

Figure 19.17. Pulse-shaping waveforms for OMSK, MSKTC, RCTC and SQOTC modulations, which are compared with that of QORC.

5.2   Band-Limited Nonlinear Channel Model

The QO modulation system model considering the non-linear channel is illustrated in Figure (19.18), where a binary input data stream $\{c_k\}$, with bit duration T, is demultiplexed by a serial-to-parallel converter into two independent binary data sequences $\{a_k\}$ and $\{b_k\}$. The data sequences $\{a_k\}$ and $\{b_k\}$ (each in non-return-to-zero (NRZ) format taking the values of ± 1) then pass through pulse-shaping filters, $p(t)$ and $q(t)$, and are modulated onto the quadrature carriers, resulting in the following transmitted signal

$$
\begin{aligned}
s(t) \;=\; & \Big[\sum_{n=-\infty}^{\infty} a_n p(t - 2nT) \Big] \cos \omega_c t \\
& + \Big[\sum_{n=-\infty}^{\infty} b_n q(t - 2nT) \Big] \sin \omega_c t \qquad (a_n, b_n = \pm 1) \quad (19.55)
\end{aligned}
$$

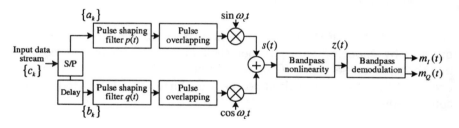

Figure 19.18.   A generic system model for studying spectral efficiency of any quadrature modulation schemes in band-limited nonlinear channels, valid for overlapped or non-overlapped (without two *pulse overlapping* elements) and staggered or non-staggered (without the *delay* element) modulations.

The spectral characteristics of the modulation are primarily determined by the pulse-shaping filters $p(t)$ and $q(t)$ in I and Q channels, respectively. Pulse-shaping can be implemented in two ways; overlapped and non-overlapped, each of which can be either staggered or non-staggered, depending on the relative phase shift between I and Q channel bit streams. The non-overlapped modulations include QPSK, offset QPSK and MSK modulations. As for the overlapped modulations, two consecutive bit pulses in both I and Q channels are overlapped by just half of the pulse duration $(2T)$, as in the QORC and QOTC modulations. In staggered modulation schemes, the Q channel bit stream is delayed by T with respect to that in the I channel. In fact, all four new modulations concerned in this section are the staggered (offset) QO modulations and the duration of a symbol pulse in either I or Q channels is $4T$, where T is the bit duration of the input data stream $\{c_k\}$.

Although the pulse-shape functions of the four new QO modulations, given in (19.51) to (19.54), look quite complicated, they can be realized using digital filter technology. Because the pulse shaping waveforms can be viewed as the output from a filter, whose input is the binary data waveform (normally, antipodal square pulses taking the levels of -1 or +1), the impulse response of the pulse-shaping filter can be easily obtained with the help of digital filter design theory. Two such pulse-shaping filters are needed in a QO modem, one for the I channel and another for the Q channel. Due to the overlapping of consecutive bit symbols in a QO modem, a waveform buffer, together with a delay and addition unit, is required for both I and Q channels. This means that the output pulse-shaped signals can be appropriately overlapped according to the signs of the corresponding consecutive bits. The complexity of such a QO modem is slightly greater than that of a non-overlapped modem, such as MSK. However, there is no technical difficulties in implementing such a QO modem and the hardware complexity can be kept reasonably low. We have to admit that the hardware implementation of the proposed modems has some impact on the real-world spectral performance of the modems, due to certain realization problems, such as quantization errors existing in realizing digital filters, etc. Therefore, the performance analysis in this section only reflects the results of an ideal scenario without considering all those realization issues. On the other hand, as the same ideal scenario is applied to all compared modulation schemes, the relative difference in their performance also tells which one is better.

As $s(t)$ is to be transmitted over a non-linear channel, it is convenient for us to represent $s(t)$ in (19.55) in the equivalent polar form as

$$s(t) = \sum_{k=-\infty}^{\infty} R_T(t - 2kT) \cos(\omega_c t - \eta_T(t - 2kT)) \qquad (19.56)$$

where $R_T(t)$ and $\eta_T(t)$ have to be determined and are non-zero in the interval $[0,2T]$. Then, $R_T(t - 2kT)$ and $\eta_T(t - 2kT)$ are just the time-shifted versions of $R_T(t)$ and $\eta_T(t)$, respectively. For the staggered and overlapped modulations, the expressions for $R_T(t-2kT)$ and $\eta_T(t-2kT)$ are given as follows (its proof has been given in [39, 41]):

$$R_T(t - 2kT) \qquad\qquad (19.57)$$

$$= \begin{cases} \left\{ \left[\sum_{i=0}^{1} a_{k-i} p(t - 2(k-i)T)\right]^2 \\ + \left[\sum_{i=0}^{2} b_{k-i} q(t - 2(k-i)T)\right]^2 \right\}^{1/2}, & 2kT \le i \le 2(k+1)T \\ 0, & \text{e.w.} \end{cases}$$

and

$$\eta_T(t - 2kT) \tag{19.58}$$

$$= \begin{cases} \tan^{-1}\left[\dfrac{\sum_{i=0}^{2} b_{k-i}(q(t-2(k-i)T))}{\sum_{i=0}^{1} a_{k-i}(p(t-2(k-i)T))}\right], & 2kT \leq i \leq 2(k+1)T \\ 0, & \text{e.w.} \end{cases}$$

The non-linearity has equivalent AM-AM and AM-PM characteristics denoted by $\tilde{f}(\cdot)$ and $\tilde{g}(\cdot)$ [39, 41], respectively. The output signal from the non-linear channel can be written as

$$z(t) = \sum_{k=-\infty}^{\infty} \tilde{f}(R_T(t - 2kT)) \cos\left[\omega_c t - \eta_T(t - 2kT) - \tilde{g}(R_T(t - 2kT))\right]$$

$$= m_I(t) \cos \omega_c t + m_Q(t) \sin \omega_c t \tag{19.59}$$

where the in-phase component $m_I(t)$ can be written into

$$m_I(t) = \sum_{k=-\infty}^{\infty} d_I(t - 2kT; a_k(1), b_k(2)) \tag{19.60}$$

with

$$d_I(t - 2kT; a_k(1), b_k(2))$$

$$= \begin{cases} \tilde{f}(R_T(t - 2kT)) \cos[\eta_T(t - 2kT) + \tilde{g}(R_T(t - 2kT))], \\ \quad\quad 2kT \leq t \leq 2(k+1)T \\ 0, \quad otherwise \end{cases} \tag{19.61}$$

and

$$a_k(1) = [a_{k-1}, a_k] \tag{19.62}$$

$$b_k(2) = [b_{k-2}, b_{k-1}, b_k] \tag{19.63}$$

Similarly, the quadrature demodulated signal $m_Q(t)$ over the non-linear channel can also be derived. Based on these equations, we can evaluate the PSD of the QO modulated signal over a band-limited nonlinear channel in the sequel.

5.3   PSD Expression Derivation

In this subsection, we shall first derive a generic PSD expression for the signals over a non-linear channel. From the definition, the power spectral density of $m_I(t)$ is given by

$$S_m(f) = \mathcal{F}(R_m(\tau)) = \mathcal{F}(\langle \overline{m_I(t)m_I(t - \tau)} \rangle) \tag{19.64}$$

where $R_m(\tau)$ is the autocorrelation function of the signal $m_I(t)$, $\mathcal{F}$ denotes the Fourier transform operator, the over-bar denotes the statistical average and the symbol "$\langle\rangle$" denotes the time average. Substituting equation (19.60) into (19.64), we have

$$S_m(f) = \tag{19.65}$$

$$\mathcal{F}\{[\sum_{j=-\infty}^{\infty} \sum_{k=-\infty}^{\infty} \overline{d(t - 2jT; a_j(1), b_j(2))d(t - 2kT - \tau; a_k(1), b_k(2))}]\}$$

Denoting

$$v(t - 2jT) = d(t - 2jT; a_j(1), b_j(2)) \tag{19.66}$$
$$v(t - 2kT - \tau) = d(t - 2kT - \tau; a_k(1), b_k(2)) \tag{19.67}$$

we thus obtain the following equation to calculate the time average for (19.65)

$$\langle \sum_{j=-\infty}^{\infty} \sum_{k=-\infty}^{\infty} v(t - 2jT)v(t - 2kT - \tau) \rangle$$

$$= \langle \cdots + v(t)v(t - \tau) + v(t)v(t - 2T - \tau) + \cdots$$
$$+ \quad v(t - 2T)v(t - \tau) + v(t - 2T)v(t - 2T - \tau) + \cdots \rangle$$

$$= \frac{1}{2T} \sum_{k=-\infty}^{\infty} \int_0^{2T} v(t)v(-(\tau - t) - 2kT)dt$$

$$= \frac{1}{2T} \sum_{k=-\infty}^{\infty} v(t) \otimes v(-\tau - 2kT) \tag{19.68}$$

where $\otimes$ denotes the time convolution operation. Thus, the power spectral density in (19.65) can be reduced to

$$S_m(f) = \frac{1}{2T}\mathcal{F}\{\sum_{k=-\infty}^{\infty} \overline{d(\tau; a_0(1), b_0(2)) \otimes d(-\tau - 2kT; a_k(1), b_k(2))}\}$$

$$= \frac{1}{2T} \sum_{k=-\infty}^{\infty} \overline{D(f; a_0(1), b_0(2))D^*(f; a_k(1), b_k(2))}e^{j4\pi fkT} \tag{19.69}$$

where the asterisk denotes the complex conjugate, and

$$D(f; a_k(1), b_k(2)) = \int_0^{\infty} d(t; a_k(1), b_k(2))e^{-j2\pi ft}dt$$

$$= \int_0^{2T} d(t; a_k(1), b_k(2))e^{-j2\pi ft}dt \tag{19.70}$$

We should bear in mind that the PSD expression given in (19.69) is valid for staggered and overlapped signals passing through a non-linear channel. When the non-linearity is modeled by a hard limiter, the AM-AM and AM-PM characteristics of (19.59) become

$$\tilde{f}(\cdot) = \gamma = \sqrt{2} \tag{19.71}$$

$$\tilde{g}(\cdot) = 0 \tag{19.72}$$

Applying Equation (19.72) to (19.61), we obtain a further simplified expression

$$d_I(t - 2kT; a_k(1), b_k(2))$$
$$= \begin{cases} \gamma \cos[\eta_T(t - 2kT)], & 2kT \le t \le 2(k+1)T \\ 0, & otherwise \end{cases} \tag{19.73}$$

In the above analysis, we have only considered the I channel signal $m_I(t)$. In fact, the same analysis can also be applied to the Q channel signal, $m_Q(t)$. Therefore, we could ignore the subscripts I and Q and (19.69) gives a generic PSD expression for the QO signals.

5.4   Numerical Evaluation of PSD

Equation (19.69), however, is not suitable for a direct PSD numerical calculation for a specific QO signal. Therefore, we want to derive a computationally efficient PSD expression suitable for any staggered or/and overlapped modulations. Using the results obtained in [39, 41], the numerical expression for the PSD of staggered and overlapped modulations (assuming that pulse-shaping waveform $p(t)$ is symmetrical about $t = 2T$) after passing through a non-linear channel is given below

$$\begin{aligned}
S_m(f) &= 2T\gamma^2 \left\{ \frac{3}{4} \sum_{k=1}^{4} R_k^2 + \frac{1}{4} \sum_{k=1}^{4} I_k^2 \right. \\
&- \frac{1}{2}(R_1 R_3 + R_2 R_4 - I_1 I_3 - I_2 I_4) \\
&+ \frac{1}{8} \cos 8\pi f T \left[\left(\sum_{k=1}^{4} R_k\right)^2 - \left(\sum_{k=1}^{4} I_k\right)^2 \right] \\
&- \frac{1}{4} \sin 8\pi f T \left[\left(\sum_{k=1}^{4} R_k\right)\left(\sum_{k=1}^{4} I_k\right) \right] \\
&+ \frac{1}{2}(R_1 + R_2 - R_3 - R_4)
\end{aligned}$$

$$\times \ \left(\sum_{k=1}^{4} R_k \cos 4\pi fT - \sum_{k=1}^{4} I_k \sin 4\pi fT \right) \right\} \quad (19.74)$$

where

$$
\begin{aligned}
R_1 &= Re[K(f;1,1)]; & I_1 &= Im[K(f;1,1)] \\
R_2 &= Re[K(f;1,-1)];, & I_2 &= Im[K(f;1,-1)] \\
R_3 &= -Re[K(f;-1,1)];, & I_3 &= -Im[K(f;-1,1)] \\
R_4 &= -Re[K(f;-1,-1)];, & I_4 &= -Im[K(f;-1,-1)]
\end{aligned}
\qquad (19.75)
$$

and

$$K(f; x_k, y_k) \qquad (19.76)$$

$$= \frac{1}{2T} \int_0^T \frac{[p(t+2T) + x_k p(t)]\, e^{-j2\pi ft} dt}{\left\{ [p(t+2T) + x_k p(t)]^2 + [q(t+2T) + y_k q(t+4T)]^2 \right\}^{1/2}}$$

We have also defined x_k and y_k as

$$
\begin{aligned}
y_k &= b_k b_{k-1} = b_k/b_{k-1} = b_{k-1}/b_k \\
x_k &= a_k a_{k-1} = a_k/a_{k-1} = a_{k-1}/a_k
\end{aligned}
\qquad (19.77)
$$

where $\{a_k\}$ and $\{b_k\}$ take the values of ± 1 with equal likelihood. Then x_k and y_k are identically independently distributed (*i.i.d.*) sequences, taking values of ± 1 with equal probability. In the next subsection, (19.74) will be used to study the spectral efficiency of the four new QO modulations over the non-linear channels.

5.5   Numerical Results

Equation (19.74) can be used to numerically evaluate the spectral efficiency for various modulations (either overlapped or non-overlapped, staggered or non-staggered). Here we are particularly interested in the staggered QO modulations over both linear and non-linear channels.

5.5.1   PSD over Linear Channels.   The PSD's of the QO signals transmitted over a linear channel are obtained by performing Fourier transform of the autocorrelation functions of transmitted signals. The pulse-shaping waveforms of staggered QORC, OMSK, MSKTC, RCTC and SQOTC modulations have been shown in Figure (19.17); whereas their corresponding spectra are given in Figure (19.19).

It can be observed in Figure (19.19) that the new modulation schemes over the linear channel have faster side lobe reduction rates than the

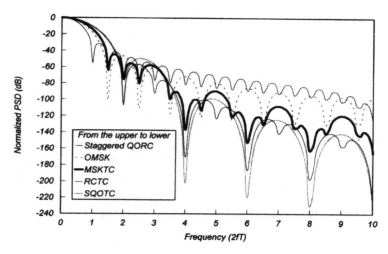

Figure 19.19. PSD functions of staggered QORC, OMSK, MSKTC, RCTC and SQOTC modulated signals in a linear channel without bandpass filter.

staggered QORC modulation, which already possesses better spectral characteristics than the OQPSK and MSK modulations [9, 17]. The PSD main lobe widths of the OMSK and MSKTC signals remain the same as that of MSK, which is 1.5 times wider than that of staggered QORC modulation. As for the RCTC and SQOTC signals, their PSD main lobe widths are twice as wide as the staggered QORC modulation. However, they exhibit faster side lobe reduction rate than the QORC modulation.

It is interesting to note that somehow there is a relation between the PSD and the shaping-pulse appearance. The flatter the central part of the bit pulse, the narrower the PSD main lobe; whereas the flatter the part around the edges, the faster the side lobe levels drop. It is verified in Figures (19.17) and (19.19) that the RCTC and SQOTC pulses with flatter edges have faster side lobe reduction rates, and the staggered QORC pulse has a narrower main lobe due to its flat central portion of the pulse shape.

5.5.2   PSD over Nonlinear Channels.   Using Equation (19.74), we can obtain the PSD's of staggered and overlapped modulations (staggered QORC, OMSK, MSKTC, RCTC and SQOTC modulations) after passing through the hard-limited channel. The power spectra of the OMSK, MSKTC,RCTC and SQOTC modulations, prior to and after hard-limiting, are shown in Figures (19.20) and (19.21) respectively. After hard-limiting, the spectral side lobes of the OMSK and MSKTC modulations can maintain their respective levels prior to hard-limiting. However, the side lobe levels regenerated due to the hard-limiting effect for the RCTC and SQOTC modulations are higher than those of the OMSK and MSKTC modulations, although they are still better than the QORC modulation.

The power spectra after bandpass and hard-limiting channel of the OQPSK, MSK, staggered QORC and the four new modulations are compared in Figures (19.22) and (19.23), respectively. As expected, the spectra of constant-envelope OQPSK and MSK modulations change little after hard-limiting (their PSDs are derived in [39, 41]). These modulations suffer from very high side lobe levels even prior to hard-limiting. On the other hand, the new QO modulations can keep relatively low side lobe levels (by taking into account the side lobe regeneration effect), if compared with those for OQPSK, MSK and staggered QORC modulations. From Figures (19.22) and (19.23), we can see that the MSKTC modulation yields the lowest side lobe level after bandpass hard-limiting channel. Figure (19.19) shows that RCTC and SQOTC modulations have the lowest side lobe levels before hard-limiting, because the PSD side lobes of the SQOTC and RCTC modulations are regenerated due to the hard-limiting effect. In general, the new modulations can offer a better spectral efficiency than that of their counter-part modulations OQPSK, MSK and staggered QORC.

6.   Robustness Against Sampling Timing Jitter

It is well-known that QORC modulation [17] is more bandwidth-efficient than QPSK and MSK due mainly to the application of its pulse shaping and overlapping before quadrature carrier modulation. With the same transmission rate, the pulse duration in a QO modulation can be made twice as wide as that for QPSK or MSK [9, 15, 17]. In a QO modulation there should be no inter-symbol interference if sampling to received signal can be perfectly synchronized at $t = 2nT$ in both I and Q channels at a receiver, where n is any integer, T is the bit duration of the input signal to the modulator and raised-cosine pulse is defined over $(0, 4T)$ as given in (19.11). However, when timing for sampling is jittered because of either non-perfect bit synchronization or carrier phase tracking problems or other interference, the sampling from adjacent bits will degrade the bit error rate (BER) performance. Therefore, the shape

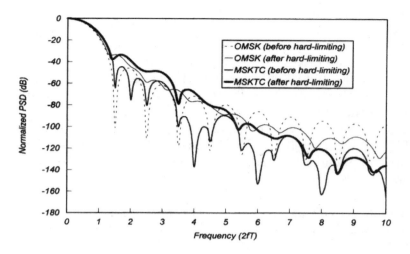

Figure 19.20. PSD functions of staggered OMSK and MSKTC modulated signals before and after bandpass hard-limiting channel.

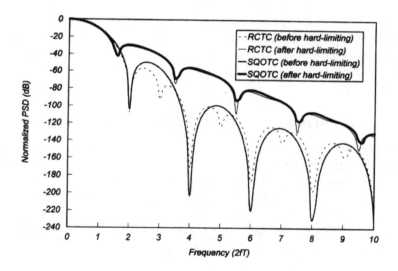

Figure 19.21. PSD functions of staggered RCTC and SQOTC modulated signals before and after bandpass hard-limiting channel.

of the pulse shaping waveform in a quadrature-overlapped modulation plays an important role in BER when timing jitter is present.

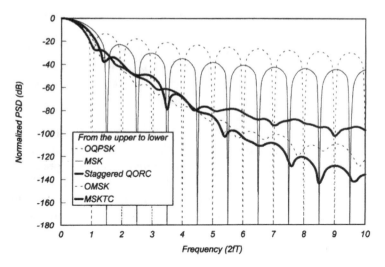

Figure 19.22. PSD functions of staggered OMSK and MSKTC modulated signals, which are compared to those for OQPSK and MSK modulated signals, after bandpass hard-limiting channel.

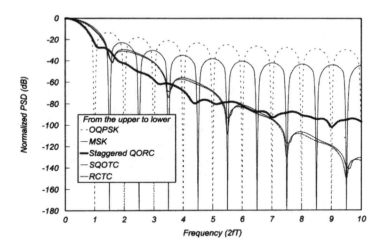

Figure 19.23. PSD functions of staggered RCTC and SQOTC modulated signals, which are compared to those for OQPSK and MSK modulated signals, after bandpass hard-limiting channel.

In this section we would like to study the impact of receiver timing jitter on the bit error rate (BER) of a pulse-shaped QO modulation.

We are interested in *Raised-Cosine and Triangular-Cosine* (RCTC) modulation, whose PSW has been given in (19.53). In fact, the RCTC waveform in (19.53) is generated with the help of the TDC method by time convolution between a raised-cosine (19.45) and a triangular-cosine (19.37) waveforms, both having a $2T$-length. The resultant RCTC PSW, together with that of raised-cosine waveform, has been shown in Figure (19.17).

It can be seen from Figure (19.17) that the edges of the RCTC pulse (the regions nearby either $t=0$ or $4T$) are flatter than those of the raised-cosine pulse, thus a lower sampling interference level is expected when timing jitter exists. To concentrate on BER for QORC and RCTC modulations with timing jitter, inter-symbol interference (ISI) due to limited bandwidth is not considered here. The transmitted signal can be written as

$$s(t) = I(t) \cos \omega t + Q(t) \sin \omega t \tag{19.78}$$

where

$$I(t) = A \sum_{n=-\infty}^{\infty} a_n g(t - 2nT) \tag{19.79}$$

$$Q(t) = A \sum_{n=-\infty}^{\infty} b_n g(t - 2nT - T) \tag{19.80}$$

are baseband signals in the I and Q channels, respectively, a_n and b_n take either 1 or -1, A is the signal amplitude and $g(t)$ is the pulse shaping waveform defined within $(0, 4T)$. After corruption by Gaussian noise $n(t) = n_c \cos \omega t - n_s \sin \omega t$ with zero mean and variance σ^2 and coherent demodulation, the recovered signal becomes $y_I(t) = I(t) + n_c$ in the I channel. Because the I and Q channels are symmetrical, we need only consider $y_I(t)$ at $(0, 4T)$ when the sampling time is $2T \pm \Delta t$ where Δt represents the timing jitter, or

$$y_I(2T \pm \Delta t) = I(2T \pm \Delta t) + n_c \tag{19.81}$$

where

$$I(2T \pm \Delta t) = A\left[a_{-1}g(4T \pm \Delta t) + a_0 g(2T \pm \Delta t) + a_1 g(\pm\Delta t)\right] \tag{19.82}$$

Without losing generality, we take the positive sign in (19.81), thus $g(4T + \Delta t) = 0$ and $g(\Delta t) \neq 0$. Therefore (19.81) can be rewritten as

$$y_I(2T + \Delta t) = A[a_0 g(2T + \Delta t) + a_1 g(\Delta t)] + n_c \tag{19.83}$$

It can be shown that the average power for a signal given by (19.53) is $P_{av} = 2A^2\mu$, where μ is the power of a single pulse $\mu = \frac{1}{4T}\int_0^{4T} g^2(t)dt$. The signal to noise ratio is $r = \frac{2A^2\mu}{\sigma^2}$. After calculating the statistical average over a_0 and a_1 in (19.83), the BER can be shown to be

$$
\begin{aligned}
P_e &= \frac{1}{4}erfc\left(\frac{1}{2}\sqrt{\frac{r}{u}}\,[g(2T+\Delta t)+g(\Delta t)]\right) \\
&+ \frac{1}{4}erfc\left(\frac{1}{2}\sqrt{\frac{r}{u}}\,[g(2T+\Delta t)-g(\Delta t)]\right)
\end{aligned}
\tag{19.84}
$$

where $erfc(x)$ is the complementary error function of x. The BER's for QORC and RCTC modulations with $\Delta t = 0.01T$, $0.25T$ and $0.5T$ are plotted in Figure (19.24). Figure (19.25) shows BER against timing jitter Δt for both modulations with r as a parameter. From Figure (19.24), it is seen that RCTC outperforms QORC in BER for different Δt with the SNR gain ranging from 1.5 dB for $\Delta t= 0.5T$ to 2.2 dB for $t=0.01T$, for $P_e = 10^{-6}$. Figure (19.25) shows four pairs of curves for $r=0$, 5, 10 and 15 dB, respectively. When r is large, the performance gap between the two modulations also becomes large. However, if Δt becomes very large, such as $0.5T$, the BER for different signals deteriorates rapidly and the BER gap tends to shrink.

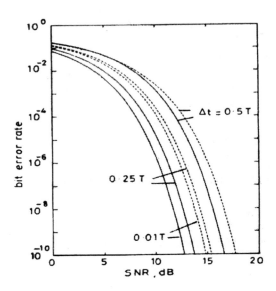

Figure 19.24. Bit error rate for QORC and RCTC modulations under sampling timing jitter Δt, where dotted lines are for QORC and solid lines are for RCTC.

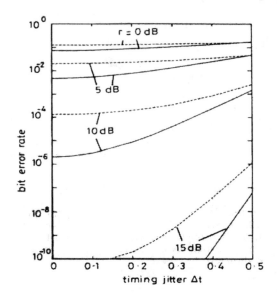

Figure 19.25. Bit error rate for QORC and RCTC modulations with respect to sampling timing jitter Δt, where dotted lines are for QORC and solid lines are for RCTC.

We can summarize from the above analysis about the RCTC modulation and its BER performance under timing jitter as follows. The proposed RCTC modulation performs much better than reported QORC modulation [17] in noisy channels with timing jitter at the receiver. The tolerance of RCTC modulation to timing jitter makes it a suitable modulation scheme for some undesirable working environments, where severe interference may be present to make bit synchronization difficult to track perfectly. It can also be shown that many other pulse-shaped modulations proposed in this chapter have also similar feature, being extremely robust against synchronization timing jitter.

7.   Summary

In this chapter we have addressed the issues on pulse-shaping waveform design for bandwidth efficient digital modulations. The chapter starts with a review of the history of symbol waveform design techniques under the context of digital modulations with a high bandwidth efficiency. The importance of the pulse shaping technology and its role in bandwidth efficient digital modems is discussed in section 1. Three major pulse shaping waveform design methodologies have been discussed in section 2, which include intuitive methods, piece-wise fitting methods and time-domain convolution (TDC) methods. The intuitive methods were widely used in the early research on digital modulations, such as QPSK, OQPSK, MSK, GMSK and QORC, etc. The method is primitive in nature but useful in improving the bandwidth efficiency of a digital modulation scheme. The piece-wise fitting methods employ segmented curves to fit a particular waveform as an effort to enhance the overall spectral characteristics of a digital modem. However, both intuitive methods and piece-wise fitting methods are non-systematic approaches for pulse shaping waveform design, all of which usually involve a very time-consuming try-and-error process to obtain a useful pulse shape for a particular modulation scheme. On the other hand, the TDC method offers several important advantages over the previous two methods in terms of its flexibility and performance. The method itself offers a great degree-of-freedom to control several important spectral parameters of obtained signal, such as the PSD side lobe reduction rate, the width of the PSD main lobe, the time-domain appearance of the pulse-shape, etc. Thus, it is a very promising pulse shaping waveform design method, which can be found very useful in many applications of digital modulation optimization as well as newly emerging UWB technology, where pulse shape design is also of significance in determining the bandwidth efficiency.

In section 3, we have introduced various merit parameters for a pulse-shaping waveform, such as envelope fluctuation ratio ρ, the average power of the PSW μ, the quarter sample point level $g(T)$ and etc., which are closely associated with its bandwidth efficiency in different applications.

Section 4 of the chapter has demonstrated how to analyze bandwidth efficiency of a digital modem under bandpass nonlinear channels, which can always be found in many practical applications, such as satellite systems and mobile cellular transceivers. We have taken the four new QO modulations as examples to show how effectively they could improve their bandwidth efficiency in bandpass nonlinear channels, if compared with conventional modems, such as OQPSK, MSK and QORC, etc. Finally, we have also addressed the issues on robustness against sampling timing jitter for pulse-shaped digital modems in section 5.

Throughout this chapter very much attention has been paid to quadrature-overlapped (QO) modulations due to their great advantages in improving the overall bandwidth-efficiency of a digital modem, which has become an increasingly important issue recently as wireless applications prevail around the world. Several new QO modulation schemes have been proposed in this chapter based on the new pulse-shaping waveform design method, namely the TDC method, which can be used to replace many traditional digital modulation schemes for their excellent bandwidth efficiency and other attractive characteristics.

Acknowledgments

The author wishes to thank the Editor of this book, Prof. Mohsen Guizani, for his encouragement and support in writing this chapter. He would like also to thank Mr. Yu-Hsin Lin, Mr. Chien-Yao Chao and Mr. Zhe-Min Lin, Institute of Communications Engineering, National Sun Yat-Sen University, Taiwan, for their help in preparing some figures and coding programs for some analysis presented in this chapter. The author would like also to thank National Science Council (NSC), Taiwan, for the following research grants, NSC 89-2213-E-005-032, NSC 89-2213-E-005-029, NSC 89-2213-E-005-032, NSC 90-2213-E-110-054, NSC 91-2213-E-110-032, NSC 92-2213-E-110-015, NSC 92-2213-E-110-047, NSC 92-2213-E-110-048 and NSC 93-2213-E-110-012, which have partially supported the related research work.

References

[1] Baker, P. A., (1962) Phase Modulation Data Sets for Serial Transmission at 2000 and 2400 bits per second, Part I, *AIEE Trans on Communications Electronics*, July, 1962.

[2] Bussgang, J. & Leiter, M., (1964) Error rate approximations for differential phase-shift keying, *IEEE Trans on Communications Systems*, vol. CS-12, pp. 18-27, Mar 1964.

[3] Bussgang, J. & Leiter, (1965) Error performance of a quarature phase shift keying system, The MITRE Corp., Bedford, Mass., *MITRE Tech. Memo.* 04236, May 14, 1965.

[4] Bussgang, J. & Leiter, (1966) Phase-shift keying with a transmitted reference, *IEEE Trans. Communication Technology*, vol. COM-14, pp. 14-22, February 1966.

[5] Bussgang, J. & Leiter, M., (1968) Error Performance of Quadrature Pilot Tone Phase-Shift Keying, *IEEE Transactions on Communications*, pp.526-529, vol: 16, Issue: 4, Aug 1968.

[6] R. DeBuda, (1972) Coherent demodulation of frequency-shift keying with low deviation ratio, *IEEE Transactions on Communications*. vol. COM-20, pp. 429-435, June 1972.

[7] S. A. Rhodes, (1972) Effects of hard limiting on bandlimited transmissions with conventional and offset QPSK modulation, *Conf. Rec. of 1972 IEEE National Telecommunications Conf.* 1972, CHO601-5-NTC, pp. 2OF-1, Dec. 1972.

[8] Mathwich, H., Balcewicz, J. & Hecht, M., (1974) The Effect of Tandem Band and Amplitude Limiting on the Eb/No Performance of Minimum (Frequency) Shift Keying (MSK), *IEEE Transactions on Communications*, pp.1525-1540, Vol: 22, Issue: 10, Oct 1974.

[9] Gronemeyer, S. & McBride, A. (1976) MSK and Offset QPSK Modulation, *IEEE Transactions on Communications*, pp.809-820, Vol: 24, Issue: 8, Aug 1976.

[10] Vasant K. Prabhu, (1977) PSK-type modulation with overlapped baseband pulses, *IEEE Transactions on Communications*, vol. COM-25, No. 9, pp. 980-990, Sept. 1977.

[11] Oetting, J., (1979) A Comparison of Modulation Techniques for Digital Radio, *IEEE Transactions on Communications*, page(s): 1752-1762, Vol: 27, Issue: 12, Dec 1979.

[12] Morais, D. & Feher, K., (1979) Bandwidth Efficiency and Probability of Error Performance of MSK and Offset QPSK Systems, *IEEE Transactions on Communications*, pp. 1794-1801, Vol: 27, Issue: 12, Dec 1979.

[13] F. Amoroso, (1979) The use of quasi-bandlimited pulses in MSK transmission, *IEEE Transactions on Communications*, vol. COM-27, No. 10, pp. 1616-1624, Oct. 1979.

[14] Korn, I., (1980) Generalized minimum shift keying, *IEEE Transactions on Information Theory*, pp. 234-238, Vol: 26, Issue: 2, Mar 1980.

[15] D. H. Morais & K. Feher, (1980) The effects of filtering and limiting on the performance of QPSK, offset QPSK, and MSK systems, *IEEE Transactions on Communications*, vol. COM-28, No. 12, pp. 1999-2009, Dec. 1980.

[16] Murota, K. & Hirade, K., (1981) GMSK Modulation for Digital Mobile Radio Telephony, *IEEE Transactions on Communications*, pp. 1044-1050, Vol: 29, Issue: 7, Jul 1981.

[17] Mark C. Austin & Ming U. Chang, (1981) Quadrature overlapped raised-cosine modulation, *IEEE Transactions on Communications*, vol. COM-29, No. 3, pp. 237-249, March 1981.

[18] Milstein, L., Pickholtz, R. & Schilling, D., (1982) Comparison of Performance of Digital Modulation Techniques in the Presence of Adjacent Channel Interference, *IEEE Transactions on Communications*, pp. 1984-1993, Vol: 30, Issue: 8, Aug 1982.

[19] D. Divsalar & M. K. Simon, (1982) The power spectral density of digital modulations transmitted over nonlinear channels, *IEEE Transactions on Communications*, vol. COM-30, No. 1, pp. 142-151, Jan. 1982.

[20] A. Yongacoglu & K. Feher, (1985) DCTPSK: An efficient modulation technique for differential detection, *Proceedings of the IEEE ICC*, June, 1985.

[21] Ali, A.A. & Al-Kadi, I. (1988) A comparison of multipath effect of digital modulations for mobile channels,1988. *Conference Proceedings on Area Communication*, EUROCON 88., 8th European Conference on Electrotechnics, Stockholm, Sweden, pp. 291-294, 13-17 Jun 1988.

[22] Kim, K. & Polydoros, A., (1988) Digital modulation classification: the BPSK versus QPSK case, 1988. *MILCOM 88, Conference record*, pp. 431-436, vol.223-26, Oct 1988.

[23] Cohen, D.J., (1988) Necessary bandwidth of digital modulation, *Symposium Record of IEEE 1988 International Symposium on Electromagnetic Compatibility*, page(s): 247-251, Seattle, USA, 2-4 Aug 1988.

[24] D. S-Dias & K. Feher, (1991) Baseband pulse shaping techniques for $\pi/4$-DQPSK in non-linearly amplified land mobile channels, *Proceedings of INFOCOM 1991*, pp. 759-764, 1991.

[25] S. B. Slimane & T. Le-Ngoc, (1991) Performance of quadrature pulse-overlapping modulated signals in Rayleigh fading channels, *Record of GLOBECOM 1991*, pp. 51.3.1-51.3.5, 1991.

[26] M. E. Fox & M. W. Marcellin, (1991) Shaped BPSK and the 5 KHz UHF SatCom channels, *Record of MILCOM 1991*, pp. 15.2.1-15.2.7.

[27] H. H. Chen & J. Oksman, (1992) MSKRC quadrature-overlapping modulation and its performance analysis, *Record of the 3rd international symposium on personal, indoor and mobile radio communications* (PIMRC 1992), pp. 605-609, Boston, USA, October 19-21, 1992.

[28] S. B. Slimane & T. Le-Ngoc, (1992) Maximum Likelihood sequence estimation of quadrature pulse-overlapping modulated signals for portable/mobile satellite communications, *IEEE Journal on Selected Areas in Communications*, Vol. 10, No. 8, October 1992, pp. 1278-1288.

[29] H. H. Chen, (1993) Quadrature-overlapped convoluted raised-cosine (QOCRC) modulation and its BER analysis over nonlinear satellite channels, *Record of 1993 international symposium on communications* (ISCOM 1993), National Chiao Tung University, Hsinchu, Taiwan, Vol. 2, pp. 21-28/35, 7-10 December, 1993.

[30] H. H. Chen, (1993) A novel quadrature-overlapped modulation scheme with immunity to timing jitters at coherent receiver, *Record of 1993 international symposium on personal communications* (ISPC 1993), Nanjing, China, pp. C4. 1 - C4. 6, October 26-28, 1993.

[31] H. H. Chen & FVC Mendis, (1993) A new bandwidth-efficient digital modulation scheme-MSKRC modulation and its performance analysis in band limited nonlinear channels, *Record of the 5th international symposium on IC technology, system & applications* (ISIC-93), Nanyang Technological University, Singapore, pp. 500-504, 15-17 September 1993.

[32] H. H. Chen & J. Oksman, (1993) Performance of quadrature-overlapped convoluted triangular-cosine (QOCTC) modulation over nonlinear land mobile satellite channels, *Record of the 4th international symposium on personal, indoor and mobile radio communications* (PIMRC 1993), YoKohama, Japan, pp. 444-447, September 9-11, 1993.

[33] H. H. Chen, (1993) Convoluted raised-cosine and triangular-cosine (RCTC) modulation and its performance against timing jitters over noisy channels, *IEE Electronics Letters*, Vol. 29, No. 14, pp. 1239-1240, 8th July 1993.

[34] Klovsky, D., Periyalwar, S.S. & Fleisher, S. (1993) Comments on multi-carrier and single-carrier digital modulation in a multipath radio channel, *Canadian Conference on Electrical and Computer Engineering*, Vancouver, Canada, pp. 381-384, vol.1, 14-17 Sep 1993.

[35] H. H. Chen & J. Oksman, (1994) A robust quadrature-overlapped modulation scheme & its performance over nonlinear band-limited channels, *Record of GLOBECOM 1994*, San Francisco, US, pp. 431-436, Vol. II, November 27-December 1, 1994.

[36] H. H. Chen & W. H. Ong, (1994) Performance analysis of quadrature overlapping modulations in non-linear fading channels, *Proceedings of ICCS 1994*, pp. 691-695, Vol. II, Singapore, November 14-18, 1994.

[37] H. H. Chen & J. Oksman, (1996) A Quasi-constant envelope quadrature overlapped modulation & its performance over nonlinear bandlimited satellite channels, *International Journal of Satellite Communications*, John Wiley & Sons, Vol. 14, pp. 351-359, 1996.

[38] H. H. Chen, (1996) Quadrature overlapping modulation system, *US Patent, No. US5509033*, Patent granted date: April 1996.

[39] H. H. Chen & S. Y. (1997) Wong, Four novel quadrature overlapping modulations and their spectral efficiency analysis over band-limited non-linear channels, *International Journal of Satellite Communications*, John Wiley & Sons, vol. 15, pp. 117-127, 1997.

[40] H. H. Chen & X. D. Cai, (1997) Optimization of transmitter and receiver filters for the OQAM-OFDM systems by using nonlinear programming algorithms, *IEICE Trans. on Communications*, vol. E80-B, No. 11, November 1997.

[41] H. H. Chen & S. Y. Wong, (1997) Spectral efficiency analysis of new quadrature overlapped modulations over band-limited non-linear channels, *International Journal of Communication Systems*, John Wiley & Sons, vol. 10, 1-11, pp.1-11, 1997.

[42] H. H. Chen & X. D. Cai, (1997) Waveform optimization for OQAM-OFDM systems by using nonlinear programming algorithm, *Proceedings of VTC*, 1997, Phoenix, USA, June 1997.

[43] Parr, B.; ByungLok Cho; Wallace, K.; Zhi Ding; (2003) A novel ultra-wideband pulse design algorithm, Communications Letters, IEEE, Volume: 7 Issue: 5, Page(s): 219 -221, May 2003.

[44] Cheng Yonghong; Chen Xiaolin; Jiang Yan; Lu Liro; Xie Hengkun; (2001) A fingerprint analyzing method of ultra-wideband partial discharge time-domain waveform, Electrical Insulating Materials, 2001. (ISEIM 2001). Proceedings of 2001 International Symposium on, Page(s): 107-110, 19-22 Nov. 2001.

[45] Jeongwoo Han; Cam Nguyen; (2002) A new ultra-wideband, ultra-short monocycle pulse generator with reduced ringing, Microwave and Wireless Components Letters, IEEE [see also IEEE Microwave and Guided Wave Letters], Volume: 12 Issue: 6, Page(s): 206 -208, June 2002.

[46] Yue Yansheng; Long Teng; (2000) An ultra-wide band orthogonal digital signal generator, Signal Processing Proceedings, 2000. WCCC-ICSP 2000. 5th International Conference on, Volume: 3, Page(s): 2086-2090 vol.3, 21-25 Aug. 2000.

[47] Xiaojing Huang; Yunxin Li; (2001) Generating near-white ultra-wideband signals with period extended PN sequences, Vehicular Technology Conference, 2001. VTC 2001 Spring. IEEE VTS 53rd, Volume: 2, Page(s): 1184 -1188 vol.2, 6-9 May 2001.

[48] Mesyats, G.A.; Rukin, S.N.; Shpak, V.G.; Yalandin, M.I.; (1998) Generation of high-power subnanosecond pulses, Ultra-Wideband Short-Pulse Electromagnetics 4, 1998, Page(s): 1-9, 14-19 June 1998.

[49] Smithson, G., (1998) Introduction to digital modulation schemes, *IEE Colloquium on The Design of Digital Cellular Handsets*, London , UK, page(s): 2.1-2.9, 4 Mar 1998.

[50] Hill, F. & Won Lee, (1974) PAM Pulse Generation Using Binary Transversal Filters, *IEEE Transactions on Communications*, page(s): 904-913, Vol: 22, Issue: 7, Jul 1974.

[51] Civanlar, M.R. & Nobakht, R.A. (1988) Optimal pulse shape design using projections onto convex sets, *1988 International Conference on Acoustics, Speech, and Signal Processing*, page(s): 1874-1877 vol.3, New York, USA, 11-14 Apr 1988.

[52] Koch, D.B., (1990) Analysis of a generalized coding/decoding method using FIR digital filters for radar waveform design, *Twenty-Second Southeastern Symposium on System Theory*, page(s): 652-656, Cookeville,USA, 11-13 Mar 1990.

[53] Dzung, D. & Braun, W.R. (1990) Generalized CPM and DPM: digital angle modulation schemes with improved bandwidth efficiency, *IEEE Transactions on Communications*, page(s): 1971-1979, Vol: 38, Issue: 11, Nov 1990.

[54] Sollenberger, N.R., (1990) Pulse design and efficient generation circuits for linear TDMA modulation, *1990 IEEE 40th Vehicular Technology Conference*, page(s): 616-621, Orlando, USA, 6-9 May 1990.

[55] Salmasi, A. & Gilhousen, K.S., (1991) On the system design aspects of code division multiple access (CDMA) applied to digital cellular and personal communications networks, *41st IEEE Vehicular Technology Conference*, 1991, 'Gateway to the Future Technology in Motion', page(s): 57-62, St. Louis, USA, 19-22 May 1991.

[56] Madhusudhana Rao, G., Reddy, K.V.V.S. & Reddy, E.M.D.E.A. (1993) Waveform design of FH-BFSK communication system useful for mobile telephone subscribers, *Proceedings of the IEEE 1993 National Aerospace and Electronics Conference*, Dayton, USA, page(s): 447-453 vol.1, 24-28 May 1993.

[57] Stott, G.F., (1994) Digital modulation for radar jamming, *IEE Colloquium on Signal Processing in Electronic Warfare*, London , page(s): 1/1-1/6, UK, 31 Jan 1994.

[58] Tortoli, P. Guidi, F. & Atzeni, C. (1994) Digital vs. SAW matched filter implementation for radar pulse compression, *Proceedings of 1994 IEEE Ultrasonics Symposium*, Cannes, France, 1-4 Nov 1994.

[59] Schock, S.G., Leblanc, L.R. & Panda, S. (1994) Spatial and temporal pulse design considerations for a marine sediment classification sonar, *IEEE Journal of Oceanic Engineering*, page(s): 406-415, Vol: 19, Issue: 3, Jul 1994.

[60] Nikookar, H. & Prasad, R. (1997) Optimal waveform design for multicarrier transmission through a multipath channel, *1997 IEEE 47th Vehicular Technology Conference*, page(s): 1812-1816, vol.3, Phoenix, USA, 4-7 May 1997.

[61] Chan, Y.T., Plews, J.W. & Ho, K.C. (1997) Symbol rate estimation by the wavelet transform, *Proceedings of 1997 IEEE International Symposium on Circuits and Systems*, Hong Kong, page(s): 177-180 vol.1, 9-12 Jun 1997.

[62] Lallo, P.R.U., (1999) Signal classification by discrete Fourier transform, *1999 IEEE Military Communications Conference Proceedings*, page(s): 197-201 vol.1, Atlantic City, USA, Oct 31 - Nov 3, 1999.

[63] Cvetkovic, Z., (1999) Modulating waveforms for OFDM, *Proceedings of 1999 IEEE International Conference on Acoustics, Speech, and Signal Processing*, page(s): 2463-2466 vol.5, Phoenix, USA, Mar 15-19 1999.

[64] Wern-Ho Sheen, Chun-Chieh Tseng & Chia-Shu Wang, (2000) On the diversity, bandwidth, and performance of digital transmission over frequency-selective slow fading channels, *IEEE Transactions on Vehicular Technology*, page(s): 835-843, Vol: 49, Issue: 3, May 2000.

[65] Xinjun Zhang, Xin Jiang, Wentao Song & Hanwen Luo (2002) A novel direct waveform synthesis technique with carrier frequency programmable, *2002 IEEE Wireless Communications and Networking Conference*, page(s): 150- 154 vol.1, 17-21 Mar 2002.

Chapter 20

MULTIPLE ANTENNAS

Ezio Biglieri
Politecnico di Torino, Italy
biglieri@polito.it

Giorgio Taricco
Politecnico di Torino, Italy
taricco@polito.it

Abstract   In this chapter we investigate transmission systems where more than one antenna is used at both ends of the radio link. The use of multiple transmit and receive antennas allows one to reach capacities that cannot be obtained with any other technique using present-day technology. After computing these capacities, we show how "space–time" codes can be designed, and how suboptimum architectures can be employed to simplify the receiver.

Keywords: Multiple antennas, Fading channels, Space–time codes

1.   Preliminaries

The use of multiple antennas for transmission and reception is a major technological advance that is expected to make it possible to increase the data rate in wireless networks by orders of magnitude. While the use of more than one antenna at the receiver side is quite common, only recently was a communication theory of multiple transmit and receive antenna developed. This chapter focuses on the main aspects of this theory.

Here we consider transmission with t antennas and reception with r antennas (Fig. 20.1). Assuming two-dimensional (i.e., complex) elementary constellations throughout, the channel model becomes

$$y = Hx + z \qquad (20.1)$$

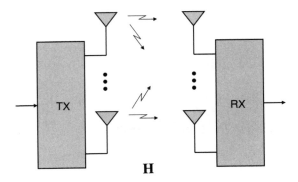

Figure 20.1. Transmission and reception with multiple antennas. The channel gains are described by the $r \times t$ matrix $\mathbf{H}$.

where $\mathbf{x} \in \mathbb{C}^t$, $\mathbf{y} \in \mathbb{C}^r$, $\mathbf{H} \in \mathbb{C}^{r \times t}$ (i.e., $\mathbf{H}$ is an $r \times t$ complex matrix whose entries h_{ij} describe the gains of each transmission path to a receive from a transmit antenna), and $\mathbf{z}$ is circularly-symmetric complex Gaussian noise. We also assume that

$$\mathbb{E}[\mathbf{z}\mathbf{z}^\dagger] = N_0\mathbf{I}_r \qquad (20.2)$$

that is, the noise components affecting the different receivers are independent with variance N_0, and the signal energy is constrained by $\mathbb{E}[\mathbf{x}^\dagger\mathbf{x}] = tE_s$. Assuming that $\mathbb{E}[|h_{i,j}|^2] = 1$ for all i, j, the average signal-to-noise ratio (SNR) at receiver is

$$\rho \triangleq \frac{tE_s}{N_0}$$

where we assume that the symbol rate equals the signal bandwidth at the Nyquist signalling rate.

The ith component x_i, $i = 1, \ldots, t$, of vector $\mathbf{x}$ is the signal transmitted from antenna i; the jth component y_j, $j = 1, \ldots, r$, of vector $\mathbf{y}$ is the signal received by antenna j. Explicitly, we have from (20.1):

$$y_j = \sum_{i=1}^{t} h_{ij}x_i + z_j, \qquad j = 1, \ldots, r \qquad (20.3)$$

which shows how every component of the received signal includes a linear combination of the signals emitted by each antenna. We say that $\mathbf{y}$ is affected by *spatial interference* from the signals transmitted by the various antennas. This interference has to be removed, or controlled in some way, in order to sort out the transmitted signals. We shall see in the following how this can be done: for the moment we may just observe that the problem of removing interference

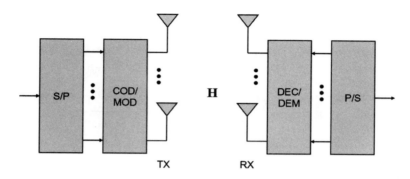

Figure 20.2. Spatial multiplexing through transmission and reception with multiple antennas.

makes the tools for the analysis of multiple-antenna transmission have much in common with those used in the study of digital equalization (which deals with intersymbol interference) and of multiuser detection (which deals with multiple-access interference).

The upsides of using multiple antennas can be summarized by defining two types of gain. As we shall see in the following, in the presence of fading a multiplicity of transmit antennas creates a set of parallel channels, that can be used to potentially increase the data rate up to a factor of $\min\{t, r\}$ (with respect to single-antenna transmission) and hence generate a *rate gain*. This corresponds to the *spatial multiplexing* depicted in Fig. 20.2. Here the serial-to-parallel S/P converter distributes the stream of data across the transmit antennas; after reception, the original stream is reconstituted by the parallel-to-serial converter P/S. The other gain, called *diversity gain*, is due to the combination of received signals that are independently faded replicas of a single transmitted signal, which allows a more reliable reception. We hasten to observe here that these two gains are not independent, but there is a fundamental tradeoff between the two [ZheTse03]. Actually, it can be said that the problem of designing a multiple-antenna systems is based on this tradeoff. As an example, Fig. 20.3 illustrates the diversity–rate tradeoff for a MIMO system with $t = 2$ transmit and $r = 2$ receive antennas. Fig. (a) assumes the channels are orthogonal so that the rate is maximum (twice as much as each channel rate) but there is no diversity gain. Fig. (b) assumes the transmitter replicates the same signal over the two channels so that there is no rate gain but the diversity increases to four.

The problems we address in this chapter are the following:

- What is the limiting performance (channel capacity) of this multiple-antenna system?

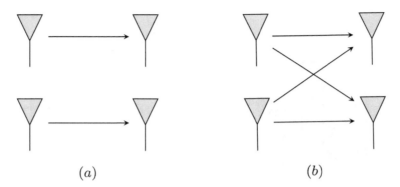

(a) (b)

Figure 20.3.   Spatial multiplexing through transmission and reception with multiple antennas.

- How can we design codes matched to the channel structure (*space–time codes*)?

- How can we design architectures allowing simple decoding of space–time codes, and what is their performance?

2.   Channel capacity

In this section we evaluate the capacity of a multiple-antenna, or MIMO (multiple-input, multiple-output) transmission system as described by (20.1). Unless otherwise specified, capacity will be always measured in bit/dimension pair, corresponding to bit/s/Hz at the Nyquist signalling rate. Several models for the matrix $\mathbf{H}$ can be considered:

(a) $\mathbf{H}$ is deterministic.

(b) $\mathbf{H}$ is a random matrix with a given joint probability distribution of its entries, and each channel use (viz., the transmission of one symbol from each of the t transmit antennas) corresponds to an *independent* realization of $\mathbf{H}$ (ergodic channel).

(c) $\mathbf{H}$ is a random matrix, but once it is chosen it remains fixed for the whole transmission.

When $\mathbf{H}$ is random (cases (b) and (c) above) we assume here that its entries are $\sim \mathcal{N}_c(0, 1)$, i.e., independent, identically distributed (iid) Gaussian with zero-mean, independent real and imaginary parts, each with variance $1/2$. Equivalently, each entry of $\mathbf{H}$ has uniform phase and Rayleigh magnitude. This choice models frequency-flat Rayleigh fading with enough separation within antennas such that the fades for each TX/RX antenna pair are independent. We also assume, for most of our treatment, that the channel state information (CSI, that is, the realization of **H**) is known at the receiver, while the distribution of **H** is known at the transmitter (the latter assumption is necessary for capacity computations, since the transmitter must choose an optimum code for that specific channel).

Model (c) is applicable to an indoor wireless data network or a personal communication system with mobile terminals moving at walking speed, so that the channel gain, albeit random, varies so slowly with time that it can be assumed as constant along transmission of a long block of data (see [ChuTseKah98,Foschini96,FosGan98,RalCio98,TarSesCal98,Telatar99]). More generally, fading blocks can be thought of as separated in time (e.g., in a time-division system [OzaShaWyn94]), as separated in frequency (e.g., in a multicarrier system), or as separated both in time and in frequency (e.g., with slow time-frequency hopping [CaiKnoHum98, Knopp97, KnoHum00]). This model, the block-fading (BF) channel [BigProSha98], allows the incorporation of delay constraints, which imply that, even though very long code words are transmitted, perfect (i.e., infinite-depth) interleaving cannot be achieved. With this fading model, the channel may or may not be ergodic (ergodicity is lost when there is a delay constraint). Without ergodicity, the average mutual information over the ensemble of channel realizations cannot characterize the achievable transmission rates. Since the instantaneous mutual information of the channel turns out to be a random variable, the quantity that characterizes the quality of a channel when coding is used is not channel capacity, but rather the *information outage probability*, i.e., the probability that this mutual information is lower than the rate of the code used for transmission [OzaShaWyn94]. This outage probability is closely related to the code word error probability, as averaged over the random coding ensemble and over all channel realizations; hence, it provides useful insight on the performance of a delay-limited coded system [CaiKnoHum98, Knopp97, KnoHum00, MalLei97, MalLei99]. An additional important definition related to outage probability is that of *zero-outage capacity*, sometimes also referred to as *delay-limited capacity*. This is the maximum rate for which the minimum outage probability is zero under a given power constraint [BigCaiTar01, CaiTarBig98, CaiTarBig99, TseHanly98].

2.1 Deterministic channel

Capacity is the maximum of the average mutual information $I(\mathbf{x}; \mathbf{y})$ between input and output of the channel over the choice of the distribution of **x**.

Singular-value decomposition (SVD) of matrix **H** yields

$$\mathbf{H} = \mathbf{U}\mathbf{D}\mathbf{V}^{\dagger} \qquad (20.4)$$

where $\mathbf{U} \in \mathbb{C}^{r \times r}$ and $\mathbf{V} \in \mathbb{C}^{t \times t}$ are unitary, and $\mathbf{D} \in \mathbb{R}^{r \times t}$ is diagonal. We can write

$$\mathbf{y} = \mathbf{U} \mathbf{D} \mathbf{V}^\dagger \mathbf{x} + \mathbf{z} \tag{20.5}$$

Premultiplication of (20.5) by $\mathbf{U}^\dagger$ shows that the original channel is equivalent to the channel described by the input-output relationship

$$\tilde{\mathbf{y}} = \mathbf{D} \tilde{\mathbf{x}} + \tilde{\mathbf{z}} \tag{20.6}$$

where $\tilde{\mathbf{y}} \triangleq \mathbf{U}^\dagger \mathbf{y}$, $\tilde{\mathbf{x}} \triangleq \mathbf{V}^\dagger \mathbf{x}$ (so that $\mathbb{E}[\tilde{\mathbf{x}}^\dagger \tilde{\mathbf{x}}] = \mathbb{E}[\mathbf{x}^\dagger \mathbf{x}]$), and $\tilde{\mathbf{z}} \triangleq \mathbf{U}^\dagger \mathbf{z} \sim \mathcal{N}_c(0, N_0 \mathbf{I}_r)$. Now, the rank of $\mathbf{H}$ is at most $m \triangleq \min\{t, r\}$, and hence at most m of its singular values are nonzero. Denote these by $\sqrt{\lambda_i}$, $i = 1, \ldots, m$, and rewrite (20.6) componentwise in the form

$$\tilde{y}_i = \sqrt{\lambda_i} \tilde{x}_i + \tilde{z}_i, \qquad i = 1, \ldots, m \tag{20.7}$$

which corresponds to a set of m independent parallel additive Gaussian noise channels. The remaining components of $\tilde{\mathbf{y}}$ (if any) are equal to the corresponding components of the noise vector $\tilde{\mathbf{z}}$: we see that, for $i > m$, $\tilde{y}_i$ is *independent of the transmitted signal*, and $\tilde{x}_i$ does not play any role.

Defining ρ_i the SNR constraint for the ith channel, the overall channel capacity is

$$C = \sum_{i=1}^{m} \log(1 + \lambda_i \rho_i) \tag{20.8}$$

Using the constraint $\sum_i \rho_i \leq \rho$ we get the "water-filling" solution [CovTho91, Gal68]

$$\rho_i(\mu) = (\mu - \lambda_i^{-1})_+ \tag{20.9}$$

where μ is obtaied by solving the equation

$$\rho = \sum_{i=1}^{m} (\mu - \lambda_i^{-1})_+ \tag{20.10}$$

The capacity can then be written as

$$C = \sum_{i=1}^{m} (\log(\mu \lambda_i))_+ \tag{20.11}$$

and is attained by choosing $\tilde{x}_i \sim \mathcal{N}_c(0, N_0(\mu - \lambda_i^{-1})_+)$ for $i = 1, \ldots, m$, i.e., iid circularly-symmetric zero-mean complex Gaussian random variables with variances $N_0(\mu - \lambda_i^{-1})_+$.

Example 1. Consider as $\mathbf{H}$ the all-1 matrix. The SVD of $\mathbf{H}$ yields

$$\mathbf{H} = \mathbf{UDV}^\dagger = \begin{bmatrix} 1 & \cdots & 1 \\ \vdots & \ddots & \vdots \\ 1 & \cdots & 1 \end{bmatrix} = \frac{1}{\sqrt{r}} \begin{bmatrix} 1 \\ \vdots \\ 1 \end{bmatrix} [\sqrt{rt}] \frac{1}{\sqrt{t}} [1, \ldots, 1]$$

Since there is only one singular value we have the trivial solution $\rho_1 = \rho$ yielding the channel capacity

$$C = \log(1 + rt\rho) \qquad (20.12)$$

The signals achieving this capacity can be described as follows. Since $\tilde{x}_1 \sim \mathcal{N}_c(0, \rho N_0)$, the vector $\mathbf{x} = \mathbf{V}\tilde{x}_1$ has all equal components $\sim \mathcal{N}_c(0, \rho N_0/t)$. Thus, the TX antennas all send the same signal with power $\rho N_0/t$. Because of the structure of $\mathbf{H}$ the signals *add coherently* at the receiver, so that at each receiver we have the voltage $t\sqrt{\rho N_0/t}$, and hence the power $t^2 \times \rho N_0/t = \rho N_0 t$. Since each receiver sees the same signal, and the noises are uncorrelated, the overall SNR is $rt\rho$, as shown by the capacity formula (20.12). □

Example 2. Take $t = r = m$, and $\mathbf{H} = \mathbf{I}_m$. Due to the structure of $\mathbf{H}$, there is no spatial interference here, and transmission occurs over m parallel AWGN channels, each having SNR ρ/m and hence capacity $\log(1 + \rho/m)$. Thus,

$$C = m \log(1 + \rho/m) \qquad (20.13)$$

Notice how the capacity here increases faster with the number of antennas than in Example 1: this is due to the absence of spatial interference. Notice also that as m grows to infinity the capacity converges to the finite limit $\rho \log e$. □

2.2 Ergodic Rayleigh fading channel

We assume here that $\mathbf{H}$ is independent of both $\mathbf{x}$ and $\mathbf{z}$, with entries $\sim \mathcal{N}_c(0, 1)$. We also assume that for each channel use an *independent* realization of $\mathbf{H}$ is drawn, so that the channel is memoryless. If the receiver has perfect knowledge of the realization of $\mathbf{H}$ (the "channel state information," or CSI), the channel capacity is achieved by a transmitted signal $\mathbf{x} \sim \mathcal{N}_c(0, (\rho/t)\mathbf{I}_t)$, and takes value [Telatar99]

$$C = \mathbb{E}\left[\log\det\left(\mathbf{I}_r + \frac{\rho}{t}\mathbf{HH}^\dagger\right)\right] \qquad (20.14)$$

Note that for fixed r and as $t \to \infty$, the strong law of large numbers yields

$$\frac{1}{t}\mathbf{HH}^\dagger \to \mathbf{I}_r \qquad \text{a.s.} \qquad (20.15)$$

Thus, as $t \to \infty$ the capacity tends almost surely to

$$\log \det (\mathbf{I}_r + \rho \mathbf{I}_r) = r \log(1 + \rho) \tag{20.16}$$

so that the capacity increases *linearly* with r. This result should be compared with (20.12), where C increases only *logarithmically* with r.

The expectation in (20.14) can be computed explicitly, yielding [Telatar99]

$$C = \int_0^\infty \log \left(1 + \frac{\rho}{t}\lambda\right) \sum_{k=0}^{m-1} \frac{k!}{(k-n+m)!} \left[L_k^{n-m}(\lambda)\right]^2 \lambda^{n-m} e^{-\lambda} \, d\lambda \tag{20.17}$$

where $m \triangleq \min\{t, r\}$, $n \triangleq \max\{t, r\}$, and $L_k^\ell(\lambda)$ is the associated Laguerre polynomial defined as [Szego39]:

$$L_k^\ell(\lambda) \triangleq \frac{1}{k!} e^x x^{-\ell} \frac{d^k}{dx^k} [e^{-x} x^{k+\ell}] \tag{20.18}$$

Capacity values for $\rho = 20$ dB are plotted in Fig. 20.4 and 20.5. Some special cases as well as asymptotic aproximations to the values of C are examined in the examples that follow.

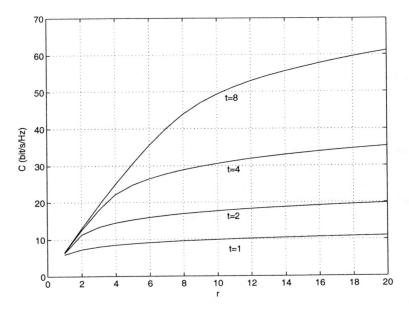

Figure 20.4. Capacity of the ergodic Rayleigh MIMO channel with $\rho = 20$ dB.

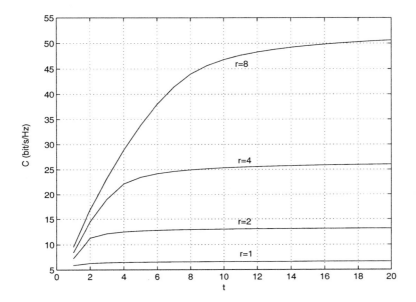

Figure 20.5. Capacity of the ergodic Rayleigh MIMO channel with $\rho = 20$ dB.

Example 3 $(t = 1, r \to \infty)$. Consider first the case $t = 1$, so that $m = 1$ and $n = r$. Since by definition $L_0^{n-m}(\cdot) = 1$, application of (20.17) yields

$$C = \frac{1}{(r-1)!} \int_0^\infty \log(1 + \rho\lambda)\lambda^{r-1}e^{-\lambda} d\lambda$$

This is plotted in Fig. 20.6. An asymptotic expression of C as $r \to \infty$ is $C \sim \log(1 + \rho r)$. This approximation to the capacity is also plotted in Fig. 20.6. We see here that if $t = 1$ the capacity increases only logarithmically as the number of receive antennas is increased—a quite inefficient way of boosting capacity. □

Example 4 $(r = 1, t \to \infty)$. Consider first $r = 1$, so that $m = 1$ and $n = t$. Application of (20.17) yields

$$C = \frac{1}{(t-1)!} \int_0^\infty \log\left(1 + \frac{\rho}{t}\lambda\right) \lambda^{t-1}e^{-\lambda} d\lambda$$

This is plotted in Fig. 20.7. An asymptotic expression of C as $t \to \infty$ is $C \sim \log(1 + \rho)$. This approximation to the capacity is also plotted in Fig. 20.7. □

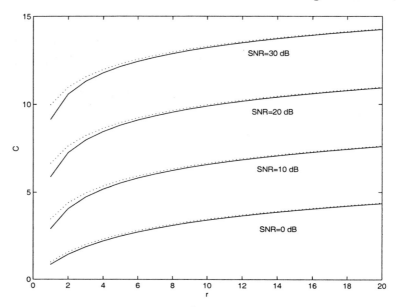

Figure 20.6.   Capacity of the ergodic Rayleigh MIMO channel with $t = 1$ (continuous line). The asymptotic approximation $C \sim \log(1 + \rho r)$ is also shown (dotted line).

Example 5 $(r = t)$.   With $r = t$ we have $m = n = r$, so that application of (20.17) yields

$$C = \int_0^\infty \log\left(1 + \frac{\rho}{r}\lambda\right) \sum_{k=0}^{r-1} [L_k(\lambda)]^2 e^{-\lambda} \, d\lambda$$

where $L_k \triangleq L_k^0$ is a Laguerre polynomial of order k (see eq. (20.18)). The capacity is plotted in Fig. 20.8.   □

A general asymptotic approximation to (20.14) can be derived by using results from random-matrix theory [Sil95]. We have

$$\frac{C}{m} \to \left(\log(w_+\rho) + (1 - \alpha)\log(1 - w_-) - \frac{w_-\alpha}{\ln 2}\right) \max(1, 1/\alpha) \quad (20.19)$$

where

$$w_\pm \triangleq (w \pm \sqrt{w^2 - 4/\alpha})/2 \quad (20.20)$$

and

$$w \triangleq 1 + \frac{1}{\alpha} + \frac{1}{\rho} \quad (20.21)$$

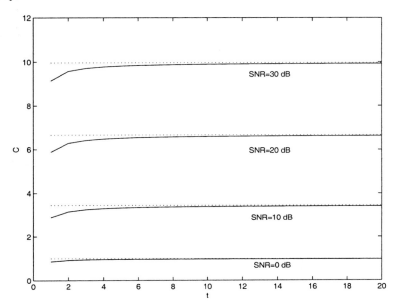

Figure 20.7. Capacity of the ergodic Rayleigh MIMO channel with $r = 1$ (continuous line). The asymptotic approximation $C \sim \log(1 + \rho)$ is also shown (dotted line).

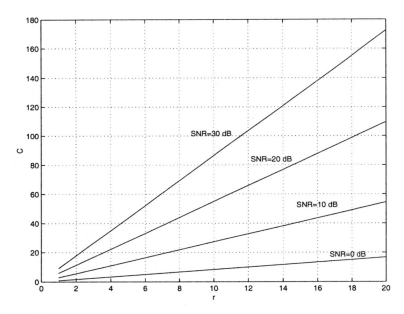

Figure 20.8. Capacity with independent Rayleigh fading and $r = t$ antennas.

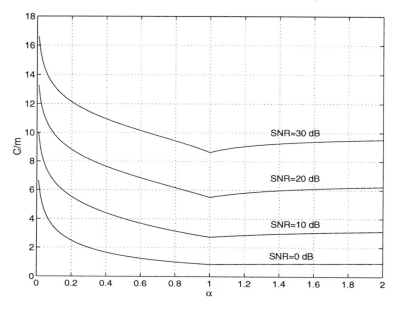

Figure 20.9. Capacity per antenna as $t, r \to \infty$ with independent Rayleigh fading and $t/r \to \alpha$.

This asymptotic result (plotted in Fig. 20.9) can be used to approximate the value of C for finite r, t.

The Appendix shows a simple way to derive the above result. By setting $\alpha = t/r$, (20.19) yields C as a function of t and r, and provides values very close to the true capacity even for small r, t. Tables 20.1–20.3 show some examples of exact and asymptotic values of capacity for different values of α and of the SNR ρ.

Table 20.1. Exact values of C/m for $t/r = 1/3$ and different values of SNR and of t.

ρ (dB)	$t = 1$	$t = 2$	$t = 4$	$t = 8$	$t \to \infty$
0 dB	1.87	1.87	1.86	1.86	1.86
10 dB	4.72	4.71	4.70	4.70	4.70
20 dB	7.98	7.97	7.96	7.96	7.96
30 dB	11.30	11.28	11.28	11.28	11.28

Table 20.2. Exact values of C/m for $t/r = 1/2$ and different values of SNR and of t.

ρ (dB)	$t = 1$	$t = 2$	$t = 4$	$t = 8$	$t \to \infty$
0 dB	1.44	1.43	1.43	1.43	1.43
10 dB	4.06	4.02	4.01	4.01	4.01
20 dB	7.27	7.23	7.22	7.22	7.22
30 dB	10.58	10.54	10.53	10.53	10.53

Table 20.3. Exact values of C/m for $t/r = 2$ and different values of SNR and of t.

ρ (dB)	$t = 2$	$t = 4$	$t = 8$	$t = 16$	$t \to \infty$
0 dB	0.92	0.92	0.91	0.91	0.91
10 dB	3.17	3.14	3.13	3.13	3.13
20 dB	6.28	6.24	6.23	6.23	6.23
30 dB	9.58	9.54	9.53	9.53	9.53

Observation 1. For the validity of (20.19), it is not necessary to assume that the entries of $\mathbf{H}$ are Gaussian: a sufficient condition is that $\mathbf{H}$ have iid entries with unit variance.

Observation 2. If $C(r, t, \rho)$ denotes the capacity of a channel with r receive antennas, t transmit antennas, and total SNR ρ, we have

$$C(a, b, \rho b) = C(b, a, \rho a) \qquad (20.22)$$

Thus, for example, $C(r, 1, \rho) = C(1, r, r\rho)$, which shows that with TX rather than RX diversity we need r times as much transmit power to achieve the same capacity.

Observation 3. Choose $t = r = 1$ as the baseline; this yields *one* more bit/s/Hz for every 3 dB of SNR increase. In fact, for large ρ,

$$C = \log(1 + \rho) \sim \log \rho \qquad (20.23)$$

For multiple antennas with $t = r$, for every 3 dB of SNR increase we have t more bit/s/Hz.

A critique. The previous results derived under the assumption $r \to \infty$ should be accepted *cum grano salis*. The assumption that the entries of the channel-gain matrix $\mathbf{H}$ are independent random variables becomes increasingly questionable as r increases. In fact, for this assumption to be justified the antennas should be separated by some multiple of the wavelength, which cannot be obtained when a large number of antennas is packed in a finite volume. Furthermore, if the variance of the elements of $\mathbf{H}$ does not depend on r, increasing r leads to an increased total received power, which is physically unacceptable. These observations lead to the consideration of a multiple-antenna system where not only the total transmit power remains constant as t increases, but also the average received power remains constant when r increases. This is obtained by rescaling $\mathbf{H}$ by a factor $r^{-1/2}$, so that the capacity (20.14) becomes

$$C = \mathbb{E}\left[\log\det\left(\mathbf{I}_r + \frac{\rho}{rt}\mathbf{H}\mathbf{H}^\dagger\right)\right] \tag{20.24}$$

As t and r grows to infinity, we obtain [ChiRimTel01]

$$C \to \rho\log e \tag{20.25}$$

a conclusion in contrast with our previous result that capacity increases linearly with the number of antennas.

2.3   Nonergodic Rayleigh fading channel

When $\mathbf{H}$ is chosen randomly at the beginning of the transmission, and held fixed for all channel uses, average capacity has no meaning, because the channel is nonergodic. In this case the quantity to be evaluated is, rather than capacity, outage probability. Under these conditions, the mutual information of the channel is the random variable

$$C(\mathbf{H}) = \log\det\left(\mathbf{I}_r + \frac{\rho}{t}\mathbf{H}\mathbf{H}^\dagger\right) \tag{20.26}$$

and the outage probability is given by

$$P_{\text{out}}(R) = \mathbb{P}\big(C(\mathbf{H}) < R\big) \tag{20.27}$$

The evaluation of (20.27) can be done by Monte Carlo simulation. However, one can profitably use an asymptotic result from matrix theory which states that, as t and r grow to infinity, the capacity $C(\mathbf{H})$ tends to a Gaussian random variable. This result has been recently obtained independently by a variety of authors under slightly different assumptions [HocMarTar02, SenMit02, SmiSha02].

The expected value of $C(\mathbf{H})$ is the ergodic capacity discussed in previous section. Its variance was evaluated independently in the form of an integral for finite r and t in [SmiSha02] and [WanGia02] (the latter reference derives the moment generating function of $C(\mathbf{H})$, and hence all of its moments). Moreover, it was observed in [SmiSha02] that $C(\mathbf{H})$ follows very closely a normal distribution even for small values of r and t. Thus, if μ_C and σ_C^2 denote the asymptotic mean and variance of $C(\mathbf{H})$, respectively, then a close approximation of the outage probability can be obtained in the form

$$P_{\text{out}}(R) \approx Q \left(\frac{\mu_C - R}{\sigma_C} \right) \tag{20.28}$$

where $Q(\cdot)$ is the Gaussian tail function. Refs. [SenMit02, MouSimSen03] derive results that can be specialized to our problem to yield

$$\sigma_C^2 = -\log e \log \left(1 - \frac{\gamma^2 \delta^2}{\beta} \right) \tag{20.29}$$

with $\beta \triangleq \alpha^{-1}$, and γ, δ are solutions of the equation pair

$$\gamma = \frac{\beta}{w + \delta} \qquad\qquad \delta = \frac{1}{w + \gamma} \tag{20.30}$$

with $w \triangleq \sqrt{1/\rho}$, which in our case yields

$$\gamma \triangleq \frac{\beta - 1 - w^2 + \sqrt{(\beta - 1 - w^2)^2 + 4\beta w^2}}{2w} \tag{20.31}$$

$$\delta \triangleq \frac{1 - \beta - w^2 + \sqrt{(1 - \beta - w^2)^2 + 4w^2}}{2w} \tag{20.32}$$

An alternative expression in the form of an integral for the asymptotic capacity variance was obtained in [BaiSil02]. This integral can be evaluated numerically, and results are shown in Fig. 20.10; the limiting value of the variance as $\rho \to \infty$ is

$$\lim_{\rho \to \infty} \sigma_C^2 = \log e \cdot \log(1 - \min(\alpha, \alpha^{-1})) \tag{20.33}$$

Fig. 20.11, which plots P_{out} versus R for $r = t = 4$ and two values of SNR, shows the quality of the Gaussian approximation for a Rayleigh channel.

Based on these results, we can evaluate the outage probabilities as in Figs. 20.12 and 20.13. These figures show the rate R that can be supported by the channel for a given SNR and a given outage probability, that is, from (20.28):

$$R = \mu_C - \sigma_C Q^{-1}(P_{\text{out}})$$

Notice how as r, t increase the outage probabilities curves come closer to each other: in fact, as r and t grow to infinity the channel tends to an ergodic channel. Finally, Fig. 20.14 shows the outage capacity (at $P_{\text{out}} = 0.01$) of an independent Rayleigh fading MIMO channel.

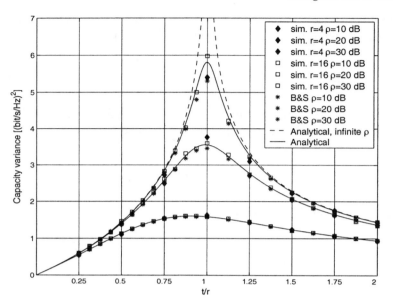

Figure 20.10. Variance of the nonergodic capacity. Continuous line: Asymptotic variance from [SenMit02, MouSimSen03]. (∗): Numerical integration of Bai-Silverstein expression for the variance [BaiSil02]. (♦) and (□): Monte Carlo simulation.

Block-fading channel. We now choose a block-fading channel model as shown in Fig. 20.15. The channel is described by the F matrices $\mathbf{H}_k$, $k = 1, \ldots, F$, each describing the fading gains in a single block. These gains are independent of each other. The channel input-output equation is

$$\mathbf{y}_k[\ell] = \mathbf{H}_k \mathbf{x}_k[\ell] + \mathbf{z}_k[\ell] \tag{20.34}$$

for $k = 1, \ldots, F$ (block index) and $\ell = 1, \ldots, N$ (symbol index along a block), $\mathbf{y}_k, \mathbf{z}_k \in \mathbb{C}^r$, and $\mathbf{x}_k \in \mathbb{C}^t$.

It is convenient to use the singular-value decomposition

$$\mathbf{H}_k = \mathbf{U}_k \mathbf{D}_k \mathbf{V}_k^\dagger \tag{20.35}$$

where $\mathbf{D}_k$ is a $r \times t$ matrix whose main-diagonal elements are the ordered singular values $\sqrt{\lambda_{k,1}} \geq \cdots \geq \sqrt{\lambda_{k,m}}$, with $\lambda_{k,i}$ the ith largest eigenvalue of the Hermitian matrix $\mathbf{H}_k \mathbf{H}_k^\dagger$, and $m \triangleq \min(r, t)$. Since $\mathbf{U}_k$ and $\mathbf{V}_k$ are unitary, by premultiplying $\mathbf{y}_k[\ell]$ by $\mathbf{U}_k^\dagger$ the input-output relation (20.34) can be rewritten in the form

$$\tilde{\mathbf{y}}_k[\ell] = \mathbf{D}_k \tilde{\mathbf{x}}_k[\ell] + \tilde{\mathbf{z}}_k[\ell] \tag{20.36}$$

where $\tilde{\mathbf{y}}_k[\ell] \triangleq \mathbf{U}_k^\dagger \mathbf{y}_k[\ell]$, $\tilde{\mathbf{x}}_k[\ell] \triangleq \mathbf{V}_k^\dagger \mathbf{x}_k[\ell]$, $\tilde{\mathbf{z}}_k[\ell] \triangleq \mathbf{U}_k^\dagger \mathbf{z}_k[\ell]$, and $\tilde{\mathbf{z}}_k[\ell] \sim \mathbf{z}_k[\ell] \sim \mathcal{N}_c(0, N_0)$.

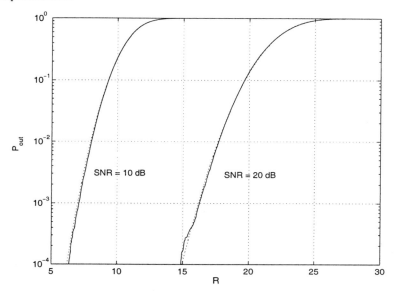

Figure 20.11. Outage probability for $r = t = 4$ and a nonergodic Rayleigh channel vs. R, the transmission rate in bit/s/Hz. The continuous line shows the results obtained by Monte Carlo simulation, while the dashed line shows the normal approximation.

No delay contraints. Since the random matrix process $\{\mathbf{H}_k\}_{k=1}^{F}$ is iid, as $F \to \infty$ the channel is ergodic and the average capacity is the relevant quantity. When perfect CSI is available to the receiver only, this is given by

$$C = \sum_{i=1}^{m} \mathbb{E}\left[\log\left(1 + \frac{\rho}{t}\lambda_i\right)\right] \tag{20.37}$$

If perfect CSI is available to transmitter and receiver, we proceed as in Section 2.1 and obtain

$$C = \mathbb{E}\left[\sum_{i=1}^{m} (\log(\mu\lambda_i))_+\right] \tag{20.38}$$

where μ is the solution of the "water-filling" equation

$$\sum_{i=1}^{m}(\mu - 1/\lambda_i)_+ = \rho \tag{20.39}$$

For all block lengths $N = 1, 2, \ldots$, the capacities (20.37) and (20.38) are achieved by code sequences with length FNt with $F \to \infty$. Capacity (20.37)

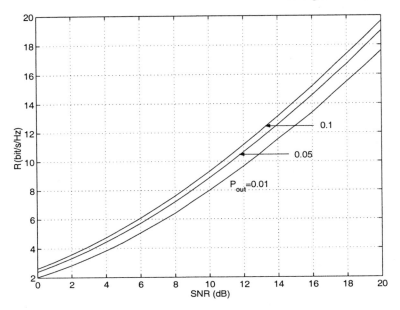

Figure 20.12. Transmission rate that can be supported with $r = t = 4$ and a give outage probability by a nonergodic Rayleigh channel. The results are based on the Gaussian approximation.

is achieved by random codes whose symbols are iid complex $\sim \mathcal{N}_c(0, \rho/t)$. Thus, all antennas transmit the same average energy per symbol. capacity (20.38) can be achieved by generating a random code with iid components $\sim \mathcal{N}_c(0, 1)$ and having each code word split into F blocks of N vectors $\tilde{\mathbf{x}}_k[\ell]$ with t components each. For block k, the optimal linear transformation

$$\mathbf{W}_k = \mathbf{V}_k \operatorname{diag}\left(\sqrt{\rho_{k,1}}, \ldots, \sqrt{\rho_{k,m}}, \underbrace{0, \ldots, 0}_{t-m}\right) \qquad (20.40)$$

is computed, where $\rho_{i,k} \triangleq (\xi - 1/\lambda_{k,i})_+$. The vectors $\mathbf{x}_k[\ell] = \mathbf{W}_k \tilde{\mathbf{x}}_k[\ell]$ are transmitted from the t antennas. This optimal scheme can be viewed as the concatenation of an optimal encoder for the unfaded AWGN channel, followed by a beamformer with weighting matrix $\mathbf{W}_k$ varying from block to block [Big-CaiTar01].

Delay constraints.   Consider now a delay constraint that forces F to take on a finite value. The random capacity is given by

$$C(\mathbf{\Lambda}, \mathbf{\Gamma}) = \frac{1}{F} \sum_{k=1}^{F} \sum_{i=1}^{m} \log(1 + \lambda_{k,i}\rho_{k,i}) \qquad (20.41)$$

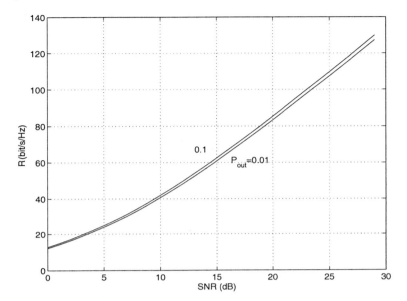

Figure 20.13. Transmission rate that can be supported with $r = t = 16$ and a give outage probability by a nonergodic Rayleigh channel. The results are based on the Gaussian approximation.

which corresponds to the outage probability

$$P_{\text{out}}(R) = \mathbb{P}\left(C(\boldsymbol{\Lambda}, \boldsymbol{\Gamma}) < R\right) \tag{20.42}$$

The outage probability is minimized under a long-term average power constraint by setting a threshold SNR^* such that if the SNR per block necessary to avoid an outage exceeds SNR^* then transmission is turned off and an outage is declared, while if it is below SNR^* transmission is turned on, and the power is allocated to the blocks according to a rule depending on the fading statistics only through the threshold value SNR^* (see [BigCaiTar01]).

2.4 Influence of channel-state information

A crucial factor in determining the performance of a multi-antenna system is the availability of the channel-state information (CSI), that is, the knowledge of the values of the fading gains in each one of the transmission paths. As we have seen, in a system with t transmit and r receive antennas and an ergodic Rayleigh fading channel modeled by an $t \times r$ matrix with random i.i.d. complex Gaussian entries the average channel capacity with perfect CSI at the receiver is about $m \triangleq \min(t, r)$ times larger than that of a single-antenna system for the same transmitted power and bandwidth. The capacity increases by about

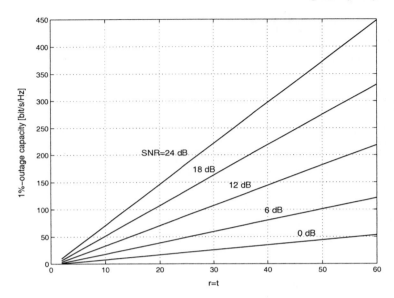

Figure 20.14. Outage capacity (at $P_{\text{out}} = 0.01$) with independent Rayleigh fading and $r = t$ antennas.

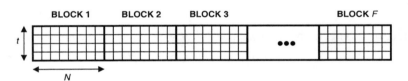

Figure 20.15. One code word in an F-block fading channel.

m bit/s/Hz for every 3-dB increase in SNR. Due to the assumption of perfect CSI available at the receiver, this result can be viewed as a fundamental limit for coherent multiple-antenna systems [ZheTse02].

In a fixed wireless environment, the fading gains can be expected to vary slowly, so their estimate can be obtained by the receiver with a reasonable accuracy, even in a system with a large number of antennas. One way of obtaining this estimate is by periodically sending pilot signals on the same channel used for data signals: if the channel is assumed to remain constant for I symbol periods, then we may write $I = I_r + I_d$, where I_r is the number of pilot symbols, while data transmission occupies I_d symbols. Since the transmission of pilot symbols lowers the information rate, there is a tradeoff between system performance and transmission rate.

Perfect CSI at the receiver. The most commonly studied situation is that of perfect CSI available at the receiver, which is the assumption under which we described multiple-antenna systems above.

No CSI. Fundamental limits of non-coherent communication, i.e., one taking place in an environment where estimates of the fading coefficients are not available, were derived in [MarHoc99, HocMar98]. In this channel model the fading gains are iid Rayleigh and remain constant for I symbol periods before changing to new independent realizations. Under these assumptions, further increasing the number of transmit antennas beyond I cannot increase capacity. Zheng and Tse [ZheTse02] derive an explicit formula for the high-SNR average channel capacity for $t = r$ and $I \geq 2r$, and characterize the rate at which capacity increases with SNR for $I < 2r$. By defining $K \triangleq \min(r, \lfloor I/2 \rfloor)$, [ZheTse02] shows that the capacity gains $K(1 - K/I)$ bits per second per Hz for every 3-dB SNR increase.

Imperfect CSI at the receiver. In real world the receiver has an imperfect knowledge of the CSI. In [TarBig03] some answers are provided to two questions: 1) How long should the training interval I_r be for satisfactory operation? 2) What is the performance degradation caused by imperfect estimation of CSI?

It can be shown that we must have $I_r \geq t$, i.e., the duration of the training interval must be at least as great as the number of transmit antennas. Moreover, the optimum training signals are orthogonal with respect to time among the transmit antennas, and each transmit antenna is fed equal energy.

CSI at the transmitter and at the receiver. It is also possible to envisage a situation in which channel state information is known to the receiver and to the transmitter: the latter can take the appropriate measures to counteract the effect of channel attenuations by suitably modulating its power. To assure causality, the assumption of CSI available at the transmitter is valid if it is applied to a multicarrier transmission scheme in which the available frequency band (over which the fading is selective) is split into a number of subbands, as with OFDM. The subbands are so narrow that fading is frequency-flat in each of them, and they are transmitted simultaneously, via orthogonal subcarriers. From a practical point of view, the transmitter can obtain the CSI either from a dedicated feedback channel (some existing systems already implement a fast power-control feedback channel) or by time-division duplex, where the uplink and the downlink time-share the same subchannels and the fading gains can be estimated from the incoming signal.

Ref. [BigCaiTar01] derives the performance limits of a channel with additive white Gaussian noise, delay and transmit-power constraints, and perfect channel-state information available at both transmitter and receiver. Because of a delay constraint, the transmission of a code word is assumed to span a finite (and typically small) number of independent channel realizations; therefore, the channel is nonergodic, and the relevant performance limits are the information outage probability and the delay-limited capacity. The coding scheme that minimizes the information outage probability was also derived. This scheme can be interpreted as the concatenation of an optimal code for the AWGN channel without fading to an optimal linear beamformer, whose coefficients change whenever the fading changes. For this optimal scheme we evaluated minimum-outage probability and delay-limited capacity. Among other results, [BigCaiTar01] proves that, for the fairly general class of *regular* fading channels, the asymptotic delay-limited capacity slope, expressed in bit/s/Hz per dB of transmit SNR, is proportional to $m \triangleq \min(t, r)$ and independent of the number of fading blocks F. Since F is a measure of the time diversity (induced by interleaving) or of the frequency diversity of the system, this result shows that, if channel-state information is available also to the transmitter, very high rates with asymptotically small error probabilities are achievable without need of deep interleaving or high frequency diversity. Moreover, for a large number of antennas the delay-limited capacity approaches the ergodic capacity. Finally, the availability of CSI at the transmitter makes transmit-antenna diversity equivalent, in terms of capacity improvement, to receive-antenna diversity, in the sense that reciprocity holds.

3.   Coding for multiple-antenna systems

Given that considerable gains are achievable by a multi-antenna system, the challenge is to design coding schemes that perform close to capacity: space–time trellis codes, space–time block codes, and layered space–time codes have been advocated (see, e.g., [GueFitBelKuo99, ShiKah99, ShiKah99a, SteDum99, TarSesCal98, TarJafCal99]).

A space–time code with block length N is described by the $t \times N$ matrix $\mathbf{X} \triangleq (\mathbf{x}[1], \ldots, \mathbf{x}[N])$. The code, which we denote $\mathcal{X}$, has $|\mathcal{X}|$ words. The row index of $\mathbf{X}$ indicates space, while the column index indicates time: to wit, the ith component of the t-vector $\mathbf{x}[\ell]$, denoted $x_i[\ell]$, is a complex number representing the two-dimensional signal transmitted by the ith antenna at discrete time ℓ, $\ell = 1, \ldots, N$, $i = 1, \ldots, t$. The received signal is the $r \times N$ matrix

$$\mathbf{Y} = \mathbf{HX} + \mathbf{Z} \qquad (20.43)$$

where $\mathbf{Z}$ is matrix of zero-mean complex Gaussian random variables (RV) with zero mean and independent real and imaginary parts with the same variance $N_0/2$ (i.e., circularly-distributed). Thus, the noise affecting the received signal is spatially and temporally independent, with $\mathbb{E}[\mathbf{ZZ}^\dagger] = NN_0\mathbf{I}_r$, where $\mathbf{I}_r$ denotes the $r \times r$ identity matrix and $(\cdot)^\dagger$ denotes Hermitian transposition.

The channel is described by the $r \times t$ matrix $\mathbf{H}$. Here we assume that $\mathbf{H}$ is independent of both $\mathbf{X}$ and $\mathbf{Z}$, it remains constant during the transmission of an entire code word, and its realization (the CSI) is known at the receiver.

3.1 Maximum likelihood detection

Under the assumption of CSI perfectly known by the receiver, and of additive white Gaussian noise, maximum-likelihood decoding corresponds to choosing the code word $\mathbf{X}$ which minimizes the squared Frobenius norm $\|\mathbf{Y} - \mathbf{HX}\|^2$. Explicitly, ML detection and decoding corresponds to the minimization of the quantity

$$\|\mathbf{Y} - \mathbf{HX}\|^2 = \sum_{i=1}^{r} \sum_{n=1}^{N} \left| y_{in} - \sum_{j=1}^{t} h_{ij} x_{jn} \right|^2 \tag{20.44}$$

Pairwise error probability. For computations, we may resort to the union bound to error probability

$$P(e) \leq \frac{1}{|\mathcal{X}|} \sum_{\mathbf{X} \in \mathcal{X}} \sum_{\widehat{\mathbf{X}} \neq \mathbf{X}} P(\mathbf{X} \to \widehat{\mathbf{X}}) \tag{20.45}$$

where the pairwise error probability (PEP) $P(\mathbf{X} \to \widehat{\mathbf{X}})$ is given by [BenBig99]

$$P(\mathbf{X} \to \widehat{\mathbf{X}}) = \mathbb{E}\left[Q\left(\frac{\|\mathbf{H\Delta}\|}{\sqrt{2N_0}} \right) \right] \tag{20.46}$$

where $\mathbf{\Delta} \triangleq \mathbf{X} - \widehat{\mathbf{X}}$. By writing

$$\|\mathbf{H\Delta}\|^2 = \text{Tr}\left(\mathbf{H}^\dagger \mathbf{H} \mathbf{\Delta} \mathbf{\Delta}^\dagger \right) \tag{20.47}$$

we see that the exact pairwise error probability, and hence the union bound to $P(e)$, depends on the $t \times r$ matrix $\mathbf{H}^\dagger \mathbf{H}$. This matrix can be interpreted as representing the effect of spatial interference on error probability: in particular, if $\mathbf{H}^\dagger \mathbf{H} = \mathbf{I}_t$ then (20.46) becomes

$$P(\mathbf{X} \to \widehat{\mathbf{X}}) = Q\left(\frac{\|\mathbf{\Delta}\|}{\sqrt{2N_0}} \right) \tag{20.48}$$

which is the PEP we would obtain on a set of t parallel independent AWGN channels, each transmitting a code formed by a row of $\mathbf{X}$.

A simple and useful upper bound to the pairwise error probability (20.46) can be computed by bounding above the Q function:

$$Q\left(\frac{\|\mathbf{H\Delta}\|}{\sqrt{2N_0}} \right) \leq \exp\left(-\|\mathbf{H\Delta}\|^2 / 4N_0 \right) \tag{20.49}$$

Under the assumption of independent Rayleigh fading, that is, when the entries h_{ij} of $\mathbf{H}$ are independent complex Gaussian random variables, circularly distributed with variance of their real and imaginary parts equal to $1/2$, the exact expectation of the RHS of the above can be computed:

$$P(\mathbf{X} \rightarrow \widehat{\mathbf{X}}) \leq \det \left[\mathbf{I}_t + \mathbf{\Delta}\mathbf{\Delta}^\dagger / 4N_0 \right]^{-r} \qquad (20.50)$$

The rank-and-determinant criterion.   Since the determinant of a matrix is equal to the product of its eigenvalues, (20.50) yields

$$P(\mathbf{X} \rightarrow \widehat{\mathbf{X}}) = \prod_{j=1}^{t} (1 + \lambda_j / 4N_0)^{-r} \qquad (20.51)$$

where λ_j denotes the jth eigenvalue of $\mathbf{\Delta}\mathbf{\Delta}^\dagger$. We can also write

$$P(\mathbf{X} \rightarrow \widehat{\mathbf{X}}) \leq \prod_{j \in \mathcal{J}} (\lambda_j / 4N_0)^{-r} \qquad (20.52)$$

where $\mathcal{J}$ is the index set of the nonzero eigenvalues of $\mathbf{\Delta}\mathbf{\Delta}^\dagger$. Denoting by ν the number of elements in $\mathcal{J}$, and rearranging the indexes so that $\lambda_1, \ldots, \lambda_R$ are the nonzero eigenvalues, we have

$$P(\mathbf{X} \rightarrow \widehat{\mathbf{X}}) \leq \left[\prod_{j=1}^{\nu} \lambda_j \right]^{-r} \gamma^{-r\nu} \qquad (20.53)$$

where $\gamma \triangleq 1/4N_0$. From this expression we see that the total diversity order (the "diversity gain") of the coded system is $r\nu_{\min}$, where $\nu_{\min}$ is the minimum rank of $\mathbf{\Delta}\mathbf{\Delta}^\dagger$ across all possible pairs $\mathbf{X}$, $\widehat{\mathbf{X}}$. In addition, the pairwise error probability depends on the rth power of the product of eigenvalues of $\mathbf{\Delta}\mathbf{\Delta}^\dagger$. This does not depend on the SNR (which is proportional to γ), and displaces the error probability curve instead of changing its slope. This is called the "coding gain." Thus, for high enough SNR we can choose, as a criterion for designing a space–time code, the maximization of the coding gain as well as of the diversity gain.

Notice that if $\nu_{\min} = t$, i.e., $\mathbf{\Delta}\mathbf{\Delta}^\dagger$ is full-rank for all the code word pairs, we have

$$\prod_{j=1}^{t} \lambda_j = \det \left(\mathbf{\Delta}\mathbf{\Delta}^\dagger \right) \qquad (20.54)$$

A necessary condition for $\mathbf{\Delta}\mathbf{\Delta}^\dagger$ to be full-rank is that $N \geq t$, i.e., the code block length must be at least equal to the number of transmit antennas.

The Euclidean-distance criterion.   Observe now that the term in the upper bound (20.50) can be written as

$$\det\left(\mathbf{I}_t + \gamma\boldsymbol{\Delta}\boldsymbol{\Delta}^\dagger\right) = 1 + \gamma\text{Tr}\left(\boldsymbol{\Delta}\boldsymbol{\Delta}^\dagger\right) + \ldots + \gamma^t\det\left(\boldsymbol{\Delta}\boldsymbol{\Delta}^\dagger\right) \qquad (20.55)$$

where $\boldsymbol{\Delta} \triangleq \mathbf{X} - \widehat{\mathbf{X}}$. We see that if $\gamma \ll 1$ then the LHS of (20.55), and hence the PEP, depends essentially on $\text{Tr}\left(\boldsymbol{\Delta}\boldsymbol{\Delta}^\dagger\right)$, which is the Euclidean distance between $\mathbf{X}$ and $\widehat{\mathbf{X}}$, while if $\gamma \gg 1$ it depends essentially on $\det\left(\boldsymbol{\Delta}\boldsymbol{\Delta}^\dagger\right)$, that is, on the product of the eigenvalues of $\boldsymbol{\Delta}\boldsymbol{\Delta}^\dagger$.

A different perspective can be obtained by allowing the number r of receive antennas to grow to infinity. To do this, we first renormalize the entries of $\mathbf{H}$ so that their variance is now $1/r$ rather than 1: this prevents the total receive power to diverge as $r \to \infty$. We obtain the following new form of (20.50):

$$P(\mathbf{X} \to \widehat{\mathbf{X}}) \le \det\left[\mathbf{I}_t + \boldsymbol{\Delta}\boldsymbol{\Delta}^\dagger/4rN_0\right]^{-r} \qquad (20.56)$$

which yields, in lieu of (20.55):

$$\det\left[\mathbf{I}_t + (\gamma/r)\boldsymbol{\Delta}\boldsymbol{\Delta}^\dagger\right] = 1 + (\gamma/r)\text{Tr}\left(\boldsymbol{\Delta}\boldsymbol{\Delta}^\dagger\right) + \ldots + (\gamma/r)^t\det\left(\boldsymbol{\Delta}\boldsymbol{\Delta}^\dagger\right) \quad (20.57)$$

This shows that as $r \to \infty$ the rank-and-determinant criterion is appropriate for a SNR increasing as fast as r, while the Euclidean-distance criterion is appropriate for relatively small SNRs. This situation is illustrated in the example of Fig. 20.16, which shows the union upper bound (obtained numerically) on the word-error probability of the space–time code obtained by splitting evenly the code words of the $(24, 8, 12)$ extended Golay code between two transmit antennas. This space–time code has the minimum rank of $\boldsymbol{\Delta}$ equal to 1, and hence a diversity gain equal to r. Now, it is seen from Fig. 20.16 how the slope predicted by (20.50), and exhibited by a linear behavior in the $P(e)$-vs.-E_b/N_0 chart, may be reached only for very small values of error probability (how small, that depends on the weight structure of the code under scrutiny). To justify this behavior, observe from Fig. 20.16 that for a given value of r the error-probability curve changes its behavior from a "waterfall" shape (for small to intermediate SNR) to a linear shape (high SNR). As the number of receive antennas grows, this change of slope occurs for values of $P(e)$ that are smaller and smaller as r increases. Thus, to study the word-error-probability curve in its waterfall region it makes sense to examine its asymptotic behavior as $r \to \infty$. The case $r \to \infty$, $t < \infty$ can be easily dealt with by using the law of large numbers: this yields $\mathbf{H}^\dagger\mathbf{H} \to \mathbf{I}_t$, $\mathbf{I}_t$ the $t \times t$ identity matrix. We can see that, as $r \to \infty$,

$$\|\mathbf{H}\boldsymbol{\Delta}\|^2 \to \|\boldsymbol{\Delta}\|^2 \qquad (20.58)$$

and hence

$$P(\mathbf{X} \to \widehat{\mathbf{X}}) = Q\left(\frac{\|\boldsymbol{\Delta}\|}{\sqrt{2N_0}}\right) \qquad (20.59)$$

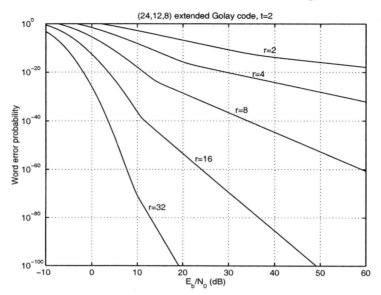

Figure 20.16. Performance of the binary $(24, 8, 12)$ extended Golay code with binary PSK over a channel with $t = 2$ transmit antennas and r receive antennas with ML interface.

This result shows that as the number of receiving antennas grows large the union bound on the error probability of the space–time code depends only on the Euclidean distances between pairs of code words. This is the result one would get with a transmission occurring over a non-fading additive white Gaussian noise (AWGN) channel whose transfer matrix $\mathbf{H}$ has orthogonal columns, i.e., is such that $\mathbf{H}^\dagger\mathbf{H}$ is a scalar matrix. In this situation the smallest error probability, at the expense of a larger complexity, can be achieved by using a single code, optimized for the AWGN channel, whose words of length tN are equally split among the transmit antennas. Within this framework, the number of transmit antennas does not affect the PEP, but only the transmission rate which, expressed in bits per channel use, increases linearly with t.

For another example, observe Fig. 20.17. This shows how for intermediate SNRs the Euclidean criterion may yield codes better than the rank-and-determinant criterion. It compares the simulated performances, in terms of frame-error rate, of the 4-state, rate-1/2 space–time code of [TarLo98] and a comparable space–time code obtained by choosing a good binary, 4-state, rate-2/4 convolutional code [ChaHwaLin97] and mapping its symbols onto QPSK. The frame length N is 130 symbols for both codes, including 1 symbol for trellis termination. It is seen that in the error-probability interval of these two figures the "standard" convolutional code generally outperforms the space–time code of [TarLo98] even for small values of r.

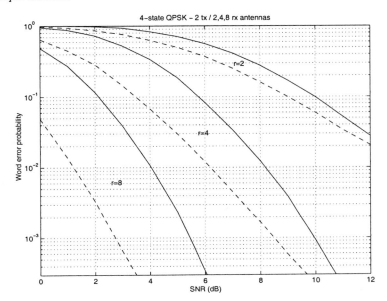

Figure 20.17. Frame-error rates of two space–time codes with 4 states, rate $1/2$, and QPSK. Number of transmit antennas: $t = 2$; number of receive antennas: $r = 2, 4, 8$. Continuous line: Code from [TarLo98]. Dashed line: Code obtained from a binary convolutional code good for the AWGN channel [ChaHwaLin97].

Decoding space–time codes. Decoding of space–time codes is performed by using Viterbi algorithm whose branch metrics at time n are, from (20.44):

$$\sum_{i=1}^{r} \left| y_{in} - \sum_{j=1}^{t} h_{ij} x_{jn} \right|^2 \qquad (20.60)$$

Computation of (20.60) requires knowledge of the fading gains h_{kj}, i.e., CSI at the receiver.

4. Space–time coding schemes

4.1 Alamouti scheme

This allows transmission of one signal per channel use, and removes the spatial interference by orthogonalization. We describe it by first considering the simple case $t = 2, r = 1$, which yields the scheme illustrated in Fig. 20.18.

The code matrix $\mathbf{X}$ has the form

$$\mathbf{X} = \begin{bmatrix} x_1 & -x_2^* \\ x_2 & x_1^* \end{bmatrix} \qquad (20.61)$$

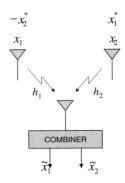

Figure 20.18. Alamouti TX-diversity scheme with $t = 2$ and $r = 1$.

This means that during the first symbol interval, the signal x_1 is transmitted from antenna 1, while signal x_2 is transmitted from antenna 2. During the next symbol period, antenna 1 transmits signal $-x_2^*$, and antenna 2 transmits signal x_1^*. Thus, the signals received in two adjacent time slots are

$$y_1 = h_1 x_1 + h_2 x_2 + z_1$$

and

$$y_2 = -h_1 x_2^* + h_2 x_1^* + z_2$$

where h_1, h_2 denote the path gains from the two TX antennas to the RX antenna. The combiner of Fig. 20.18, which has perfect CSI and hence knows the values of the path gains, generates the signals

$$\tilde{x}_1 = h_1^* y_1 + h_2 y_2^*$$

and

$$\tilde{x}_2 = h_2^* y_1 - h_1 y_2^*$$

so that

$$
\begin{aligned}
\tilde{x}_1 &= h_1^*(h_1 x_1 + h_2 x_2 + z_1) + h_2(-h_1^* x_2 + h_2^* x_1 + z_2^*) \\
&= (|h_1|^2 + |h_2|^2) x_1 + (h_1^* z_1 + h_2 z_2^*)
\end{aligned}
\tag{20.62}
$$

and similarly

$$\tilde{x}_2 = (|h_1|^2 + |h_2|^2) x_2 + (h_2^* z_1 - h_1 z_2^*) \tag{20.63}$$

Thus, we have separated x_1 from x_2. This scheme has the same performance as one with $t = 1$, $r = 2$, and maximal-ratio combining (provided that each TX antenna transmits the same power as the single antenna for $t = 1$). To prove the last statement, observe that if the signal x_1 is transmitted, the two receive antennas observe $h_1x_1 + n_1$ and $h_2x_1 + n_2$, respectively, and after maximal-ratio combining the decision variable is

$$h_1^*(h_1x_1 + z_1) + h_2^*(h_2x_1 + z_2) = (|h_1|^2 + |h_2|^2)x_1 + (h_1^*z_1 + h_2z_2^*) = \tilde{x}_1$$

This scheme can be generalized to other values of r. For example, with $t = r = 2$ and the same transmission scheme as before (see Fig. 20.19), we

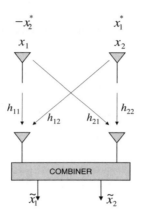

Figure 20.19. Alamouti TX-diversity scheme with $t = 2$ and $r = 2$.

have, if $y_{11}, y_{12}, y_{21}, y_{22}$ denote the signals received by antenna 1 at time 1, by antenna 1 at time 2, by antenna 2 at time 1, and by antenna 2 at time 2, respectively,

$$\begin{bmatrix} y_{11} & y_{12} \\ y_{21} & y_{22} \end{bmatrix} = \begin{bmatrix} h_{11} & h_{12} \\ h_{21} & h_{22} \end{bmatrix} \begin{bmatrix} x_1 & -x_2^* \\ x_2 & x_1^* \end{bmatrix} + \begin{bmatrix} z_{11} & z_{12} \\ z_{21} & z_{22} \end{bmatrix}$$

$$= \begin{bmatrix} h_{11}x_1 + h_{12}x_2 + z_{11} & -h_{11}x_2^* + h_{12}x_1^* + z_{12} \\ h_{21}x_1 + h_{22}x_2 + z_{21} & -h_{21}x_2^* + h_{22}x_1^* + z_{22} \end{bmatrix}$$

The combiner generates

$$\tilde{x}_1 = h_{11}^*y_1 + h_{12}y_2^* + h_{21}^*y_3 + h_{22}y_4^*$$

and

$$\tilde{x}_2 = h_{12}^*y_1 - h_{11}y_2^* + h_{22}^*y_3 - h_{21}y_4^*$$

which yields

$$\tilde{x}_1 = (|h_{11}|^2 + |h_{12}|^2 + |h_{21}|^2 + |h_{22}|^2)x_1 + \text{noise}$$

and

$$\tilde{x}_2 = (|h_{11}|^2 + |h_{12}|^2 + |h_{21}|^2 + |h_{22}|^2)x_2 + \text{noise}$$

As above, it can be easily shown that the performance of this $t = 2$, $r = 2$ scheme is equivalent to that of a $t = 1$, $r = 4$ scheme with maximal-ratio combining (again, provided that each TX antenna transmits the same power as with $t = 1$).

A general scheme with $t = 2$ and a general value of r can be exhibited: it has the same performance of a single-TX-antenna scheme with $2r$ RX antennas and maximal-ratio combining.

4.2 Alamouti scheme revisited: Orthogonal designs

We can rewrite the transmitted signal in Alamouti scheme with $t = 2$ and $r = 1$ in the following equivalent form:

$$\begin{bmatrix} y_1 \\ y_2^* \end{bmatrix} = \begin{bmatrix} h_1 & h_2 \\ h_2^* & -h_1^* \end{bmatrix} \begin{bmatrix} x_1 \\ x_2 \end{bmatrix} + \begin{bmatrix} z_1 \\ z_2 \end{bmatrix} \tag{20.64}$$

Now, if we define

$$\check{\mathbf{H}} \triangleq \begin{bmatrix} h_1 & h_2 \\ h_2^* & -h_1^* \end{bmatrix}$$

we see that

$$\check{\mathbf{H}}^\dagger \check{\mathbf{H}} = (|h_1|^2 + |h_2|^2)\mathbf{I}_2 \tag{20.65}$$

This, in conjunction with (20.46) and (20.47), shows that the error probability for this Alamouti scheme is the same as without spatial interference, and with a received power increased by a factor $(|h_1|^2 + |h_2|^2)$. For this reason Alamouti scheme is called an *orthogonal design*. There are also orthogonal designs with $t > 2$. For example, with $t = 3$, $r = 1$, and $N = 4$ we have

$$\mathbf{X} = \begin{bmatrix} x_1 & -x_2^* & -x_3^* & 0 \\ x_2 & x_1^* & 0 & -x_3^* \\ x_3 & 0 & x_1^* & x_2^* \end{bmatrix}$$

so that the equation $\mathbf{Y} = \mathbf{HX} + \mathbf{Z}$ can be rewritten in the equivalent form

$$\begin{bmatrix} y_1 \\ y_2^* \\ y_3^* \\ y_4^* \end{bmatrix} = \check{\mathbf{H}} \begin{bmatrix} x_1 \\ x_2 \\ x_3 \end{bmatrix} \tag{20.66}$$

where

$$\check{\mathbf{H}} \triangleq \begin{bmatrix} h_1 & h_2 & h_3 \\ h_2^* & -h_1^* & 0 \\ h_3^* & 0 & h_1^* \\ 0 & h_3^* & -h_2^* \end{bmatrix} \tag{20.67}$$

In this case we can verify that

$$\check{\mathbf{H}}^\dagger\check{\mathbf{H}} = (|h_1|^2 + |h_2|^2 + |h_3|^2)\mathbf{I}_3$$

Notice that with this code we transmit three signals in four time intervals, so that its rate is $3/4$ signals per channel use, less than the original Alamouti schemes, which transmit 1 signal per channel use. In fact, orthogonal designs with $t > 2$ have rates that cannot exceed $3/4$ [WanGia02]. To avoid the rate loss of orthogonal designs, algebraic codes can be designed that, for any number of transmit and receive antennas, achieve maximum diversity, as Alamouti codes, while the rate is t symbols per channel use (see [MaGian02] and references therein).

4.3   Trellis space–time codes

Trellis space–time codes are TCM schemes in which every transition among states, as described by a trellis, is labeled by t signals, each being associated with a transmit antenna [TarSesCal98].

Example 7.   An example of a space–time code is shown in Fig. 20.20 through its trellis. This has $t = 2$, four states, and transmits 2 bit/channel use by using 4PSK, whose signals are denoted $0, 1, 2, 3$. With $r = 1$ its diversity is 2. Label xy means that signal x is transmitted by antenna 1, while signal y is simultaneously transmitted by antenna 2.   □

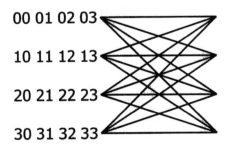

Figure 20.20.   A space–time coding scheme with $t = 2$.

4.4   Space–time codes when CSI is not available

In a rapidly-changing mobile environment, or when long training sequences are not allowed, the assumption of perfect CSI at the receiver may not be valid. In the absence of CSI at the receiver, [HocMar00,MarHocHas00] advocate *unitary space–time modulation*, a technique which circumvents the use of training symbols (which for maximum throughput should occupy half of the transmission interval, as seen before). Here the information is carried on the subspace that is spanned by orthonormal signals that are sent. This subspace survives multiplication by the unknown channel-gain matrix. A scheme based on *differential unitary space–time* signals is advocated in [HocSwe00]. High-rate constellations with excellent performance, obtained via algebraic techniques, are described in [HasHocShoSwe01].

5.   Suboptimum receiver interfaces

From the capacity results described above we can see that extremely large spectral efficiencies can be achieved on a wireless link if the number of transmit and receive antennas is large. However, as t and r increase, the complexity of space–time coding with maximum-likelihood detection may become too large. This motivates the design of suboptimal receiver interfaces whose complexity is lower than ML, and yet perform close to what predicted by the theory. We describe here some of these schemes, categorized as *linear* and *nonlinear*.

5.1   Linear interfaces

A linear interface operates a linear transformation $\mathbf{A}$ of the received signal, with $\mathbf{A} = \mathbf{A}(\mathbf{H})$ a $t \times r$ matrix chosen so as to allow a simplification of the metrics used in the Viterbi algorithm employed for soft decoding of $\mathcal{X}$.

Zero-forcing interface.   This consists of choosing $\mathbf{A} = \mathbf{H}^+$, where the superscript $^+$ denotes the Moore-Penrose pseudoinverse of a matrix [HorJoh85]. If we assume $r \geq t$, then $\mathbf{H}^\dagger \mathbf{H}$ is invertible with probability 1, and we have

$$\mathbf{H}^+ = (\mathbf{H}^\dagger \mathbf{H})^{-1} \mathbf{H}^\dagger \tag{20.68}$$

so that

$$\mathbf{H}^+ \mathbf{Y} = \mathbf{X} + \mathbf{H}^+ \mathbf{Z} \tag{20.69}$$

which shows that the spatial interference in completely removed from the received signal, while the noise is made nonwhite. The metric used with this interface is $\|\mathbf{H}^+ \mathbf{Y} - \mathbf{X}\|^2$. The conditional PEP is given by [BigTarTul02]

$$P(\mathbf{X} \to \widehat{\mathbf{X}}) = \mathbb{E}\left[Q\left(\frac{\|\mathbf{\Delta}\|^2}{\sqrt{2\,N_0 \mathrm{Tr}\left[(\mathbf{H}^\dagger \mathbf{H})^{-1} \mathbf{\Delta}\mathbf{\Delta}^\dagger\right]}} \right) \right] \tag{20.70}$$

where $\mathbf{\Delta} \triangleq \mathbf{X} - \widehat{\mathbf{X}}$.

Linear MMSE interface.   Here we choose the matrix $\mathbf{A}$ so as to minimize the mean-square value of the spatial interference plus noise. We have

$$\mathbf{A} = (\mathbf{H}^{\dagger}\mathbf{H} + \delta_s\mathbf{I}_t)^{-1}\mathbf{H}^{\dagger}$$

where $\delta_s \triangleq N_0/E_s$, E_s the average energy per coded symbol. We have [Big-TarTul02]

$$P(\mathbf{X} \to \widehat{\mathbf{X}}) = \mathbb{E}\left[Q\left(\frac{\|\mathbf{\Delta}\|^2 + 2(((\mathbf{H}^{\dagger}\mathbf{H} + \delta_s\mathbf{I}_t)^{-1}\mathbf{H}^{\dagger}\mathbf{H} - \mathbf{I}_t)\mathbf{X}, \mathbf{\Delta})}{\sqrt{2N_0\|\mathbf{H}(\mathbf{H}^{\dagger}\mathbf{H} + \delta_s\mathbf{I}_t)^{-1}\mathbf{\Delta}\|^2}}\right)\right]$$

$$(20.71)$$

where $\mathbf{\Delta} \triangleq \mathbf{X} - \widehat{\mathbf{X}}$. Notice that as $\delta_s \to 0$ the right-hand side of (20.71) tends to the PEP of the zero-forcing detector, as it should.

Performance of linear interfaces.　　Asymptotic analysis carried out for a large number of antennas [BigTarTul02] shows that these linear interfaces exhibit a PEP close to that of ML interface only for $r \gg t$; otherwise the performance loss may be substantial. As an example, Fig. 20.21 shows the error probability of a multiple-antenna system where the binary $(8, 4, 4)$ Reed-Muller code is used by splitting its code words evenly between 2 transmit antennas. The word error probabilities shown are obtained through Monte Carlo simulation. Binary PSK is used, and the code rate is 1 bit per channel use. It is seen that for $r = 2$ both MMSE and ZF interface exhibit a considerable performance loss with respect to ML, while for $r = 8$ the losses are very moderate.

5.2　　Nonlinear interfaces

The task of reducing the spatial interference affecting the received signal can be accomplished by first processing $\mathbf{Y}$ linearly, then subtracting from the result an estimate of the spatial interference obtained from preliminary decisions $\widehat{\mathbf{X}}$ on the transmitted code word $\mathbf{X}$. The metric used for decoding is $\|\widetilde{\mathbf{Y}} - \mathbf{X}\|$, where

$$\widetilde{\mathbf{Y}} \triangleq \mathbf{G}\mathbf{Y} - \mathbf{L}\widehat{\mathbf{X}} \qquad (20.72)$$

for a suitable choice of the two matrices $\mathbf{G}$ and $\mathbf{L}$.

BLAST interface.　　One nonlinear interface is called BLAST (this stands for Bell Laboratories Layered Space–Time Architecture). Several implementations of the basic BLAST idea are possible, two of them being the *vertical MMSE-BLAST* interface and the *vertical ZF-BLAST* interface: these arise from the minimization of the mean-square error of the disturbance $\widetilde{\mathbf{Y}} - \mathbf{X}$ with or without considering noise, respectively.

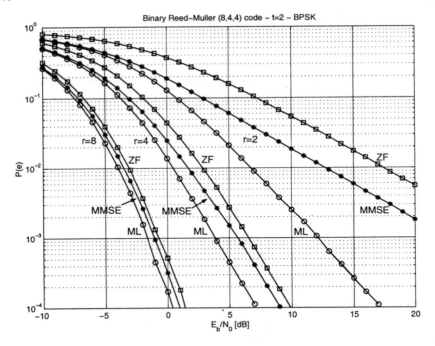

Figure 20.21. Performance of the binary $(8, 4, 4)$ Reed-Muller code with binary PSK over a channel with $t = 2$ transmit antennas and r receive antennas with ML, MMSE, and ZF interfaces.

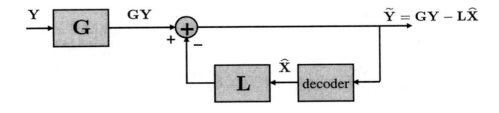

Figure 20.22. General structure of a nonlinear interface.

Zero-forcing BLAST.   Here we have

$$\mathbf{G} = \text{diag}^{-1}(\mathbf{R})\mathbf{Q}^{\dagger} \tag{20.73}$$

$$\mathbf{L} = \text{diag}^{-1}(\mathbf{R})\mathbf{R} - \mathbf{I}_t \tag{20.74}$$

where $\mathbf{Q}$ and $\mathbf{R}$ come from the QR factorization $\mathbf{H} = \mathbf{QR}$ [HorJoh85]

where $\mathbf{Q}$ and $\mathbf{R}$ come from the QR factorization $\mathbf{H} = \mathbf{QR}$ [HorJoh85]

Here $\widetilde{\mathbf{Y}}$ can be given the form

$$\widetilde{\mathbf{Y}} = \underbrace{\mathbf{X}}_{①} + \underbrace{[\mathrm{diag}^{-1}(\mathbf{R})\mathbf{R} - \mathbf{I}_t]\boldsymbol{\Delta}}_{②} + \underbrace{\mathrm{diag}^{-1}(\mathbf{R})\mathbf{Q}^{\dagger}\mathbf{Z}}_{③} \qquad (20.75)$$

The three terms in expression above can be recognized as: ① useful term (free from spatial interference: this justifies the name "zero-forcing"); ② interference due to past wrong decisions; and ③ noise. Notice that the entries of the noise matrix are independent but *not* identically distributed (i.e., the noise is *colored*).

MMSE-BLAST. Here

$$\begin{cases} \mathbf{G} = \mathrm{diag}^{-1}(\mathbf{S})\mathbf{S}^{-\dagger}\mathbf{H}^{\dagger} \\ \mathbf{L} = \mathrm{diag}^{-1}(\mathbf{S})\mathbf{S} - \mathbf{I}_t \end{cases} \qquad (20.76)$$

where $\mathbf{S}$ is an upper triangular matrix derived from the Cholesky factorization [HorJoh85] $\mathbf{H}^{\dagger}\mathbf{H} + \delta_s\mathbf{I}_t = \mathbf{S}^{\dagger}\mathbf{S}$, with $\delta_s \triangleq N_0/E_s$ as before. As a result, the soft estimate $\widetilde{\mathbf{Y}}$ can be written as

$$\begin{aligned} \widetilde{\mathbf{Y}} &= \mathrm{diag}^{-1}(\mathbf{S})\mathbf{S}^{-\dagger}\mathbf{H}^{\dagger}\mathbf{Y} - (\mathrm{diag}^{-1}(\mathbf{S})\mathbf{S} - \mathbf{I}_t)\widehat{\mathbf{X}} \\ &= \underbrace{(\mathbf{I}_t - \mathrm{diag}^{-1}(\mathbf{S})\mathbf{S}^{-\dagger})\mathbf{X}}_{①} + \underbrace{(\mathrm{diag}^{-1}(\mathbf{S})\mathbf{S} - \mathbf{I}_t)\boldsymbol{\Delta}}_{②} \\ &\quad + \underbrace{\mathrm{diag}^{-1}(\mathbf{S})\mathbf{S}^{-\dagger}\mathbf{H}^{\dagger}\mathbf{Z}}_{③} \qquad (20.77) \end{aligned}$$

where the three terms in the last expression are: ① the (biased) useful term; ② the interference due to past wrong decisions; and ③ colored noise.

Appendix: derivation of (20.19)

Using results from [Sil95] we obtain the asymptotic value of the mean capacity as

$$\frac{C}{t} \to C(\alpha, \rho) \triangleq \int_a^b \log(1 + \alpha^{-1}\rho x) \frac{\sqrt{(x-a)(b-x)}}{2\pi\alpha x}\, dx \qquad (20.78)$$

Then, setting $x = 1 + \alpha + 2\sqrt{\alpha}\cos\theta$, we obtain

$$C(\alpha, \rho) = \frac{1}{\pi}\int_0^{2\pi} \log\left(1 + \alpha^{-1}\rho(1+\alpha+2\sqrt{\alpha}\cos\theta)\right)\frac{\sin^2\theta}{1+\alpha+2\sqrt{\alpha}\cos\theta}\, d\theta \qquad (20.79)$$

Now, we factor the logarithm argument as a polynomial in $e^{j\theta}$ and obtain:

$$1 + \alpha^{-1}\rho(1+\alpha+2\sqrt{\alpha}\cos\theta) = \alpha^{-1/2}\rho e^{-j\theta}(e^{j\theta} + \alpha^{1/2}w_+)(e^{j\theta} + \alpha^{1/2}w_-)$$

where we defined $w \triangleq 1 + \alpha^{-1} + \rho^{-1}$ and $w_\pm \triangleq (w \pm \sqrt{w^2 - 4/\alpha})/2$. From the inequality

$$(\alpha^{1/2}w_+ - 1)(\alpha^{1/2}w_- - 1) = -\alpha^{-1/2}(1 - \alpha^{1/2})^2 - \alpha^{1/2}\rho^{-1} < 0$$

we have that $\alpha^{1/2}w_- < 1$ and $\alpha^{1/2}w_+ > 1$ so that we can write the logarithm argument as

$$1 + \alpha^{-1}\rho(1+\alpha+2\sqrt{\alpha}\cos\theta) = w_+\rho\left(1 + \frac{\exp(j\theta)}{\alpha^{1/2}w_+}\right)\left(1 + \frac{\alpha^{1/2}w_-}{\exp(j\theta)}\right)$$

Thus, we obtain the following series expansion

$$\ln(1 + \alpha^{-1}\rho(1+\alpha+2\sqrt{\alpha}\cos\theta))$$
$$= \ln(w_+\rho) + \sum_{k=1}^\infty \frac{(-1)^{k-1}}{k}\left[\left(\frac{\exp(j\theta)}{\alpha^{1/2}w_+}\right)^k + \left(\frac{\alpha^{1/2}w_-}{\exp(j\theta)}\right)^k\right] \qquad (20.80)$$

which is uniformly convergent in θ and can be integrated term-by-term. Now, we need to calculate

$$I_k \triangleq \frac{1}{\pi}\int_0^{2\pi} e^{jk\theta}\frac{\sin^2\theta}{1+\alpha+2\sqrt{\alpha}\cos\theta}\, d\theta \qquad (20.81)$$

for every integer k. After some algebra, we obtain

$$I_k = \begin{cases} 1 & k = 0 \\[2mm] -\dfrac{1}{2}\sqrt{\alpha} & |k| = 1 \\[3mm] (-1)^{k+1}\dfrac{1-\alpha}{2\alpha}\alpha^{|k|/2} & |k| > 1 \end{cases} \qquad (20.82)$$

The termwise integration of the series yields

$$
\begin{aligned}
C(\alpha, \rho) &= \ln(w_+\rho) - \frac{1}{2\alpha}\left[w_+^{-1} + \alpha w_-\right] + \frac{1-\alpha}{2\alpha}\sum_{k=1}^{\infty}\frac{1}{k}\left[w_+^{-k} + (\alpha w_-)^k\right] \\
&= \ln(w_+\rho) - w_- - \frac{1-\alpha}{2\alpha}\left[\ln(1 - w_+^{-1}) + \ln(1 - \alpha w_-)\right] \\
&= \ln(w_+\rho) - w_- - \frac{1-\alpha}{\alpha}\ln(1 - w_+^{-1}) \\
&= \frac{1}{\alpha}\ln(w_+\rho) - w_- - \frac{1-\alpha}{\alpha}\ln((w_+ - 1)\rho) \\
&= \frac{1}{\alpha}\ln(w_+\rho) - w_- + \frac{1-\alpha}{\alpha}\ln(1 - w_-) \qquad (20.83)
\end{aligned}
$$

Both series expansions in brackets converge because $w_+ > 1$ and $\alpha w_- < 1$ since

$$
(w_+ - 1)(\alpha w_- - 1) = 2 - w_+ - \alpha w_- = -\frac{(w_+ - 1)^2}{w_+} < 0 \qquad (20.84)
$$

Notations and Acronyms

☞ a.s., almost surely

☞ AWGN, Additive white Gaussian noise

☞ CSI, Channel state information

☞ iid, Independent and identically distributed

☞ LHS, Left-hand side

☞ ln, Natural logarithm

☞ log, Logarithm in base 2

☞ ML, Maximum-likelihood

☞ pdf, Probability density function

☞ RHS, Right-hand side

☞ RV, Random variable

☞ SNR, Signal-to-noise ratio

☞ TCM, Trellis-coded modulation

☞ a^*, Conjugate of the complex number a

☞ $(a)_+ \triangleq \max(0, a)$, Equal to a if $a > 0$, equal to 0 otherwise.

☞ $\mathbf{A}^\dagger$, Conjugate (or Hermitian) transpose of matrix $\mathbf{A}$

☞ $\|\mathbf{A}\| \triangleq \sqrt{\sum_{i=1}^{m}\sum_{j=1}^{n}|a_{ij}|^2}$, Frobenius norm of the $m \times n$ matrix $\mathbf{A} = (a_{ij})$, $i = 1, \dots, m, j = 1, \dots, n$

☞ $\mathbb{C}$, The set of complex numbers

☞ $\mathbb{E}[X]$, Expectation of the RV X

☞ $\mathbf{I}_n$, The $n \times n$ identity matrix

☞ $\Im$, Imaginary part

☞ $Q(x) \triangleq (2\pi)^{-1/2}\int_x^{\infty}\exp(-z^2/2)\,dz$, The Gaussian tail function

☞ $\Re$, Real part

☞ $\mathbb{R}$, The set of real numbers

☞ $\triangleq$, Equal by definition

☞ $X \sim \mathcal{N}(\mu, \sigma^2)$, X is a real Gaussian RV with mean μ and variance σ^2

☞ $X \sim \mathcal{N}_c(\mu, \sigma^2)$, X is a circularly-distributed complex Gaussian RV with mean μ and $\mathbb{E}[|X|^2] = \sigma^2$

References

[BaiSil02]   Z. D. Bai and J. W. Silverstein, "CLT for linear spectral statistics of large dimensional sample covariance matrices," *Annals of Probability*, to be published, 2003.

[BenBig99]   S. Benedetto and E. Biglieri, *Principles of Digital Transmission with Wireless Applications.* New York: Kluwer/ Plenum, 1999.

[BigTarTul02]   E. Biglieri, G. Taricco, and A. Tulino, "Performance of space–time codes for a large number of antennas," *IEEE Trans. Inform. Theory,* Vol. 48, No. 7, pp. 1794–1803, Jul. 2002.

[BigCaiTar01]   E. Biglieri, G. Caire, and G. Taricco, "Limiting performance of block-fading channels with multiple antennas," *IEEE Trans. Inform. Theory*, Vol. 47, No. 4, pp. 1273– 1289, May 2001.

[BigProSha98]   E. Biglieri, J. Proakis, and S. Shamai (Shitz), "Fading channels: Information-theoretic and communication aspects," *IEEE Trans. Inform. Theory*, Vol. 44, No. 6, *50th Anniversary Issue*, pp. 2619–2692, October 1998.

[CaiKnoHum98]   G. Caire, R. Knopp and P. Humblet, "System capacity of F-TDMA cellular systems," *IEEE Trans. Commun.,* Vol. 46, No. 12, pp. 1649–1661, Dec. 1998.

[CaiTarBig98]   G. Caire, G. Taricco and E. Biglieri, "Optimal power control for minimum outage rate in wireless communications," *Proc. of IEEE ICC '98*, Atlanta, GA, June 1998.

[CaiTarBig99]   G. Caire, G. Taricco and E. Biglieri, "Optimal power control for the fading channel," *IEEE Trans. Inform. Theory,* Vol. 45, No. 5, pp. 1468–1489, July 1999.

[ChaHwaLin97]   J.-J. Chang, D.-J. Hwang, and M.-C. Lin, "Some extended results on the search for good convolutional codes," *IEEE Trans. Inform. Theory*, Vol. 43, No. 5, pp. 1682–1697, Sept. 1997.

[ChiRimTel01]   N. Chiurtu, B. Rimoldi, and E. Telatar, "Dense multiple antenna systems," *Proc. of IEEE ITW 2001*, Cairns, Australia, pp. 108–109, Sept. 2–7, 2001.

[ChuTseKah98]   C.-N. Chuah, D. Tse, and J. M. Kahn, "Capacity of multi-antenna array systems in indoor wireless environment," *Proc. of IEEE GLOBECOM '98*, Sydney, Australia, Nov. 8–12, 1998.

[CovTho91]   T. M. Cover and J. A. Thomas, *Elements of Information Theory*. New York: Wiley, 1991.

[Foschini96]   G. J. Foschini, "Layered space–time architecture for wireless communication in a fading environment when using multi-element antennas," *Bell Labs Tech. J.*, Vol. 1, No. 2, pp. 41–59, Autumn 1996.

[FosGan98]   G. J. Foschini and M. J. Gans, "On limits of wireless communications in a fading environment when using multiple antennas," *Wireless Personal Communications*, Vol. 6, No. 3, pp. 311–335, March 1998.

[Gal68]   R. G. Gallager, *Information Theory and Reliable Communication*. New York: J. Wiley & Sons, 1968.

[GueFitBelKuo99] J.-C. Guey, M. P. Fitz, M. R. Bell, and W.-Y. Kuo, "Signal design for transmitter diversity wireless communication systems over Rayleigh fading channels," *IEEE Trans. Commun.*, Vol. 47, No. 4, pp. 527–537, April 1999.

[HasHocShoSwe01] B. Hassibi, B. M. Hochwald, A. Shokrollahi, and W. Sweldens, "Representation theory for high-rate multiple-antenna code design," *IEEE Trans. Inform. Theory*, Vol. 47, No. 6, pp. 2335–2367, Sept. 2001.

[HocMar98]   B. Hochwald and T. Marzetta, "Space–time modulation scheme for unknown Rayleigh fading environment," *Proc. of 36th Annual Allerton Conference on Communication, Control and Computing*, Allerton House, Monticello, IL, September 1998.

[HocMar00]   B. Hochwald and T. Marzetta, "Unitary space–time modulation for multiple-antenna communication in Rayleigh flat-fading," *IEEE Trans. Inform. Theory*, Vol. 46, No. 2, pp. 543–564, March 2000.

[HocMarTar02]   B. M. Hochwald, T. L. Marzetta, and V. Tarokh, "Multi-antenna channel-hardening and its implications for rate feedback and scheduling," *IEEE Trans. Inform. Theory*, submitted for publication, May 2002.

[HocSwe00]   B. Hochwald and W. Sweldens, "Differential unitary space time modulation," *IEEE Trans. Commun.*, Vol. 48, No. 12, pp. 2041–2052, Dec. 2000.

[HorJoh85]   R. Horn and C. Johnson, *Matrix Analysis*. New York: Cambridge University Press, 1985.

[Knopp97] R. Knopp, *Coding and Multiple-Accessing over Fading Channels,* PhD dissertation, EPFL, Lausanne (Switzerland), and Institut Eurécom, Sophia Antipolis (France), 1997.

[KnoHum00] R. Knopp and P. A. Humblet, "On coding for block-fading channels," *IEEE Trans. Inform. Theory,* Vol. 46, No. 1, pp. 189–205, Jan. 2000.

[MaGian02] X. Ma and G. B. Giannakis, "Full-diversity full-rate complex-field space–time coding," *IEEE Trans. Sig. Proc.*, Vol. 51, No. 11, pp. 2917–2930, Nov. 2003.

[MalLei97] E. Malkamäki and H. Leib, "Coded diversity schemes on block fading Rayleigh channels," *Proc. of IEEE ICUPC '97,* San Diego, CA, Oct. 1997.

[MalLei99] E. Malkamäki and H. Leib, "Coded diversity on block-fading channels," *IEEE Trans. Inform. Theory*, Vol. 45, No. 2, pp. 771–781, March 1999.

[Marzetta99] T. L. Marzetta, "BLAST training: Estimating channel characteristics for high capacity space–time wireless," *Proc. of 37th Annual Allerton Conference on Communication, Control and Computing*, Allerton House, Monticello, IL, pp. 958–966. Sept. 22–24, 1999.

[MarHoc99] T. L. Marzetta and B. M. Hochwald, "Capacity of a mobile multiple-antenna communication link in Rayleigh flat fading," *IEEE Trans. Inform. Theory*, Vol. 45, No. 1, pp. 139–157, January 1999.

[MarHocHas00] T. L. Marzetta, B. M. Hochwald, and B. Hassibi, "New approach to single-user multiple-antenna wireless communication," *Proc. of CISS 2000*, Princeton University, pp. WA4-16–WA4-21, March 15–17, 2000.

[MouSimSen03] A.L. Moustakas, S.H. Simon, A.M. Sengupta, "MIMO Capacity Through Correlated Channels in the Presence of Correlated Interferers and Noise: A (Not So) Large N Analysis," *IEEE Trans. Inform. Theory* Vol. 49, No. 10, pp. 2545–2561, Oct. 2003.

[OzaShaWyn94] L. Ozarow, S. Shamai, and A. D. Wyner, "Information theoretic considerations for cellular mobile radio," *IEEE Trans. Vehic. Tech.,* Vol. 43, No. 2, pp. 359–378, May 1994.

[RalCio98]   G. Raleigh and J. Cioffi, "Spatio-temporal coding for wireless communication," *IEEE Trans. Commun.*, Vol. 46, No. 3, pp. 357–366, March 1998.

[SenMit02]   A. M. Sengupta and P. P. Mitra, "Capacity of multivariate channels with multiplicative noise: Random matrix techniques and large-N expansions for full transfer matrices," *submitted for publication*, 2002.

[ShiKah99]   D. Shiu and J. M. Kahn, "Design of high-throughput codes for multiple-antenna wireless systems," Submitted to *IEEE Trans. Inform. Theory*, January 1999.

[ShiKah99a]   D. Shiu and J. M. Kahn, "Layered space–time codes for wireless communications using multiple transmit antennas," *Proc. of IEEE ICC'99*, Vancouver, BC, June 6–10, 1999.

[Sil95]   J. W. Silverstein, "Strong convergence of the empirical distribution of eigenvalues of large dimensional random matrices," *Journal of Multivariate Analysis*, Vol. 55, pp. 331–339, 1995.

[SmiSha02]   P. J. Smith and M. Shafi, "On a Gaussian approximation to the capacity of wireless MIMO systems," *Proc. of IEEE ICC 2002*, pp. 406–410, New York, April 28–May 2, 2002.

[SteDum99]   A. S. Stefanov and T. M. Duman, "Turbo coded modulation for systems with transmit and receive antenna diversity," *Proc. of IEEE GLOBECOM '99*, Rio de Janeiro, Brazil, pp. 2336–2340, Dec. 5–9, 1999.

[Szego39]   G. Szegö, *Orthogonal Polynomials*. Americal Mathematical Society, Providence, RI, 1939.

[TarBig03]   G. Taricco and E. Biglieri, "Space-time decoding with imperfect channel estimation," *IEEE Trans. Wireless Communications*, submitted for publication, August 2003.

[TarLo98]   V. Tarokh and T. K. Y. Lo, "Principal ratio combining for fixed wireless applications when transmitter diversity is employed," *IEEE Commun. Letters*, Vol. 2, No. 8, pp. 223–225, August 1998.

[TarSesCal98] V. Tarokh, N. Seshadri, and A. R. Calderbank, "Space–time codes for high data rate wireless communication: Performance criterion and code construction," *IEEE*

Trans. Inform. Theory Vol. 44, No. 2, pp. 744–765, March 1998.

[TarJafCal99]   V. Tarokh, H. Jafarkhani, and A. R. Calderbank, "Space–time block codes from orthogonal designs," IEEE Trans. Inform. Theory, Vol. 45, No. 5, pp. 1456–1467, July 1999.

[Telatar99]   E. Telatar, "Capacity of multi-antenna Gaussian channels," European Trans. Telecomm., Vol. 10, No. 6, pp. 585–595, November–December 1999.

[TseHanly98]   D. Tse and V. Hanly, "Multi-access fading channels—Part I: Polymatroid structure, optimal resource allocation and throughput capacities," IEEE Trans. Inform. Theory, Vol. 44, No. 7, pp. 2796–2815, November 1998.

[WanGia02]   Z. Wang and G. B. Giannakis, "Outage mutual information of space–time MIMO channels," Fortieth Annual Allerton Conference on Communication, Control, and Computing, Monticello, IL, October 2–4, 2002.

[ZheTse02]   L. Zheng and D. N. C. Tse, "Communication on the Grassman manifold: A geometric approach to the noncoherent multiple-antenna channel," IEEE Trans. Inform. Theory, Vol. 48, No. 2, pp. 359–383, February 2002.

[ZheTse03]   L. Zheng and D. N. C. Tse, "Diversity and multiplexing: A fundamental tradeoff in multiple antenna channels," IEEE Trans. Inform. Theory, Vol. 49, No. 5, pp. 1073–1096, May 2003.

Chapter 21

DIAGONAL STBC'S FOR FADING ISI CHANNELS: CODE DESIGN AND EQUALIZATION

Robert Schober

Department of Electrical and Computer Engineering
University of British Columbia
rschober@ece.ubc.ca

Wolfgang H. Gerstacker

Institute for Mobile Communication
University of Erlangen–Nuernberg
gersta@LNT.de

Lutz H.–J. Lampe

Department of Electrical and Computer Engineering
University of British Columbia
llampe@ece.ubc.ca

Subbarayan Pasupathy

Department of Electrical and Computer Engineering
University of Toronto
pas@comm.utoronto.ca

Abstract   In this chapter, we design and optimize matrix–symbol–based space–time block codes (STBC's) for transmission over fading intersymbol interference (ISI) channels. We show that STBC's employing diagonal code matrices exclusively facilitate the successful application of suboptimum equalization techniques for the practically important case when only a single receive antenna is available. Three different types of diagonal STBC's are optimized for fading ISI channels and their performances are compared for decision–feedback equalization (DFE) and decision–feedback sequence estimation (DFSE), respectively. The robustness of the designed codes against variations of the characteristics of the fading ISI

channel and the dependence of the equalizer performance on the STBC structure
are investigated.

It is shown that for data rates of $R \leq 2$ bits/(channel use) and typical multipath fading channels diagonal STBC's outperform Alamouti's code if suboptimum equalization schemes are adopted.

Keywords: Space–time block codes, multiple antennas, frequency–selective fading channels, equalization.

21.1   Introduction

It is well known that space–time (ST) codes can significantly improve the
performance of wireless communication systems, e.g. [1–4]. Initially, the investigation of ST coding was limited to flat fading channels. However, since
most wireless channels are impaired by intersymbol interference (ISI), more
recently considerable research effort has been dedicated to ST coding for
frequency–selective channels, cf. [5–14].

A fundamental problem of ST coding for fading ISI channels is to find
ST codes which achieve a high performance and at the same time allow the
application of suboptimum low–complexity equalization techniques such as
decision–feedback equalization (DFE) or reduced–state sequence estimation
(RSSE) [15, 16]. This is especially true if multiple transmit antennas along
with only a single receive antenna are applied. In this chapter, we will focus
on this important scenario which is likely to occur in down–link transmission
(base station to mobile station).

For ST trellis codes [3], reduced–state decoding with high performance is
only possible if the cascade of ST encoder and ISI channel can be interpreted as
a delay diversity scheme [17–19]. Recently, several ST block coding schemes
based on processing of entire bursts have been proposed, e.g. [5, 7, 8, 10,
11, 14]. The scheme in [8] is a generalization of Alamouti's ST block code
(STBC) [2], whereas the schemes in [5, 7, 11] and [10, 14] are based on orthogonal frequency division multiplexing (OFDM) and on single–carrier transmission combined with frequency domain equalization, respectively. These
burst–based STBC's achieve very good performance for block fading channels with constant or very slowly varying impulse response inside each burst
and, in addition, in most cases low–complexity equalization methods exist,
cf. [21] for a recent overview and comparison of several schemes. However, it
is not clear how these schemes can be extended to channels with time–variant
behavior. Therefore, in this contribution we adopt a different approach and
investigate the applicability of (matrix–) symbol–based STBC's for transmission over fading ISI channels, since in this case, an extension to time–variant
channels can be easily accomplished by adapting the equalizer filters using
least–mean–square (LMS) or recursive least–squares (RLS) algorithms.

The only matrix–symbol–based STBC that has been considered for fading ISI channels so far is Alamouti's code [2]. It has been found that with simple conventional equalizers a good performance can only be achieved if at least two receive antennas are available, e.g. [6, 9, 13]. More recently, however, suboptimum symbol–by–symbol [22] and trellis–based [23] equalizers, which are based on widely linear (WL) signal processing [24], have been proposed, see also [12]. These WL equalizers can achieve a high performance with a single receive antenna and will serve as a benchmark for the scheme proposed here.

In this chapter, we *optimize* ST block codes for fading ISI channels under the constraint that suboptimum equalization is used at the receiver. This naturally leads to *diagonal* STBC's where only one transmit antenna is active at any time. For diagonal STBC's employing N_T transmit antennas and one receive antenna an equivalent multiple–input multiple–output (MIMO) system with N_T inputs and N_T outputs exists. This equivalent MIMO system facilitates the application of both DFE and efficient RSSE techniques. Diagonal STBC's were reported first by DaSilva and Sousa [25][1]. Later, Hughes [26] and Hochwald and Sweldens [27] proposed diagonal STBC's which enable differential encoding. For optimization and comparison of different diagonal STBC's, we adapt the general design criterion for STBC's for fading ISI channels introduced in [28, 29] to the problem at hand.

Our simulation results show that for data rates of $R \leq 2$ bits/(channel use) diagonal STBC's combined with suboptimum equalization allow significant performance improvements over conventional (single antenna) transmission at the expense of a moderate increase in receiver complexity and compare favorably with the scheme proposed in [12].

This chapter is organized as follows. In Section 21.2, the transmission model is introduced, the adopted design criterion for STBC's is briefly reviewed, and the reasons for the application of diagonal STBC's are explained in detail. Various different diagonal STBC's are optimized and compared in Section 21.3. In Section 21.4, an equivalent MIMO model for diagonal STBC's is established and suboptimum equalizers are discussed. Simulation results are presented in Section 21.5 and some conclusions are drawn in Section 21.6.

21.2 Preliminaries

21.2.1 Notation

Bold upper case (X) and lower case (x) letters denote matrices and vectors, respectively. $\det(\cdot)$, $\mathrm{tr}(\cdot)$, $(\cdot)^T$, $(\cdot)^H$, and $(\cdot)^*$ refer to the determinant and the

---

[1]DaSilva and Sousa refer to their transmit diversity scheme as "fading–resistant modulation" [25]. Nevertheless, this method can be interpreted as a diagonal STBC.

trace of a matrix, transposition, Hermitian transposition, and complex conjugation, respectively. $\boldsymbol{I}_X$ and $\boldsymbol{0}_X$ are the $X \times X$ identity matrix and the all–zero column vector of size X, respectively. $\mathcal{E}\{\cdot\}$, mod, $j \stackrel{\triangle}{=} \sqrt{-1}$, and $\lfloor X \rfloor$ denote expectation, the modulo operator, the imaginary unit, and the largest integer smaller than or equal to X, respectively. Throughout this chapter, all signals are represented by their complex–baseband equivalents.

21.2.2   Transmission Model

In this chapter, we focus on ST block coded transmission employing N_T, $N_T \geq 2$, transmit and $N_R = 1$ receive antennas. The STBC matrix in its most general form is given by

$$\boldsymbol{C}[k_N] = \begin{pmatrix} c_0[Nk_N] & \cdots & c_{N_T-1}[Nk_N] \\ \vdots & \ddots & \vdots \\ c_0[Nk_N + N - 1] & \cdots & c_{N_T-1}[Nk_N + N - 1] \end{pmatrix} \quad (2.1)$$

with matrix discrete–time index $k_N \stackrel{\triangle}{=} \lfloor k/N \rfloor$ ($k \in \mathbb{Z}$: symbol discrete–time index), i.e., at time $k = Nk_N + \kappa, 0 \leq \kappa \leq N - 1$, the symbol $c_\nu[k]$ is transmitted from antenna ν, $0 \leq \nu \leq N_T - 1$. Thereby, $\boldsymbol{C}[\cdot]$ is taken from a set $\mathcal{C}$ of 2^{N_b} code matrices $\boldsymbol{C}^\alpha$, $0 \leq \alpha \leq 2^{N_b} - 1$. $c_\nu^\alpha[\kappa]$, $0 \leq \nu \leq N_T - 1$, $0 \leq \kappa \leq N - 1$, denotes the element of $\boldsymbol{C}^\alpha$ in column ν and row κ. $N_b = RN$ information bits are mapped to the elements of $\mathcal{C}$, where R is the data rate in bits per channel use. We adopt the normalization $\sum_{\nu=0}^{N_T-1} \mathcal{E}\{|c_\nu[\kappa]|^2\} = 1$ in order to make the transmitted energy independent of the number of transmit antennas N_T.

The ST block coded signal is transmitted over a fading ISI channel and after receive filtering and T–spaced sampling (T denotes the symbol interval) the received signal $r[k]$ is modeled as

$$r[k] = \sum_{\nu=0}^{N_T-1} \sum_{l=0}^{L-1} h_\nu[l]c_\nu[k - l] + n[k], \quad (2.2)$$

where $h_\nu[l]$ and $n[k]$ refer to the lth coefficient of the discrete–time overall channel impulse response (CIR) between the νth transmit antenna and the receive antenna and discrete–time additive white Gaussian noise (AWGN), respectively. The discrete–time CIR's $h_\nu[\cdot]$, $0 \leq \nu \leq N_T - 1$, are truncated to the same length L, where $\sum_{l=-\infty}^{-1} \mathcal{E}\{|h_\nu[l]|^2\} + \sum_{l=L}^{\infty} \mathcal{E}\{|h_\nu[l]|^2\} \approx 0$ is assumed. Furthermore, $h_\nu[\cdot]$ includes the combined effects of transmit filter, multipath Rayleigh fading channel, and (square–root Nyquist) receiver input filter. Therefore, for a given ν the discrete–time CIR coefficients $h_\nu[l]$ are mutually correlated zero–mean Gaussian random variables. For simplicity,

for different transmit antennas the CIR coefficients are assumed to be mutu-
ally uncorrelated, i.e., $\mathcal{E}\{h_{\nu_1}[l_1]h_{\nu_2}^*[l_2]\} = 0$, $\nu_1 \neq \nu_2$, $\forall l_1, l_2$, but to have
identical statistical properties. Nevertheless, most concepts introduced in this
chapter can be extended in a straightforward way to mutually correlated CIR's.
We assume that $h_\nu[\cdot]$ is constant over one burst but varies from burst to burst,
e.g. due to frequency hopping (block fading model). Such a model is realistic
for moderate burst lengths and moderate fading velocities. As customary in
practice, we adopt a (suboptimum) fixed square–root Nyquist receiver input
filter instead of the optimum matched filter. This facilitates implementation
and usually causes only a negligible loss in performance [30].

Because of an appropriate normalization, $\sum_{l=0}^{L-1} \mathcal{E}\{|h_\nu[l]|^2\} = 1$, $0 \leq \nu \leq$
$N_T - 1$, is valid and the noise variance is given by $\sigma_n^2 \triangleq \mathcal{E}\{|n[k]|^2\} =$
$N_0/(RE_b)$, where N_0 and E_b denote the single–sided power spectral density of
the underlying continuous–time passband noise process and the mean received
energy per bit, respectively.

21.2.3   Design Criterion for STBC's

In [28, 29], a lower bound for the pairwise error probability (PEP) $P_e(\alpha, \beta)$,
i.e., the probability that C_α is transmitted and C_β, $\beta \neq \alpha$, is detected, is given
for the transmission scheme described in the previous subsection. This bound
is related to the classical matched filter bound, e.g. [31], since it was obtained
by considering the transmission of a single STBC matrix. A Chernoff bound
on this PEP is given by [29]

$$P(\alpha, \beta) \leq \frac{1}{\prod\limits_{q=1}^{Q} \lambda_q \left(\dfrac{1}{4\sigma_n^2}\right)^Q}, \tag{2.3}$$

where Q and λ_q, $1 \leq q \leq Q$, are the rank and the non–zero eigenvalues of
matrix $C_m^{\alpha|\beta} \mathbf{\Phi}_{hh} (C_m^{\alpha|\beta})^H$, respectively. Thereby, $\mathbf{\Phi}_{hh}$ is the autocorrelation
matrix (ACM) of all CIR coefficients and is defined as

$$\mathbf{\Phi}_{hh} \triangleq \mathcal{E}\{\mathbf{h}\mathbf{h}^H\} \tag{2.4}$$

with

$$\mathbf{h} \triangleq [\mathbf{h}_0^T \ \ldots \ \mathbf{h}_{N_T-1}^T]^T \tag{2.5}$$

$$\mathbf{h}_\nu \triangleq [h_\nu[0] \ \ldots \ h_\nu[L-1]]^T. \tag{2.6}$$

Matrix

$$C_m^{\alpha|\beta} \triangleq C_m^\beta - C_m^\alpha \tag{2.7}$$

denotes the difference of two *modified* STBC matrices $\boldsymbol{C}_m^{x}$, $x \in \{\alpha, \beta\}$, which are uniquely specified by the original STBC matrices $\boldsymbol{C}^x$ via

$$\boldsymbol{C}_m^x \triangleq [\boldsymbol{C}_0'^x \; \dots \; \boldsymbol{C}_{N_T-1}'^x], \tag{2.8}$$

$$\boldsymbol{C}_\nu'^x \triangleq [\boldsymbol{c}_{1\nu}^x \; \dots \; \boldsymbol{c}_{L\nu}^x], \tag{2.9}$$

$$\boldsymbol{c}_{l\nu}^x \triangleq [\boldsymbol{0}_{l-1}^T \; \boldsymbol{c}_\nu^x[0] \; \dots \; \boldsymbol{c}_\nu^x[N-1] \; \boldsymbol{0}_{L-l}^T]^T. \tag{2.10}$$

From Eq. (2.3) it has been deduced in [28, 29] that the STBC set $\mathcal{C}$ should be selected to maximize the diversity order Q. For most practical channels it can be assumed that a STBC can be found for which matrix $\boldsymbol{C}_m^{\alpha|\beta} \boldsymbol{\Phi}_{hh} (\boldsymbol{C}_m^{\alpha|\beta})^H$ has full rank for all pairs (α, β), $\alpha \neq \beta$, provided that the CIR truncation length L and N (see next subsection) are chosen appropriately. Therefore, we may define the distance measure

$$d^2(\alpha, \beta) \triangleq \det \left(\boldsymbol{C}_m^{\alpha|\beta} \boldsymbol{\Phi}_{hh} (\boldsymbol{C}_m^{\alpha|\beta})^H \right). \tag{2.11}$$

If $\boldsymbol{C}_m^{\alpha|\beta} \boldsymbol{\Phi}_{hh} (\boldsymbol{C}_m^{\alpha|\beta})^H$ is not full rank, in Eq. (2.11) the determinant of the matrix has to be replaced by the product of the Q non–zero eigenvalues.

Now, a reasonable STBC design criterion for fading ISI channels is to maximize the minimum distance[2]

$$d_{min}^2 \triangleq \underset{\substack{(\alpha, \beta) \\ \beta \neq \alpha}}{\text{argmin}} \left\{ d^2(\alpha, \beta) \right\}. \tag{2.12}$$

The optimum code among all STBC's which achieve the maximum possible diversity order is that one which maximizes d_{min}^2. Note that for $L = 1$ and $N = N_T$ the above design rule is identical to the conventional design rule for flat fading channels [3].

21.2.4   Why Diagonal STBC's?

A primary goal of this chapter is to find STBC's which also provide a performance improvement compared to single–antenna transmission if suboptimum (low–complexity) equalization strategies are applied. The straightforward approach to this problem would be to optimize the STBC given the suboptimum equalization scheme (e.g. DFE) used. For this purpose we would need to derive a corresponding design criterion. Unfortunately, this is a very difficult if not impossible task and a practical design criterion is not likely to exist. On the other hand, the unconstrained maximization of the minimum distance

---

[2]A related design criterion for ST trellis codes has been proposed independently by Liu et al. [32].

d_{min}^2 will lead to STBC's which perform very well for computationally complex maximum–likelihood sequence detection (MLSD) [28, 29] but may be not suitable at all for application of low–complexity suboptimum equalization techniques. Therefore, we have to impose additional constraints on the structure of the STBC matrix to get viable solutions.

First, we should be aware that, in general, the $N_T N$ elements of the transmitted STBC matrix $C[\cdot]$ are not correlated. If we consider e.g. Alamouti's code with $N = N_T = 2$,

$$C[k_2] = \frac{1}{\sqrt{2}} \begin{pmatrix} b_0[k_2] & b_1[k_2] \\ -b_1^*[k_2] & b_0^*[k_2] \end{pmatrix} \tag{2.13}$$

is valid, where $b_0[\cdot]$ and $b_1[\cdot]$ are independently taken from the same M–ary phase–shift keying (MPSK) or M–ary quadrature amplitude modulation (MQAM) signal alphabet [2]. Clearly, in this case, for rotationally invariant complex signal constellations such as 4PSK or 16QAM the elements of $C[\cdot]$ are mutually uncorrelated. Since the four elements of $C[\cdot]$ are transmitted in two time steps, a separation of these elements at the receiver using equalizers which rely on second–order statistics is only possible if $N_R \geq 2$ receive antennas are employed. This problem could be slightly relaxed of course if oversampling of the received continuous–time signal was applied, cf. e.g. [33]. However, since, in general, the polyphase components of the oversampled continuous–time overall CIR are strongly correlated, the resulting receivers would be very sensitive to channel noise. For the special case of Alamouti's code a solution to this problem is given in [12] using widely linear signal processing [24]. However, for a general STBC the $N_T N$ elements of matrix $C[\cdot]$ are not separable at the receiver. To circumvent the above mentioned problems, we transmit in N time steps only N symbols since in this case the (non–zero) elements of $C[\cdot]$ can be separated at the receiver.

It has been shown in [28] that for mutually uncorrelated CIR coefficients the maximum possible diversity order $Q = LN_T$ can only be achieved, if N is chosen as $N = (N_T - 1)L + 1$. However, our investigations have shown that if only one transmit antenna is active at any time, $N = N_T$ yields the best performance for bit error rates (BER's) of practical interest, while $N > N_T$ is only beneficial at very low BER's. Therefore, for simplicity of presentation, in the rest of this chapter we restrict our attention to the case $N = N_T$, i.e., to diagonal STBC matrices $C[\cdot]$.

21.3 Design of Diagonal STBC's

In this section, three different types of diagonal STBC's are discussed and optimized for the widely accepted hilly terrain (HT) channel model [34] using the criterion given in Eq. (2.12). The obtained designs are then compared based on an approximation for the BER for MLSD [28].

For diagonal STBC's the only non–zero elements of the code matrix C^α are $c_\nu^\alpha[\nu]$, $0 \le \nu \le N_T - 1$. Therefore, for the following, we introduce the simpler notation $c_\nu^\alpha \triangleq c_\nu^\alpha[\nu]$.

21.3.1   Repetition Codes (RC's)

The simplest diagonal STBC's are repetition codes which will serve as a benchmark for more sophisticated codes. For N_T transmit antennas and RC's of rate R, $c_\nu^\alpha = b^\alpha$, $0 \le \nu \le N_T - 1$, is valid, where b^α are MPSK or MQAM symbols with $M = 2^{N_T R}$. If we take in Eq. (2.12) into account that all c_ν^α are identical, we obtain $d_{min}^2 = x_h \cdot d_{b,min}^2$, where x_h is a constant which depends only on the channel correlation matrix Φ_{hh} and $d_{b,min}^2$ is the minimum squared Euclidean distance of the adopted MPSK or MQAM signal constellation. Therefore, as one would intuitively expect, in order to maximize d_{min}^2 for given R and N_T a scalar signal constellation which has the maximum minimum squared Euclidean distance has to be selected. In addition, a Gray labeling of the scalar signal constellation constitutes also a Gray labeling for the RC.

21.3.2   Hochwald/Sweldens Codes (HS–C's) [27]

HS–C's have been introduced in [27] for differential ST transmission. In this work, we focus on coherent transmission, i.e., differential encoding is not applied. Thus, the transmitted symbols are given by

$$c_\nu^\alpha = \exp\left(j\frac{2\pi u_\nu \alpha}{2^{N_T R}}\right), \tag{3.1}$$

$0 \le \nu \le N_T - 1$, $0 \le \alpha \le 2^{N_T R} - 1$. The parameters u_ν, $0 \le u_\nu \le 2^{N_T R} - 1$, are integers[3] which are optimized in [27] for differential detection and flat Rayleigh fading channels. Here, we have to optimize the parameter vector $u \triangleq [u_0 \ \ldots \ u_{N_T-1}]$ for coherent reception and Rayleigh fading ISI channels. We note that since we assume that the CIR coefficients pertaining to different transmit antennas are mutually uncorrelated, Φ_{hh} is a block diagonal matrix. Therefore, for HS–C's the distance $d^2(\alpha, \beta)$ defined in Eq. (2.11) depends only on the difference $(\alpha - \beta)\mathrm{mod}(2^{N_T R})$. Hence, the optimum parameter vector u_{opt} can be obtained from

$$u_{opt} = \underset{0 \le u_0, \ldots, u_{N_T-1} \le 2^{N_T R} - 1}{\mathrm{argmax}} \left\{ \underset{\alpha = 1, \ldots, 2^{N_T R} - 1}{\min} \left\{ d^2(\alpha, 0) \right\} \right\}, \tag{3.2}$$

---

[3]Since we do not employ differential encoding, we could also allow non–integer values for the u_ν. However, this complicates the code search and is expected to yield only marginal performance improvements.

i.e., we have to search over $2^{N_T^2 R}$ possible parameter vectors $\boldsymbol{u}$. Using similar methods as in [27] the search space may be reduced to some extent. However, we do not expand on this topic since HS–C's are mainly beneficial for low data rates and a small number of transmit antennas implying a moderate code search complexity.

21.3.3   Rotated MASK (R–MASK) and MQAM (R–MQAM) Constellations

For the flat fading channel, diagonal STBC's have been designed in [25] by rotating MASK constellations[4]. For this, we define the vector $\boldsymbol{b}^\alpha \triangleq [b_0^\alpha \; \cdots \; b_{N_T-1}^\alpha]^T$ of MASK symbols and the rotated vector $\boldsymbol{c}^\alpha \triangleq [c_0^\alpha \; \cdots \; c_{N_T-1}^\alpha]^T$, which is obtained from

$$\boldsymbol{c}^\alpha = \boldsymbol{R}\boldsymbol{b}^\alpha, \tag{3.3}$$

where $\boldsymbol{R}$ is a real orthogonal matrix. Note that because of the orthogonality of $\boldsymbol{R}$, the transformation according to Eq. (3.3) preserves the Euclidean distance. A canonical representation of $\boldsymbol{R}$ is given by [35, 25]

$$\boldsymbol{R} = \prod_{\nu=0}^{N_T-1} \prod_{\mu=\nu+1}^{N_T-1} \boldsymbol{T}_{\nu\mu}, \tag{3.4}$$

where the elementary rotation matrix $\boldsymbol{T}_{\nu\mu}$ differs from the identity matrix only in four elements: the main diagonal elements in row ν and row μ are $\cos(\phi_{\nu\mu})$, respectively, while the elements in row ν, column μ and row μ, column ν are $-\sin(\phi_{\nu\mu})$ and $\sin(\phi_{\nu\mu})$, respectively. Hence, matrix $\boldsymbol{R}$ is uniquely specified by $N_T(N_T-1)/2$ phases $\phi_{\nu\mu}$ reflecting the $N_T(N_T-1)/2$ degrees of freedom of an $N_T \times N_T$ orthogonal matrix [35]. In [25] these $N_T(N_T-1)/2$ phases are optimized for maximization of the minimum product distance.

Here, an approach more general than the one in [25] is considered. First, in general, we adopt complex signal constellations (MQAM) and, second, we allow complex unitary matrices $\boldsymbol{R}$. In particular, motivated by Eq. (3.4), we propose the construction

$$\boldsymbol{R} = \prod_{\nu=0}^{N_T-1} \prod_{\mu=\nu+1}^{N_T-1} \boldsymbol{T}_{\nu\mu}^0 \boldsymbol{T}_{\nu\mu}^1, \tag{3.5}$$

where $\boldsymbol{T}_{\nu\mu}^0$ is defined in the same way as $\boldsymbol{T}_{\nu\mu}$ but the corresponding phase is denoted by $\phi_{\nu\mu}^0$. $\boldsymbol{T}_{\nu\mu}^1$ also differs from the identity matrix only in four

---

[4]Similar methods have also been used for construction of modulation diversity schemes, e.g. [35, 36].

elements: the main diagonal elements in row ν and row μ are $\cos(\phi^1_{\nu\mu})$, respectively, while the elements in row ν, column μ and row μ, column ν are $j\sin(\phi^1_{\nu\mu})$, respectively. The adopted construction allows for $N_T(N_T - 1)$ degrees of freedom (parameter vectors $\boldsymbol{\phi}^\mu \triangleq [\phi^\mu_{01} \cdots \phi^\mu_{0\,N_T-1} \ \phi^\mu_{12} \cdots \phi^\mu_{N_T-2\,N_T-1}]$, $\mu \in \{0, 1\}$), while general unitary matrices have N_T^2 degrees of freedom [37, Appendix D]. The "missing" N_T degrees of freedom can be realized by multiplying $\boldsymbol{R}$ (as given in Eq. (3.5)) by a diagonal matrix with main diagonal entries $\exp(j\phi^2_\nu)$, $0 \leq \nu \leq N_T - 1$. However, because of the block diagonal structure of $\boldsymbol{\Phi}_{hh}$, this would not change the distance $d^2(\alpha, \beta)$ according to Eq. (2.11). For an alternative parameterization of unitary matrices we refer to [38, Appendix].

The optimum vectors $\boldsymbol{\phi}^0_{opt}$ and $\boldsymbol{\phi}^1_{opt}$ (and consequently the optimum rotation matrix $\boldsymbol{R}_{opt}$) should maximize the minimum distance d^2_{min} (Eq. (2.12)). However, as outlined in [25] for a much simpler distance measure, a closed–form optimization seems to be not possible. Therefore, we optimized $\boldsymbol{\phi}^0$ and $\boldsymbol{\phi}^1$ by searching in discretized intervals. In order to speed up the search, we used several iterations where the interval size and the discretization size were reduced successively. The obtained parameter vectors are not necessarily the optimum ones, but, in general, they ensure a high performance.

21.3.4   Results and Discussion

In this section, we present some optimized diagonal STBC's and discuss their performance. For this, we need to specify the channel. Here, we adopt the HT channel model [34]. For transmit and receive filtering a linearized Gaussian minimum–shift keying (GMSK) impulse [39] and a square–root raised cosine filter with roll–off factor 0.3 are adopted, respectively. This choice is motivated by the GSM/EDGE (Global System for Mobile Communications/Enhanced Data Rates for GSM Evolution) system, e.g. [30]. The discrete–time CIR was truncated to $L = 7$.

For the comparison of the optimized diagonal STBC's we use an approximation of the BER for MLSD, which is based on a "quasi" matched filter bound for the PEP (cf. comments in Section 21.2). The approximation for BER is obtained by taking the union bound over the lower bound for PEP for all possible pairs (α, β), $\alpha \neq \beta$, $0 \leq \alpha, \beta \leq 2^{N_T R} - 1$ [28, 29].

All HS–C's and R–MASK's/R–MQAM's presented in the following were obtained as described in the previous two subsections, respectively. For all codes the labeling of the elements of the code set $\mathcal{C}$ was optimized with respect to the adopted distance measure (Eq. (2.11)).

In Fig. 21.1a) and b), we consider STBC's with $N_T = 2$ and $N_T = 3$, respectively. The rate is $R = 1$ bit/(channel use) and for comparison also results for BPSK with and without receive diversity are shown. Since in Fig. 21.1

the E_b/N_0 ratio is normalized by N_R, the curves for receive diversity constitute (attainable) lower bounds for the achievable BER for transmit diversity.

For $N_T = 2$ the optimum HS–C parameter vector is $\boldsymbol{u}_{opt} = [1\ 1]$, i.e., the HS–C is identical to the QPSK–RC. For R–2ASK we found the parameter vectors $\boldsymbol{\phi}^0 = [0^o]$ and $\boldsymbol{\phi}^1 = [45^o]$ (not necessarily optimum because of the suboptimum search method described in the previous subsection). Both HS–C and R–2ASK approach BPSK with $N_R = 2$, i.e., it can be expected that for MLSD they perform equally well. For $N_T = 3$, the optimum HS–C parameter vector is $\boldsymbol{u}_{opt} = [1\ 2\ 3]$, but the vector $\boldsymbol{u} = [1\ 1\ 3]$ yields only a slightly inferior performance. The performance of the proposed R–2ASK constellation with $\boldsymbol{\phi}^0 = [22^o\ 27^o\ 11^o]$ and $\boldsymbol{\phi}^1 = [46^o\ 28^o\ 45^o]$ is not much worse. On the other hand, at high signal–to–noise ratios (SNR's) 8PSK–RC imposes a performance penalty of about 3.5 dB compared to the optimum HS–C.

Diagonal STBC's with rates of $R = 2\,\text{bits}/(\text{channel use})$ and $N_T = 2$ are studied in Fig. 21.2. For comparison QPSK with and without receive diversity is also considered. Now, R–4QAM with $\boldsymbol{\phi}^0 = [59^o]$ and $\boldsymbol{\phi}^1 = [45^o]$ yields the best performance. The optimum HS–C parameter vector is $\boldsymbol{u}_{opt} = [1\ 7]$. The poorer performance of the HS–C compared to R–4QAM can be explained by the fact that for HS–C's all transmitted symbols $c_\nu[k]$ have unit magnitude which imposes additional constraints (besides the diagonal STBC matrix structure) and limits performance for higher rates. As can be observed, at high SNR's 16QAM–RC performs about 3 dB worse than R–4QAM.

A comparison of Figs. 21.1 and 21.2 reveals that the gap between receive diversity and the best diagonal STBC is larger for $R = 2\,\text{bits}/(\text{channel use})$ than for $R = 1\,\text{bit}/(\text{channel use})$. It can be expected that this gap further increases for higher rates and a larger number of transmit antennas. In these cases, the constraints imposed by the restriction to diagonal STBC matrices may be too stringent to yield high power efficiency.

21.3.5   Robustness of the Designed Codes

In the previous subsections, we designed codes for special fading ISI channels. A natural question is how these codes will perform if the underlying channel does not have exactly the properties we assumed for code design, which is likely to be the case in practice. Thus, in Fig. 21.3 we compare HS–C's ($R = 1\,\text{bit}/(\text{channel use})$) with $\boldsymbol{u} = [1\ 1\ 3]$ and $\boldsymbol{u} = [1\ 2\ 3]$, respectively. In Fig. 21.3a), we consider a channel whose discrete–time CIR's $h_\nu[\cdot]$, $0 \leq \nu \leq 2$, have $L = 2$ taps of equal variance. The taps are mutually correlated with correlation coefficient $r \overset{\triangle}{=} \mathcal{E}\{h_\nu[0]h_\nu^*[1]\}/\mathcal{E}\{|h_\nu[0]|^2\}$. Our code search showed that as r tends to 1, $\boldsymbol{u} = [1\ 1\ 3]$ becomes optimum, whereas otherwise $\boldsymbol{u} = [1\ 2\ 3]$ is the optimum choice. In Fig. 21.3b), fading ISI channels with L mutually uncorrelated taps of equal variance are consid-

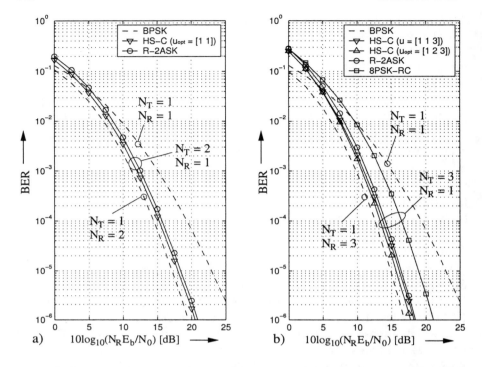

Figure 21.1. Approximate BER for MLSD vs. $10\log_{10}(N_R E_b/N_0)$ for various diagonal STBC's with a) $N_T = 2$ and b) $N_T = 3$. The rate is $R = 1$ bit/(channel use) and the HT channel is adopted. For comparison also the results for BPSK with and without receive diversity are shown.

ered. For $L = 1$ and $L > 1$ the parameter vectors $\boldsymbol{u} = [1\ 1\ 3]$ and $\boldsymbol{u} = [1\ 2\ 3]$ are optimum, respectively. Both Figs. 21.3a) and b) show that although the optimum STBC depends on the underlying channel, the designed codes are quite robust against variations of the channel characteristics. This claim is also supported by the fact that although the HS–C's in [27] were optimized for the flat fading channel (and differential encoding), for $N_T = 2$ and rates of $R = 1$ bit/(channel use) and $R = 2$ bits/(channel use) the same codes were found as in this chapter for the HT profile (and coherent transmission). Thus, in practice, we might design the STBC for a typical power delay profile and high performance can be expected even if the actual channel profile deviates to some extent from the assumed one.

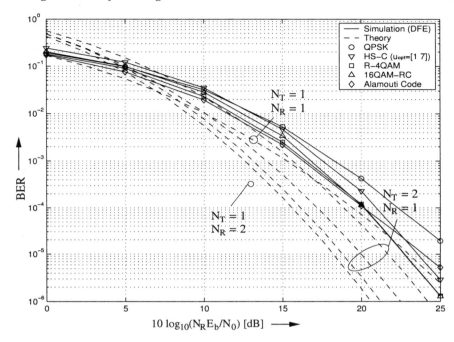

Figure 21.2. Approximate BER for MLSD vs. $10 \log_{10}(N_R E_b/N_0)$ for various diagonal STBC's with $N_T = 2$. The rate is $R = 2$ bits/(channel use) and the HT channel is adopted. For comparison also the results for QPSK with and without receive diversity are shown.

21.4 Equalization

In this section, suboptimum equalization strategies for diagonal STBC's are proposed. However, before we proceed, an equivalent $N_T \times N_T$ MIMO channel model is introduced.

21.4.1 MIMO Model

The received signal $r[k]$ according to Eq. (2.2) is cyclostationary with period $N = N_T$. Therefore, a corresponding equalizer would have to be time–variant. In order to circumvent this, we may consider the vector signal $\boldsymbol{r}[k] \triangleq [r[N_T k]\ r[N_T k + 1]\ \ldots\ r[N_T k + N_T - 1]]^{T\,5}$, which is stationary [40]. Taking the special structure of diagonal STBC's into account, from Eq. (2.2) we

---

[5]To be fully consistent with Section 21.2, $\boldsymbol{r}[k]$ would have to be replaced by $\boldsymbol{r}[k_{N_T}]$. However, in the following we simplify our notation and use k instead of k_{N_T}.

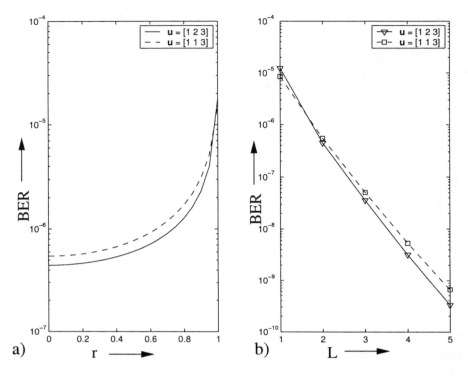

Figure 21.3. Approximate BER for MLSD vs. a) correlation coefficient r and b) L for two different HS–C's with $N_T = 3$. $R = 1$ bit/(channel use) and $10 \log_{10}(E_b/N_0) = 20$ dB is valid.

obtain

$$r[k] = \sum_{l=0}^{\tilde{L}-1} H[l]c[k-l] + n[k] \tag{4.1}$$

with the MIMO channel length $\tilde{L} \triangleq \lfloor (L-2)/N_T \rfloor + 2$ and the definitions

$$H[l] \triangleq \begin{pmatrix} h_0[N_T l] & \cdots & h_{N_T-1}[N_T l - N_T + 1] \\ h_0[N_T l + 1] & \ddots & \vdots \\ \vdots & \ddots & \vdots \\ h_0[N_T l + N_T - 1] & \cdots & h_{N_T-1}[N_T l] \end{pmatrix} \tag{4.2}$$

$$c[k] \triangleq [c_0[N_T k]\ c_1[N_T k + 1]\ \ldots\ c_{N_T-1}[N_T k + N_T - 1]]^T \tag{4.3}$$

$$n[k] \triangleq [n[N_T k]\ n[N_T k + 1]\ \ldots\ n[N_T k + N_T - 1]]^T. \tag{4.4}$$

Observe that in $\boldsymbol{H}[\cdot]$, $h_\nu[l] = 0$ for $l < 0$ and $l \geq L$ is valid. $\boldsymbol{r}[\cdot]$ may be interpreted as the output of an $N_T \times N_T$ MIMO system with (matrix) impulse response $\boldsymbol{H}[\cdot]$. Time–invariant linear equalization (LE) and DFE of such a system is possible with mild conditions on $\boldsymbol{H}[\cdot]$, e.g. [40–42]. On the other hand, a general non–diagonal STBC corresponds to a MIMO system with $N_T N$ inputs and N outputs, and consequently LE and DFE lead to very poor results for $N_T > 1$.

21.4.2   Decision–Feedback Equalization

DFE appears to be a favorable equalization scheme for the above MIMO system. In a MIMO DFE the vector signal $\boldsymbol{y}[k]$ at the input of the decision device is given by (cf. Fig. 21.4)

$$\boldsymbol{y}[k] = \sum_{l=0}^{L_F-1} \boldsymbol{W}[l]\boldsymbol{r}[k-l] - \sum_{l=1}^{L_B} \boldsymbol{B}[l]\hat{\boldsymbol{c}}[k-l-k_0], \qquad (4.5)$$

where $\boldsymbol{W}[l]$, $0 \leq l \leq L_F - 1$, and $\boldsymbol{B}[l]$, $1 \leq l \leq L_B$, are the $N_T \times N_T$ matrix filter coefficients of the feedforward (FF) and the feedback (FB) filter, respectively. L_F, L_B, and k_0 are the FF filter length, the FB filter length, and a suitably chosen decision delay, respectively. $\hat{\boldsymbol{c}}[\cdot]$ denotes the estimate for $\boldsymbol{c}[\cdot]$. Various methods for optimization of the FF and FB filters can be found in the literature, e.g. [40–42]. In this chapter, we adopt the finite impulse response (FIR) minimum mean–squared error (MMSE) MIMO DFE proposed in [41, Scenario 1]. This approach has the advantage that fast algorithms for closed–form filter calculation are available [41] (for this, the channel noise variance σ_n^2 and the CIR coefficients $h_\nu[l]$, $0 \leq l \leq L - 1$, $0 \leq \nu \leq N_T - 1$, have to be known) and alternatively, the filters may be calculated recursively using adaptive algorithms, e.g. [43].

If the FF and FB filters are adjusted properly, $\boldsymbol{y}[k]$ can be modeled as

$$\boldsymbol{y}[k] = \boldsymbol{c}[k - k_0] + \boldsymbol{e}[k], \qquad (4.6)$$

where $\boldsymbol{e}[\cdot]$ denotes the error signal vector. Although it can be expected that $\boldsymbol{e}[k]$ is a temporally white process provided that the FF and FB filters are sufficiently long [44], in general, for a given k the elements of $\boldsymbol{e}[k]$ are mutually correlated. The error autocorrelation matrix

$$\boldsymbol{\Phi}_{ee} \triangleq \mathcal{E}\{\boldsymbol{e}[k]\boldsymbol{e}^H[k]\} \qquad (4.7)$$

can be calculated in closed–form from [41, Eq. (16)] or estimated using standard techniques, e.g. [45].

Although $\boldsymbol{e}[k]$ is not exactly Gaussian distributed, in general, for long FF and FB filters its probability density function (pdf) may be well approximated as a multivariate Gaussian distribution. Consequently, the vector $c[k - k_0]$ which contains the diagonal elements of the transmitted STBC matrix may be obtained from

$$\hat{c}[k - k_0] = \underset{c[k-k_0]}{\mathrm{argmin}} \left\{ (y[k] - c[k - k_0])^H \Phi_{ee}^{-1} (y[k] - c[k - k_0]) \right\}. \quad (4.8)$$

In order to reduce the computational complexity, the *Cholesky factorization*

$$\Phi_{ee}^{-1} \triangleq L^H \Phi_{dd}^{-1} L \quad (4.9)$$

may be used. Here, L is a lower triangular matrix, whose main diagonal elements are all ones, and Φ_{dd} is a diagonal matrix with main diagonal entries d_ν^2, $0 \leq \nu \leq N_T - 1$. If we precompute

$$\begin{aligned} \bar{y}[k] &\triangleq [\bar{y}[N_T k] \ \ldots \ \bar{y}[N_T k + N_T - 1]]^T \\ &\triangleq L y[k] \end{aligned} \quad (4.10)$$

and all possible vectors

$$\begin{aligned} \bar{c}[k] &\triangleq [\bar{c}_0[N_T k] \ \ldots \bar{c}_{N_T-1}[N_T k + N_T - 1]^T \\ &\triangleq L c[k], \end{aligned} \quad (4.11)$$

Eq. (4.8) simplifies to

$$\hat{\bar{c}}[k - k_0] = \underset{\bar{c}[k-k_0]}{\mathrm{argmin}} \left\{ \sum_{\nu=0}^{N_T-1} \frac{1}{d_\nu^2} |\bar{y}[N_T k + \nu] - \bar{c}_\nu[N_T(k - k_0) + \nu]|^2 \right\}, \quad (4.12)$$

where the estimate for the original signal vector can be obtained from $\hat{c}[k - k_0] = L^{-1}\hat{\bar{c}}[k - k_0]$.

Another (suboptimum) possibility to reduce computational complexity in Eq. (4.8) is to neglect the mutual correlations of the elements of $e[\cdot]$. In this case, also Eq. (4.12) can be used for determination of $\hat{c}[k - k_0]$, however, now $L \triangleq I_{N_T}$ is valid and d_ν^2 has to be replaced by the νth main diagonal element of the error ACM Φ_{ee}. The associated total error variance relevant for the performance of the MIMO DFE is

$$\sigma_e^2 \triangleq \mathrm{tr}\{\Phi_{ee}\}, \quad (4.13)$$

whereas the relevant error variance for the first approach is

$$\sigma_d^2 \triangleq \mathrm{tr}\{\Phi_{dd}\}. \quad (4.14)$$

Since the Cholesky factorization accomplishes the (spatial) whitening of $e[k]$, i.e., the components of $Le[k]$ are mutually uncorrelated,

$$\sigma_d^2 \leq \sigma_e^2 \tag{4.15}$$

is valid (with equality if and only if the components of $e[k]$ are mutually uncorrelated, e.g. [45]).

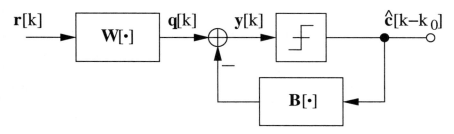

Figure 21.4. Block diagram of receiver structure for MIMO DFE.

21.4.3 Decision–Feedback Sequence Estimation (DFSE)

The performance gap between DFE and MLSD can be bridged by RSSE techniques at the expense of additional complexity [15]. Here, we focus on DFSE [16]. In general, DFSE can yield high performance only if the overall MIMO channel transfer function is minimum phase, i.e., an appropriate prefilter has to be applied to transform the original MIMO channel transfer function

$$H(z) \stackrel{\triangle}{=} \sum_{l=0}^{\tilde{L}-1} H[l]z^{-l} \tag{4.16}$$

into its minimum–phase equivalent $H_{min}(z)$. Similar to the single–input single–output (SISO) case [46], also for the MIMO case the FIR–MMSE–DFE FF filter appears to be a favorable choice for the prefilter since it accomplishes an approximate minimum–phase transformation and can be computed easily.

For DFSE the absolute metric

$$\Lambda[k+1] = \Lambda[k] + \lambda[k] \tag{4.17}$$

can be minimized by applying the Viterbi algorithm (VA) [47, 48] on a trellis with $Z = 2^{N_T RK}$ states. The design parameter K, $0 \leq K \leq L_B$, allows to adjust the desired trade–off between complexity and power efficiency. The DFSE branch metric $\lambda[k]$ is given by

$$\lambda[k] = \left(q[k] - \sum_{l=0}^{K} B[l]\check{c}[k - k_0 - l] - \sum_{l=K+1}^{L_B} B[l]\hat{c}[k - k_0 - l] \right)^H \Phi_{ee}^{-1}$$

$$\cdot \left(q[k] - \sum_{l=0}^{K} \boldsymbol{B}[l]\tilde{\boldsymbol{c}}[k - k_0 - l] - \sum_{l=K+1}^{L_B} \boldsymbol{B}[l]\hat{\boldsymbol{c}}[k - k_0 - l] \right) \quad (4.18)$$

where

$$\boldsymbol{q}[k] = \sum_{l=0}^{L_F-1} \boldsymbol{W}[l]\boldsymbol{r}[k - l] \quad (4.19)$$

is the output of the DFE FF filter and the definition $\boldsymbol{B}[0] \stackrel{\triangle}{=} \boldsymbol{I}_{N_T}$ is used. $\tilde{\boldsymbol{c}}[\cdot]$ and $\hat{\boldsymbol{c}}[\cdot]$ denote equalizer trial symbols (defined by the states and the transitions of the trellis) and state–dependent feedback symbols, respectively. Note that for the limiting cases $K = 0$ and $K = L_B$ DFSE is identical to DFE and MLSD, respectively. An alternative representation of the branch metric $\lambda[k]$, analogous to Eq. (4.12), is straightforward.

21.4.4   Suitability of STBC for Suboptimum Equalization

The optimality criterion adopted for code design in Section 21.3 guarantees high performance for MLSD. Since the power efficiency of suboptimum equalizers is bounded by that of optimum MLSD and the applicability of suboptimum equalizers is facilitated by the diagonal structure of the code matrices, this is a reasonable approach. However, suppose there are two STBC's which have a similar performance for MLSD, which one should we adopt if suboptimum equalization is to be used?

Let us first consider DFE. Here, the receiver performance is not only governed by the distance properties of the STBC but also by the total error variance σ_d^2. On the other hand, as can be observed from [41, Eqs. (14), (16)], the total error variance depends on the ACM $\boldsymbol{\Phi}_{cc} \stackrel{\triangle}{=} \mathcal{E}\{\boldsymbol{c}[k]\boldsymbol{c}^H[k]\}$ of the MIMO input vector $\boldsymbol{c}[k]$, i.e., on the second–order statistics of the STBC signal set. In order to illustrate this dependence, we consider the special cases

$$\boldsymbol{\Phi}_{cc} = \begin{pmatrix} 1 & \rho \\ \rho & 1 \end{pmatrix} \quad (4.20)$$

and

$$\boldsymbol{\Phi}_{cc} = \begin{pmatrix} 1 & \rho & 0 \\ \rho & 1 & 0 \\ 0 & 0 & 1 \end{pmatrix} \quad (4.21)$$

for $N_T = 2$ and $N_T = 3$, respectively, where ρ, $0 \leq \rho \leq 1$, is the correlation coefficient. In Figs. 21.5a) and b) we depict the corresponding average total error variance $\overline{\sigma_d^2}$ vs. ρ for $10 \log_{10}(E_b/N_0) = 10$ dB and $10 \log_{10}(E_b/N_0) = 20$ dB, respectively. Thereby, $L_F = 10$, $L_B = 3$ and $L_F = 7$, $L_B = 2$ has been used for $N_T = 2$ and $N_T = 3$, respectively. In addition, $R =$

1 bit/(channel use) is valid and $\overline{\sigma_d^2}$ is obtained by averaging over 1000 realizations of an HT channel ($L = 7$). For comparison, also the average total error variance $\overline{\sigma_e^2}$ is shown, which would be relevant if spatial noise whitening was not performed. From Fig. 21.5 we observe that the total error variance can be considerably reduced by spatial noise whitening for both $N_T = 2$ and $N_T = 3$, and for all values of ρ. Another very interesting observation is that both $\overline{\sigma_d^2}$ and $\overline{\sigma_e^2}$ decrease with increasing correlation coefficient ρ. This suggests that correlations among the signals transmitted from different antennas are beneficial for DFE. Indeed, we will see in Section 21.5 that DFE can approach MLSD closely for STBC's with strong (spatial) correlations, whereas a loss in performance of about 3–4 dB results if there are no correlations. As an example, we may consider the HS–C and R–2ASK codes discussed in Section 21.3 for $N_T = 2$ and $R = 1$ bit/(channel use), which yield a similar performance for MLSD, cf. Fig. 21.1a). It can be easily checked that the ACM's for the HS–C and the R–2ASK are given by Eq. (4.20) with $\rho = 1$ and $\rho = 0$, respectively. As expected, for DFE the HS–C outperforms the R–2ASK by about 3 dB. For DFSE this gap reduces with increasing K. As a second example, we consider the two HS–C's proposed in Section 21.3 for $N_T = 3$ and $R = 1$ bit/(channel use). The ACM's for $\boldsymbol{u} = [1\ 1\ 3]$ and $\boldsymbol{u} = [1\ 2\ 3]$ are given by Eqs. (4.21) with $\rho = 1$ and $\rho = 0$, respectively. Although Fig. 21.1b) suggests that $\boldsymbol{u} = [1\ 2\ 3]$ yields a superior performance for MLSD, our simulations showed that $\boldsymbol{u} = [1\ 1\ 3]$ performs better for DFE.

The above findings can be explained as follows. If the symbols emitted by the transmit antennas are mutually uncorrelated, the DFE, which relies on second–order statistics only, treats them as originating from independent sources and consequently has to perform interference suppression to separate them. On the other hand, if the transmitted symbols are strongly mutually correlated, the DFE essentially can combine the contributions originating from different transmit antennas and thus, can yield a lower total error variance.

21.4.5   Complexity Issues

In this section, we briefly discuss the complexity of the proposed receivers and compare it with that of SISO equalization and with that of the equalizers reported in [12]. A comparison with other receivers for STBC's, e.g. [6, 9], seems to be difficult since these schemes require more than one receive antenna.

For the MIMO DFE scheme presented here, there are $N_T^2(L_F + L_B)$ scalar FF and FB filter taps, and the decision rules according to Eqs. (4.8), (4.12) demand $2^{N_T R}/(N_T R)$ metric calculations per decided bit. For the SISO case only $L_F + L_B$ filter taps and $2^R/R$ metric calculations per bit decision are necessary. However, if we take into account that for long SISO channels (i.e.,

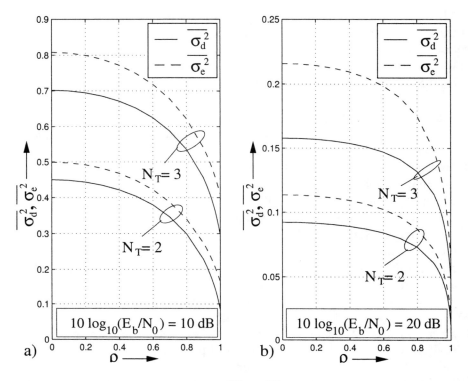

Figure 21.5. Average total error variances $\overline{\sigma_d^2}$ and $\overline{\sigma_e^2}$ vs. ρ for a) $10\log_{10}(E_b/N_0) = 10$ dB and b) $10\log_{10}(E_b/N_0) = 20$ dB. The HT channel model is adopted and $R = 1$ bit/(channel use) is valid.

$L \gg N_T$) the effective channel length of the corresponding MIMO channel is $\tilde{L} \approx L/N_T$, and consequently, the MIMO DFE FF and FB filters can be made by a factor of N_T shorter than the SISO DFE filters, the number of scalar filter taps for the ST coded system is approximately by a factor of N_T larger than for single antenna transmission.

For the scheme in [12], cf. also [22, 23], which is tailored for Alamouti's code (i.e., $N_T = 2$), not only the received signal vector $r[k]$ but also $r^*[k]$ is processed. Thus, $(2N_T)^2(L_F + L_B)$ scalar FF and FB filter taps have to be calculated. If symmetry relations are exploited, half of the calculated taps can be discarded and $2N_T^2(L_F + L_B)$ taps are necessary for filtering. Due to the orthogonal structure of Alamouti's code, only $2^R/R$ metric calculations per bit decision have to be carried out. For low rates and a small number of transmit antennas, the approach presented in this chapter requires a lower computational complexity than the scheme presented in [12] because of the lower number of filter taps. For higher rates the number of metric calculations becomes more important and the scheme in [12] has a complexity advantage. However, it should be noted that diagonal STBC's are mainly beneficial for rates $R \leq 2$ bits/(channel use) anyway.

Finally, it is worth mentioning that for both the DFSE scheme proposed here for diagonal STBC's and that proposed in [12] for Alamouti's code $Z = 2^{N_T RK}$ states are necessary if a (MIMO) channel memory of K symbols is taken into account, whereas for SISO DFSE only $Z = 2^{RK}$ states are required.

21.4.6   Extensions

There are several useful extensions to the proposed suboptimum equalizers. As a first example, instead of simple DFSE more sophisticated RSSE techniques relying on set partitioning [15] might be used. It can be expected that similar to the SISO case, e.g. [30], this would allow to reduce complexity without compromising performance. Furthermore, as an alternative approach to RSSE the MIMO model presented here also enables the application of impulse response truncation techniques [49].

In this work, we restrict our attention to coherent equalization. Nevertheless, since HS–C's can be differentially encoded [27], an interesting field for future research is the design of noncoherent equalizers for ST block coded transmission. Similar to differential PSK transmission in the single antenna case, e.g. [50, 51], it is expected that noncoherent equalizers would enable a favorable trade–off between power efficiency and robustness against phase noise and frequency offsets.

21.5   Simulation Results

In this section, we present some simulation results for the diagonal STBC's proposed in Section 21.3 and the suboptimum equalization schemes discussed in Section 21.4. We adopt the transmit and receiver input filters as described in Section 21.3, and for Figs. 21.6–21.8 the HT channel model is used, whereas for Fig. 21.9 the typical urban (TU) channel model [34] is valid. For the HT channel ($L = 7$) we adopted $L_F = 20$ ($L_B = 6$), $L_F = 10$ ($L_B = 3$), and $L_F = 7$ ($L_B = 2$), for $N_T = 1$, $N_T = 2$, and $N_T = 3$, respectively, whereas for the TU channel ($L = 5$) we adopted $L_F = 20$ ($L_B = 4$), $L_F = 10$ ($L_B = 2$), and $L_F = 7$ ($L_B = 1$), for $N_T = 1$, $N_T = 2$, and $N_T = 3$, respectively. For all simulations perfect knowledge of the CIR's and the noise variance at the receiver is assumed.

In Fig. 21.6 BER vs. $10 \log_{10}(E_b/N_0)$ is depicted for $R = 1$ bit/(channel use) and BPSK ($N_T = 1$), HS–C's ($N_T = 2$ and $N_T = 3$), and Alamouti's code ($N_T = 2$). Besides the simulation results for DFE and DFSE also the approximation for the BER for MLSD (dashed lines) is shown, cf. Section 21.3,

[28, 29]. For the reasons discussed in Section 21.4 for $N_T = 2$ we consider the HS–C rather than R–2ASK, and for $N_T = 3$ the HS–C parameter vector $\mathbf{u} = [1\ 1\ 3]$ is adopted instead of $\mathbf{u}_{opt} = [1\ 2\ 3]$, which is optimum for MLSD but has an inferior performance for DFE. Indeed, it can be observed that for $N_T = 2$ the HS–C with DFE suffers only a small performance penalty of 0.5 dB compared to the approximation for MLSD. For $N_T = 3$ the gap for DFE is about 1 dB but can be narrowed by DFSE with $Z = 8$ states ($K = 1$). On the other hand, Alamouti's code suffers from a large loss in performance if widely linear (WL) DFE [12, 22] is adopted instead of MLSD. The BER curve for WL–DFE is essentially parallel to the curves for BPSK, i.e., it appears that Alamouti's code cannot take advantage of the additional diversity offered by two transmit antennas if suboptimum equalizers are used. This might be attributed to the additional interference which is caused if two antennas transmit simultaneously and which has to be suppressed by the WL–DFE.

Fig. 21.7 shows simulation results for DFE for QPSK, Alamouti's code, and the diagonal STBC's already considered in Fig. 21.2. The rate is $R = 2$ bits/(channel use) and for comparison also the corresponding approximate BER's for MLSD are depicted. Although at high SNR's the approximate BER for MLSD for 16QAM–RC is about 3 dB worse than that for R–4QAM, both STBC's have a similar performance if DFE is applied. This behavior can again be explained by the findings in Section 21.4 since the symbols transmitted over the two antennas are fully mutually correlated and uncorrelated for 16QAM–RC and R–4QAM, respectively. At low SNR's Alamouti's code with WL–DFE performs as good R–4QAM with DFE, whereas the diagonal STBC's perform better at high SNR's, where the curve for Alamouti's code is again essentially parallel to that for $N_T = 1$ (QPSK). At BER = 10^{-5}, R–4QAM outperforms Alamouti's code and QPSK by 1.3 dB and 3.8 dB, respectively. As can be observed from Fig. 21.8, the gap between DFE and MLSD can be closed by application of DFSE for both HS–C and R–4QAM at the expense of an increase in computational complexity. It is also interesting to note that the performance gap between different diagonal STBC's for DFE and DFSE can be accurately predicted by the performance gap of the respective approximate BER curves for MLSD as long as the STBC's have the same correlation properties.

So far, we considered only the HT channel model. In order to illustrate the possible performance gains for a different power delay profile, we adopt the TU channel model for Fig. 21.9. For TU and $R = 1$ bit/(channel use) the same HS–C's as for HT are optimum. Again, we consider the parameter vector $\mathbf{u} = [1\ 1\ 3]$ instead of $\mathbf{u}_{opt} = [1\ 2\ 3]$ since this leads to a better performance for DFE. Fig. 21.9 shows that for the TU channel both the HS–C for $N_T = 2$ and that for $N_T = 3$ perform close to the approximate BER for MLSD. The small gap between DFE and MLSD for $N_T = 3$ can be attributed to short feedback filter length ($L_B = 1$ is valid). Alamouti's code with WL–DFE is clearly outperformed by the diagonal STBC's and, at high SNR's, it performs even worse than BPSK.

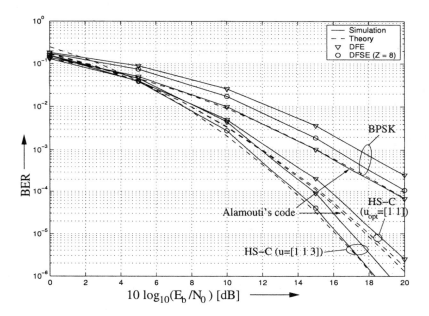

Figure 21.6.   BER vs. $10\log_{10}(E_b/N_0)$ for BPSK ($N_T = 1$), HS–C's ($N_T = 2$ and $N_T = 3$), and Alamouti's code ($N_T = 2$). The rate is $R = 1$ bit/(channel use) and the HT channel is adopted.

21.6   Conclusions

In this chapter, diagonal STBC's for transmission over fading ISI channels have been optimized based on a bound for the PEP for MLSD and their performance has been studied. An equivalent MIMO channel model enabled the successful application of suboptimum equalization techniques although only a single receive antenna was employed. It has been shown that for DFE also the mutual correlation of the symbols transmitted over different antennas plays an important role but becomes less important for DFSE as the number of states is increased. Simulations have shown that diagonal STBC's with DFE clearly outperform Alamouti's code with WL–DFE [12, 22] for rates $R \leq 2$ bit/(channel use). For higher rates the power efficiency of diagonal STBC's decreases due to the constraints imposed by the restriction to diagonal code matrices. Here, the design of more power efficient STBC's allowing for suboptimum equalization is a promising field for future research. Thereby, a similar approach as in this chapter may be adopted, i.e., the general STBC

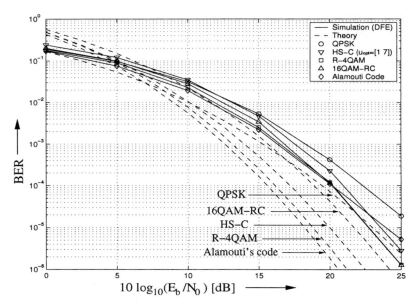

Figure 21.7.   BER vs. $10 \log_{10}(E_b/N_0)$ for QPSK ($N_T = 1$) and various STBC's ($N_T = 2$). The rate is $R = 2$ bits/(channel use) and the HT channel is adopted.

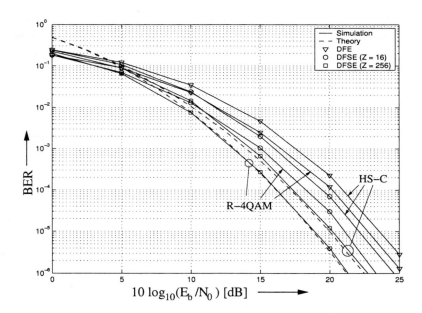

Figure 21.8.   BER vs. $10 \log_{10}(E_b/N_0)$ for HS–C and R–4QAM with DFE and DFSE. The rate is $R = 2$ bits/(channel use) and the HT channel is adopted.

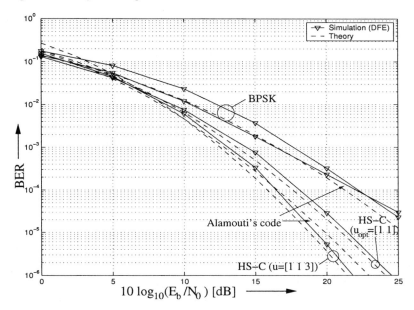

Figure 21.9. BER vs. $10 \log_{10}(E_b/N_0)$ for BPSK ($N_T = 1$), HS–C's ($N_T = 2$ and $N_T = 3$), and Alamouti's code ($N_T = 2$).The rate is $R = 1$ bit/(channel use) and the TU channel is adopted.

design criterion can be used while constraints on the structure of the STBC matrix ensure the applicability of suboptimum (WL) equalization.

References

[1] A. Wittneben. A New Bandwidth Efficient Transmit Antenna Modulation Diversity Scheme for Linear Digital Modulation. In *Proceedings of the International Conference on Communications*, pages 1630–1634, Geneva, May 1993.

[2] S.M. Alamouti. A Simple Transmitter Diversity Scheme for Wireless Communications. *IEEE Journal on Selected Areas in Communications*, SAC-16:1451–1458, October 1998.

[3] V. Tarokh, N. Seshadri, and A.R. Calderbank. Space–Time Codes for High Data Rate Wireless Communication: Performance Criterion and Code Construction. *IEEE Transactions on Information Theory*, IT-44:744–765, March 1998.

[4] J.-C. Guey, M.P. Fitz, M.R. Bell, and W.-Y. Kuo. Signal Design for Transmitter Diversity Wireless Communication Systems Over Rayleigh Fading Channels. *IEEE Transactions on Communications*, COM-47:527–537, April 1999.

[5] D. Agrawal, V. Tarokh, A. Naguib, and N. Seshadri. Space–Time Coded OFDM for High Data–Rate Wireless Communication Over Wideband Channels. In *Proceedings of the IEEE Vehicular Technology Conference*, pages 2232–2236, Ottawa, May 1998.

[6] W.-J. Choi and J.M. Cioffi. Space–Time Block Codes over Frequency Selective Rayleigh Fading Channels. In *IEEE Vehicular Technology Conference*, pages 2541–2545, November 1999.

[7] Y. Li, J.C. Chung, and N.R. Sollenberger. Transmitter Diversity for OFDM Systems and its Impact on High–Rate Data Wireless Networks. *IEEE Journal on Selected Areas in Communications*, SAC-17:1233–1243, July 1999.

[8] E. Lindskog and A. Paulraj. A transmit diversity scheme for channels with intersymbol interference. In *Proceedings of the IEEE International Conference on Communications*, New Orleans, June 2000.

[9] N. Al-Dhahir, A.F. Naguib, and A.R. Calderbank. Finite–Length MIMO Decision–Feedback Equalization for Space–Time Block–Coded Signals Over Multipath–Fading Channels. *IEEE Transactions on Vehicular Technology*, VT-50:1176–1182, July 2001.

[10] N. Al-Dhahir. Single–Carrier Frequency–Domain Equalization for Space–Time Block–Coded Transmission Over Frequency–Selective Fading Channels. *IEEE Communications Letters*, 5:304–306, July 2001.

[11] Z. Liu, G.B. Giannakis, S. Barbarossa, and A. Scaglione. Transmit–Antennae Space–Time Block Coding for Generalized OFDM in the Presence of Unknown Multipath. *IEEE Transactions on Communications*, COM-49:1352–1364, July 2001.

[12] W.H. Gerstacker, F. Obernosterer, R. Schober, A. Lehmann, A. Lampe, and P. Gunreben. Receiver Concepts for Space–Time Block–Coded Transmission over Frequency–Selective Fading Channels. *Revised version submitted to IEEE Transactions on Communications*, September 2003.

[13] L. Li, H. Li, and Y.-D. Yao. Transmit Diversity and Equalization for Frequency Selective Fading Channels. In *IEEE Vehicular Technology Conference*, pp. 1673–1677, Rhodes, Greece, May 2001.

[14] S. Zhou and G.B. Giannakis. Single–Carrier Space–Time Block–Coded Transmission Over Frequency–Selective Fading Channels. *IEEE Transactions on Information Theory*, IT-49:164–179, January 2003.

[15] M.V. Eyuboglu and S.U. Qureshi. Reduced–State Sequence Estimation with Set Partitioning and Decision Feedback. *IEEE Transactions on Communications*, COM-36:13–20, January 1988.

[16] A. Duel-Hallen and A. C. Heegard. Delayed Decision–Feedback Sequence Estimation. *IEEE Transactions on Communications*, COM-37:428–436, 1989.

[17] M.J. Heikkilae, K. Majonen, and J. Lilleberg. Decoding and Performance of Space–Time Trellis Codes in Fading Channels with Intersymbol Interference. In *Proceedings of the IEEE International Symposium on Personal, Indoor and Mobile Radio Communications (PIMRC)*, pages 1077–1082, London, September 2000.

[18] A.F. Naguib. Equalization of Transmit Diversity Space–Time Coded Signals. In *IEEE Global Telecommunications Conference*, San Francisco, December 2000.

[19] W. Younis and N. Al-Dhahir. Joint Prefiltering and MLSE Equalization of Space–Time–Coded Transmission Over Frequency–Selective Channels. *IEEE Transactions on Vehicular Technology*, VT-51:144–154, January 2002.

[20] V. Tarokh, H. Jafarkhani, and A.R. Calderbank. Space–Time Block Codes from Orthogonal Designs. *IEEE Transactions on Information Theory*, IT-45:1456–1467, July 1999.

[21] N. Al-Dhahir. Overview and Comparison of Equalization Schemes for Space–Time–Coded Signals With Application to EDGE. *IEEE Transactions on Signal Processing*, SP-50:2477–2488, October 2002.

[22] W.H. Gerstacker, F. Obernosterer, R. Schober, A. Lehmann, A. Lampe, and P. Gunreben. Widely Linear Equalization for Space–Time Block–Coded Transmission Over Fading ISI Channels. In *Proceedings of the IEEE Vehicular Technology Conference*, pages 238–242, Vancouver, Canada, September 2002.

[23] W.H. Gerstacker, F. Obernosterer, R. Schober, A. Lehmann, A. Lampe, and P. Gunreben. Symbol–by–Symbol and Trellis–Based Equalization with Widely Linear Processing for Space–Time Block–Coded Transmission over Frequency–Selective Channels. In *Proceedings of the IEEE Global Telecommunications Conference*, Taipei, Taiwan, December 2002.

[24] B. Picinbono and P. Chevalier. Widely linear estimation with complex data. *IEEE Transactions on Signal Processing*, 43:2030–2033, August 1995.

[25] V.M. DaSilva and E.S. Sousa. Fading–Resistant Modulation Using Several Transmitter Antennas. *IEEE Transactions on Communications*, COM-45:1236–1244, October 1997.

[26] B.L. Hughes. Differential Space–Time Modulation. *IEEE Transactions on Information Theory*, IT-46:2567–2578, November 2000.

[27] B.M. Hochwald and W. Sweldens. Differential Unitary Space–Time Modulation. *IEEE Transactions on Communications*, COM-48:2041–2052, December 2000.

[28] R. Schober, W. H. Gerstacker, and L.H.-J. Lampe. Performance Analysis and Design of STBC's for Frequency–Selective Fading Channels. *To appear in IEEE Transactions on Wireless Communications*, 2003.

[29] R. Schober, W. H. Gerstacker, and L.H.-J. Lampe. Performance Analysis and Design of STBC's for Fading ISI Channels. *Proceedings of the IEEE International Conference on Communications*, pp. 1451–1455, New York, May 2002.

[30] W. H. Gerstacker and R. Schober. Equalization Concepts for EDGE. *IEEE Transactions on Wireless Communications*, TW-01:190-199, January 2002.

[31] M.V. Clark, L.J. Greenstein, W.K. Kennedy, and M. Shafi. Matched Filter Performance Bounds for Diversity Combining Receivers in Digital Mobile Radio. *IEEE Transactions on Vehicular Technology*, VT-41:356–362, November 1992.

[32] Y. Liu, M.P. Fitz, and O.Y. Takeshita. Space–time codes performance criteria and design for frequency selective fading channels. In *Proceedings of the IEEE International Conference on Communications*, Helsinki, June 2001.

[33] N.W.K. Lo, D.D. Falconer, and A.U.H. Sheikh. Adaptive Equalization for Co-Channel Interference in a Multipath Fading Environment. *IEEE Transactions on Communications*, COM-43:787–794, February-April 1995.

[34] *GSM Recommendation 05.05: "Propagation Conditions", Vers. 5.3.0, Release 1996.*

[35] D. Rainish. Diversity Transform for Fading Channels. *IEEE Transactions on Communications*, COM-44:1653–1661, December 1996.

[36] J. Boutros and E. Viterbo. Signal Space Diversity: A Power- and Bandwidth-Efficient Diversity Technique for the Rayleigh Fading Channel. *IEEE Transactions on Information Theory*, IT-44:1453–1467, July 1998.

[37] P.P. Vaidyanathan, T.Q. Nguyen, Z. Doganata, and T. Saramaki. Improved Technique for Design of Perfect Reconstruction FIR QMF Banks with Lossless Polyphase Matrices. *IEEE Transactions on Acoustics, Speech, and Signal Processing*, 37:1042–1056, July 1989.

[38] D. Agrawal, T.J. Richardson, and R.L. Urbanke. Multiple–Antenna Signal Constellations for Fading Channels. *IEEE Transactions on Information Theory*, IT-47:2618–2626, September 2001.

[39] *ETSI, Tdoc SMG2 WPB 325/98, November 1998.*

[40] A. Duel-Hallen. Equalizers for Multiple Input/Multiple Output Channels and PAM Systems with Cyclostationary Input Sequences. *IEEE Journal on Selected Areas in Communications*, SAC-10:630–639, April 1992.

[41] N. Al-Dhahir and A.H. Sayed. The Finite–Length Multi–Input Multi–Output MMSE–DFE. *IEEE Transactions on Signal Processing*, SP-48:2921–2936, October 2000.

[42] C. Tidestav, A. Ahlen, and M. Sternad. Realizable MIMO Decision Feedback Equalizers: Structure and Design. *IEEE Transactions on Signal Processing*, SP-49:121–133, January 2001.

[43] A. Maleki-Tehrani, B. Hassibi, and J.M. Cioffi. Adaptive Equalization of Multiple–Input Multiple–Output (MIMO) Channels. In *IEEE International Conference on Communications*, pages 1670–1674, New Orleans, June 2000.

[44] J. M. Cioffi, G. P. Dudevoir, M. V. Eyuboglu, and G. D. Forney Jr. MMSE Decision–Feedback Equalizers and Coding – Part I: Equalization Results. *IEEE Transactions on Communications*, COM-43:2582–2594, October 1995.

[45] S. Haykin. *Adaptive Filter Theory*. Prentice-Hall, Upper Saddle River, New Jersey, Third Edition, 1996.

[46] W. Gerstacker and J. Huber. Improved Equalization for GSM Mobile Communications. In *Proceedings of International Conference on Telecommunication*, Istanbul, April 1996.

[47] G.D. Forney, Jr. The Viterbi Algorithm. *IEEE Proceedings*, 61:268–278, 1973.

[48] W. Van Etten. Maximum–Likelihood Receiver for Multiple Channel Transmission Systems. *IEEE Transactions on Communications*, COM-76:276–283, February 1976.

[49] N. Al-Dhahir. FIR Channel–Shortening Equalizers for MIMO ISI Channels. *IEEE Transactions on Communications*, COM-49:213–218, February 2001.

[50] R. Schober and W. H. Gerstacker. Adaptive Noncoherent DFE for MDPSK Signals Transmitted over ISI Channels. *IEEE Transactions on Communications*, COM-48:1128–1140, July 2000.

[51] R. Schober, W. H. Gerstacker, and J.B. Huber. Adaptive Noncoherent Linear Minimum ISI Equalization for MDPSK and MDAPSK Signals. *IEEE Transactions on Signal Processing*, SP-49:2018–2030, September 2001.

Chapter 22

FAST ROUTING AND RECOVERY PROTOCOLS IN HYBRID AD-HOC CELLULAR NETWORKS

Mostafa Bassiouni, Wei Cui, and Bin Zhou
School of Computer Science
University of Central Florida
Computer Science Building
Orlando, Florida 32816
{bassi, willy, bzhou}@cs.ucf.edu

Abstract   A hybrid ad-hoc cellular network (HACN) is based on the idea of replacing the stationary cellular base stations with mobile routers. Communications among the mobile routers is achieved using satellite links or high bandwidth wireless channels. In this chapter, we present and evaluate efficient recovery and routing protocols for mobile routers in HACN. In the dual backup recovery protocol, a standby router is mapped (dedicated) to each primary router. A more flexible protocol, called the distributed recovery protocol, is obtained by relaxing this one to one mapping and scattering the backup routers geographically among the primary routers. The effectiveness of the distributed recovery protocol is demonstrated by simulation results. A simple analytical model is also presented for computing the blocking probability of new calls in the presence of router failures. The chapter is concluded by presenting an efficient location-based routing protocol for HACN mobile routers. Performance results of the routing protocol are presented.

Keywords: Cellular networks, ad-hoc wireless networks, fault tolerance, recovery protocols, location based routing protocols, terminal mobility.

1.   Introduction

The hybrid ad-hoc cellular network (HACN) is a hybrid mobile wireless architecture that combines the advantages of both the ad-hoc and the cellular models. This hybrid network provides rapidly deployable communications capability and communications-on-the- move services with enhanced flexibility and scalability. In HACN, specialized mobile

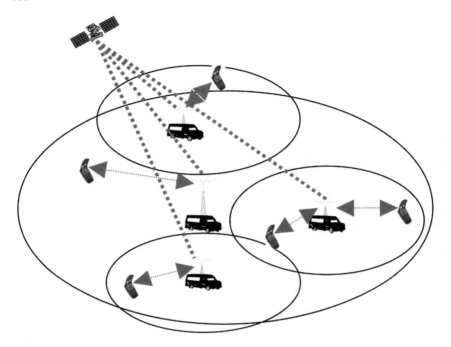

Figure 22.1.   Example of a two-tier HACN

routers are used to achieve continued connectivity and fast forwarding. A mobile router has functionality similar to a cellular base station, but has no wired connections. In its simplest form, the mobile router could be a truck-mounted transceiver box with rechargeable battery. Satellite links or backbone wireless channels are used for communications among the mobile routers while short-hop wireless channels are used for communications between each mobile router and its users (mobile hosts). When the mobile hosts move, the mobile routers would also move to ensure the continuity of coverage and improve the quality of service for active connections. In one-tier (flat) HACN, the coverage areas (cells) of different mobile routers are mostly non-overlapping. A hierarchical version of HACN is obtained by adding another tier of mobile routers, called umbrella routers. An umbrella router has a more powerful transmitter and its coverage area usually overlaps that of several one-tier routers. Conceptually, a two-tier HACN is similar to the hierarchical macro-micro architecture used in cellular networks. Figure 22.1 1 shows a two-tier HACN with one umbrella cell and three regular mobile routers.

The performance of one-tier HACN and its two-tier hierarchical counterpart has been investigated in [3]. In general, the two-tier model has the potential to give better performance and better geographical coverage than the one-tier counterpart. This is especially true when the degree of the randomness of the movement of mobile terminals is high. However, under the constraint of equal power consumption and for mobility patterns with low randomness, the one tier system could provide higher throughput.

2.   Recovery Protocols for HACN Mobile Routers

To improve the reliability and resiliency of HACN, mechanisms must be put in place to ensure that the operations of the mobile routers can survive environmental hazards or hostile attacks. A mobile router can become immobilized due to a flat tire, failed automotive engine, or some type of road obstruction. Although the mobility of the router is compromised, the wireless transceiver in this case is intact and can continue to provide service in a stationary mode. A more serious scenario is the failure or total destruction of the router's transceiver. This condition forces the termination of all active connections served by the failed router. In order to be able to handle these faults when they occur, the network must be equipped with robust fault tolerance and recovery protocols. These protocols normally entail adding redundant hardware and incurring some extra overhead during normal operations. Understandably, there is a tradeoff between the cost of the extra hardware and the level of robustness and gracefulness by which the faults can be handled. Below we elaborate on a new recovery protocol, called the distributed recovery protocol, that we have designed and evaluated.

In the dual backup restoration protocol [2], an extra backup router is assigned to each mobile router. The backup router exchanges state-update messages with the primary router in order to be able to accurately mirror its state. The backup router also follows the movement of its primary as close as possible. If the primary router malfunctions, its backup can take over promptly and re-establish links to the hosts served by the failed primary.

The dual backup protocol is simple and provides definite performance gains in face of router failures and hostile attacks. In particular, the dual backup protocol provides the fastest recovery when a backup router survives the destruction of its primary router. However, one disadvantage of the dual backup protocol is that the primary router and its backup are likely to be destroyed together (e.g., they enter a mine field together).

Furthermore, the dual backup protocol does not provide optimal utilization of the backup resources in the long run. As more primary routers get knocked out and are replaced by their backups, several cells will operate without backups, which ultimately degrades the gracefulness of future recovery.

Rather than strictly dedicating a backup router to each primary mobile router, the backup routers are carefully scattered among the primary mobile routers. When a primary router fails, one or more of the nearby backup routers is dispatched to provide coverage for the set of mobile terminals originally covered by the failed router. This protocol is called the distributed recovery protocol. It is flexible and does not require dedicating a backup router for each primary router.

The evaluation of the recovery protocols for HACN has been primarily done by extensive simulation tests. An analytical model has also been developed for certain useful cases. The model assumes that the time to dispatch a backup router to the location of the failed router is small compared to the time needed to repair the failed router as well as the mean time between failures. Inter-cell traffic is assumed to be of low volume, i.e., new call requests represent the dominant source of requests for mobile routers. It is possible to extend the model in order to relax most or all of the above assumptions, but we will focus in this chapter on the basic model without the added complexity of removing these assumptions.

2.1 Analysis of the Distributed Recovery Protocol

Assume that the time between failures (TBF) of mobile routers is exponentially distributed with a mean ω_p for a primary router and ω_s for a standby (backup) router. The repair time of a failed router is exponentially distributed with mean $\mathfrak{R}$. The interarrival times of new calls to a primary router are exponentially distributed with mean λ and the duration of each call is exponentially distributed with mean μ. Let M be the number of cells served by the HACN network and let N be the initial number of mobile routers deployed in the network, where $M \leq N \leq 2M$. Thus initially M routers are dispatched as primary routers and $N - M$ routers are used as standby routers.

Let j represent the total number of operational routers (both primary and standby) in the HACN network. The system can be represented by a birth-death Markov process where the transition rate from state j to state $j + 1$ is given by

$$\alpha_j = (N - j) * \Re \qquad\qquad 0 \le j \le N$$

and the transition rate form state j to state $j - 1$ is given by

$$\gamma_j = j * \omega_p \qquad\qquad \text{for} \quad 0 \le j \le M$$
$$\gamma_j = M * \omega_p + (j - M) * \omega_s \quad \text{for} \quad M + 1 \le j \le N$$

The stationary distribution π_j of the occupancy of state j can be obtained by solving the above balance of flow equations. The blocking rate of new calls in state j is given by

$$\Phi_j = 1 - (j/M) + j * B/M \quad \text{for} \quad 0 \le j \le M - 1$$
$$\Phi_j = B \qquad\qquad\qquad \text{for} \quad M \le j \le N$$

where B is calculated using the well-known Erlang-B formula

$$B = \frac{\frac{(\lambda/\mu)^c}{c!}}{\sum_{j=1}^{c} \frac{(\lambda/\mu)^j}{j!}}$$

In the above equation, c is the maximum link capacity of the mobile router. The steady state blocking rate of new calls in the HACN network can be computed as follows

$$\Phi = \sum_{j=0}^{N} \pi_j \Phi_j$$

The above analysis enables us to evaluate the performance of distributed recovery in a hybrid mobile network that initially has M primary routers and $N - M$ backup routers.

2.2 Performance Results of the Recovery Protocols

A detailed simulation model written in C++ was used in the evaluation of the recovery protocols for HACN. In the performance results reported in this chapter, the simulation tests used a one-tier network with 36 active mobile routers serving 1800 mobile terminals. The number of backup routers was zero for the simple (no backup) protocol, 36 for the dual backup recovery protocol, and either 18 or 36 for the distributed recovery protocol. Each mobile router has the capacity to serve 20 calls simultaneously. Call durations were exponentially distributed with mean of 180 seconds. The mean times between failures (MTBF) and repair times were also exponentially distributed with a wide range of values.

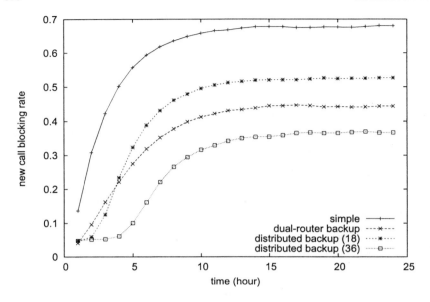

Figure 22.2. New call blocking rates for different recovery options

Figure 22.2 shows the new call blocking rate versus time for the different recovery options. In all options (including the simple protocol that does not use backups), the failed router is brought up back to operational status after a repair period. Figure 22.2 shows that the distributed recovery protocol (with 36 backup routers) outperforms the dual backup protocol which uses the same number of backup routers.

Figure 22.3 shows the new call blocking rate for the different protocols as the mean time between failures is increased from 2 to 16 hours. The results from the analytical model closely agree with those obtained by simulation. Both results confirm that the distributed protocol is quite effective and outperforms the dual router protocol under equal number of backups.

3.   Location-based Routing for HACN Mobile Routers

Although HACN is a hybrid ad-hoc cellular architecture, routing messages from one mobile router to the other is basically a multi-hop ad-hoc routing problem. In this section, we present a new location-based algorithm for mobile routers different from those previously proposed in the literature [6, 8, 9].

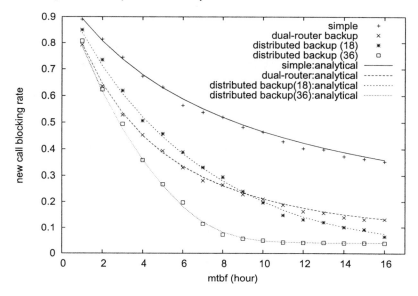

Figure 22.3. New call blocking rates vs MTBF

Location-based services (LBS) is becoming increasingly popular. New location positioning products continue to be deployed for a variety of applications with improved reliability and lower cost. Mobile phones with location positioning capabilities are rapidly spreading around the globe and it is projected that a large majority of digital wireless phones will have some position location capability in the near future. In [1], we used GPS positioning services in the design of a novel handoff prioritization technique for cellular networks and in [12] we presented a position-based multicasting (i.e., geocasting) algorithm for mobile wireless networks. In this chapter, we use similar positioning information services to improve the reachability [11] of routing algorithms in HACN. The new algorithm proposed in this chapter assumes that every mobile router can determine its position information with reasonable accuracy. This information can be easily obtained by GPS [10]. As in [7], we also assume that the source mobile router knows the approximate position of the destination mobile router. This is an acceptable assumption in the HACN environment since the number of mobile routers is an order of magnitude smaller than the number of mobile terminals and a geographical location service (GLS) that provides a translation from address to geographical location of the mobile routers can be cost-effective.

In our protocol, every mobile router periodically sends summary information about its status to its neighbors. By using the exchanged information, our algorithm seeks to reduce the number of intermediate hops while improving the chances of successful route discovery. At a given moment, the HACN network is modeled as a graph $G = (V, E)$, where V is the set of vertices (mobile routers) and E is the set of edges. An edge between one mobile router and the other means that the two routers are neighbors, i.e., they are within the transmission range of each other. The flooding algorithm is a well known routing protocol in which the routing message is sent to all neighbors and the process is repeated until the destination is reached or the algorithm fails. The flooding protocol gives the best outcome for routing success (i.e., the best reachability) since it guarantees finding the destination if there exists a feasible path to it in the network. However, this is done at the expense of higher cost of bandwidth and power consumption [4, 5]. In our performance tests, we implemented an ideal version of the flooding protocol such that the search is immediately halted once the destination is reached. The implementation of this ideal flooding algorithm is not practically feasible, but we use it in our tests as a benchmark for evaluating our location-based routing protocol for HACN.

In our algorithm, every node sends its status only to its neighbors (the terms *mobile router* and *node* are used interchangeably in the description of the routing protocol). The status information consists of the position of the node and its degree (number of neighbors within the node's transmission range). When a search is initiated from source node s to a destination node d, each node in the established path keeps track of the previous node and the next node in the path. Backtracking to the previous node is possible when the search via the next node fails and no other unvisited neighbor exists for the current node. The algorithm given in [7] is denoted GRA (geographical routing algorithm) and can be briefly described as follows. For a source node s and a destination node d, s sends its request message to the neighbor that has the shortest distance to d. That neighbor then sends the message to its neighbor with the shortest distance to d, and so on until the destination node is reached or the algorithm is stuck at some node and cannot advance. In the latter case, a route discovery protocol is used to find a new path. Basically, a depth first search algorithm is used to find an acyclic path and a routing table is needed to save the routing information.

Below we give the definitions and the notation used in our routing algorithm.

Useful degree $(\ddot{U}_v)$ of a neighbor: For a current node c and its neighbor v, the useful degree of v is a measure of the number of nodes reachable from v which can be useful in continuing the search for the destination d. In the tests reported in this chapter, the useful degree of v is equal to half the number of neighbors of v, i.e., half the number of mobile routers that are within the transmission range of v. The rationale of this method is that if the nodes are randomly located, then half of the neighbors of v will be closer than v to the destination d.

Deviation angle (α_v): Consider a current mobile router c, a destination mobile router d and a neighboring mobile router of c, say v. The deviation angle of v, α_v, is defined as the angle between vector cd and cv.

Direct distance $(\check{D}_v)$: Consider a current mobile router c, a destination mobile router d and a neighbor router of c, say v. The direct distance of v with respect to c and d is the length of the projection of vector cv on vector cd.

Routing weight (ω_v): For a current mobile router c, one of its neighbor mobile routers v and a destination mobile router d, the routing weight of v, ω_v, is defined as:

$$\omega_v = \check{D}v * \delta(\ddot{U}_v)/L + \ddot{U}_v + f(\alpha_v) \tag{22.1}$$

Where,

$$\delta(X) = \begin{cases} 1, & X \leq 1 \\ 0, & \text{otherwise} \end{cases}$$

Equation (22.1) has three primary parameters: the direct distance, the useful degree and the deviation angle of node v. The parameter L is tunable and is used to scale the contribution of the direct distance with respect to the useful degree. We refer to parameter L as the "level selector" (or simply the level) since it determines the level of contribution of the direct distance $\check{D}$ on the routing weight. Smaller values of L produce higher weights for $\check{D}$ in the selection of the next hop. The function f in equation (22.1) is the angle weight function and is used to scale the contribution of the deviation angle. This angle could have a value between 0 and π (measured symmetrically on both sides of the line cd). The contribution of the deviation angle to the routing weight of a node should decrease when the angle becomes larger. This is because a larger deviation angle represents a less favorable choice with respect to reaching the destination d.

In our simulation tests, we evaluated several choices of the value of the level parameter L and the angle weight function f. In general, we have found that the performance is improved when the direct distance $\check{D}$ and the useful degree $\ddot{U}$ have the same effective weight in equation

(22.1). The tests have shown that a good choice of the angle weight function f is a stair-wise function yielding a maximum weight that is approximately equal to the average value of $\bar{U}$ (this maximum weight is obtained when α_v approaches 0).

3.1   Performance Results of the Routing Protocol

In our simulation tests, we used a model similar to that described in [5]. We assume the mobile routers are distributed randomly in a 1000×1000 unit area. The number of mobile routers in our tests ranged from 20 to 70. The value of the transmission radius R ranged from 200 to 500 and the level parameter L ranged from 1 to 100. For each test situation, we generated 10000 randomly distributed scenarios to get the reported results.

Reachability metric: For every test situation, we compute the fraction of scenarios in which our algorithm fails to find a route to the destination but the flooding algorithm succeeds to do so. We call this metric the reachability difference and is computed as follows: Let $Total_L$ be the total number of tests for a given setting and let $Diff_L$ be the difference between the number of cases of failure of our algorithm and that of the ideal flooding algorithm, then the reachability difference of our algorithm is defined as:

$$reachability\_difference = \frac{Diff_L}{Total_L} \times 100\%$$

Another metric shown in our graphs is the number of search nodes. A search node is any node that has been visited by the search algorithm during the search process. The number of search nodes reflects the total cost (bandwidth, computation and power consumption) of the search process at all nodes. The flooding algorithm has a high number of search nodes since the search request propagates concurrently to all neighbors.

Figure 22.4 gives comparisons of our protocol, denoted HACN, with the GRA algorithm presented in [7]. The value of L was set to 60 and the transmission range was set to 200. The figure shows that the reachability of our protocol is better than that of GRA. Figure 22.5 shows that the improvement in reachability is achieved with a very small increase in the number of search nodes. The number of search nodes for HACN is almost the same as that of GRA, and they are both much smaller than the corresponding number of the ideal flooding method. The proposed protocol significantly improves the reachability of the routing algorithm without incurring any significant overhead. We are currently investigating ways to improve the selection of the algorithm's parameters in order to further improve its performance.

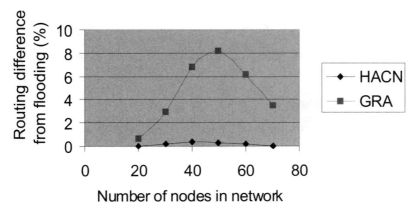

Figure 22.4.   Comparison of routing reachability

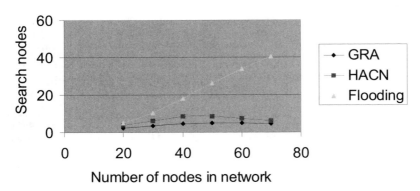

Figure 22.5.   Comparison of search overhead

4. Conclusions

In this chapter, we presented a distributed recovery protocol for the mobile routers in hybrid ad-hoc cellular networks. We also presented a location based routing algorithm for these mobile routers. Simulation tests showed that both protocols are quite effective. Further research is planned for improving and fine tuning both algorithms.

Acknowledgment

This work has been supported by ARO under Grant Number DAAD19-01-1-0502 and by NSF under Grant Number 0086251.

References

[1] M. Chiu and M. Bassiouni, "Predictive Schemes for Handoff Prioritization in Cellular Networks Based on Mobile Positioning," *IEEE Journal on Selected Areas in Communications*, vol.18, no.3, pp.510-522, March 2000.

[2] W. Cui and M. Bassiouni, "Channel Planning and Fault Recovery in Hierarchical Hybrid cellular Networks with Mobile Routers," *Proceedings of IEEE Wireless Local Networks – 26th LCN Conference*, pp.646-652, November 2001.

[3] W. Cui and M. Bassiouni "Analysis of Hierarchical Cellular Networks with Mobile Base Stations," *Journal of Wireless Communications and Mobile Computing*, John Wiley & Sons Publishing, vol.2, pp.131-149, March 2002.

[4] Min-Te Sun, Wuchi Feng and Ten-Hwang Lai, "Location aided broadcast in wireless ad hoc networks," *IEEE Global Telecommunications Conference (GLOBECOM)*, vol.5, pp.2842-2846, 2001.

[5] Young-Bae Ko and Nitin H. Vaidya, "Location-aided routing (LAR) in mobile ad hoc networks," *Wireless Networks*, vol.6, Issue4, pp.307-321, July 2000.

[6] V. Rodoplu and T. H. Meng, "Position based CDMA with multiuser detection (P- CDMA/MUD) for wireless ad hoc networks," *IEEE Sixth International Symposium on Spread Spectrum Techniques and Applications*, vol.1, pp.336 -340, 2000.

[7] R. Jain and A. Puri and R. Sengupta, "Geographical routing using partial information for wireless ad hoc networks," *IEEE Personal Communications*, vol.8, no.1, pp.48-57, Feb. 2001.

[8] Yu-Liang Chang and Ching-Chi Hsu, "Routing in wireless/mobile ad-hoc networks via dynamic group construction," *Mobile Networks and Applications*, vol.5, no.1, pp.27-37, March 2000.

[9] Yu-Chee Tseng, Sze-Yao Ni, Yuh-Shyan Chen and Jang-Pang Sheu, "The broadcast storm problem in a mobile ad hoc network," *Wireless Networks*, vol.8, Issue2/3, pp.153-167, March 2002.

[10] G. Dommety and R. Jain, "Potential networking applications of global positioning systems (GPS)," *Technical Report TR-24*, The Ohio State University, 1996, http://www.cis.ohio-state.edu/~jain/papers/gps.htm.

[11] C-K Toh, "Wireless ATM and ad-hoc networks," Kluwer academic publishers, ISBN 079239822-x, 1997.

[12] H. El-Aarag and M. Bassiouni, "A Reliable Congestion Control Mechanism for Geocasting in Mobile Wireless Networks," *International Journal of Network Management*, John Wiley & Sons Publishing, vol.13, no.5, pp.375-387, September 2003.

INDEX

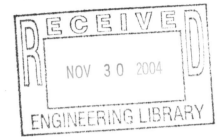